Transglutaminases

Kiyotaka Hitomi • Soichi Kojima • Laszlo Fesus
Editors

Transglutaminases

Multiple Functional Modifiers and Targets for New Drug Discovery

Springer

Editors
Kiyotaka Hitomi
Graduate School of Pharmaceutical Sciences
Nagoya University
Nagoya, Japan

Laszlo Fesus
Department of Biochemistry and Molecular Biology
University of Debrecen
Debrecen, Hungary

Soichi Kojima
Micro-Signaling Regulation Technology Unit
RIKEN Center for Life Science Technologies (CLST)
Wako, Saitama, Japan

RIKEN Molecular and Chemical Somatology
Tokyo Medical and Dental University
Bunkyo-ku, Tokyo, Japan

ISBN 978-4-431-55823-1 ISBN 978-4-431-55825-5 (eBook)
DOI 10.1007/978-4-431-55825-5

Library of Congress Control Number: 2015958519

Springer Tokyo Heidelberg New York Dordrecht London

Printed on acid-free paper

Springer Japan KK is part of Springer Science+Business Media (www.springer.com)

Preface

Posttranslational modification, which refers to covalent and generally enzymatic modification of proteins during or after protein biosynthesis, is a highly important subject of the next-generation, post-genome research. Among various posttranslational modifications, enzymes that catalyze protein–protein cross-linking reactions are transglutaminases conserved from microorganisms to mammals, forming covalent isopeptide bonds between peptide-bound glutamine and lysine residues under strict regulatory conditions. Transglutaminases also catalyze attachment of primary amine (transamidation) or replacement to glutamic acid residue (deamidation) at the glutamine residues. In mammals, this enzyme family consists of eight isoforms (isozymes) differentially expressed in various tissues.

Although the first description on transglutaminase was made almost 60 years ago and numerous studies have been performed to characterize these enzymes and their physiological and pathological roles, there are many unresolved issues. Newer and newer mysteries have evolved, and once the answers to these are found, the challenge of application of novel knowledge to practical use remains for the future. Intriguing transglutaminase features include diverse functions and sometimes opposing activities of transglutaminase family members, their multiple and changing localization inside and outside of cells, their non-enzymatic interactions and scaffolding activity, the mechanism of their secretion, and so on. By transglutaminase-catalyzed reactions the structure, function, and stability of the substrate proteins are altered, associating with a number of biological phenomena. The unique catalytic reactions and multiple interactions of transglutaminases have fascinated many scientists during the last six decades so that they have achieved biological significance in a wide scientific area and applications in chemical, food, cosmetic, and pharmaceutical industries. Because the aberrant activity or ectopic expression of the enzymes causes several diseases, inhibitory and regulatory molecules have been developed as promising new drugs. Additionally, the dramatic advances in molecular life science technologies have brought much progress in all the transglutaminase research areas.

Considering current trends and advances, we planned to publish this review book to cover basic knowledge and novel findings in transglutaminase research. We have attempted to review structures, expression, functions, and regulatory mechanisms from the scope of enzymology, biochemistry, physiology, pathology, pharmacology, chemistry, and applied bioscience. Particularly, we focused on diseases and drug development related to the enzymes' role in various pathologies.

At the Gordon Research Conference on Transglutaminases in Human Disease Processes held in Italy in 2014, we discussed with the authors, who have been engaged in prominent transglutaminase work, the purpose and scope of the book's content and decided to devote more space to basic knowledge underlying the theme of each chapter, in addition to the new, cutting-edge findings, so that "newcomers" can obtain useful information and technical insight, and most importantly an interest in studying these enigmatic enzymes in the future.

The book could not have been achieved without the full dedication of each contributing author and the people who have supported us. We hope that it will develop a further basis of new collaborations by stabilizing "cross-linking" of researchers and newcomers in the future.

Nagoya, Japan — Kiyotaka Hitomi

Wako, Japan — Soichi Kojima

Debrecen, Hungary — Laszlo Fesus

Contents

Chapter 1
Structure of Transglutaminases: Unique Features Serve Diverse Functions

Máté Á. Demény, Ilma Korponay-Szabó, and László Fésüs

Abstract Understanding the diverse functions and pathologies of transglutaminases requires detailed analysis and interpretation of their structures. This chapter is an attempt to describe in detail how these enzymes are folded into functional domains, what type of catalytic and scaffolding functions have been gained as the result of their evolution, how their regulation is achieved through unique Ca^{2+} and purine nucleotide binding sites, redox changes and specific proteolytic actions, and by influencing the equilibrium of open-close configurations. The importance of structural motifs in pathologies is underlined by the celiac epitopes of transglutaminase 2, responsible for autoimmune reactions.

Keywords Domain organization • Crystallography • Catalytic mechanism • Regulation • Ca^{2+} • Purine nucleotides • Proteolysis • Redox • Open-close conformation • Substrates • Interactions • Celiac epitope

1.1 Introduction

Transglutaminases (Tgases) are a large family of enzymes canonically responsible for amidation of protein and non-protein amines. They are ubiquitous in higher organisms but have also been identified in lower life forms, including archea, bacteria, plants, worms and insects. Their structural core and catalytic residues are strongly related to proteases and other hydrolases. Their common ancestor evolved to utilize acyl acceptors other than water by changing the active site so that it would be less accessible for water by a mechanism that is perfected in the vertebrate enzymes. The simple ancient Tgases, similar to the present microbial enzymes, acquired additional domains to serve regulatory, interacting and new enzymatic functions. In some instances these additions rendered them zymogens, which show differential activity based on proteolytic activation. They evolved to be

---

M.Á. Demény • L. Fésüs (✉)
Department of Biochemistry and Molecular Biology, University of Debrecen, Debrecen, Hungary
e-mail: fesus@med.unideb.hu

I. Korponay-Szabó
Department of Pediatrics, University of Debrecen, Debrecen, Hungary

K. Hitomi et al. (eds.), *Transglutaminases*, DOI 10.1007/978-4-431-55825-5_1

responsive to effector molecules, which are characteristic of the majority of the group.

Structural information has accumulated in the last 25 years and year by year we learn important facts. This chapter intends to summarize what we know today, focusing particularly on vertebrate and human enzymes. The mammalian transglutaminase family contains eight enzymatically active members (FXIIIa, TGM1-7) and an inactive member (erythrocyte protein band 4.2 (EPB42)). They have an absolute requirement for calcium ions for their transglutaminase activity, while other regulatory mechanisms are more particular to specific members. These signals cue the enzymes to undergo dramatic conformational rearrangements selecting from their available enzymatic activities and substantially determining their interactions with other proteins.

1.2 Primary Structure, X-Ray Diffraction and Homology Models of Secondary and Tertiary Structure, Domain Organization

The first Tgases isolated from different animal tissues differed significantly in terms of not only catalytic activity but also proteolytic sensitivity and size suggesting the existence of multiple forms of the enzymes. Cloning of TGM2 in 1980 revealed the amino acid sequence and led to the identification of other isoenzymes. Before the appearance of the first crystallographic model of FXIIIa in 1994, structural insight was limited to the amino acid sequence, molecular size, subunit composition of activated zymogenic isoforms, awareness of catalytically important residues, affinities of the enzymes for metal ions and other allosteric effectors and the tendency of the enzymes for conformational change. This insight could be derived from chemical reactivity, electrophoretic mobility, chromatographic behavior, proteolytic sensitivity, thermal denaturation, small angle X-ray scattering and infrared spectroscopy (Folk and Cole 1965, 1966a, b, c; Folk et al. 1967; Bergamini 1988; Tanfani et al. 1993; Pedersen et al. 1994).

Our knowledge of Tgase structure was to be advanced vastly by the large number of models developed primarily – with a single exception – of the vertebrate enzymes from X-ray crystallographic data. Several attempts have been made at crystallizing the same enzyme in order to answer questions of activation and conformational change. Human FXIII was crystallized in the absence of bound calcium in zymogenic (PDB ID:1GGT) and thrombin activated forms (PDB ID:1FIE) (Yee et al. 1994, 1995), in the presence of bound calcium, ytterbium and strontium (PDB IDs:1GGU; 1GGY; 1QRK)(Fox et al. 1999) and an irreversible inhibitor (PDB ID:4KTY) (Stieler et al. 2013). The structure of the FXIII mutant Trp279Phe was also resolved and deposited in the Protein Data Bank (PDB ID:1EX0). Human TGM2 has been crystallized in GDP- and GTP-bound forms (PDB IDs:1KV3 and 4PYG), in complex with ATP (PDB ID:3LY6), and in complex with four different irreversible inhibitors (PDB IDs:2Q3Z; 3S3J; 3S3P; 3S3S) (Liu et al. 2002; Pinkas et al. 2007; Han et al. 2010; Jang et al. 2014). The crystals have been resolved at 2.00–3.14 Å. Human TGM3 was crystallized as a zymogen in complex with a single calcium ion (PDB ID: 1L9M) and in the partially cleaved form with two additional calcium ions (PDB ID: 1L9N) (Ahvazi et al. 2002).

To construct a homology model for human TGM1, crystal structures of TGM2 and 3, and the model of TGM5 were used (Boeshans et al. 2007). As the N-terminus could not be modeled from other Tgases, the segment Pro24-Ala55 was modeled on a segment of Cd-6 metallothionenin (PDB ID:1DMF) and residues Asp66-Leu109 on human fascin (PDB ID: 1DFC). For Met1-Ser23 no template was found. The isolated second β-barrel of human TGM1 comprising residues 693–787 has been also crystallized and resolved at 2.3 Å (PDB ID:2XZZ). An illustration made of a homology model of TGM5 was presented in Pietroni et al.; however, no technical detail was made public (Pietroni and von Hippel 2008). For TGM6 two homology models were generated, one based on the activated and calcium-bound form of TGM3 (PDB ID:1L9N), the other based on the GDP-bound TGM2 to infer the structures of both the calcium binding sites and the guanine nucleotide binding pocket (PDB ID:1KV3) (Thomas et al. 2013). The total energies calculated from these models were −32,300 and 17,400 kJ/mol, respectively, reflecting, possibly, the stabilizing effect of calcium binding and the significantly higher level (50 %) of sequence conservation between TGM6 and TGM3. No model has been published of the 3D structures of TGM4 and 7. The crystal of red sea bream (Pargus major) liver transglutaminase (PDB ID:1G0D) was resolved at 2.50 Å resolution (Noguchi et al. 2001). The models of all vertebrate Tgases reveal the same four folded domains (Fig. 1.1) which are similar in organization and structure: an N-terminal β-sandwich domain, a catalytic core domain, and two C-terminal β-barrel domains (Kiraly et al. 2011). These domains are also conserved in TGM1 with the exception that according to its predicted model it contains an additional fifth domain at its N-terminus (Boeshans et al. 2007). All except FXIIIa are monomeric. FXIIIa works as a dimer. As will be discussed later the enzymes are thought to exist in closed and open forms.

The structure of *Streptoverticillium mobaraense* transglutaminase was determined at 2.4 Å resolution (PDB ID:1 IU4) (Kashiwagi et al. 2002). As can be expected from the sequence dissimilarity and the different sizes of the molecules the overall structure of the *Streptoverticillium mobaraense* transglutaminase is completely different from that of the vertebrate transglutaminases. The microbial Tgase consists of a single domain that folds into a plate-like shape, and has one deep cleft at the edge, which contains the active site (Kashiwagi et al. 2002).

The shape of the closed forms of the vertebrate enzymes is reminiscent of a flattened triangle and has roughly the dimensions of 110x65x50 Å. Secondary structural elements are conserved enough so that a description of TGM2 will give an idea about all isoforms. The *N-terminal β-sandwich* domain is 130–140 amino acids long and is folded into 9 β-strands and an α-helix. The *catalytic core domain*

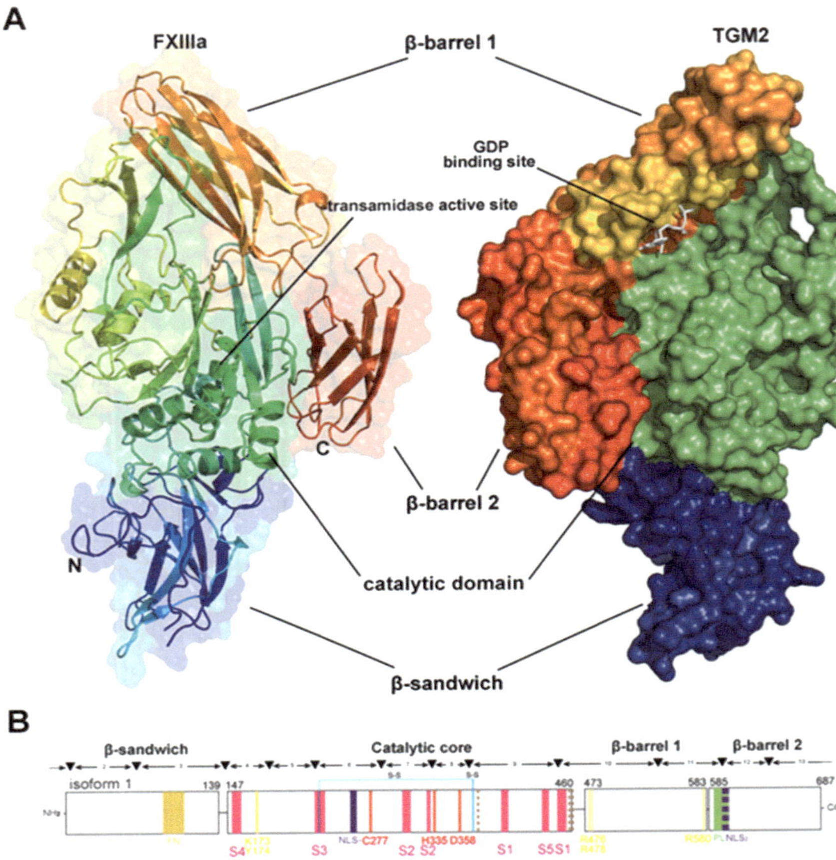

Fig. 1.1 General domain organization of vertebrate Tgases illustrated by human FXIIIa and TGM2. (**a**) Folded 3D organization of the FXIIIa (*left*) and TGM2 (*right*) isoforms (PDB 1FIE and 1KV3) with the characteristic four structural domains. (**b**) The four domains of the TGM2 protein are represented with *rectangles*. Exon boundaries within the coding sequence of the *TGM2* gene are marked with *arrowheads*. Functional regions and amino acid positions are shown as follows: *red*, amino acids of the catalytic triad; *yellow*, key residues involved in GTP/GDP-binding; FN, *khaki*, fibronectin binding site, WTATVVDQQDCTLSLQLTT(88–106) (Hang et al. 2005); NLS_1 and NLS_2, *violet*, putative nuclear localization signals (Peng et al. 1999); S1–S5, *magenta*, calcium binding regions (Kiraly et al. 2009); *dashed brown*, matrix metalloproteinase 2 cleavage sites C-terminal to Pro375, Arg458 and His461 (Belkin et al. 2004); PL, *green*, phospholipid binding motif KIRILGEPKQRKK(590–602) (Zemskov et al. 2011); *blue*, Cys230, Cys370 and Cys371 are capable of forming disulfide bonds in the closed and open conformations, respectively (Stamnaes et al. 2010)

could be encased within a 35 × 35 Å box with a large triangular protrusion that has been called the 'flap'. It contains ~330 amino acids and consists of 12 or 13 β-sheets interspersed with 9 or 10 α-helices and a 3_{10} helix. (Secondary structure assignment can be subjective in borderline cases. The description given here follows the predictions in the DSSP database (Kabsch and Sander 1983).) The longest α-helix is located in the center of the molecule and carries the catalytic cysteine toward its end. β14, β16 and β17-18 seem to be the remainder of the ancient β-barrel characteristic of the so called 'transglutaminase fold' with the catalytic histidine and aspartate sitting on the adjacent β-strands, β16 and β17. If one compares it with the blueprint given by Anantharaman, half of this barrel appears to have been disrupted by insertions of helical segments in the connecting loops (Anantharaman and Aravind 2003). β19-20 were pushed away with respect to the rest of the barrel. The base of the hairpin made of β21 and 22 being dragged with it, the hairpin twisted and turned downward changing the orientation of the strands to the opposite with respect to the rest. The hairpin containing β15, which constitutes the 'flap' together with β19–20, seems to be a later evolutionary addition (Makarova et al. 1999). The active site is shielded from contact with the solvent because it is buried within a tunnel that is covered by the side chains of two conserved tryptophans, one of which is part of the same β-sheet as the catalytic histidine and the other sitting in a loop of the α-helical bundle. The residues connecting the last α-helical segment of the core domain to the β1-strand of the first β-barrel form a flexible solvent-exposed loop that is the site of proteolytic cleavage in TGM1, 3, 5 and 6. At the same time, this *connecting region* constitutes the hinge around which the domain movements opening the enzymes are thought to occur, as has been demonstrated for TGM2 and FXIIIa (Pinkas et al. 2007; Stieler et al. 2013). *The two β-barrels* are composed of ~100–110 amino acids, each folded into β-sheets that are organized into two four- and three- antiparallel stranded plates. The first barrel contains a short α-helical segment between β-sheets 5–6. In the closed form of the enzymes the two β-barrels are closely appositioned to the surface of the catalytic domain in the region of the active site. The loop between β-sheets 3 and 4 in barrel 1 harbors a conserved tyrosine residue that protrudes into the catalytic site area in the closed form and is hydrogen bonded to the active site cysteine completing the occlusion of the catalytic residue.

There are two non-proline and one proline conserved *cis peptide bonds* in the red sea bream and human TGM2, the zymogen and active crystals of human TGM3 and FXIIIa (Noguchi et al. 2001; Ahvazi et al. 2002; Weiss et al. 1998). The bonds are present in the same locations within conserved sequence stretches in the isoforms and much heed has been formerly paid to their relevance for stabilization, activation or mechanism of action of the enzymes. *Cis* peptide bonds are energetically not favorable and are found only in 0.03 % of proteins. Their stabilization is attributed to extensive hydrogen bonding with neighbor atoms (Jabs et al. 1999). This network of bonds may weave the cis peptide bond containing regions together, potentially holding the active site containing residues in the proper orientation (Ahvazi et al. 2002). Alternatively, they may isomerize to *trans* in association with the conformational change and activation of the enzymes. However, the *cis* bonds were found in the substrate-bound, open forms of both TGM2 and FXIIIa, which supposedly represent the active enzymes and the proteolytically activated, nevertheless, closed form of TGM3, likely dismissing the hypothesis that the energy stored in these bonds is necessary for activation and supporting the idea that they are important for stability. The question remains unresolved.

1.3 Catalytic Mechanism – Organization of the Active Site

The essence of the catalytic mechanism of Tgases was understood long before the first atomic scale structural models became available thanks to classical enzymology studies, homology with papain-like proteases and experimental observations obtained from site-directed mutagenesis and chemical compound screening. However, to grasp the nuances of the catalytic mechanism it was necessary to reveal and derive the sense of the structural organization of the active site region of these enzymes (Keillor et al. 2015).

The reaction catalyzed by the Tgases has been described as a *modified ping-pong mechanism* (Folk 1969). The ping-pong mechanism is used by enzymes which sequentially react with two substrates. They also exist in two states, E and E', the latter being a relatively stable reaction intermediate, a form chemically modified as a result of interaction with the first reactant. When this substrate dissociates from the modified enzyme, E' can react with the second substrate. In the Tgase reaction (Fig. 1.2) the first substrate is the protein- or peptide-bound glutamine donor and the second substrate is a primary amine. The reaction may deviate from this scheme and is called 'modified' inasmuch as there is an alternate second substrate, water, which can hydrolyze the acyl-enzyme intermediate (E') and direct the reaction to deamidation. In the Tgase reaction the carboxamido group of the substrate glutamine is attacked by a nucleophilic cysteine (Cys314 in Hs FXIIIa/Cys277 in Hs TGM2/Cys272 in HsTGM3) located in the active site, that is made reactive by partial deprotonation via a charge relay system composed of a nearby histidine (His373 in Hs FXIIIa/His335 in Hs TGM2/His330 in HsTGM3) and an aspartate (Asp396 in Hs FXIIIa/Asp358 in Hs TGM2/Asp353 in HsTGM3). One of the imidazole nitrogens acting as a base deprives the sulfhydryl residue of the cysteine of a proton generating an imidazolium thiolate. The nucleophilic thiolate reacts with the carbonyl leading to a tetrahedral intermediate. The histidine imidazolium acts as a base catalyst of the decomposition of this intermediate to one molecule of released ammonia and an enzyme-substrate thioester. The reactive enzyme can be regenerated from this acylated complex by either amino- or hydrolysis. In the case of aminolysis the histidine again functions as a base to deprotonate the neutral amine, which will undertake a nucleophilic attack on the thiolester carbonyl forming a second tetrahedral intermediate. The decomposition of this releases the transacylated product and regenerates the catalytically competent thiolate cysteine. If there is no suitable acyl-acceptor amine but water has access to the active site, the acyl-enzyme intermediate is hydrolyzed to thiolate and glutamic acid. In this scheme the histidine functions as a general base, which is rendered more electronegative by the interaction of the appropriately situated aspartate residue with the other nitrogen of the imidazole ring.

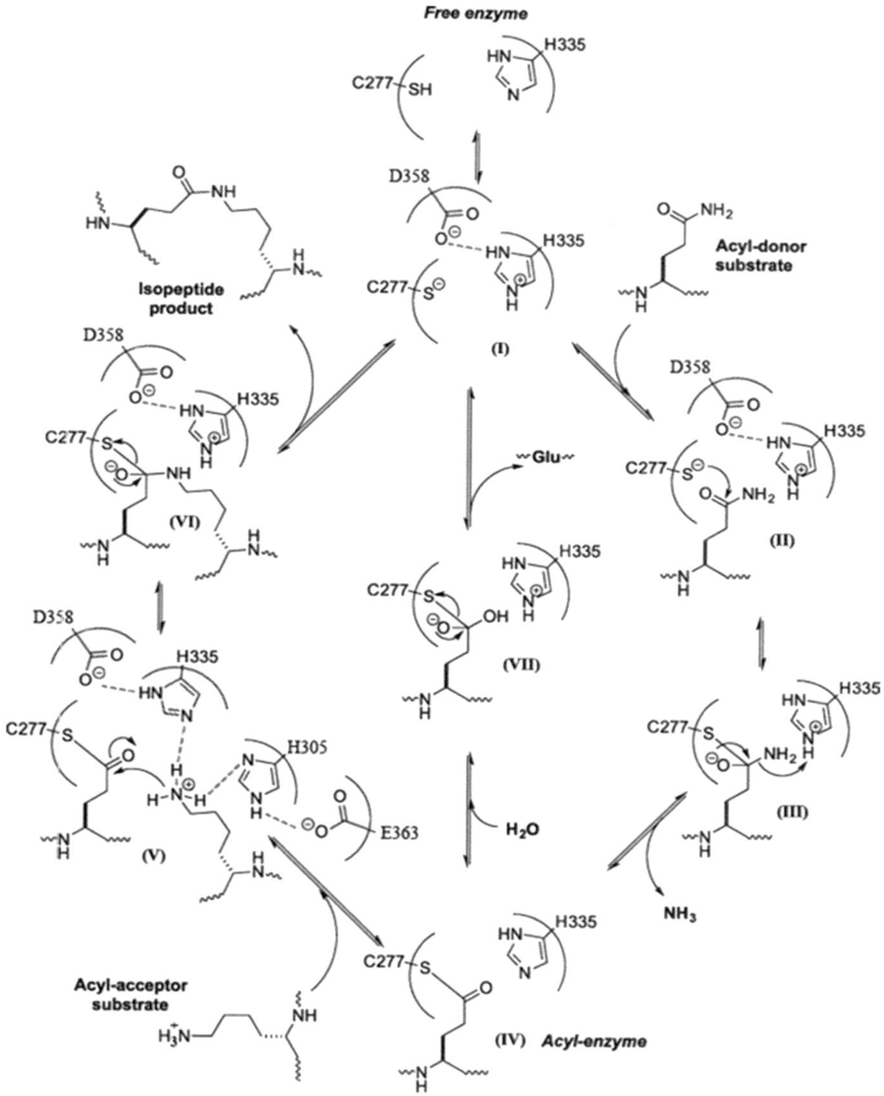

Fig. 1.2 Unified mechanism of Tgase-mediated acyl-transfer reactions on the example of TGM2. (Adapted from Keillor et al. 2014). The active form of the enzyme is the imidazolium thiolate shown as form I. The side chain of a select Gln residue is bound in a tunnel leading down to active site (see Michaelis complex II). Nucleophilic attack by the active site thiolate on the amide substrate carbonyl leads to formation of tetrahedral intermediate III. The subsequent decomposition of III gives acyl enzyme IV and results in expulsion of one equivalent of product ammonia, presumably with general acid assistance by imidazolium. Acyl enzyme IV now has two fates – either aminolysis or hydrolysis, the selectivity of which is discussed below. If a suitable amine acyl-acceptor substrate (such as the side chain of an accessible Lys residue) is present, it is bound in a second binding tunnel to form Michaelis complex *V*. Two imidazole groups near the active site organized in charge relays with acidic residues now function as general bases to deprotonate the amine, which is positively charged at physiological pH, during its attack on the thiolester carbonyl.

The catalytic cysteine is located near the N-terminus of an α-helix. This arrangement of the position of the nucleophile is also observed in cysteine proteases, subtilisin proteases, and α/β hydrolases, in microbial transglutaminase and in glutamine specific N-terminal amidase and peptide N-terminal glycanase (Pedersen et al. 1994). Mammalian transglutaminases and some cysteine proteases, such as papain and actinidin, also share a similar segment of α-helix and β-sheet containing the catalytic triad (Drenth et al. 1976; Baker 1980).

Although Tgases have an identical catalytic triad in a very similar configuration to thiol proteases, a significant difference lies in the surrounding residues. The active site of proteases needs to be well accessible so that a polypeptide segment containing the targeted peptide bond within a substrate protein could be easily accommodated in it without steric clash between the enzyme and its substrate. The active site also needs to be penetrable to water, which in this case is the primary second substrate that is supposed to hydrolize the bond. Therefore the active sites of proteases resemble a groove. In Tgases the E', acyl-enzyme intermediate needs to be protected for a sufficiently long time for a second substrate much less abundant than water to react with it. The first catalytic phase until the point of formation of the acyl-enzyme intermediate is quick and the nucleophilic attack by the amine substrate is considered the rate limiting step after the amine has formed a saturable Michaelis-Menten complex with the enzyme. This means that the intermediate needs to be guarded from water. To help the much less concentrated amine substrate in winning this race against the outnumbering and much more diffusible water molecules *the active site* (Fig. 1.3) *is shielded by hydrophobic residues and is located in a channel*, which the substrates to be connected need to approach from two opposing directions. In the active vertebrate Tgase isoforms *two tryptophan residues* arch over the catalytic cysteine containing a groove and the indole rings form an apolar roof. In human TGM2 five out of nine tryptophan residues sit within 20 Å of the active site cysteine, which may be involved in locking out water from the active site area (Nemes et al. 2005).

In TGM2 Trp241 was shown to stabilize the two tetrahedral oxyanion intermediates by hydrogen bonding of its indole nitrogen with the carbonyl. The corresponding tryptophans in active vertebrate Tgases are therefore also considered direct catalytic residues (Iismaa et al. 2003). This tryptophan is conspicuously absent from EBP42, the inactive member of the Tgase family. Modeling experiments suggested that Tyr 516 in TGM2 plays the same role. This idea was based on

---

Fig. 1.2 (continued) This attack, forming tetrahedral intermediate *VI*, has been shown to be rate limiting for a variety of acyl-donor and acyl-acceptor substrates. The decomposition of VI results in the release of transamidation product and the expulsion of thiolate, regenerating the free enzyme in its catalytically competent form. In the absence of suitable acyl-acceptor substrate, water can infiltrate the active site of acyl-enzyme IV. Subsequent activation by the general base imidazole, as observed in the cysteine proteases, leads to attack on thiolester to form tetrahedral intermediate *VII*. The decomposition of VII expels thiolate to regenerate the active site and releases the deamidation product glutamic acid

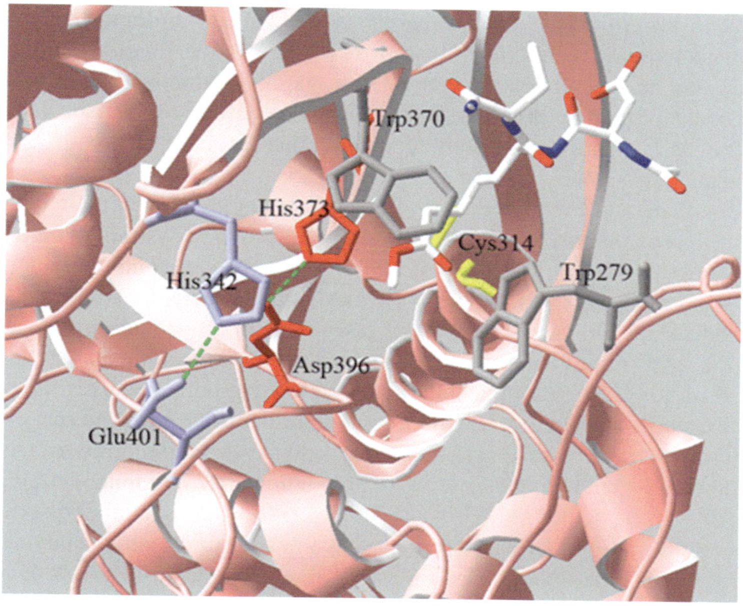

Fig. 1.3 Catalytically important residues in the active site region of FXIIIa° (PDB 4TYK). The active site is viewed from the side of the access of the amine donor. The catalytic cysteine (Cys314) is shown in *yellow*. The other two members of the catalytic triad (His373 and Asp396), which form a charge relay to deprotonate the cysteine, are shown in *red*. The other charge relay diad (His342 and Glu401), which play a newly identified role in deprotonating the attacking amine are rendered in *purple*. Large aromatic residues (Trp279, Trp370) shield the catalytic cysteine from water. In *white*: a fragment of the irreversible inhibitor, ZED1301, bonded to Cys314

the then existing GDP-bound closed TGM2 structure, in which to fit the substrate to the active site Trp241 had to be rotated out of its way (Chica et al. 2004). With the insight from the open enzyme structures it is clear that Trp241 remains in place and Tyr516 is removed from the vicinity of the active site together with the β-barrels, which opens access for the substrate from a different direction. The conserved or even increased activity of Tyr516 mutants and the deleterious effect of Trp241 mutations are consistent with these assumptions. Tyr516 is not directly involved in transamidase catalysis, but by forming a bond with Cys277 it stabilizes the compact conformation.

Recently, an *additional diad of essential catalytic residues* has been proposed based on the structure of activated calcium-complexed FXIIIa° (Stieler et al. 2013). A conserved histidine and a glutamate at positions corresponding to His342 in Hs FXIIIa/His305 in Hs TGM2 and Glu401 in FXIIIa/Glu363 in Hs TGM2 would facilitate the nucleophilic attack of the amine substrate at the acyl-enzyme intermediate. Mutation of this histidine did not affect the Km for the amine substrate significantly, but drastically reduced crosslink formation (Hettasch and Greenberg 1994). To act as a nucleophile in the reaction scheme the amine group, which is expected to be protonated and positively charged at physiological pH, when it enters the active site tunnel, must lose two protons. Formerly, the histidine of the catalytic triad has been evoked to answer for one of the deprotonation events, but the fate of the remaining proton was unknown. The new His-Glu diad is suggested to act like the His-Asp charge relay to soak up this second proton from the amino head group.

The food and textile industries have recently taken great interest in *Streptoverticillium mobaraense* transglutaminase for its different catalytic properties. The active site of this enzyme is located in a deep cleft at its edge. The catalytic residue, Cys64, sits at the bottom of the cleft. The secondary structure frameworks around these residues are similar to vertebrate Tgases. The most striking difference is that although Cys64, Asp255, and His274 superimpose well on the catalytic Cys-His-Asp triad of the vertebrate transglutaminases the asparagine occupies the position of the histidine and vice versa. Unlike in mammalian transglutaminases Cys64 in the bacterial enzyme is exposed to the solvent and can promptly react with the acyl-donor substrate eliminating the level of regulation that lies with the controlled accessibility of the active site. This may be one reason for the higher reaction rate. The microbial transglutaminase seems to have developed a novel catalytic mechanism by circular permutation of the original active site residues, which is not unprecedented and can also be observed in the NlpC/p60 superfamily, where the positions of the catalytic cysteine and histidine are swapped (Anantharaman et al. 2001). Here, in contrast the Cys64-Asp255 diad replaces functionally the Cys-His-Asp triad and His274 – although present – is considered enzymatically less essential.

1.4 Evolutionary Origins and Relationships of Vertebrate Tgases

The essence of an enzyme is its catalytic core, whereas the rest of its structure provides framework and stability to the catalytic residues, solvation, interaction surfaces with substrates and other proteins and regulation. In all known active Tgases the catalytic core is composed of an amino acid diad or triad formed by a cysteine, a histidine and less obligatorily, an acidic residue, usually an aspartate. Phylogenetic relationships show that this catalytic triad was derived once in evolution and was later co-opted for different mechanistically similar reactions such as transamidation, acyl transfer and peptide-bond hydrolysis in thiol proteases (c.f. previous and later sections).

Identification of novel transglutaminases progressed by way of sequence comparison with known mammalian Tgases and also via biochemical purification of novel proteins with de novo identified transglutaminase activities. The latter usually yielded *non-homologous proteins in bacteria, plants, ascidians and arthropods, which only share the active site residues* supporting an identical catalytic mechanism (Weraarchakul-Boonmark et al. 1992; Tokunaga et al. 1993; Cariello et al. 1997). PSI-BLAST searching of sequence databases initiated with human transglutaminases as a query unveiled an extensive class of phage, archeal, bacterial, yeast, nematode and higher eukaryotic proteins – termed transglutaminase-like (TGL) proteins. The sequence homology comprises three motifs around the three active site residues with conservation of the small residues two positions upstream from the catalytic cysteine, of the aromatic residues two positions down from the catalytic histidine, and an aromatic residue at the N-terminal side of the catalytic polar amino acid (Makarova et al. 1999; Zhang and Aravind 2012).

When X-ray crystallographic structures of transglutaminases became available they confirmed the *relationship between Tgases and papain-like thiol proteases* prompting their classification in the same superfamily in the structural Classification of Proteins (SCOP) database (Murzin et al. 1995). This kinship is based on spatiology, corresponding 3D topography of their active sites and their vicinities often called the transglutaminase-like or papain-like protease fold, while sequelogy between the two families is not high enough to be detected by search engines. The essential framework of the transglutaminase fold comprises an α-helix carrying the catalytic cysteine at its amino terminus packed against a successive three-stranded β-sheet with the second and third strands harboring the histidine and its polar partner to appropriately position them for a charge relay arrangement (Anantharaman and Aravind 2003). This kernel can be traced in an extremely broad functional and phylogenetic spectrum of proteins among which papain-like proteases, amino-group acetyltransferases, adenoviral proteases and ubiquitin carboxyl terminal hydrolases are the best known. In transglutaminases, papain-like proteases, and NH_2-acetyltransferases the three-stranded sheet is incorporated into a β-barrel (Anantharaman and Aravind 2003). Diversification of the TGL proteins ensued as phyletic change extended this kernel configuration by inserting various structural elements in the loops between the strands for purposes of creating interaction surfaces, regulatory elements and substrate specificity, as has been demonstrated by comparison of a homology model of the methanobacterial protein MTH795 with the core structure of FXIIIa (Makarova et al. 1999; Anantharaman and Aravind 2003). Most bacterial, archeal, yeast and nematode TGL proteins have short inserts, consistent with the lack of counterparts to the additional conserved elements of the Tgases (Makarova et al. 1999).

The advent of sequence-structure threading algorithms also yielded *enzymes, which are non-homologous to Tgases, but catalyze mechanistically similar reactions* and possess very similar active sites arranged into the thiol-protease fold. Examples include peptide N-terminal glycanases, glutamine specific N-terminal amidase, and the NlpC/p60 superfamily including bacterial autolysins, phage cell-wall hydrolases, and eukaryotic lecithin-retinol acyl-transferase (Anantharaman and Aravind 2003).

Other transglutaminase fold containing proteins lack the predicted catalytic residues and may have arisen during evolution from active ancestors by elimination of the enzymatic activity. These include a family of cyanobacterial and eukaryotic potential cytoskeletal proteins, typified by yeast Cyk3 and mouse Ky and the

PNGase-related Rad4/Xp-C, which are involved in nucleotide excision repair in yeast and humans (Anantharaman et al. 2001). They are thought to have been exempted for protein-protein interaction at different points of evolutionary development not unlike erythrocyte protein band 4.2.

The only primitive transglutaminase characterized in full structural detail was isolated from *Streptoverticillium mobaraense*. Its structure has been determined by X-ray crystallography (PDB ID:1 IU4) (Kashiwagi et al. 2002). In contrast to higher order Tgases this protein contains a single domain, its activity is Ca^{2+}-independent, the reaction rate and the substrate specificity for the acyl donor are higher and lower, respectively, and the deamidation activity is weaker than those of the vertebrate Tgases implying that it is difficult for a water molecule to play the role of an acyl acceptor (Ando et al. 1989; Ohtsuka et al. 2000). This protein is secreted from the cytoplasm as a zymogen and is activated by proteolytic processing like several vertebrate Tgases (Pasternack et al. 1998).

1.5 Regulation of Transglutaminases

1.5.1 Regulatory Role of Calcium

The actual role of calcium in the activity is somewhat less enigmatic today than it was for a long time, but still not fully understood. All studies are consistent in that every known vertebrate transglutaminase requires calcium or other metal ions. Experiments with other lanthanide trivalent cations (Er^{3+}, Sm^{3+}, Tb^{3+}, Lu^{3+}) have shown that calcium is the preferred cation although differences exist (Ahvazi et al. 2002). TGM3 showed similar activity with 1 mol Ca^{2+} + 2 moles Er^{3+} or 2 moles Ca^{2+} + 1 mol Yb^{3+}/mol enzyme, whereas FXIIIa is inhibited by Yb^{3+}. The *Streptoverticillium mobaraense* Tgase, however, is calcium independent (Ando et al. 1989). Interestingly, the transacylation of primary amines and the hydrolysis of the carboxamido groups of glutamine show widely different calcium ion requirements (Folk et al. 1968). The apparent k_D for calcium in the case of TGM2 is 90 μM (Bergamini 1988). The summed calcium affinity of FXIIIa is on a par with that of TGM2 with $k_D \approx 100$ μM. TGM2 could bind 6 moles of calcium per mole enzyme in equilibrium dialysis experiments (Bergamini 1988). The FXIIIa zymogen and FXIIIa bound equally 8 moles of calcium per mole enzyme out of which two had high and six had lower affinities (Lewis et al. 1978). NMR studies and equilibrium dialysis on an extended free calcium concentration range suggested the existence of low affinity Ca^{2+} binding sites on both human FXIIIa and human TGM2 in addition to high affinity ones in accordance with surface polarity analysis identifying high numbers of negatively charged clusters (Ambrus et al. 2001). Site-directed mutagenesis also suggested that TGM2 has at least five calcium binding sites, one of which corresponds to the high affinity site in TGM3 (Kiraly et al. 2009).

Suggested by the structural similarity in the involved regions the same high affinity metal ion binding sites are shared by probably all mammalian Tgase enzymes. The amino acids critical for calcium coordination are conserved in the entire transglutaminase family (Stieler et al. 2013). All calcium binding sites are located in the catalytic core domain. Depending on the circumstances of crystallization different numbers of these sites are occupied by calcium ions in different isoforms. Equilibrium dialysis experiments and $^{43}Ca^{2+}$ NMR usually indicate the presence of a higher number of binding sites because they are sensitive to low affinity associations with polar surface patches. The first resolved model of zymogenic TGM3 in the presence of calcium shows one tightly bound calcium ion. This represents a high affinity interaction site that is, probably, permanently occupied by calcium, which strongly contributes to the stability of the enzyme. The proteolytically processed active TGM3 was shown to take up two more calcium ions. Calcium ion chelation was highly exothermic and is therefore also expected to contribute to protein stability (Ahvazi et al. 2002). None of the numerous TGM2 crystals, not even those of activated TGM2, contained calcium (Pinkas et al. 2007 and PDB IDs:2Q3Z; 3S3J; 3S3P; 3S3S). The firstly solved structures of zymogenic inactive FXIIIa with a single or no calcium ion did not differ much (Yee et al. 1994; Fox et al. 1999). The most recently published 1.98 Å model of a proteolytically equally unprocessed yet active FXIIIa (FXIIIa°) in complex with an irreversible inhibitor contains three calcium ions. The so far best elaboration of the presumable activation mechanism stemmed from comparison of these three structures of FXIIIa. The differences between the inactive and active structures were translated into a plausible string of events suggested to unfold in sequence upon calcium binding.

The numbering of calcium binding sites in TGM3 is different from those assigned in FXIIIa with site 1 referring to site 3, site 2 to site 1, and site 3 to site 2. Unfortunately neither of the numbering schemes follows the order based on the polypeptide sequence. The reason for this is that when the inactive FXIIIa and the latent TGM3 structures were resolved both contained a single calcium ion, but at different physical sites that came to be referred to as site 1 in the respective literatures (Fox et al. 1999; Ahvazi et al. 2002). The description below follows the nomenclature adopted for FXIIIa (Fig. 1.4). As homology modeling of TGM1 and TGM6 was primarily based on the TGM3 models the calcium binding sites in these enzymes follows the TGM3 scheme (Boeshans et al. 2007; Thomas et al. 2013). While TGM2 also follows the TGM3 regime, it has been suggested to have three additional sites (Bergamini 1988; Kiraly et al. 2009).

Site 1 was identified as an unique calcium binding site in the X-ray structure of non-activated FXIIIa near the end of the loop connecting the catalytic core to the first β-barrel domain (Fox et al. 1999). This sole calcium ion is coordinated by the carbonyl group of Ala457 and five water molecules. Upon activation the coordination sphere will be completed and the calcium ion will also be bound by the side chains of Asn436, Glu485 and Glu490. This results in the movement of a loop and an α-helix closer to the ion. Calcium is bound to the equivalent physical area in

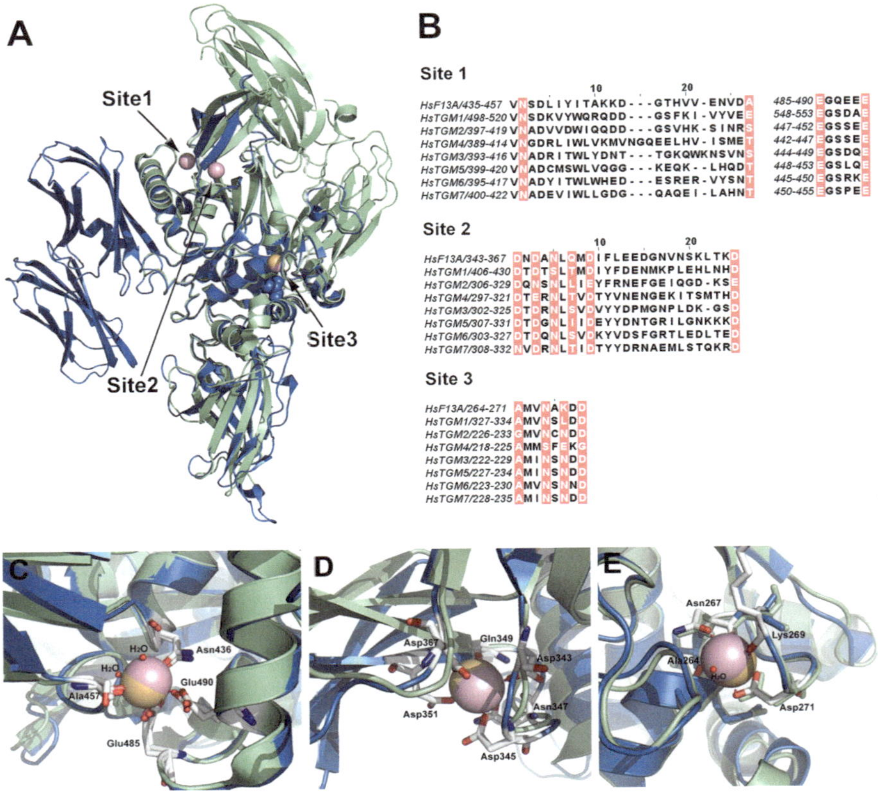

Fig. 1.4 Calcium binding sites of FXIIIa and activated TGM3. (**a**) The positions of the Ca^{2+}-binding sites on the overlay of FXIIIa° (PDB 4TKY) and TGM3 (PBD 1L9N); blue - FXIIIa°, green - TGM3, pink - Ca^{2+} ions belonging to FXIIIa, light orange - Ca^{2+} ions belonging to TGM3. (**b**) The calcium coordinating sequences are well conserved in the active members of the human Tgase family. (**c**) At site 1 virtually the same coordination geometry is found in both structures. (**d**) Compared to the active state of FXIIIa, the coordination polyhedral around the calcium is not yet fully established as the bond with Gln349 is lacking and the coordination site of Asp345 is not occupied by a water molecule in the TGM3 structure. As a consequence, the 'flap' (its β-strands are visible on the *left*) is not shifted and rotated and the hydrophobic substrate binding cavity is not formed. (**e**) The third calcium ion binding site in TGM3 shows similar geometry as the site in active FXIIIa° (Reproduced from Stieler et al. 2013)

TGM3 upon proteolytic cleavage of the zymogen (Ahvazi et al. 2002). The site has similar configurations in FXIII, FXIIIa° and latent or active TGM3 and is thought to adopt an EF-hand-like conformation (Fox et al. 1999; Stieler et al. 2013). Binding of a calcium ion to site 1 promotes no (FXIIIa) or only minor (TGM3) changes in local topography that do not in any directly obvious way affect conformations of residues near the active site. It has been suggested that in TGM2 occupation of the corresponding putative site by calcium would drag the peptide stretch Ile416-Ser419 away from stabilizing the first β-barrel through hydrogen bonding with its first β-sheet and could weaken the affinity of the enzyme for GDP/GTP, thus facilitating an activity disposed state (Liu et al. 2002).

Site 2 seems to be responsible for most of the inducible effects of calcium binding. In the TGM3 structure occupation of the corresponding site creates a transverse channel connecting the front and rear sides of the enzyme by pulling on the loop Asp320-Ser325 in the vicinity of the active site (Ahvazi et al. 2002). This change exposes Trp236 and Trp327, two tryptophans thought to control access to the active site. As this channel in the model of TGM3 is bound on one side by the first β-barrel domain, the corresponding regions of the open enzymes look totally different. In FXIIIa° binding of a calcium ion to Asp367, Asp351, Glu345, Asn347 and Asp343 at this site induces the rotation of a three-stranded β-sheet and leads to the opening of a hydrophobic cavity near the active site that will be crucial in accommodating non-polar residues of the substrate. One of the β-strands contains Trp370 whose side chain is rotated due to these movements at a right angle and will be pointing at Trp279. The side chains of the two tryptophans together form the roof of the tunnel across the active site. Structural developments at this site seem to be the most responsible for adaptation of the catalytic domain to accepting the substrates.

Virtually the same coordination geometry is found in FXIIIa° and active TGM3 at *site 3* (site 1 in TGM3) (Stieler et al. 2013). In TGM3 the tightly bound calcium ion at this site is essential but not sufficient for activity. It may be required to maintain the correct three-dimensional structure of the active site region (Kanchan et al. 2013). In FXIIIa occupation of this site results in the reorientation of a loop near the protein surface and this may affect the binding of the lysine-containing substrate.

Beyond the conserved three binding sites, TGM2 has been suggested to have three more (Bergamini 1988). Mutagenesis of the putative calcium sites in human TGM2 raised the possibility of cooperative interactions between the sites, since mutation of one binding site led to loss of up to four calcium ions per protein molecule. Two of the non-conserved sites were putatively allocated to the negatively charged surface patches, Asp151-Glu155 and Asp434-Asp438 (*sites 4 and 5*). Multiplex mutations in each of these peptide stretches led to loss of three calcium ions per TGM2 molecule (Kiraly et al. 2009).

Based on the TGM1 homology model it has been suggested that calcium binding at sites 1 and 2 serves to anchor together β-strands 7 and 8 and β-strands 12 and 13 of the core domain, respectively, with neighboring α-helices. Clustering these structural elements together forms what has been termed the 'flap-motif', a β-stranded protrusion on the core domain. The 'flap' overlays the first β-barrel and is connected to it with hydrogen bonds in the closed conformation. Although the strands of the 'flap'-motif are common with other transglutaminases, the sequence homology is the lowest in this region and their orientations are different, which is consistent with the observation that the high <B> temperature factor values of the TGM2 and TGM3 crystal structures reveal variations and possible flexibility in this region (Boeshans et al. 2007). Indeed, comparison with the GDP-bound structure after deuterium exchange revealed that this region in

TGM2 becomes stabilized upon calcium (and inhibitor) binding (Iversen et al. 2014).

Importantly, structural changes elicited by binding of calcium ions do not provide complete explanation for the hinge motion of the two β-barrel domains. The calcium-bound crystal structures revealed no structural rearrangements (of FXIIIa and TGM3) that would point at a strikingly evident mechanism for this rearrangement. TGM3 has been crystallized with three bound calcium ions in a closed conformation suggesting that calcium binding is not orthogonally antagonistic with adopting a closed state, although it probably makes it not favored. Whether the Tgases rendered in the models binding one to three calcium ions are active or binding of additional lower affinity ions would be necessary for their activation is not certain.

1.5.2 Regulation by Purine Nucleotide Binding

The regulation of transglutaminases is accomplished at several levels. All known vertebrate transglutaminases require calcium for their activity, while FXIIIa and TGM1 and 3 (and potentially also TGM5 and 6) are produced as zymogens and require proteolytic activation. Negative regulation by purine nucleotide concentration represents an additional level of protection against unsolicited transglutaminase activation in the intracellular milieu (Lai et al. 2007). The basis of the inhibitory effect is binding of the nucleotides to a pocket contributed by the catalytic core and the first β-barrel domains.

Besides TGM2, TGM3, 5 and 6 have also been shown to be inhibited by purine nucleotides (Achyuthan and Greenberg 1987; Im et al. 1990; Boeshans et al. 2007; Candi et al. 2004; Thomas et al. 2013). To reveal the structural basis of nucleotide inhibition TGM2 has been co-crystallized with GDP, GTP and ATP, and TGM3 with bound GMP (although the article reporting the latter structure was later retracted) (Liu et al. 2002; Han et al. 2010; Jang et al. 2014). The purine nucleotide binding pockets of TGM2 and 3 are related to each other in their positions and several key amino acids, but they are very superficially related to those described in G-proteins. Mg^{2+} is not required for guanine nucleotide binding to Tgase proteins. Homology modeling suggests that TGM1, 5, 6 and 7 may also bind GDP/GTP (Boeshans et al. 2007). Rat TGM4 has been shown to bind to GTP-agarose resin (Mariniello et al. 2003). Although a binding pocket is predicted in TGM1 based on homology modeling after TGM3, GTP does not inhibit TGM1 activity in the range of 20–500 μM in the presence of 0.5 mM calcium, which has been shown to drastically reduce the activities of TGM2 and 3 and less so, but still significantly, those of TGM5 and 6 (Candi et al. 2004; Thomas et al. 2013). The authors of this study left it undecided whether GTP affinity may be diminished or GTP binds but does not have an effect. The position of the putative binding pocket on the interface of the catalytic and the first β-barrel domains akin to TGM2 and 3, with the nucleotide pivoting the two together, makes it unlikely that in a nucleotide-bound state the active site could be accessible. TGM2 has been shown to be fifty times more sensitive to GTP mediated inhibition than TGM6 (Thomas et al. 2013). TGM2 has been found to hydrolyze both ATP and GTP with equal k_{cat}, TGM3 is a 100-fold faster GTPase than ATPase, while TGM5 has a measurable hydrolase activity only toward GTP (Iismaa et al. 1997; Candi et al. 2004). Protein band 4.2 in contrast with the active members of the protein family binds ATP but not GTP (Azim et al. 1996).

The reason for the apparently diversified responsiveness of the otherwise closely related Tgases to purine nucleotides, which is unlike the similar calcium sensitivity, may be a surprisingly low level of sequence conservation in the nucleotide binding groove. Although the groove is fitted with aromatic residues (Tyr174 and Tyr583 in TGM2) in each isoform to hold the purine ring in place and positively charged side chains to interact with the phosphates and to neutralize the extra anionic charges arising during hydrolysis, they are not in corresponding positions (Fig. 1.5). An important exception, Arg579 in rat TGM2, is a conserved residue in human TGM2 and also in TGM3 and conservatively substituted with a lysine in TGM4-7. It has been suggested that unlike other residues in the nucleotide binding cleft, which contribute to bolting the core and barrel 1 together when purine nucleotides are present, it destabilizes the closed form. When it is mutated to alanine the enzyme tends to adopt a semi-compact form based on its behavior in nPAGE (Begg et al. 2006). The purpose of GTP binding would in part be to neutralize the intrinsic destabilizing effect of this arginine.

It is confounding that the nucleotide hydrolase activity of TGM2 has been consistently localized to an N-terminal part of the molecule comprising the β-sandwich and part of the transamidase catalytic domain extending to at least amino acid 185 (Lai et al. 1996). This is in concert with the surprising observation that short splice variants of the enzyme lacking most or all of the β-barrel domains and, therefore, most of the amino acids forming the nucleotide binding pocket have elevated GTPase activity (Fraij 1996).

TGM2 was shown to be identical with the 74 to 80 kD $G\alpha_h$ subunit of the dimeric G-protein, Gh, that functions in concert with the α_{1B}-adrenergic, α_{1D}-adrenergic, oxitocine, FSH and thromboxane A_2 receptors to activate phospholipase C (Nakaoka et al. 1994; Baek et al. 1993, 1996; Vezza et al. 1999; Lin et al. 2010). TGM2 shows no consensus sequences with heterotrimeric G-proteins or small molecular size GTP-binding proteins. Despite their conserved overall structure and similar behavior toward purine nucleotides, no G-protein function has been described for other Tgase family members.

The hallmark of both tripartite and small GTP-binding proteins, which function in cellular signaling, is their ability to undergo structural changes in response to alternate binding of GDP and GTP. The resulting structural transition is essential for recognizing different protein partners and is at the heart of the mechanism by which these factors relay signals between ligand-bound receptors and their intracellular executors. It is also important to recognize that these structural changes must not occur spontaneously, but need to be prompted exclusively by nucleotide exchange, otherwise the regulated signaling pathway would go awry. GEFs have a

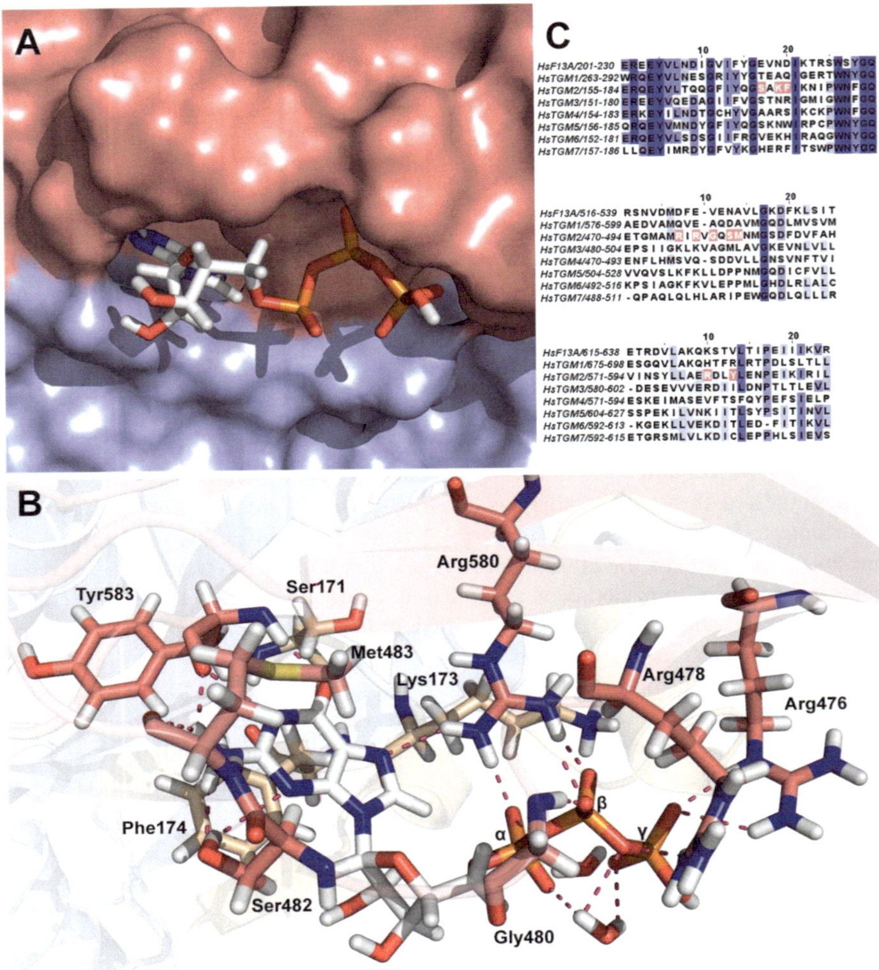

Fig. 1.5 Environment of the GTP binding site in TGM2 as revealed by the structure PDB:4PYG. (**a**) The GTP molecule snugly fits into a pocket formed by the catalytic (*purple*) and the first β-barrel (*salmon*) domains. (**b**) The amino acid residues and water molecules involved in interaction with GTP are shown as viewed from the *top* in (**a**). H-bonds are shown as *pink dashed lines*. (**c**) The GDP/GTP-binding pocket shows surprisingly low sequence level conservation (the amino acids involved in direct contact with the nucleotide are highlighted in pink)

dual activity: they destabilize the strong interaction with GDP and stabilize the nucleotide-free G-protein. The GEF-G-protein complex is in turn dissociated by GTP binding, which – present at a tenfold higher concentration in the cell – outcompetes GDP in this step. The interaction surface with adrenoreceptor α1B, a GEF of TGM2, was localized to a tripartite motif consisting of Leu547-Ile561, Arg564-Asp581 and Gln633-Glu646 (Feng et al. 1999). The Val665-Lys672 motif near the C-terminus in the second β-barrel domain was identified as the interaction site with phospholipase Cδ1, the downstream effector of $G\alpha_h$ mediated adrenergic signaling (Hwang et al. 1995).

Interestingly, in TGM2 the GDP and GTP-bound structures do not show any striking conformational difference (Liu et al. 2002; Jang et al. 2014). In conclusion, from a structural point of view, at the current state of understanding of the nucleotidase activity of transglutaminases and the role of TGM2 as a G-protein, several questions remain open.

1.5.3 Regulation by Proteolysis

Under proteolytic activation of Tgases we need to speak about two apparently different processing steps. FXIIIa and TGM1 are longer than the other members of the family and contain an N-terminal extension. The two proteins share no similarity in these regions, but in the case of both TGM1 and FXIIIa they are inhibitory to the enzymes. TGM1, 3, 5 and 6, on the other hand, are cleaved in the intrinsically disordered segment between the catalytic and the first β-barrel domains.

The *N-terminus of FXIIIa*, also called the activation peptide, is cleaved off by thrombin between Arg37 and Gly38 as part of the physiological blood coagulation cascade (Muszbek et al. 2011). The structure of thrombin-cleaved FXIIIa is remarkably similar to that of the zymogen: there are no large conformational changes in the protein and the 37 residue amino terminus activation peptide remains associated with the rest of the molecule (Yee et al. 1995).

In the case of TGM1 the inhibitory effect of the N-terminal tail is not complete and allows for a low residual activity. The N-terminus contains a cysteine cluster (Cys47-Cys53) that is subject to esterification with myristoic and palmytoic acids through which TGM1 is tethered to the inner leaflet of the plasma membrane in maturing keratinocytes in the upper spinosum and granulosum strata of the epidermis, most probably to concentrate the enzyme to the site of cornified envelope formation. *In human TGM1 the proteolytic cleavage sites*, which are associated with maturation, are after Gly93 and Gly573 and TGM1 is present in keratinocytes as combinations of 106, 67, 33 and 10 kD fragments – the 106 kD form being the full-length TGM1 (Kim et al. 1992; Steinert et al. 1996a, b). The 67 kD fragment corresponds to residues 94–573 comprising the β-sandwich and the core domains and the 33 kD fragment to residues 574–816 comprising the two β-barrel domains. Processing by the calcium-dependent intracellular proteases, m-calpain and μ-calpain, removes the N-terminal 10 kD membrane-anchored fragment. The rest of TGM1 is either released from the membrane as a soluble, 67/33 kD form or remains associated with it through secondary interactions between the 10/67/33 kD fragments. The 67 kD and the 106 kD forms are also present in the soluble cytosolic fractions of the cells. The soluble full-length form has roughly tenfold higher specific activity than its membrane-anchored zymogen counterpart. Proteolytic cleavage to the membrane-bound 67/33/10 kD form increases its activity

100–200 times. The soluble 67/33 kD form is around half as active as with the 10 kD fragment and the dissociation of the 33 kD fragment reduces the activity of the 67 kD form even further by a factor of 2 to 3. Recombinant full-length TGM1 also has lower specific activity compared to the processed form even in the absence of membrane anchorage. Engineered N-terminal deletions that render the protein about the same size as the FXIIIa zymogen (TGM1Δ1-61) or thrombin activated FXIIIa (TGM1Δ1-97) resulted in 10–15 and 3–6 fold increases in specific activity, whereas a more radical truncation of the entire N-terminal domain (TGM1Δ1-123) annihilated the activity (Kim et al. 1994).

TGM3 proteolytic digestion, although not absolutely necessary since the zymogen possesses activity at un-physiologically high calcium concentrations, tremendously increases its responsiveness to calcium (Kim et al. 1990). TGM3 is processed into 27 and 55 kD fragments (Kim et al. 1990). Proteolytic cleavage at Ser469 in the hinge region that connects the catalytic domain to the first β-barrel facilitates the acquisition of two calcium ions in addition to the one already tightly bound (Ahvazi et al. 2002). This would provide only a limited mechanistic explanation for how proteolytic cleavage can activate the enzyme and the exact structural rearrangement that leads to the higher calcium sensitivity is still unclear. In the zymogenic TGM3 crystal the density for the flexible loop containing the cleavage site could not be resolved, and in the active enzyme crystal an even longer region is missing, probably because it retains its malleable dynamic nature. If it lies near the protein surface the intact twenty amino acid loop may occlude binding of calcium ions to sites 2 and 3. Although the cleaved fragment containing the N-terminal β-sandwich and the catalytic domains is equally active, the proteolytic fragments remain together in a complex that strongly resembles the closed conformation of either TGM3 or other transglutaminases (Ahvazi et al. 2002).

C-terminal deletion experiments showed that the removal of the two β-barrel domains of TGM2 stripped the enzyme of transamidating activity. However, the purified 55 kD subunit of TGM3, which is cleaved at Ser469 in the loop connecting the core and the first β-barrel domain, is fully active. It is possible that the C-terminal deletion mutants of TGM2 were not properly folded although a compensatory increase in the GTPase activity was recorded implying a functional conformation (Lai et al. 1996).

TGM5 has been shown to be processed into 28 and 53 kD fragments in insect epithelial cells used for baculoviral protein expression and in transfected HEK-293 and HaCaT cells by unknown proteases (Pietroni and von Hippel 2008). The 53 kD fragment comprises the N-terminal β-sandwich and the catalytic core domains, and the smaller fragment is composed of the two β-barrels. *TGM6* also has high specific activity even in the absence of proteolytic processing and the function of cleavage in the loop connecting the core and the first β-barrel domain is unknown (Thomas et al. 2013).

1.5.4 Regulation by Redox Potential

Early investigations with TGM2 and -SH reagents or copper in the presence of calcium suggested that the reduced state of -SH groups other than the catalytic cysteine were also essential for activity of the enzyme (Folk and Cole 1996a, b, c; Boothe and Folk 1969). Although conditions in the extracellular milieu are predicted to favor activation, extracellular TGM2 was shown to only become active in response to inflammation (Siegel et al. 2008). This phenomenon, which is very important with regard to the central role of trans- and deamidation of gliadin peptides in the celiac immune response, has also been tied to the old observation that TGM2 is susceptible to oxidation (Boothe and Folk 1969; Connellan and Folk 1969; Chung and Folk 1970; Roth et al. 1986). TGM2 contains 20 cysteines. Cys230, Cys370 and Cys371 have been shown to form a triad (Fig. 1.6) in which Cys370 can be disulfide bonded to either of the other two cysteines under oxidative conditions (Stamnaes et al. 2010). This cysteine triad is specific to TGM2 and is not present in other isoforms. Calcium protects against formation of the Cys230-Cys370 and Cys370-Cys371 disulfide bonds. Cys230 is located within calcium binding site 3.

Cys370, Cys371 and Asp358 of TGM2 are situated within the same loop as the two cysteines sitting directly opposite the catalytic asparagine in the rising and returning strands of the same loop. In the TGM2 structure in complex with Ac-P (DON)LPF-NH_2 the vicinal cysteines form the disulfide bond (Pinkas et al. 2007). The distortion from the sharp kink in the peptide backbone is evident in that the loop is twisted and coils downward. The rest of the loop carrying Asp358 is in good overlay in the structures of closed and open TGM2 (1KV3 and 2Q3Z); the distances between Asp358, His335 and the catalytic cysteine are not altered much suggesting that the disulfide bond does not significantly influence the arrangement of the catalytic triad.

Cys230 of TGM2 is located within calcium binding site 3, which based on the similarity with TGM3 and FXIIIa comprises Ala226-Asp233. The presence of the Cys230-Cys370 disulfide bond, then, may also interfere with calcium binding. A Cys230-Cys370 disulfide bond was found in both the ATP and the GTP complex crystal structures, but not in the GDP-bound structure (Liu et al. 2002; Han et al. 2010; Jang et al. 2014).

The cysteine triad is not replicated in TGM6 as the residue corresponding to Cys230 of TGM2 is a serine in TGM6. Nevertheless, TGM6 is profoundly sensitive to oxidative inactivation implying that oxidation of residues other than the proposed cysteine triad may be involved in the redox regulation of transglutaminases (Thomas et al. 2013).

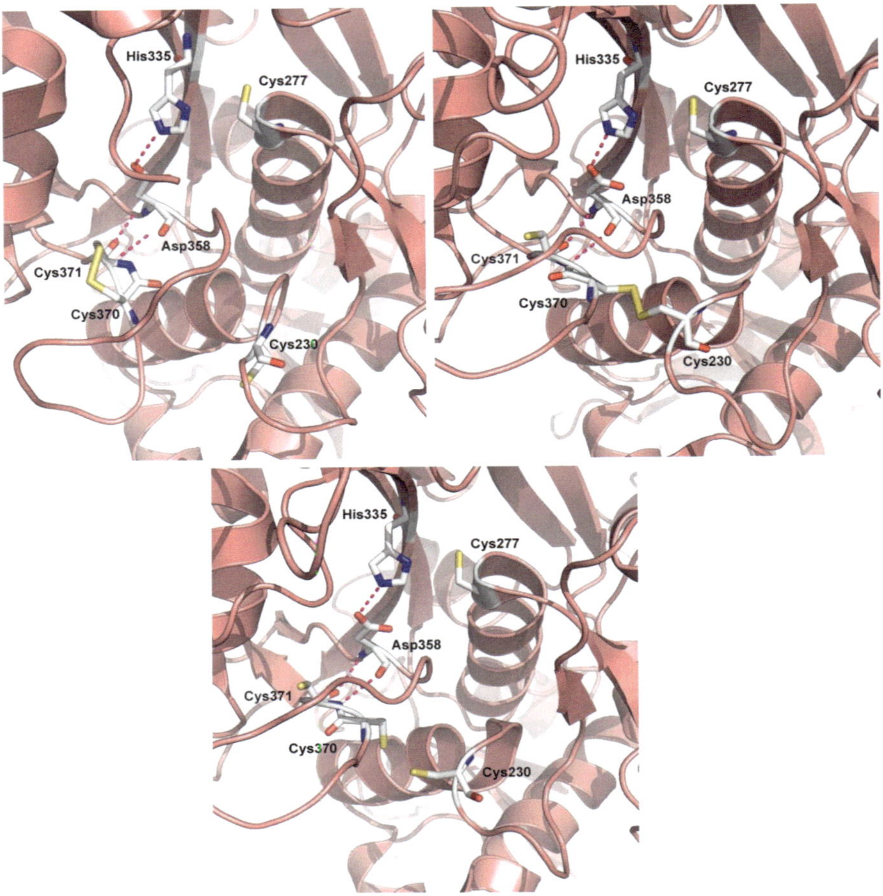

Fig. 1.6 Comparison of the oxidation state of the Cys230-Cys370-Cys371 triad in available crystal structures of TGM2. *Upper left*: 2Q3Z inhibitor-bound open; *upper right*: 4PYG GTP-bound; *bottom*: 1KV3 GDP-bound. In the GDP-bound structural model Cys230, Cys370 and Cys371 are reduced. In the GTP-bound model Cys230 forms a disulfide bond with Cys370, while Cys371 is reduced. In the open, inhibitor-bound structure of Pinkas et al., Cys370 and Cys371 form a vicinal disulfide bond and Cys230 is reduced. (The oxidized state is not a consistent, obligatory feature of the open structures as in the models, 3S3J, 3S3P, 3S3S, generated in the presence of different inhibitors, all three cysteines are reduced)

1.6 Open-Close Conformational Rearrangement and Its Relation to the Catalytic Cycle and Activation

It is didactic to describe the conformational reorganizations of FXIIIa, TGM2 and by extension probably of other vertebrate transglutaminases on at least two levels. One involves the repositioning of the domains relative to one another, and the other concerns more subtle adaptations in secondary structure, and rearrangement of local structural environment around particular structural elements. In the cases of the best studied FXIIIa and TGM2 we know of two conformational states on the first level (Fig. 1.7). The first, where the two C-terminal β-barrel domains fold back on the core around a hinge region and the proteins are compact or 'closed', is similar in the two isoforms. In the second type of conformers the N-terminal and the catalytic domains overlay nicely, but the C-terminal β-barrels are detached from the surface of the catalytic domain, in which sense the molecules are 'open'; however, in the case of TGM2 all domains come to be arranged more or less in a line and the protein is extended, whereas for FXIIIa the two β-barrels turn sidewise resulting in an horseshoe arrangement with both termini pointing to the same direction (Fig. 1.7). These changes expose ≈ 3000 $Å^2$ buried surface area to the solvent between the core and the β-barrels, which requires the investment of the equivalent of an estimated 11 kcal/mol binding energy (Weiss et al. 1998).

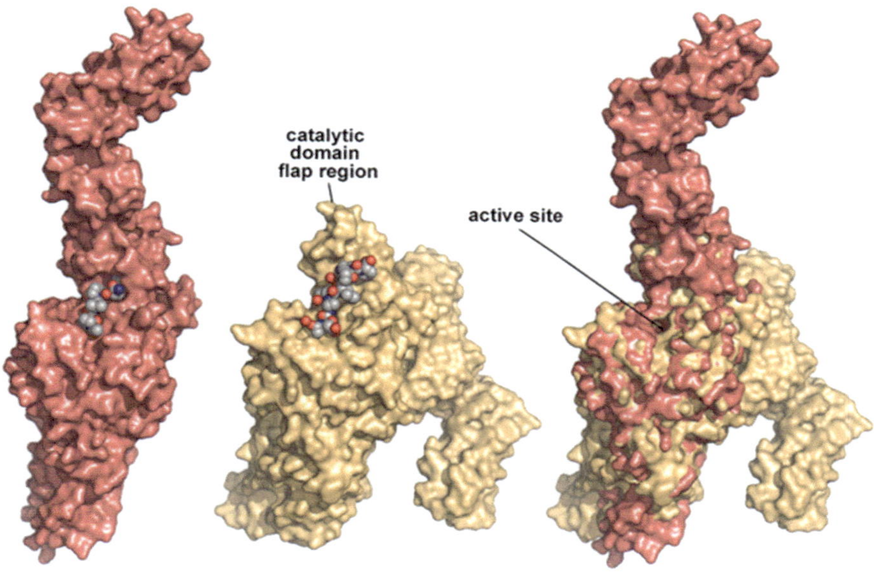

Fig. 1.7 Comparison of the open structures of TGM2 and FXIIIa° in their respective inhibitor-bound forms. 2Q3Z TGM2: *salmon*; 4KTY FXIIIa°: *mustard*; the surface areas contributed by the active site cysteines is shown in *pink*; the inhibitor molecules (Ac-P(DON)LPF-NH_2 and ZED1301) are rendered in *CPK coloring*. It is obvious that TGM2 is elongated, whereas FXIIIa° has a *horseshoe* shaped appearance. The different positions of the β-barrels and the presence of the N-terminal activation peptide in FXIIIa° are responsible for most of the space filling differences. The 'flap' is evident as a protrusion on the top of FXIIIa°. In TGM2 it is adsorbed into the first β-barrel to stabilize its new position. It is noteworthy how the paths of the substrate peptide chains on the enzyme surfaces are different due to the different conformations of the base of the 'flap' in the two isoforms. In TGM2 the inhibitor's course cannot continue in the same direction beyond the hydrophobic cavity that accommodates its phenylalanine ring and which is also present in FXIIIa°. The inhibitor peptide chain needs to turn sharply backward to curl around the density generated by a straightened repositioned β-sheet

The *transition between the open and closed states* remains controversial with respect to the actual stimuli that induces it. Firstly – although this could be a genuine theoretical possibility – opening and closing have not been suggested to occur during the catalytic cycle. TGM2 molecules in the absence of an allosteric effector populate both states, which can be separated by nPAGE (Lorand et al. 1979, 1988; Murthy et al. 2002) suggesting that they are relatively stable and equilibrate with each other slowly. Recently, it has been shown experimentally that for TGM2 the time constants for these conformational transitions in the absence of any allosteric effector ($k_{open} = 0.139\ min^{-1}$ and $k_{close} = 0.05\ min^{-1}$) are slower than the catalytic constant (k_{cat} for transamidation is in the range of 3–4 sec^{-1}) supporting the view that the open and closed forms represent metastable states respectively poised for and recalcitrant to catalysis (Kim et al. 1994; Clouthier et al. 2012). At present, it seems reasonable to argue that *opening and closing are spontaneously occurring processes and the two forms are in dynamic equilibrium with each other*. Guanine nucleotides skew this equilibrium strongly toward the closed form, obstruction of the active site and inactivation by acting as a bolt that fastens the catalytic and β-barrel domains together. Single-molecule FRET experiments performed with TGM2 tagged with donor and acceptor at its ends suggest that protonation of residues at lower pHs favors the closed conformation, whereas around pH 7 the efficiency of FRET was lower, consistent with the opening of the molecules (Tyagi et al. 2014). This may reflect the effect of redistribution of hydrogen bonds and salt bridges on the conformation and is in accord with the dominance of the slow migrating form in nPAGE.

How calcium binding per se is involved in the regulation, whether it directly induces the gross conformational transition of opening as was originally imagined (Bergamini 1988) or acts more like GDP/GTP and stabilizes a conformer that tends to arise spontaneously can be a matter of debate. The ability of the open and closed conformers to bind calcium and the affinities of these forms to the ion have not been studied yet. It nevertheless has been suggested that in TGM2 if calcium occupied the putative binding site 3 this would drag the peptide stretch Ile416-Ser419 away from stabilizing the first β-barrel through hydrogen bonding with its first β-sheet and could weaken the affinity of the enzyme for GDP/GTP, thus facilitating an activity disposed state when guanine nucleotides are present (Liu et al. 2002). This could explain decreased GDP/GTP binding in the presence of calcium. Occupation of site 2 may also diminish the contact between the core and the two C-terminal domains by inducing a twist in the four β-stranded 'flap' at the top of the catalytic domain that is H-bonded to the first β-barrel.

Finally, interaction with the substrate itself interferes with the folding of the C-terminal β-barrels back over the core and favors the open form. If an irreversible inhibitor is used this form becomes permanent because the inhibitor directly interferes with the re-closure of the enzyme (Pinkas et al. 2007; Iversen et al. 2014).

The *inhibitor-bound open TGM2 structures and inhibitor-bound FXIIIa°*, which in addition contains three calcium ions, *show marked differences* (Pinkas et al. 2007; Stieler et al. 2013). In TGM2 all domains come to be arranged more or less in a line and the protein is extended. In FXIIIa° the two β-barrel domains are lifted off the surface of the core domain and are folded to the side around the longitudinal axis of the β-sandwich and the core. The catalytic domain in both configurations shows further local adaptations to the effect that the substrate binding sites are exposed and the active site is unobstructed. Which of these changes are attributable to calcium binding and which to the irreversible inhibitors, which are the equivalent of the respective glutaminyl substrates here, is a matter of question. Which represent frozen intermediates of dynamic rearrangements captured during crystal packing is equally hard to resolve. The loop which connects the core to the first β-barrel domain is not visible in either type of open form, probably because of its pronounced flexibility. In the open FXIIIa° the loop between the sixth and seventh β-strands of the first β-barrel forms interactions with the Glu489-Tyr500 α-helix of the core domain and the connecting loop between the β3 and β4 strands of the first barrel establishes new contact with the loop connecting β8 and β9 of the catalytic domain. It follows, then, that it is not the structure of the connecting parts in the hinge region that keeps the FXIIIa° enzyme in this conformation, but the back-flipped first β-barrel on the one hand, and the catalytic domain on the other needed to provide the secondary interactions that stabilize this form. For that matter, these interactions seem rather limited and the 4KTY structure may represent just one geometry that the β-barrels are able to adopt, which may have been frozen by crystal packing effects.

Besides the conspicuous movements of the β-barrels *the region most affected by structural remodeling upon effector/substrate binding is in the vicinity of the core domain 'flap' region* (Fig. 1.8). The structural adaptation of this region is in intimate connection with the dissociation/association of the barrels and the catalytic domain. Although the strands of the 'flap' -motif are common with other transglutaminases, the sequence homology is the lowest in this region and their orientations are different, which is consistent with the observation that the high <B> temperature factor values of the TGM2 and TGM3 crystal structures reveal variations and possible flexibility in this region (Boeshans et al. 2007). Calcium binding at sites 1 and 2 clusters the structural elements around the base of the 'flap' together. Comparison of the FXIII and FXIIIa° structures reveals that in response to calcium binding to site 2 the β-strands comprising the 'flap' rotate (Stieler et al. 2013). In the closed state the 'flap' is hydrogen bonded to the first β-barrel. This rotational movement favors the disruption of these interactions and leverages the first β-barrel off the surface of the catalytic domain. Calcium occupation of this site also leads to the opening of a hydrophobic cavity near the active site that will be crucial for accommodating a non-polar residue of the substrate. In TGM2 amino acids Leu312 to Arg317 – which form the lateral β-strand of the 'flap' in the closed structure – become restructured as an α-helix when the molecule is open (Pinkas et al. 2007). This helix runs across in front of the antiparallel β-strands of barrel 1 providing support for the barrel in its new position. As a large difference from both the closed TGM2 and the FXIIIa° structures the preceding long β-strand comprising amino acids Pro294-His305, which is bent at a right angle in the closed

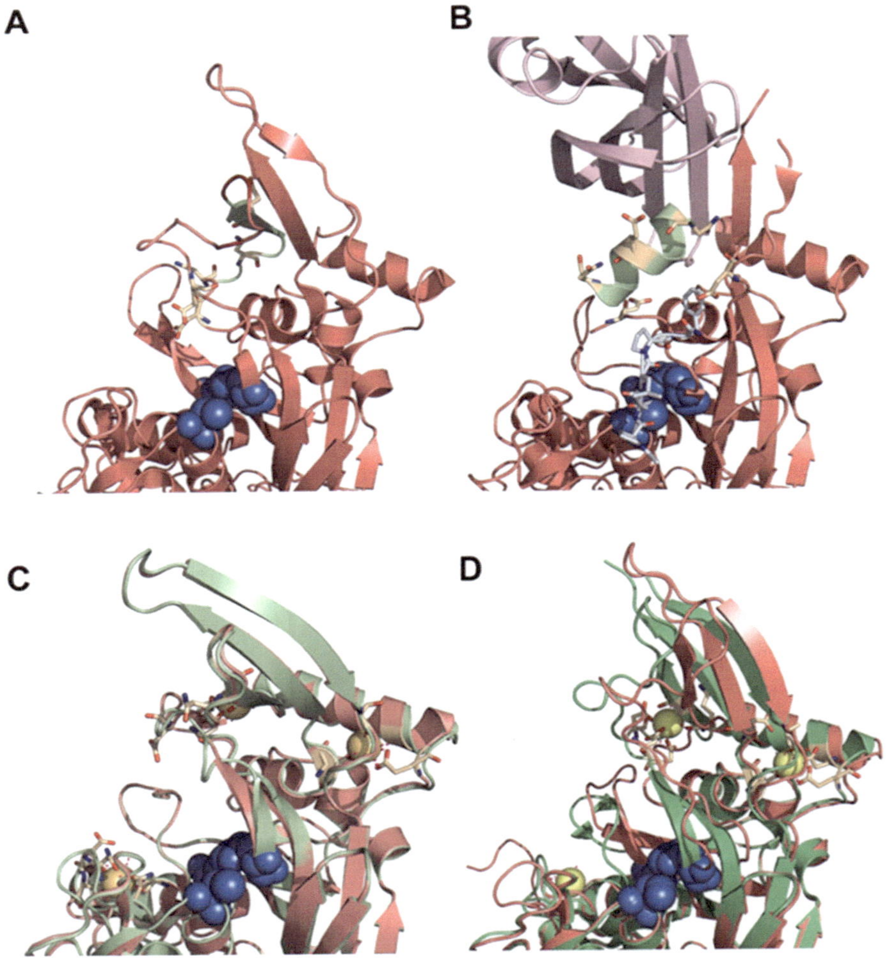

Fig. 1.8 Comparison of the active site, glutamyl substrate binding and 'flap' regions of TGM2, TGM3 and FXIIIa. (**a**) GDP-bound TGM2 (1KV3); (**b**) inhibitor-bound open TGM2 (2Q3Z); (**c**) overlay of TGM3 with a single Ca^{2+} ion (1L9M, *salmon*) and TGM3 with three Ca^{2+} ions (1L9N, *green*); (**d**) overlay of FXIIIa zymogen (1GGT, *salmon*) and FXIIIa° with three Ca^{2+} ions and irreversible inhibitor (not shown) (4KTY, *green*). The catalytic triad is rendered in *blue*. Ca^{2+} belonging to the salmon structures is shown in *light orange*, those belonging to the *green* structures are shown in *yellow*. The residues involved in calcium binding are shown in *wheat color*. The bound pentapeptide inhibitor of TGM2, Ac-P(DON)LPF-NH_2, is in *CPK coloring*. In 1KV3, 1L9M, 1L9N and 1GGT the β-barrels would block the view of the depicted regions and they have been removed. In 4KTY the barrels are not shown because they do not affect the configuration of the regions of interest. The 'flap' region of the catalytic domain is the structure formed by three β-strands and the connecting loops on top. In the inhibitor-bound open TGM2 only two of its strands are retained and tilt toward the *right aligning* to the edge of the β-sheets in the first barrel domain (*pale pink*). The remaining β-strand refolds as an α-helix (the corresponding sequences are shown in *green* in (**a**) and (**b**)). In TGM3 occupation of two more calcium binding sites does not cause a large difference in the structure of the 'flap', probably because the coordination polyhedron of the metal is not fully formed in this structure. It must not be forgotten that the surface shown here is covered by the two β-barrel domains in reality

enzyme, straightens up and the hydrophobic cavity is formed by the side chains of Leu312, Ile313 and Phe316 of the helix, Ala304, Ile331 and Leu420 from the β15 strand, which is recruited from the 'flap'. Interestingly, the hydrophobic cavity, which the substrate occupies in this arrangement, separates the amino acids on two strands, which would form calcium binding site 2. Another important difference is that the remaining two β-strands of the 'flap' align to the side of and become adsorbed into the β-sheeted plate of barrel 1. Besides the transverse helix this is the other structural element that fixes the position of the barrel. Altogether, these motifs hydrogen bonded together seem to form a much more stable structural arrangement than what is seen in FXIIIa°. In FXIIIa° – where site 2 is occupied by calcium and the 'flap' is retained – if the barrels oriented themselves in the same direction as in TGM2, they would clash with it so they probably remain detached and tethered through a flexible coil and form transient interactions with the enzyme surface.

What the difference between the two structures represents is probably that when the stabilizing calcium is lost from this site the 'flap' region rearranges. The position of the hydrophobic substrate side chain in the hydrophobic cavity behind the straightened β14 strand suggests that substrate binding is incompatible with the presence of calcium at site 2, since the binding site is cut in half. The straightened β-strand at the same time, however, is in the way of the leaving substrate polypeptide strand, which either has to bend sharply or navigate through a narrow groove to leave the enzyme surface. The open TGM2 structure, therefore, probably represents a more stable open conformation, which arises when calcium dissociates from site 2. This form requires higher flexibility from the substrate or may even react only with glutamine containing peptides and not proteins. When the calcium binding sites are saturated the 'flap' is stabilized and the 4KTY FXIIIa° structure likely captures one possible position of the barrels, which otherwise wobble freely in the sampling space around the enzyme.

Four different irreversible inhibitors have been co-crystallized with TGM2 and the open structures are in very good correspondence to one another alluding that the FXIIIa°-type lateral opening is not an alternative for TGM2. However, the crystals of FXIIIa° were grown under different conditions from those of TGM2, three of which were generated by the same laboratory probably adopting the methods of Pinkas et al., the creators of the very first crystal of the open TGM2 form. All four TGM2 models lack calcium ions, which – although necessarily present at substrate binding – probably dissociated during further purification and during crystal formation in the mother liquor. *The two different open structures, therefore, are not predilections of the two isoforms, but a consequence of the protein preparation and crystallization protocols*. Both structures open FXIIIa° and TGM2 represent frozen geometries, which the proteins can adopt. Based on the fact that the FXIIIa° crystal contains three calcium ions and that its model can accommodate both substrates and displays an active site fully compatible with the putative catalytic mechanism, the TGM2 structures may not represent a fully activated state and the FXIIIa° model may approach that situation better. It is interesting to ponder whether in a physiological context, when an activating calcium surge is on decline, Tgases may release the metal and adopt a stable conformation corresponding to the linear open form.

This may happen even in the absence of any substrate. Given the un-honed arrangement of the active site region and the more constricting topography of the substrate binding site it is dubitable whether this could engage in further catalytic cycles. These questions need to be clarified experimentally. nPAGE, where calcium is not usually present in the running buffer, may present a similar situation and the distinct slow migrating TGM2 may be identical with the linear open form. Quite surprisingly, after electrophoretic separation investigators found that this form was inactive and the fast migrating form active (Murthy et al. 1999). The answer may lie with the refractoriness of this form to another round of calcium activation.

On the strength that conformational rearrangement and activation are in absolute correspondence, TGM2 presents an ideal case for the study of its activation by fluorescence resonance energy transfer. According to the TGM2 models the two termini of the polypeptide are about 40 Å apart in the closed state and they are distanced to over 120 Å in the open conformer. Investigators have taken advantage of the strict dependence of FRET on distance within the range of 10 to 100 Å and have tagged the termini with CFP and YFP to show intracellular activation of TGM2 (Pavlyukov et al. 2012; Caron et al. 2012; Tyagi et al. 2014). The opening described for FXIIIa°, however, would lead to a much less significant increase in the end to end distance (~60 vs >120 Å) and a less impressive FRET efficiency difference.

1.7 Substrate Binding and Specificity

The acyl-transfer reaction follows a ping-pong mechanism in which the enzyme sequentially reacts with two substrates (Folk 1969). The carboxamido group of a glutamine reacts with the catalytic cysteine residue of the enzymes to form a thioester intermediate with the concomitant release of ammonia. The intermediate then reacts with the second substrate, typically a primary amine or a long-chain ω-ceramide in the case of TGM1, to yield the amide or ester product. It has been noted many times that transglutaminases exhibit a much stricter specificity for the acyl-donor substrate than for the acceptor, which can be almost any primary amine (Mycek et al. 1959; Folk 1983; Aeschlimann and Paulsson 1994). Asparagine or D-glutamine containing peptides are not substrates (Chica et al. 2004). Molecular modeling suggests that there is a hydrophobic interaction between the methylene groups of the substrate glutamine side chain and the indole residue of Trp241. This is thought to answer for why asparagine containing peptides are not substrates; the side chain of asparagine, which is shorter by one methylene group, does not establish this stabilizing contact (Chica et al. 2004).

Experiments on TGM2 have shown that the amino acid L-glutamine itself is not a substrate nor are CBz-glutaminyl-glycine or glutaminyl-glycine (Folk and Cole 1965). The minimally accepted substrate was CBz-glutaminyl-glycine. When dipeptides were tested as substrates of TGM2, in which a single amino acid was trailing CBz-glutamine, the identity of the second amino acid did not have a drastic effect either on the K_m or K_{cat}. Boc derivatized short peptides or Gly-Gln-Gly were not substrates at all showing that in minimal length substrates, substrate binding interactions of the amino-terminal group are important (Chica et al. 2004). Extension of the peptide chain beyond the minimal requirement of a single amino acid residue toward C-terminal from the glutamine increases substrate efficiency. When the glutamine is in the context of such a longer peptide, residues N-terminal from the glutamine weigh in appreciably less on substrate efficiency then the C-terminally flanking ones (Folk and Cole 1965; Keresztessy et al. 2006; Sugimura et al. 2006; Csosz et al. 2008; Sugimura et al. 2008; Yamane et al. 2010; Kuramoto et al. 2013)).

In the X-ray structures of both TGM2 and FXIIIa containing active site-bound, peptide based inhibitors the active site cysteine sits at the bottom of a tunnel pointing perpendicularly to the middle of a shallow groove. The tunnel is bridged over by the conserved tryptophans W241 and W332. Along the groove a peptide stretch of about five amino acids of the acyl donor gets appositioned to the enzyme surface. This peptide is held in place by interactions between a polar patch around the catalytic cysteine and a preferentially charged residue N-terminal to the substrate glutamine; a hydrophobic cavity at the other end of the groove accepts a hydrophobic amino acid in position +3 in the substrate. In the case of TGM2 the shape complementarity is the highest and the apolar substrate residue is best aligned with the hydrophobic cavity if the proline at position +2 assumes a trans configuration (Pinkas et al. 2007). Why the presence of a proline at position +5 in the case of TGM2 and +6 in the case of FXIIIa is favored is unknown; however, the overall dimensions of the groove are such that if it assumes a *cis* conformation it might be necessary to sharply turn the peptide chain and direct it away from the enzyme surface in order for it not to clash with the three β-stranded 'flap'. These features explain mechanistically the consensus sequences (Q-x-K/R-ϕ-x-x-x-W-P for TGM1, Q-x-P-ϕ-D-(P), Q-x-P-ϕ or Q-x-x-ϕ-D-P for TGM2, Q-S/T-K/R-ϕ for TGM3, L-Q-ϕ-P-V-W for TGM7, and Q-x-x-ϕ-W-P for FXIIIa) that studies searching for substrate motifs have identified (Keresztessy et al. 2006; Sugimura et al. 2006; Csosz et al. 2008; Sugimura et al. 2008; Yamane et al. 2010; Kuramoto et al. 2013).

To have proteins as substrates instead of short peptides can set different requirements in terms of sequence specificity as the range of interactions stabilizing the enzyme-substrate complex is broader. It has been a consistent observation that glutamines present in flexible loops are preferred substrates of vertebrate Tgases (Csosz et al. 2008).

The substrate specificity of the *Streptoverticillium mobaraense* transglutaminase is extremely broad, which makes it an exceptional tool for the food and textile industries (Gundersen et al. 2014).

Tgases are also able to function as iso-peptide hydrolases (Parameswaran et al. 1997). Their active sites are too tight for shorter true peptide bonds directly flanked by other side chain-containing, space requiring residues to fit into them, which is the reason why they do not work as proteases. The long N′ε-(γ-glutamyl)-lysine isopeptide bond (Cα-C′α >10 Å) can insert into the tunnel with the two cross-linked polypeptides remaining at the front and back entrances. The isopeptidase activity of Tgases, its structural requirements and regulation have not been extensively studied. Recent observations suggest that by manipulating the hydrophobic residues around the active site, it is possible to create Tgase variants, which are significantly deficient in cross-linking activity while having normal or increased isopeptidase activity raising the possibility that cross-links formed in proteins by a transglutaminase may be reversed by the same enzyme under specific regulatory conditions (Kiraly et al. 2015).

1.8 Interaction Motifs

Tgases interact with a swarm of other proteins; however, very few interaction sites are precisely known, thus few structurally relevant facts are revealed by this knowledge (Kanchan et al. 2015; Csosz et al. 2009). For example, allosteric effects of the interacting proteins on enzymatic activity cannot be deduced in most cases.

The interactome of TGM2 is the best characterized and several lists of interacting proteins have been assembled and even deposited in publicly available databases (Csosz et al. 2008; Kanchan et al. 2015). To focus on the most important interactions, three sequence motifs, Leu547-Ile561, Arg564-Asp581 and Gln633-Glu646, were identified as parts of the interaction surface with adrenoreceptor α1B (Feng et al. 1999). Directed mutations of multiple amino acids at a time in any of these regions severely weakened the interaction with the receptor, indicating that they are all required. An eight amino acid motif between Val665 and Lys672 near the C-terminus in the second β-barrel domain was identified as the interaction site with phospholipase C δ1 (Hwang et al. 1995). Binding to fibronectin and to heparane sulphate (HS) proteoglycans are the two most important intermolecular associations of TGM2 in the extracellular matrix and on the cell surface. The interaction surface for fibronectin was located to the N-terminal β-sandwich domain and has been shown to involve amino acids Trp88-Thr106 comprising two antiparallel β-strands (Hang et al. 2005). TGM2 binds to heparane sulphate with very high affinity (the k_D is nanomolar). The heparane sulphate proteoglycans include syndecan and glypican, which can be binding partners of extracellular TGM2 on the cell surface and perlecans that are present in basement membranes. The polyanionic HS binds to positively charged surface patches in proteins; however, there is no consensus as to which of those is engaged in interaction on TGM2. Two binding regions have been proposed. One involves the cluster of the motifs Arg262-Arg-Trp-Lys265 and Lys598-Gln-Lys-Arg-Lys602 complemented with the basic residues Arg19, Arg28 and Lys634. These positively charged patches are distant in the primary sequence, but they are brought into spatial proximity upon folding in the closed conformation of the enzyme. Mutations in either the Arg262-Lys265 or Lys598-Lys602 motif were sufficient to strongly impair HS binding. Consistent with these findings the open Ca^{2+} and inhibitor-bound form of the enzyme has severely reduced affinity to HS (Lortat-Jacob et al. 2012). The other putative HS binding region comprises Lys202 and Lys205 (Wang et al. 2012). The authors of this study also concluded that HS binds preferentially to GTP-bound closed TGM2 and poorly to a protein that has reacted with an irreversible inhibitor and is trapped in the open state. Interestingly, the positions of Lys202 and Lys205 are not different in the open and closed structures. To complicate matters even more, none of these regions are conserved in other members of the human Tgase family, although TGM1 binds to heparin with an equally high affinity, while TGM3 and FXIIIa do so with moderate affinities (Lortat-Jacob et al. 2012).

Protein-protein interactions often occur through short (3–20 amino acids) peptide segments of low amino acid complexity, often referred to as short linear motifs (SLiMs). SLiMs usually mediate highly specific protein interactions with moderate affinities, which are often exploited in signaling pathways (Davey et al. 2012). The majority of SLiMs are embedded in protein segments, which lack a well-defined three-dimensional structure (Fuxreiter et al. 2007). These protein regions are termed intrinsically disordered (ID) and are present in ~40 % of the human proteome (Ward et al. 2004). Bioinformatical analysis using prediction software has been performed on TGM2 in search for ID regions (Kanchan et al. 2015). Thirteen ID regions were identified in the molecule, which are located along a spiral surrounding the guanine nucleotide binding site in both the open (PDB:1KV3) and closed forms (PDB:2Q3Z) and overlap with multiple presumable SLiMs. Among the verified interactions, one site implicated in syndecan-4 binding lies in the Gly208-Pro217 ID region (Wang et al. 2012). The α-1B adrenoreceptor interacts with the Gly626-Glu647 ID region.

1.9 Structural Characteristics of the Celiac Epitope

Celiac disease (CD) is characterized by a specific set of predominantly IgA class anti-TGM2 autoantibodies accumulating on TGM2 present in the extracellular matrix and bound to the surface of fibers (reticulin and endomysium) giving a positive silver staining reaction. These antibodies serve as diagnostic markers and they are thought to play a pathogenic role in the disease process both in the gut leading to villous atrophy and also in inducing extracellular manifestations (hepatopathy, myocardial and placental dysfunction, glomerular disease, diabetes and others). The sensitivity and specificity of testing for autoantibodies in CD approaches 100 % making a gut mucosal biopsy no longer necessary for diagnosis in well-defined patients with high serum anti-TGM2 confirmed to react with endomysium as well (Husby et al. 2012).

The targeted epitopes (Fig. 1.9) are conformational (Seissler et al. 2001; Simon-Vecsei et al. 2012) and only accessible in vivo in a specific pattern (endomysial and reticulin pattern, not in connective tissue in general) outside the cells (Korponay-Szabo et al. 2004). These properties make celiac antibodies valuable experimental tools to study the structural and compartmentalizational changes of TGM2. Celiac anti-TGM2 antibodies are gluten-dependently produced, therefore initially they

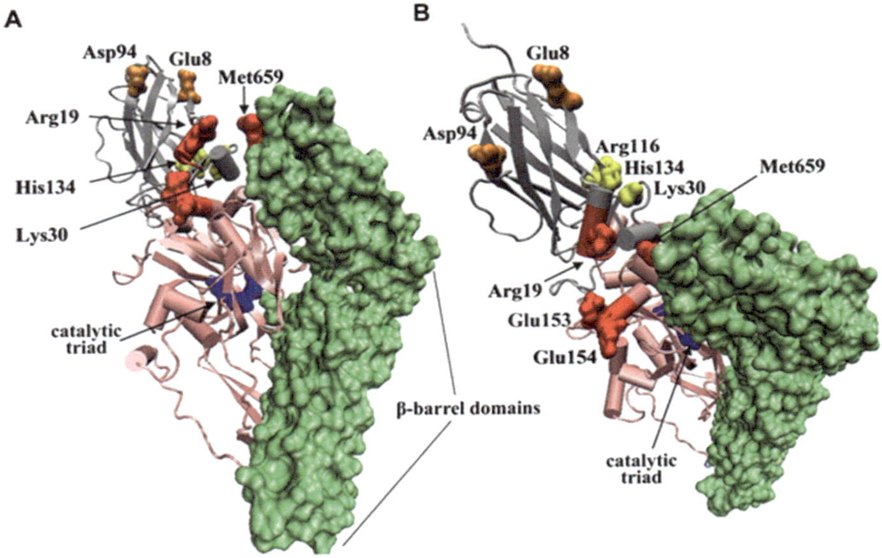

Fig. 1.9 Celiac epitope-relevant residues and fragments of TGM2. The structure shown here is based on 1KV3 crystal structure (**a**) and the same rotated 90° to look into the cleft between the N-terminus and C-terminus (**b**). The N-terminal domain is depicted in *gray* and the core domain in *pink*. Fragment 473–687 comprising the two β-barrel domains is shown in *lime*. Amino acids of Epitope 1 (Lys30, Arg116, His134) are shown in *yellow* and those of Epitope 2 (Arg 19, Glu153, Glu154, Met 658) are shown in *red*. Asp94 and Glu8 corresponding to minor antibody re-activities and to previously identified fibronectin binding sites are shown in *orange*. The catalytic triad is shown in *blue*. There is a spatial relationship of target amino acids

were thought to target neoepitopes involving or induced by cereal gluten peptides which can be used as substrates by TGM2. It is now, however, clear that recombinant TGM2 or natural TGM2 in human fetal, animal tissues and in cell culture never having been in contact with gluten before are equally well recognized by CD antibodies (Ladinser et al. 1994; Sblattero et al. 2002).

The role of Arg19 on the first α-helix of the N-terminal domain has been confirmed by several studies based on site-directed mutagenesis (Simon-Vecsei et al. 2012; Iversen et al. 2013), deuterium exchange studies (Iversen et al. 2014) and interference with heparin binding localized to this same epitope (Teesalu et al. 2012). However, Arg19 is probably only one part of a composite epitope requiring anchor amino acids also on the core domain. Mutation of Arg19 to serine leads to the complete loss of binding of cloned celiac antibodies utilizing VH3 heavy chain segments, and almost similar decrease is observed in half of patient sera comprising natural polyclonal antibodies. When the mutation of Arg19 is combined with the mutation of either Glu153 or Glu154 on the first α-helix of the core domain and situated only 12.9 Å apart from Arg19, all patient sera showed very reduced binding (Simon-Vecsei et al. 2012). Mutation of Glu158 mimics all these changes even when Arg19 is intact. Glu158 is not surface exposed, but is it situated at the base of the helix carrying Glu153 and Glu154 at its edge, and its mutation is capable of changing the axis and the position of the first α-helix of the core domain relative to the first α-helix of the N-terminal domain and thus to Arg19. The patient samples sensitive to the Arg19 mutation make a clear competition with those which are not sensitive (Simon-Vecsei et al. 2012), but the main anchor residues for the latter are not yet precisely known. A distinct epitope (called Epitope 1) was suggested by the study of other cloned celiac antibodies (Iversen et al. 2013) represented mainly by the usage of VH5 segments which did not compete with VH3 clones binding to Arg19 (also called Epitope 2). Further, Epitope 3 related to Epitope 2 and Epitope 4 related to Epitope 1 were suggested (Iversen et al. 2013). The binding of VH5 celiac antibodies to TGM2 was interfered with by fibronectin when Ca^{2+} was not present while VH3 antibodies were not sensitive to these conditions. However, in these experiments TGM2 was bound to the KCC009 inhibitor when Ca^{2+} was included and also other experimental conditions (antigen coating procedure, temperature, buffers) considerably differed from the competition studies with natural antibodies described above. KCC009 requires DMSO for dissolution and as discussed before, inhibitors may force TGM2 into a rigid conformational state reducing natural flexibility. In has also been indeed described that binding of the celiac antibodies also may alter the conformation of TGM2: in the absence of effectors and inhibitor, the protein is shifted to the open conformational state when VH3 antibodies bind, whereas binding of VH5 antibodies drives more molecules into the closed state (Iversen et al. 2014). Such effects may be inhibited and thus may lead to altered binding efficacy and the apparent lack of competition when the structure is frozen by an inhibitor. Therefore it is at present unclear to which extent Epitope 1(4) and 2(3) may overlap with each other and further studies are required for the clarification of this issue. Changes in a number of N-terminal domain residues (Lys30, Glu8 and Asp94) rather far from Arg19 also diminished the binding of VH5 antibodies, but to variable extents, and the known fibronectin binding sites of the enzyme have not been shown conclusively involved. The antibodies rather seem to get a steric hindrance which can be relieved when Ca^{2+} is present. Interestingly, all VH5 celiac antibodies are able to bind to endomysium, despite the co-expression of both fibronectin and TGM2 in this localization (Iversen et al. 2013; Cardoso et al. 2015).

Although multiple evidence shows the presence of the celiac epitope(s) on the N-terminal domain, a number of results also suggest that parts of the core domain are also indispensable for the efficient binding (Seissler et al. 2001; Sblattero et al. 2002; Nakachi et al. 2004; Simon-Vecsei et al. 2012). However, the highly conformational nature of the celiac antibody binding makes the exploration of core domain anchor points very difficult because deletions or changes in this area can heavily alter conformation. The change of Glu154 to lysine occurs at the corresponding site in factor XIII naturally without changes in the crystal structure and this change effectively reduces the binding of celiac antibodies to TGM2 (Simon-Vecsei et al. 2012). It has repeatedly been observed that small deletions of the core domain often had more profound effects on the binding than larger ones and the studies utilizing TGM2 fragments did not yield conclusive results. It was also impossible to locate these binding segments with a deuterium exchange approach, although small changes were detected also in domain fragments (Iversen et al. 2014). The binding of both natural and cloned celiac antibodies is somewhat higher to the extended conformation of TGM2 induced by various active site directed inhibitors compared to the effector free state and decreases when the protein is forced into a compacted closed state by GDP or GTP (Lindfors et al. 2011; Iversen et al. 2013; Korponay-Szabo et al. 2000). Also oxidation interferes with the binding, which, however, is not altering the N-terminal domain (Iversen et al. 2014). Taken together, these results indicate that the N-terminal amino acids alone do not represent the entire binding area.

All early studies also indicated binding of both natural and cloned antibodies to C-terminal fragments of TGM2 as well (Seissler et al. 2001; Sblattero et al. 2002; Nakachi et al. 2004). Met659 is spatially very close to both Arg19 (7.7 Å) and Glu153 (16.8 Å) in the closed crystal structure (1KV3) and its mutation had an additive effect to the mutation of Arg19 and Glu153 or Glu154 (Simon-Vecsei et al. 2012). Interestingly, the fragment 473–687 comprising only β-barrels 1 and 2 also bound celiac antibodies to some extent in immunoprecipitation studies with radiobinding assay using in vitro translated TGM2 (Tiberti et al. 2003; Nenna et al. 2013). In these and similar studies, the protein was expressed in a rabbit reticulocyte system. Clinical studies showed that under these conditions, rabbit full-length erythrocyte TGM2, which is also antigenic for a subset of celiac antibodies, may complicate the interpretation of the relative binding efficacy (Agardh et al. 2005). Nonetheless, in the 473–687 fragment, the loop containing 659 and many amino acids beneath Glu153 are present and can offer part of the common epitope surface to the antibodies (Fig. 1.9).

It is not well understood how the anti-TGM2 immune response is elicited, but the most likely hypothesis is a haptene carrier mechanism, in which the primary antigens are proteolytically resistant peptide fragments of gliadin, which get access to the mucosal interstitium. Based on the fact that TGM2 can crosslink gliadin to itself, it has been proposed that TGM2 reactive B-cells, recognizing these adducts, process and present the cross-linked peptide fragments on their MHCII molecules to gliadin reactive T-helper cells. Binding assays in vitro showed that deamidation of α-gliadin by TGM2 enhanced affinity to HLA-DQ2 25-fold and the non-deamidated PFPQPQLPY peptide is almost non-stimulatory to T-cells, whereas its deamidated counterpart, PFPQPELPY, is highly antigenic. The preference of the autoantibodies for certain epitopes is not evident from the cross-stimulatory mechanism by which they arise. It is, however, expected that the surface covered by the cross-linked gliadin would be hidden from the B-cell receptor. In fact, the clustering of the autoantibody epitopes in a restricted region of the protein is surprising in light of the mechanism by which they arise. Such a phenomenon would comply much better with exposure of a neoepitope as, for example, in atypical hemolytic uremic syndrome or autoimmune Addison's disease. The preference for these epitopes must first appear at the B-cell receptor's interaction with TGM2. There is another hypothesis, that is founded on observation of cross-reaction of anti-TGM2 antibodies with gliadin fragments (Korponay-Szabo et al. 2008). This hypothesis assumes molecular mimicry between the major antigens, gliadin and TGM2. In dermatitis herpetiformis, which is thought to be a dermatological manifestation of CD, antibodies also reactive with TGM3 are produced, while gluten ataxia autoantibodies recognize TGM6 (Sardy et al. 2002; Hadjivassiliou et al. 2008; Stamnaes et al. 2010). Antibodies against TGM6 represent a new marker associated with gluten-related neurological dysfunction (Boscolo et al. 2010; Hadjivassiliou et al. 2013; Stenberg et al. 2014).

1.10 Conclusion

Since the solution of the first Tgase crystal structure 24 years ago, structural analysis has tremendously advanced our understanding of Tgase function at the molecular level. However, despite their ubiquity and several decades of investigation, the physiological role of transglutaminases in cells and tissues is only partly revealed. The challenging questions are intimately connected to problems of their substrate specificity, allosteric regulation and molecular interactions. In the paradigms, clinically or industrially relevant, where it will be desirable to regulate their function, it is going to be inevitable to have atomic-level understanding of their operation.

References

Achyuthan KE, Greenberg CS (1987) Identification of a guanosine triphosphate-binding site on guinea pig liver transglutaminase. Role of GTP and calcium ions in modulating activity. J Biol Chem 262(4):1901–1906

Aeschlimann D, Paulsson M (1994) Transglutaminases: protein cross-linking enzymes in tissues and body fluids. Thromb Haemost 71(4):402–415

Agardh D, Dahlbom I et al (2005) Autoantibodies against soluble and immobilized human recombinant tissue transglutaminase in children with celiac disease. J Pediatr Gastroenterol Nutr 41(3):322–327

Ahvazi B, Kim HC et al (2002) Three-dimensional structure of the human transglutaminase 3 enzyme: binding of calcium ions changes structure for activation. EMBO J 21(9):2055–2067

Ambrus A, Banyai I et al (2001) Calcium binding of transglutaminases: a 43Ca NMR study combined with surface polarity analysis. J Biomol Struct Dyn 19(1):59–74

Anantharaman V, Aravind L (2003) Evolutionary history, structural features and biochemical diversity of the NlpC/P60 superfamily of enzymes. Genome Biol 4(2):R11

Anantharaman V, Koonin EV et al (2001) Peptide-N-glycanases and DNA repair proteins, Xp-C/Rad4, are, respectively, active and inactivated enzymes sharing a common transglutaminase fold. Hum Mol Genet 10(16):1627–1630

Ando Y, Imamura S et al (1989) Cross-linking of lipocortin I and enhancement of its Ca^{2+} sensitivity by tissue transglutaminase. Biochem Biophys Res Commun 163(2):944–951

Azim AC, Marfatia SM et al (1996) Human erythrocyte dematin and protein 4.2 (pallidin) are ATP binding proteins. Biochemistry 35(9):3001–3006

Baek KJ, Das T et al (1993) Evidence that the Gh protein is a signal mediator from α 1-adrenoceptor to a phospholipase C. I. Identification of α 1-adrenoceptor-coupled Gh family and purification of Gh7 from bovine heart. J Biol Chem 268:27390–27397
Baek KJ, Kwon NS et al (1996) Oxytocin receptor couples to the 80 kDa Gh α family protein in human myometrium. Biochem J 315:739–744
Baker EN (1980) Structure of actinidin, after refinement at 1.7 A resolution. J Mol Biol 141 (4):441–484
Begg GE, Carrington L et al (2006) Mechanism of allosteric regulation of transglutaminase 2 by GTP. Proc Natl Acad Sci U S A 103(52):19683–19688
Belkin AM, Zemskov EA et al (2004) Cell-surface-associated tissue transglutaminase is a target of MMP-2 proteolysis. Biochemistry 43(37):11760–11769
Bergamini CM (1988) GTP modulates calcium binding and cation-induced conformational changes in erythrocyte transglutaminase. FEBS Lett 239(2):255–258
Boeshans KM, Mueser TC et al (2007) A three-dimensional model of the human transglutaminase 1: insights into the understanding of lamellar ichthyosis. J Mol Model 13(1):233–246
Boothe RL, Folk JE (1969) A reversible, calcium-dependent, copper-catalyzed inactivation of guinea pig liver transglutaminase. J Biol Chem 244(2):399–405
Boscolo S, Lorenzon A et al (2010) Anti transglutaminase antibodies cause ataxia in mice. PLoS One 5(3):e9698
Candi E, Paradisi A et al (2004) Transglutaminase 5 is regulated by guanine-adenine nucleotides. Biochem J 381(Pt 1):313–319
Cardoso I, Stamnaes J et al (2015) Transglutaminase 2 interactions with extracellular matrix proteins as probed with celiac disease autoantibodies. FEBS J 282(11):2063–2075
Cariello L, Ristoratore F et al (1997) A new transglutaminase-like from the ascidian Ciona intestinalis. FEBS Lett 408(2):171–176
Caron NS, Munsie LN et al (2012) Using FLIM-FRET to measure conformational changes of transglutaminase type 2 in live cells. PLoS One 7(8):e44159
Chica RA, Gagnon P et al (2004) Tissue transglutaminase acylation: proposed role of conserved active site Tyr and Trp residues revealed by molecular modeling of peptide substrate binding. Protein Sci 13(4):979–991
Chung SI, Folk JE (1970) Mechanism of the inactivation of guinea pig liver transglutaminase by tetrathionate. J Biol Chem 245(4):681–689
Clouthier CM, Mironov GG et al (2012) Real-time monitoring of protein conformational dynamics in solution using kinetic capillary electrophoresis. Angew Chem Int Ed Engl 51 (50):12464–12468
Connellan JM, Folk JE (1969) Mechanism of the inactivation of guinea pig liver transglutaminase by 5,5′-dithiobis-(2-nitrobenzoic acid). J Biol Chem 244(12):3173–3181
Csosz E, Bagossi P et al (2008) Substrate preference of transglutaminase 2 revealed by logistic regression analysis and intrinsic disorder examination. J Mol Biol 383(2):390–402
Csosz E, Mesko B et al (2009) Transdab wiki: the interactive transglutaminase substrate database on web 2.0 surface. Amino Acids 36(4):615–617
Davey NE, Van Roey K et al (2012) Attributes of short linear motifs. Mol Biosyst 8(1):268–281
Drenth J, Kalk KH et al (1976) Binding of chloromethyl ketone substrate analogues to crystalline papain. Biochemistry 15(17):3731–3738
Feng JF, Gray CD et al (1999) Alpha 1B-adrenoceptor interacts with multiple sites of transglutaminase II: characteristics of the interaction in binding and activation. Biochemistry 38 (7):2224–2232
Folk JE (1969) Mechanism of action of guinea pig liver transglutaminase. VI. Order of substrate addition. J Biol Chem 244(13):3707–3713
Folk JE (1983) Mechanism and basis for specificity of transglutaminase-catalyzed epsilon-(gamma-glutamyl) lysine bond formation. Adv Enzymol Relat Areas Mol Biol 54:1–56
Folk JE, Cole PW (1965) Structural requirements of specific substrates for guinea pig liver transglutaminase. J Biol Chem 240:2951–2960

Folk JE, Cole PW (1966a) Identification of a functional cysteine essential for the activity of guinea pig liver transglutaminase. J Biol Chem 241(13):3238–3240
Folk JE, Cole PW (1966b) Mechanism of action of guinea pig liver transglutaminase. I. Purification and properties of the enzyme: identification of a functional cysteine essential for activity. J Biol Chem 241(23):5518–5525
Folk JE, Cole PW (1966c) Transglutaminase: mechanistic features of the active site as determined by kinetic and inhibitor studies. Biochim Biophys Acta 122(2):244–264
Folk JE, Mullooly JP et al (1967) Mechanism of action of guinea pig liver transglutaminase. II. The role of metal in enzyme activation. J Biol Chem 242(8):1838–1844
Folk JE, Cole PW et al (1968) Mechanim of action of guinea pig liver transglutaminase. V. The hydrolysis reaction. J Biol Chem 243(2):418–427
Fox BA, Yee VC et al (1999) Identification of the calcium binding site and a novel ytterbium site in blood coagulation factor XIII by X-ray crystallography. J Biol Chem 274(8):4917–4923
Fraij BM (1996) GTP hydrolysis by human tissue transglutaminase homologue. Biochem Biophys Res Commun 218(1):45–49
Fuxreiter M, Tompa P, Simon I (2007) Local structural disorder imparts plasticity on linear motifs. Bioinformatics 23(8):950–956
Gundersen MT, Keillor JW et al (2014) Microbial transglutaminase displays broad acyl-acceptor substrate specificity. Appl Microbiol Biotechnol 98(1):219–230
Hadjivassiliou M, Aeschlimann P et al (2008) Autoantibodies in gluten ataxia recognize a novel neuronal transglutaminase. Ann Neurol 64(3):332–343
Hadjivassiliou M, Aeschlimann P et al (2013) Transglutaminase 6 antibodies in the diagnosis of gluten ataxia. Neurology 80(19):1740–1745
Han BG, Cho JW et al (2010) Crystal structure of human transglutaminase 2 in complex with adenosine triphosphate. Int J Biol Macromol 47(2):190–195
Hang J, Zemskov EA et al (2005) Identification of a novel recognition sequence for fibronectin within the NH2-terminal beta-sandwich domain of tissue transglutaminase. J Biol Chem 280 (25):23675–23683
Hettasch JM, Greenberg CS (1994) Analysis of the catalytic activity of human factor XIIIa by site-directed mutagenesis. J Biol Chem 269(45):28309–28313
Husby S, Koletzko S et al (2012) European society for pediatric gastroenterology, hepatology, and nutrition guidelines for the diagnosis of coeliac disease. J Pediatr Gastroenterol Nutr 54 (1):136–160
Hwang KC, Gray CD et al (1995) Interaction site of GTP binding Gh (transglutaminase II) with phospholipase C. J Biol Chem 270(45):27058–27062
Iismaa SE, Chung L et al (1997) The core domain of the tissue transglutaminase Gh hydrolyzes GTP and ATP. Biochemistry 36(39):11655–11664
Iismaa SE, Holman S et al (2003) Evolutionary specialization of a tryptophan indole group for transition-state stabilization by eukaryotic transglutaminases. Proc Natl Acad Sci U S A 100 (22):12636–12641
Im MJ, Riek RP et al (1990) A novel guanine nucleotide-binding protein coupled to the alpha 1-adrenergic receptor. II. Purification, characterization, and reconstitution. J Biol Chem 265 (31):18952–18960
Iversen R, Di Niro R et al (2013) Transglutaminase 2-specific autoantibodies in celiac disease target clustered, N-terminal epitopes not displayed on the surface of cells. J Immunol 190 (12):5981–5991
Iversen R, Mysling S et al (2014) Activity-regulating structural changes and autoantibody epitopes in transglutaminase 2 assessed by hydrogen/deuterium exchange. Proc Natl Acad Sci U S A 111(48):17146–17151
Jabs A, Weiss MS, Hilgenfeld R (1999) Non-proline cis peptide bonds in proteins. J Mol Biol 286 (1):291–304
Jang TH, Lee DS et al (2014) Crystal structure of transglutaminase 2 with GTP complex and amino acid sequence evidence of evolution of GTP binding site. PLoS One 9(9):e107005

Kabsch W, Sander C (1983) Dictionary of protein secondary structure: pattern recognition of hydrogen-bonded and geometrical features. Biopolymers 22(12):2577–2637
Kanchan K, Ergulen E et al (2013) Identification of a specific one amino acid change in recombinant human transglutaminase 2 that regulates its activity and calcium sensitivity. Biochem J 455(3):261–272
Kanchan K, Fuxreiter M, Fesus L (2015) Physiological, pathological and structural implications of non-enzymatic protein-protein interactions of the multifunctional human transglutaminase 2. Cell Mol Life Sci 72(16):3009–3035
Kashiwagi T, Yokoyama K et al (2002) Crystal structure of microbial transglutaminase from Streptoverticillium mobaraense. J Biol Chem 277(46):44252–44260
Keillor JW, Clouthier CM et al (2014) Acyl transfer mechanisms of tissue transglutaminase. Bioorg Chem 57:186–197
Keillor JW, Apperley KY et al (2015) Inhibitors of tissue transglutaminase. Trends Pharmacol Sci 36(1):32–40
Keresztessy Z, Csosz E et al (2006) Phage display selection of efficient glutamine-donor substrate peptides for transglutaminase 2. Protein Sci 15(11):2466–2480
Kim HC, Lewis MS et al (1990) Protransglutaminase E from guinea pig skin. Isolation and partial characterization. J Biol Chem 265(35):21971–21978
Kim IG, McBride OW et al (1992) Structure and organization of the human transglutaminase 1 gene. J Biol Chem 267(11):7710–7717
Kim SY, Kim IG et al (1994) The structure of the transglutaminase 1 enzyme. Deletion cloning reveals domains that regulate its specific activity and substrate specificity. J Biol Chem 269 (45):27979–27986
Kiraly R, Csosz E et al (2009) Functional significance of five noncanonical Ca^{2+}–binding sites of human transglutaminase 2 characterized by site-directed mutagenesis. FEBS J 276 (23):7083–7096
Kiraly R, Demeny M et al (2011) Protein transamidation by transglutaminase 2 in cells: a disputed Ca2+–dependent action of a multifunctional protein. FEBS J 278(24):4717–4739
Kiraly R, Tangaraju K et al (2015) Isopeptidase activity of human transglutaminase 2: disconnection from transamidation and characterization by kinetic parameters. Amino Acids [7 Aug 2015; Epub ahead of print]
Korponay-Szabo IR, Sulkanen S et al (2000) Tissue transglutaminase is the target in both rodent and primate tissues for celiac disease-specific autoantibodies. J Pediatr Gastroenterol Nutr 31 (5):520–527
Korponay-Szabo IR, Halttunen T et al (2004) In vivo targeting of intestinal and extraintestinal transglutaminase 2 by coeliac autoantibodies. Gut 53(5):641–648
Korponay-Szabo IR, Vecsei Z et al (2008) Deamidated gliadin peptides form epitopes that transglutaminase antibodies recognize. J Pediatr Gastroenterol Nutr 46(3):253–261
Kuramoto K, Yamasaki R et al (2013) Phage-displayed peptide library screening for preferred human substrate peptide sequences for transglutaminase 7. Arch Biochem Biophys 537 (1):138–143
Ladinser B, Rossipal E et al (1994) Endomysium antibodies in coeliac disease: an improved method. Gut 35(6):776–778
Lai TS, Slaughter TF et al (1996) C-terminal deletion of human tissue transglutaminase enhances magnesium-dependent GTP/ATPase activity. J Biol Chem 271(49):31191–31195
Lai TS, Liu Y et al (2007) Identification of two GTP-independent alternatively spliced forms of tissue transglutaminase in human leukocytes, vascular smooth muscle, and endothelial cells. FASEB J 21(14):4131–4143
Lewis BA, Freyssinet JM, Holbrook JJ (1978) An equilibrium study of metal ion binding to human plasma coagulation factor XIII. Biochem J 169(2):397–402
Lin YF, Yeh TS et al (2010) Nonmuscle myosin IIA (myosin heavy polypeptide 9): a novel class of signal transducer mediating the activation of G alpha h/phospholipase C-delta 1 pathway. Endocrinology 151(3):876–885
Lindfors K, Koskinen O et al (2011) Serodiagnostic assays for celiac disease based on the open or closed conformation of the autoantigen, transglutaminase 2. J Clin Immunol 31(3):436–442

Liu S, Cerione RA et al (2002) Structural basis for the guanine nucleotide-binding activity of tissue transglutaminase and its regulation of transamidation activity. Proc Natl Acad Sci U S A 99 (5):2743–2747

Lorand L, Siefring GE Jr et al (1979) Dansylcadaverine specific staining for transamidating enzymes. Anal Biochem 93(2):453–458

Lorand L, Dailey JE, Turner PM (1988) Fibronectin as a carrier for the transglutaminase from human erythrocytes. Proc Natl Acad Sci U S A 85(4):1057–1059

Lortat-Jacob H, Burhan I et al (2012) Transglutaminase-2 interaction with heparin: identification of a heparin binding site that regulates cell adhesion to fibronectin-transglutaminase-2 matrix. J Biol Chem 287(22):18005–18017

Makarova KS, Aravind L et al (1999) A superfamily of archaeal, bacterial, and eukaryotic proteins homologous to animal transglutaminases. Protein Sci 8(8):1714–1719

Mariniello L, Esposito C et al (2003) N-terminus end of rat prostate transglutaminase is responsible for its catalytic activity and GTP binding. Int J Biochem Cell Biol 35(7):1098–1108

Murthy SN, Lomasney JW et al (1999) Interactions of G(h)/transglutaminase with phospholipase Cdelta1 and with GTP. Proc Natl Acad Sci U S A 96(21):11815–11819

Murthy SN, Iismaa S et al (2002) Conserved tryptophan in the core domain of transglutaminase is essential for catalytic activity. Proc Natl Acad Sci U S A 99(5):2738–2742

Murzin AG, Brenner SE et al (1995) SCOP: a structural classification of proteins database for the investigation of sequences and structures. J Mol Biol 247(4):536–540

Muszbek L, Bereczky Z et al (2011) Factor XIII: a coagulation factor with multiple plasmatic and cellular functions. Physiol Rev 91(3):931–972

Mycek MJ, Clarke DD et al (1959) Amine incorporation into insulin as catalyzed by transglutaminase. Arch Biochem Biophys 84(528):540

Nakachi K, Powell M et al (2004) Epitopes recognised by tissue transglutaminase antibodies in coeliac disease. J Autoimmun 22(1):53–63

Nakaoka H, Perez DM et al (1994) Gh: a GTP-binding protein with transglutaminase activity and receptor signaling function. Science 264(5165):1593–1596

Nemes Z, Petrovski G et al (2005) Structure-function relationships of transglutaminases – a contemporary view. Prog Exp Tumor Res 38:19–36

Nenna R, Tiberti C et al (2013) Anti-transglutaminase immunoreactivity and histological lesions of the duodenum in coeliac patients. Int Immunol 25(6):389–394

Noguchi K, Ishikawa K et al (2001) Crystal structure of red sea bream transglutaminase. J Biol Chem 276(15):12055–12059

Ohtsuka T, Ota M et al (2000) Comparison of substrate specificities of transglutaminases using synthetic peptides as acyl donors. Biosci Biotechnol Biochem 64(12):2608–2613

Parameswaran KN, Cheng XF et al (1997) Hydrolysis of gamma: epsilon isopeptides by cytosolic transglutaminases and by coagulation factor XIIIa. J Biol Chem 272(15):10311–10317

Pasternack R, Dorsch S et al (1998) Bacterial pro-transglutaminase from Streptoverticillium mobaraense – purification, characterisation and sequence of the zymogen. Eur J Biochem 257(3):570–576

Pavlyukov MS, Antipova NV et al (2012) Detection of transglutaminase 2 conformational changes in living cell. Biochem Biophys Res Commun 421(4):773–779

Pedersen LC, Yee VC et al (1994) Transglutaminase factor XIII uses proteinase-like catalytic triad to crosslink macromolecules. Protein Sci 3(7):1131–1135

Peng X, Zhang Y et al (1999) Interaction of tissue transglutaminase with nuclear transport protein importin-alpha3. FEBS Lett 446(1):35–39

Pietroni P, von Hippel PH (2008) Multiple ATP binding is required to stabilize the "activated" (clamp open) clamp loader of the T4 DNA replication complex. J Biol Chem 283 (42):28338–28353

Pinkas DM, Strop P et al (2007) Transglutaminase 2 undergoes a large conformational change upon activation. PLoS Biol 5(12):e327

Roth WJ, Chung SI et al (1986) Inactivation of alveolar macrophage transglutaminase by oxidants in cigarette smoke. J Leukoc Biol 39(6):629–644
Sardy M, Karpati S et al (2002) Epidermal transglutaminase (TGase 3) is the autoantigen of dermatitis herpetiformis. J Exp Med 195(6):747–757
Sblattero D, Florian F et al (2002) The analysis of the fine specificity of celiac disease antibodies using tissue transglutaminase fragments. Eur J Biochem 269(21):5175–5181
Seissler J, Wohlrab U et al (2001) Autoantibodies from patients with coeliac disease recognize distinct functional domains of the autoantigen tissue transglutaminase. Clin Exp Immunol 125 (2):216–221
Siegel M, Strnad P et al (2008) Extracellular transglutaminase 2 is catalytically inactive, but is transiently activated upon tissue injury. PLoS One 3(3):e1861
Simon-Vecsei Z, Kiraly R et al (2012) A single conformational transglutaminase 2 epitope contributed by three domains is critical for celiac antibody binding and effects. Proc Natl Acad Sci U S A 109(2):431–436
Stamnaes J, Pinkas DM et al (2010) Redox regulation of transglutaminase 2 activity. J Biol Chem 285(33):25402–25409
Steinert PM, Chung SI et al (1996a) Inactive zymogen and highly active proteolytically processed membrane-bound forms of the transglutaminase 1 enzyme in human epidermal keratinocytes. Biochem Biophys Res Commun 221(1):101–106
Steinert PM, Kim SY et al (1996b) The transglutaminase 1 enzyme is variably acylated by myristate and palmitate during differentiation in epidermal keratinocytes. J Biol Chem 271 (42):26242–26250
Stenberg R, Hadjivassiliou M et al (2014) Anti-transglutaminase 6 antibodies in children and young adults with cerebral palsy. Autoimmun Dis 2014:237107
Stieler M, Weber J et al (2013) Structure of active coagulation factor XIII triggered by calcium binding: basis for the design of next-generation anticoagulants. Angew Chem Int Ed Engl 52 (45):11930–11934
Sugimura Y, Hosono M et al (2006) Screening for the preferred substrate sequence of transglutaminase using a phage-displayed peptide library: identification of peptide substrates for TGASE 2 and Factor XIIIA. J Biol Chem 281(26):17699–17706
Sugimura Y, Hosono M et al (2008) Identification of preferred substrate sequences for transglutaminase 1 – development of a novel peptide that can efficiently detect cross-linking enzyme activity in the skin. FEBS J 275(22):5667–5677
Tanfani F, Bertoli E et al (1993) Structural investigation of transglutaminase by Fourier transform infrared spectroscopy. Eur J Biochem 218(2):499–505
Teesalu K, Panarina M et al (2012) Autoantibodies from patients with celiac disease inhibit transglutaminase 2 binding to heparin/heparan sulfate and interfere with intestinal epithelial cell adhesion. Amino Acids 42(2–3):1055–1064
Thomas H, Beck K et al (2013) Transglutaminase 6: a protein associated with central nervous system development and motor function. Amino Acids 44(1):161–177
Tiberti C, Bao F et al (2003) Celiac disease-associated transglutaminase autoantibody target domains at diagnosis are age and sex dependent. Clin Immunol 109(3):318–324
Tokunaga F, Muta T et al (1993) Limulus hemocyte transglutaminase. cDNA cloning, amino acid sequence, and tissue localization. J Biol Chem 268(1):262–268
Tyagi S, VanDelinder V et al (2014) Continuous throughput and long-term observation of single-molecule FRET without immobilization. Nat Methods 11(3):297–300
Vezza R, Habib A, FitzGerald GA (1999) Differential signaling by the thromboxane receptor isoforms via the novel GTP-binding protein, Gh. J Biol Chem 274:12774–12779
Wang Z, Collighan RJ et al (2012) Characterization of heparin-binding site of tissue transglutaminase: its importance in cell surface targeting, matrix deposition, and cell signaling. J Biol Chem 287(16):13063–13083
Ward JJ, Sodhi JS et al (2004) Prediction and functional analysis of native disorder in proteins from the three kingdoms of life. J Mol Biol 337(3):635–645

Weiss MS, Metzner HJ et al (1998) Two non-proline cis peptide bonds may be important for factor XIII function. FEBS Lett 423(3):291–296
Weraarchakul-Boonmark N, Jeong JM et al (1992) Cloning and expression of chicken erythrocyte transglutaminase. Proc Natl Acad Sci U S A 89(20):9804–9808
Yamane A, Fukui M et al (2010) Identification of a preferred substrate peptide for transglutaminase 3 and detection of in situ activity in skin and hair follicles. FEBS J 277(17):3564–3574
Yee VC, Pedersen LC et al (1994) Three-dimensional structure of a transglutaminase: human blood coagulation factor XIII. Proc Natl Acad Sci U S A 91(15):7296–7300
Yee VC, Pedersen LC et al (1995) Structural evidence that the activation peptide is not released upon thrombin cleavage of factor XIII. Thromb Res 78(5):389–397
Zemskov EA, Mikhailenko I et al (2011) Unconventional secretion of tissue transglutaminase involves phospholipid-dependent delivery into recycling endosomes. PLoS One 6(4):e19414
Zhang D, Aravind L (2012) Novel transglutaminase-like peptidase and C2 domains elucidate the structure, biogenesis and evolution of the ciliary compartment. Cell Cycle 11(20):3861–3875

Chapter 2
Control of TG Functions Depending on Their Localization

Yutaka Furutani and Soichi Kojima

Abstract One of the unique characteristics of the transglutaminase (TG) family enzymes is multiple distribution of the members. It is known that TGs exist in the cytosol, plasma membrane, nucleus, and mitochondria, as well as outside cells, being associated with the extracellular matrix and being secreted into tissue fluids or blood circulation. Depending on where TGs locate, their molecular conformation is altered, and they exert multiple biological activities. Furthermore, their localization and thus their activities are changed in pathological situations, which is highly associated with progression of many diseases, including thrombotic disease, immune diseases including celiac disease, skin diseases, neurodegenerative diseases, hepatic diseases, fibrotic diseases and cancers.

This chapter summarizes information on distribution of TGs in various intracellular compartments and extracellular milieus, control of the activity of TGs depending on where they are, and molecular mechanisms by which TGs localize to specific compartments.

Keywords Subcellular localization • Conformational change • Multifunctional activities • Pathobiology

2.1 TGs in Various Cell Compartments

Cells contain diverse intracellular organelles, such as a nucleus, as well as mitochondria, endoplasmic reticulum, Golgi apparatus, endosome, and exosome existing in the cytosol, maintaining cellular homeostasis through their respective functions. In normal cells, 73 % of intracellular transglutaminases (TGs) are found in the cytosol and the remaining 27 % localize in every cellular compartment, mainly in the plasma membrane and the nucleus. In addition, some TG enzymes are known to be released from the cells. At each location, TGs work through exertion of a variety of activities including transamidation, phosphorylation, GTPase, protein disulfide isomerization, and scaffold activities, as well as (in some cases)

Y. Furutani, Ph.D. • S. Kojima, Ph.D. (✉)
Micro-Signaling Regulation Technology Unit, RIKEN Center for Life Science Technologies, 2-1 Hirosawa, Wako, Saitama 351-0198, Japan
e-mail: skojima@postman.riken.go.jp

K. Hitomi et al. (eds.), *Transglutaminases*, DOI 10.1007/978-4-431-55825-5_2

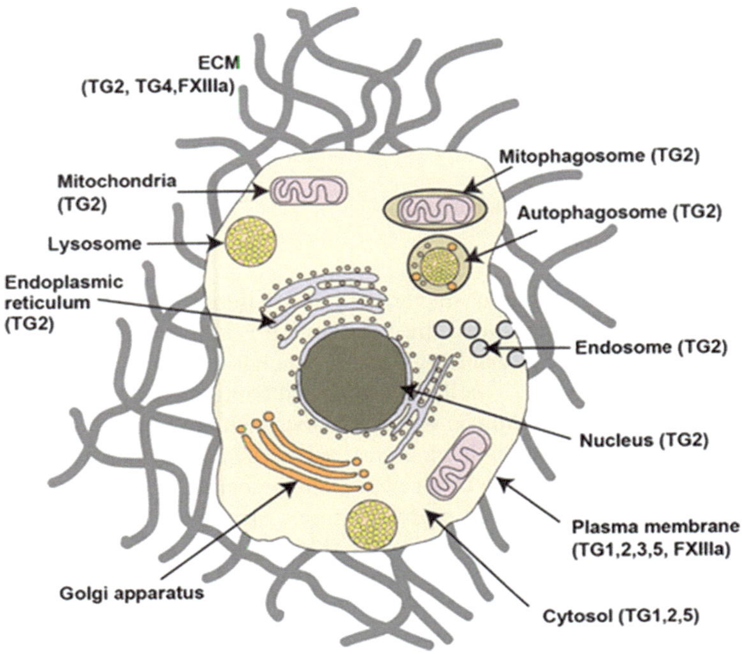

Fig. 2.1 Subcellular localization of TG family proteins. TGs are localized to the extracellular matrix, plasma membrane, cytosol, endoplasmic reticulum, endosome, and nucleus. ECM is crosslinked by TG2, TG4, and FXIIIa. In the plasma membrane, TG1, 2, 3, 5, and FXIIIa crosslink and modify the membrane. TG2 induces apoptosis dependent on mitochondria activity. Ablation of TG2 enhances autophagosome and mitophagosome formation. TG2 can be secreted through the endosome. TG2 in the nucleus mediates apoptosis induction

isopeptidase activity, and they contribute to cellular growth, differentiation, apoptosis, and adhesion, as well as keratinization and blood coagulation. Moreover, TGs change their conformation depending upon their localization. In a normal situation, cellular TGs take the closed form (negative in transamidation activity) in the cytosol, whereas they take the open form (active in transamidation activity) at both the cytosol and nucleus in cells undergoing apoptosis or outside cells (Fig. 2.1).

This chapter summarizes information on distribution of TGs in various intracellular compartments and extracellular milieu, how they function depending on where they are, and molecular mechanisms by which TGs localize to specific compartments.

2.1.1 Cytoplasm

The cytosol is defined as the water-soluble components of the cell cytoplasm, constituting the fluid portion that remains after removal of the organelles and other intracellular structures (Clegg 1984). Newly synthesized TGs go into the cytoplasm and stay there before going out of the cells or localizing to specific organelles such as the nucleus. In most cases, TGs take an inactive form (closed form) and do not exert transamidase activity (crosslinking activity) while they stay in the cytosol (Pavlyukov et al. 2012; Pinkas et al. 2007). When cells undergo apoptosis and/or a redox state, TGs (not only TG2, but also TG1, TG5, and FXIIIa) are activated and start to crosslink cytoskeleton and intermediate filament proteins such as β-actin, tubulin, cytokeratin, vimentin, and vinculin, making their structure strengthen. TG2 is known to stabilize axonal microtubules through polyamination of tubulin (Baumgartner and Weth 2007; Candi et al. 2001; Clegg 1984; Dolge et al. 2012; Hasegawa et al. 2003; Robinson et al. 2007; Song et al. 2013) (Table 2.1).

2.1.2 Extracellular Matrix (ECM)

The ECM is a complex structural entity, surrounding and supporting cells, found within mammalian tissues. ECM, often referred to as connective tissue, is composed of three major classes of biomolecules including structural proteins (such as collagen and elastin), specialized proteins (e.g. fibrillin, fibronectin, and laminin), and proteoglycans. Upon secretion to the outside cells, TG2 is largely inactivated in physiological conditions because of the oxidative state of the extracellular milieu. However, it can be activated in certain types of disease conditions. Extracellular TG2 binds fibronectin and enhances cell adhesion (Akimov and Belkin 2001; Akimov et al. 2000). TG2 also binds to heparan sulfate proteoglycans such as syndecan through a heparin-binding region at ^{202}KFLKNAGRDCSRRSSPVYVGR^{222} (Wang et al. 2012) and controls cell adhesion cooperatively with fibronectin and integrins. Furthermore, cross-linking of ECM proteins, i.e. cross-linking between laminin and nidogen (Aeschlimann and Paulsson 1991), as well as cross-linking among collagen fibers (Wang et al. 2014), contributes to stabilize ECM structures (Sane et al. 1988; van den Akker et al. 2011). Another role of extracellular TG2 is to crosslink cytokines such as midkine (Kojima et al. 1995; Mahoney et al. 2000), latent TGF-β1 (Le et al. 2001) and endostatin (Faye et al. 2010) as well as the protease inhibitors trappin-2 (Jaovisidha et al. 2006) and PAI-2 (Jensen et al. 1993), and fixes them to the ECM to modulate their activities (Nunes et al. 1997; Pfundt et al. 1996; Wang et al. 2011) (Table 2.2).

Table 2.1 Cytosol

Substrates	Type of TG	Functions	References
Actin, tubulin, cytokeratin	TG2	Transglutaminase-mediated cross-linking stabilizes the particulate material shed from the placenta	Robinson et al. (2007)
β-Actin	TG1	TG1 stabilizes β-actin in endothelial cells, correlating with stabilization of intracellular junctions	Baumgartner and Weth (2007)
	TG2	Protein disulfide isomerase in rat liver	Hasegawa et al. (2003) and Ohba et al. (1981)
β-Actin	TG1	β-Actin is crosslinked by TG1 at the synaptic ending	Dolge et al. (2012)
	TG1, TG2	TG1 and TG2 expressions are increased in AD patients	Kim et al. (1999)
	TG1	TG1 activates AKT and STAT3 pathways following oxidant injury, contributing to cell survival	Ponnusamy et al. (2009)
Calbindin-D28k, ataxin-1	TG2	TG2 crosslinks between calbindin-D28k and ataxin-1 Q82. Ataxin-1 aggregates may be toxic to neurons	Vig et al. (2007)
	TG2	TG2 knockout mice exhibited increased neutrophil phagocytic activity	Balajthy et al. (2006)
	TG2	TG2 activity is increased in R6/2 transgenic HD mice	Dedeoglu et al. (2002)
	TG2	TG2 binds GTP	Nakaoka et al. (1994) and Singh et al. (1995)
Dual leucine zipper-bearing kinase (DLK)	TG2	TG2 crosslinks DLK and downregulates its activity	Hebert et al. (2000)
	TG2	TG2 regulates the cell cycle in endothelial cells	Nara et al. (1992)
	TG2	Cytosol TG2 has a pro-apoptotic function	Milakovic et al. (2004)
Vimentin	TG5	TG5 crosslinks vimentin in vitro	Candi et al. (2001)
Vimentin, tubulin α-1, tubulin β-5, moesin, WD repeat protein 1, annexin A1, α-enolase, L-lactate dehydrogenase, pyruvate kinase, GAPDH, transketolase, elongation factor 1, THO complex,	TG2	TG2 crosslinks intermediate filaments, cytoskeletons, elongation factors, initiation factors, ribosomal proteins, kinase, and several cellular proteins in	van den Akker et al. (2011)

(continued)

Table 2.1 (continued)

Substrates	Type of TG	Functions	References
initiation factor 4A-I, ribosomal protein 53-A, elongation factor 2, heat shock protein 90, guanine nucleotide-binding protein, serum albumin precursor, multifunctional protein ADE2		a redox state in smooth muscle cells	
	TG2	TG2 changes its conformation and activates after the induction of apoptosis	Pavlyukov et al. (2012)
Filamin, vinculin	FXIIIa	FXIIIa crosslinks the cytoskeleton during platelet activation	Serrano and Devine (2002)
Polyamines	TG2	TG2 polyaminates tubulin for stabilizing axonal microtubules	Song et al. (2013)

2.1.3 Blood and Seminal Fluid (Extracellular Fluids)

FXIIIa mainly exists in blood circulation and works as a blood coagulation factor through cross-linking and stabilization of fibrin clots (See Chapter 15 by Hayashi and Kasahara). FXIIIa also crosslinks many blood proteins such as thrombospondin-1, α2-antiplasmin, plasminogen, PAI-2, and vWF to fibrin clots. Another instance of TGs present in extracellular fluids is TG4 secreted from the prostate. It crosslinks seminal proteins such as SVP and SVS secreted from seminal vesicles, and makes seminal clots (Furutani et al. 2005; Tseng et al. 2012) (Table 2.3).

2.1.4 Plasma Membrane

The plasma membrane is a biological membrane or an outer membrane of a cell, which is composed of two layers of phospholipids and embedded with proteins. TG2 exists in both the outer and inner layers of the plasma membrane and accounts for 20 % of cellular TG2 (Bruce and Peters 1983). TG2 in the outer layer crosslinks adhesion proteins such as V-CAM1 (van den Akker et al. 2011), as well as growth factor/cytokine receptors such as VEGFR2 (Dardik and Inbal 2006). TG2 present in the outer membrane also crosslinks tight-junction and intercellular-junction proteins (Baumgartner et al. 2004; Hiiragi et al. 1999). In neuronal cells, TG2 in the outer membrane contributes to formation of amyloid bodies via cross-linking of α-synuclein and pathogenesis of Alzheimer's disease (Nemes et al. 2009). On the other hand, in the inner membrane, TG2 acts as a component of the β-adrenergic receptor G_h and works to emit survival signals by virtue of its GTPase activity (Nakaoka et al. 1994). Furthermore, TG2 in the inner membrane is reported to facilitate phosphorylation of IGFBP-3 and enhance underlying signaling (Mishra

Table 2.2 Extracellular matrix

Substrates	Type of TG	Functions	References
	TG2	Collagen deposition is decreased in a TG2 knockout mice UUO model	Shweke et al. (2008)
	TG2	TG2 in ECM is activated by tissue injury	Siegel et al. (2008)
Trappin-2	TG	TG crosslinks trappin-2 in cartilage and synovial fluid in osteoarthritis	Jaovisidha et al. (2006)
	TG2	GTP-bound TG2 induces hypertrophic differentiation and calcification of chondrocytes	Johnson and Terkeltaub (2005)
Fibrin	TG2	TG2 crosslinks fibrin, promoting integrin-dependent cell adhesion	Belkin et al. (2005)
Osteopontin, bone sialoprotein, α2 HS-glycoprotein	TG2	TG2 crosslinks osteopontin, bone sialoprotein, α2 HS-glycoprotein in bone	Kaartinen et al. (2002)
Latent TGF-β1	TG2	TG2 crosslinks the latent form of TGF-β1	Le et al. (2001)
	TG2	TG2 regulates LTBP-1 deposition	Nunes et al. (1997) and Verderio et al. (1999)
Fibronectin, collagen α1 chain precursor, fibulin-2, nidogen-1		TG2 crosslinks fibronectin, collagen α1 chain precursor, fibulin-2, nidogen-1	van den Akker et al. (2011)
	TG2	Thrombospondin-2 null mice have reduced TG2 activity	Agah et al. (2005)
Fibronectin	TG2	TG2 binds fibronectin and enhances attachment, migration, and spreading	Akimov and Belkin (2001), Akimov et al. (2000), Priglinger et al. (2004) and Verderio et al. (2003)
PAI-2	TG, FXIIIa	TG crosslinks PAI-2 extracellular structures	Jensen et al. (1993)
	TG2	TG2 interacts with syndecan-4 and CD44 at macrophages to promote removal of apoptotic cells	Nadella et al. (2015)
Collagen	TG2	TG2 crosslinks collagen in corneal stroma	Wang et al. (2014)
	TG2	TG2 inhibition leads to reduced fibronectin and VEGFA depositions, suppressing angiogenesis by HUVECs	Wang et al. (2013)
SLPI	TG2, FXIIIa	TG2 crosslinks SLPI	Baranger et al. (2011)

(continued)

Table 2.2 (continued)

Substrates	Type of TG	Functions	References
Endostatin	TG2	TG2 crosslinks endostatin at ECM and stimulates angiogenesis	Faye et al. (2010)
	TG2	Higher TG2 levels are associated with collagen III and IV deposition	Fisher et al. (2009)
	TG2	TG2-mediated cross-linking occurs in human liver fibrosis	Grenard et al. (2001)
Midkine, galectin-3	TG2	TG2 crosslinks midkine and galectin-3	Kojima et al. (1995) and Mahoney et al. (2000)
Thyroglobulin	TG2	TG2 crosslinks thyroglobulin in thyrocytes	Saber-Lichtenberg et al. (2000)
Plasminogen	TG2, FXIIIa	TG2 and FXIIIa crosslink human plasminogen	Bendixen et al. (1995)
β-Amyloid	TG2	TG2 crosslinks β-amyloid	Dudek and Johnson (1994)
PAI-2	TG2, FXIIIa	TG2 and FXIIIa crosslink PAI-2	Jensen et al. (1993)
Laminin, nidogen	TG2	TG2 crosslinks laminin and nidogen complex	Aeschlimann and Paulsson (1991)
Vitronectin	TG2	TG2 crosslinks vitronectin	Sane et al. (1988)
Polyamines	TG2	TG2 crosslinks polyamines	Folk et al. (1980)
	TG4	TG4 increases adhesion of prostate cancer cells to ECM	Jiang et al. (2013)
	TG2	TG2 does not affect fibrotic matrix stability	Popov et al. (2011)
	TG2	TG2-fibronectin complex enhances cell adhesion and survival through syndecan-4 and −2 on osteoblasts	Wang et al. (2011)

Table 2.3 Extracellular fluid (blood, seminal fluid)

Substrates	Type of TG	Functions	References
Guinea pig trappin, SVP	TG4	TG4 crosslinks guinea pig trappin and SVP in seminal vesicle fluid	Furutani et al. (2005)
SVS I, II, III	TG4	TG4 crosslinks SVS I, II, III	Tseng et al. (2012)
Fibrin, thrombospondin-1, α2-antiplasmin, vWF, plasminogen, PAI-2, TAFI	FXIIIa	FXIIIa stabilizes fibrin clots	Ichinose (2005)

Table 2.4 Plasma membrane

Substrates	Type of TG	Functions	References
Annexins, integrins, placental alkaline phosphatase	TG2	Transglutaminase-mediated cross-linking stabilizes the particulate material shed from the placenta	Robinson et al. (2007)
	TG2	HSP70 prevents TG2 localization to the leading edge	Boroughs et al. (2011)
Glu-tubulin	FXIIIa	Plasma membrane FXIIIa enhances osteoblast matrix secretion and deposition	Al-Jallad et al. (2011)
α-Synuclein	TG1, 2, 5	TGs crosslink α-synuclein, promoting amyloid formation in Lewy bodies	Nemes et al. (2009)
	TG2	Cell-surface-associated TG2 is cleaved by MMP-2	Belkin et al. (2004)
	TG1	Cross-linking activity of TG1 stabilizes the intercellular junction	Baumgartner et al. (2004)
IGFBP-3	TG2	TG2 phosphorylates IGFBP-3 in breast cancer cell membrane	Mishra and Murphy (2004)
	TG2	Membrane-targeted TG2 induces anti-apoptotic function	Milakovic et al. (2004)
Hwp1	TG2	*Candida albicans* Hwp1 is a substrate for TG2 enhancing attachment to mammalian epithelial cells	Staab et al. (1999)
Loricrin, involucrin, small proline-rich proteins	TG5	TG5 crosslinks loricrin, involucrin, and small proline-rich proteins in the keratinocyte expression system	Candi et al. (2001)
Involucrin, loricrin, elafin, S100A11, keratins, filaggrin, SPR1, SPR2, desmoplakin	TG1	TG1 crosslinks involucrin, loricrin, elafin, and SPR1 and 2 in the cornified envelope	Nemes et al. (1999b), Steinert and Marekov (1995) and Robinson and Eckert (1998)
Pro-uPA	TG2	TG2 crosslinks pro-uPA to a specific protein on the surface of human monocyte-like U927 cells	Behrendt et al. (1993)
Lipocortin I (annexin A1)	TG2	TG2 crosslinks lipocortin I to plasma membrane phospholipid	Ando et al. (1991)
	TG2	TG2 is essential in receptor-mediated endocytosis	Davies et al. (1980)
VEGFR2	TG2	TG2 crosslinks VEGFR2	Dardik and Inbal (2006)

(continued)

Table 2.4 (continued)

Substrates	Type of TG	Functions	References
	TG2	TG2 inhibitor blocks migration of endothelial cells	Dardik and Inbal (2006)
	TG2	TG2 binds to LRP6 and induces β-catenin activation	Deasey et al. (2013)
SPINK6	TG1 and 3	TG1 or 3 crosslinks SPINK6 to fibronectin and protects against matrix degradation	Fischer et al. (2013)
	TG2	TG2 interacts with LDLRP5 receptor and activates β-catenin signaling	Faverman et al. (2008)
V-CAM1, voltage-dependent anion-selective channel	TG2	TG2 crosslinks V-CAM1 in smooth muscle cells	van den Akker et al. (2011)
	TG2	TG1 cross-linking activity stabilizes cadherin-based adherent junctions	Hiiragi et al. (1999)
	TG2	TG2 localized beneath the cell membrane has a transamidation active conformation	Pavlyukov et al. (2012)
	TG1	TG1 crosslinks lipid Z to involucrin	Nemes et al. (1999a)

and Murphy 2004). TG1 plays an essential role in keratinization through cross-linking of cornified envelope proteins including loricrin, involucrin, trappin-2, and small proline-rich protein (See Chapter 3 by Hitomi and Tatsukawa). Cross-linking between involucrin and lipid Z present on the outer membrane makes the skin barrier stronger (Nemes et al. 1999a). In addition to TG1, TG3 and TG5 are involved in formation of the cornified envelope (Candi et al. 2001; Hitomi et al. 2003; Sane et al. 1988) (Table 2.4).

2.1.5 Nucleus

The nucleus is a membrane-enclosed organelle containing DNA and RNA in complex with a large variety of proteins, such as histones and proteins involved in transcription, to form chromosomes. Normally, 7 % of intracellular TG2 exists in the nucleus, while further accumulation of the activated (open) form of TG2 in the nucleus is induced in cells undergoing apoptosis (Kuo et al. 2011). There, TG2 crosslinks and silences a transcription factor, Sp1, to facilitate apoptosis (Tatsukawa et al. 2009). TG2 also crosslinks histone proteins and modulates their phosphorylation, leading to regulation of transcription (Ballestar et al. 1996; Mishra et al. 2006; Nunomura et al. 2003). It is reported that TG2 gets activated in

Table 2.5 Nucleus

Substrates	Type of TG	Functions	References
Relish	Drosophila TG	TG-catalyzed cross-linking suppresses transcription factor Relish to enable immune tolerance	Shibata et al. (2013)
	TG2	Activation of calpain-induced BAX translocation, inducing apoptotic cell death	Yoo et al. (2012)
	TG2	TG2 translocates to the nucleus after ischemic stroke. TG2 suppresses upregulation of HIF-induced apoptotic genes	Filiano et al. (2010)
	TG2	TG2 activates NF-κB signaling	Kim et al. (2006)
Sp1	TG2	TG2 crosslinks Sp1 and induces apoptosis	Tatsukawa et al. (2009)
Histones	TG2	TG2 phosphorylates histones	Mishra et al. (2006)
	TG2	TG2 interacts with Rb, inducing a protective effect against apoptosis	Lee et al. (2004)
Histones	TG2	TG2 crosslinks histones in starfish sperm and chicken erythrocytes	Ballestar et al. (1996) and Nunomura et al. (2003)
Huntingtin	TG2	TG2 crosslinks huntingtin and may be involved in formation of nuclear inclusions in Huntington's disease	Karpuj et al. (1999)
	TG2	TG2 is activated in the nucleus during liver regeneration	Haddox and Russell (1981)
	TG2	Nuclear TG2 (C277S) attenuates apoptosis	Milakovic et al. (2004)

hepatocytes under liver regeneration (Haddox and Russell 1981). In neuronal cells, nuclear TG2 is known to associate with formation of aggregates of huntingtin protein (Karpuj et al. 1999) (Table 2.5).

2.1.6 Mitochondria

The mitochondrion is an organelle that carries out the process of aerobic respiration and is responsible for energy production. TG2 on the mitochondrial outer membrane crosslinks and inactivates BAX and suppresses apoptosis (Rodolfo et al. 2004). Ablation of TG2 in mouse embryonic fibroblasts induces mitophagy accompanying aggregation of LC3II, which is an autophagosomal marker, whereas overexpression of TG2 results in acceleration of mitochondria-dependent apoptosis (D'Eletto et al. 2009; Rossin et al. 2015) (Table 2.6).

Table 2.6 Mitochondria

Substrates	Type of TG	Functions	References
BAX	TG2	TG2 crosslinks BAX in the outer mitochondrial membrane	Rodolfo et al. (2004)
	TG2	TG2 ablation leads to mitophagy associated with LC3II accumulation	D'Eletto et al. (2009); Rossin et al. (2015)
	TG2	TG2 overexpression sensitizes toward apoptotic stimuli specifically toward mitochondria	Piacentini et al. (2002)
	TG2	TG2 suppress BAX, inhibiting cell death	Cho et al. (2010)
Aldehyde dehydrogenase	TG2	TG2 crosslinks aldehyde dehydrogenase	van den Akker et al. (2011)

2.1.7 *Endoplasmic Reticulum*

The endoplasmic reticulum is a series of folded membranes in the cells, which are associated with protein synthesis and storage and the movement of cellular materials. TG2 is reported to control release of Ca^{2+} from the endoplasmic reticulum via allosteric regulation of IP_3R type I through transamidation and deamidation and modulation of the open Ca^{2+} channel (Hamada et al. 2014) (Table 2.7).

2.1.8 *Endosome*

Endosomes are membrane-bound vesicles, formed via a complex family of processes collectively known as endocytosis, and found in the cytoplasm of virtually every animal cell. TG2 exists in the endosome as a transamidation inactive configuration that is not altered in cells undergoing apoptosis (Pavlyukov et al. 2012), and is released to the outside of cells together with other proteins, such as integrin, via recycling endosomes (Zemskov et al. 2011) (Table 2.8).

2.2 Molecular Mechanisms of Distribution of TG2

The molecular mechanisms of localization of TGs in cellular organelles are mostly unclear, except for the mechanism by which TG2 localizes into the nucleus (Shrestha et al. 2015) and the mechanism by which TG2 is released to the outside of cells (Chou et al. 2011; Zemskov et al. 2011).

Table 2.7 Endoplasmic reticulum

Substrates	Type of TG	Functions	References
IP_3R type I	TG2	TG2 crosslinks and deamidates IP_3R type I at the endoplasmic reticulum	Hamada et al. (2014)

Table 2.8 Endosome

Substrates	Type of TG	Functions	References
	TG2	TG2 retains the transamidation inactive conformation even at late stages of apoptosis	Pavlyukov et al. (2012)
	TG2	TG2 is secreted through recycling endosomes	Zemskov et al. (2011)

2.2.1 *Molecular Mechanism of Nuclear Localization of TG2*

Recently, it has been discovered that ^{466}AEKEETGMAMRIRV^{479} and ^{657}LHMGLHKL^{664} of human TG2 function as novel nuclear localization signal (NLS) and nuclear export signal (NES), respectively. Using these NLS and NES, TG2 shuttles between the cytoplasm and nucleus (Shrestha et al. 2015). Apoptotic stimuli either stimulate NLS function or inhibit NES function, leading to accumulation of TG2 in the nucleus. Emergence of the short form of TG2 (TG2-S) defect in NES-containing D domain has been reported in neuronal cells isolated from patients with Alzheimer's disease (Festoff et al. 2002) as well as in hepatocytes treated with alcohol and free fatty acids (Kuo et al. 2012) (Fig. 2.2).

2.2.2 *Molecular Mechanism of Secretion of TG2 from Cells*

Although the molecular mechanism of secretion of TG2 from cells has not been fully elucidated, recently Chou et al. reported that Asp94 and Asp97 present in the fibronectin-binding region within the N-terminal β-sandwich domain play an important role in secretion of TG2 from cells (Chou et al. 2011). On the other hand, Zenskov et al. reported secretion of TG2 via recycling endosomes (Zemskov et al. 2011).

2.3 Afterword

It is amazing that TGs exist in every cell compartment and exert various activities at each place, although the molecular mechanisms by which TG2 localizes into specific cell compartments are largely unknown. If these mechanisms are

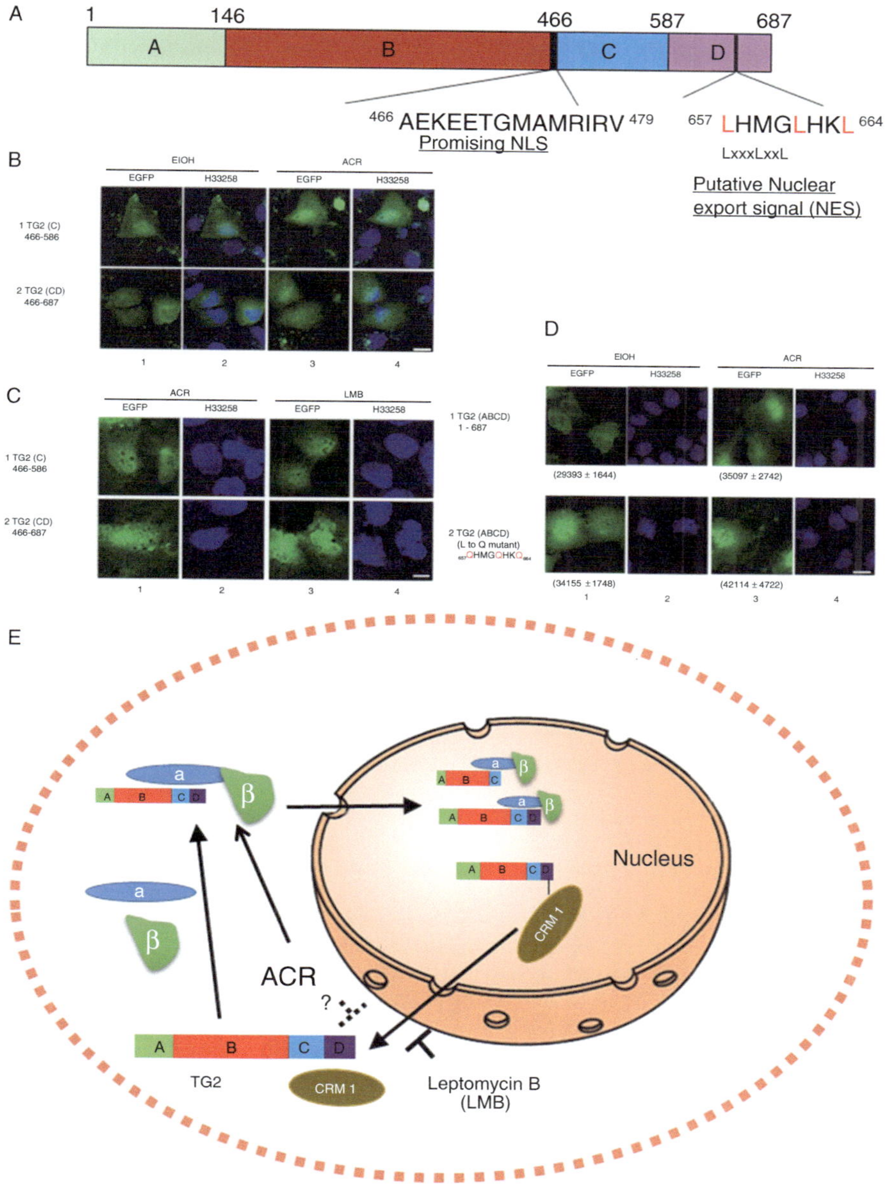

Fig. 2.2 Nuclear localization mechanism of TG2. (**a**) Schematic structure of TG2 consisting of four domains, β-sandwich shown, catalytic core, β-barrel1, and β-barrel2 shown as A, B, C, and D, respectively. The nuclear localization signal (NLS) is present at the N-terminal end of domain C. The putative nuclear export signal (NES) is found in domain D. (**b**) EGFP-tagged domain C or CD was localized to the nucleus after stimulation with EtOH or acyclic retinoid (ACR). (**c**) EGFP-tagged domain C or CD was localized to the nucleus. Nuclear export of EGFP-tagged domain CD was inhibited by leptomycin B (LMB), while EGFP-tagged domain C was not inhibited, suggesting that the NES is present in domain D. (**d**) The putative NES $_{657}$LHMGLHKL$_{664}$ in

elucidated and if we can establish techniques to control a certain activity of TG at a certain organelle(s), that will contribute to elucidation of the unsolved molecular pathogenesis of TG-related diseases and establishment of new methods to prevent or cure these diseases. To this end, identification of sites within every TG molecule responsible for localization into a specific organelle and making knockin mice will be challenging. As a defect in FXIIIa is responsible for certain bleeding disorder, administration of recombinant FXIIIa is a treatment for these patients (See Chapter 15 by Hayashi and Kasahara). Since transamidation of gluten peptides by TG2 at the intestinal membrane has been revealed to be a molecular mechanism for making autoantigens in celiac disease, great advances have been made in diagnosis and treatment of the disease (See Chapter 9 by Iversen and Sollid). However, except for these cases, the apparent molecular mechanism by which TGs are involved in many diseases are not fully elucidated, except for antifibrosis therapy in the kidney targeting TG2 (See Chapter 11 by Verderio et al.). As nucleus TG2 has been reported to be associated with the pathogenesis of neurodegenerative diseases (See Chapter 13 by Feola et al.), as well as liver diseases (Kuo et al. 2012; Tatsukawa et al. 2009), selective control of nuclear TG2 might enable us to establish new treatments for these diseases.

References

Aeschlimann D, Paulsson M (1991) Cross-linking of laminin-nidogen complexes by tissue transglutaminase. A novel mechanism for basement membrane stabilization. J Biol Chem 266(23):15308–15317

Agah A, Kyriakides TR, Bornstein P (2005) Proteolysis of cell-surface tissue transglutaminase by matrix metalloproteinase-2 contributes to the adhesive defect and matrix abnormalities in thrombospondin-2-null fibroblasts and mice. Am J Pathol 167(1):81–88. doi:10.1016/s0002-9440(10)62955-0

Akimov SS, Belkin AM (2001) Cell surface tissue transglutaminase is involved in adhesion and migration of monocytic cells on fibronectin. Blood 98(5):1567–1576

Akimov SS, Krylov D, Fleischman LF, Belkin AM (2000) Tissue transglutaminase is an integrin-binding adhesion coreceptor for fibronectin. J Cell Biol 148(4):825–838

Al-Jallad HF, Myneni VD, Piercy-Kotb SA, Chabot N, Mulani A, Keillor JW, Kaartinen MT (2011) Plasma membrane factor XIIIA transglutaminase activity regulates osteoblast matrix secretion and deposition by affecting microtubule dynamics. PLoS One 6(1):e15893. doi:10.1371/journal.pone.0015893

Ando Y, Imamura S, Owada MK, Kannagi R (1991) Calcium-induced intracellular cross-linking of lipocortin I by tissue transglutaminase in A431 cells. Augmentation by membrane phospholipids. J Biol Chem 266(2):1101–1108

Fig. 2.2 (continued) domain D was mutated to $_{657}$QHMGQHKQ$_{664}$. The mutated TG2 was accumulated in the nucleus. (**e**) TG2 has NLS and NES and interacts with importin α/β and CRM1, and then TG2 shuttles between the cytoplasm and nucleus under stimulation of ACR

Balajthy Z, Csomos K, Vamosi G, Szanto A, Lanotte M, Fesus L (2006) Tissue-transglutaminase contributes to neutrophil granulocyte differentiation and functions. Blood 108(6):2045–2054. doi:10.1182/blood-2004-02-007948

Ballestar E, Abad C, Franco L (1996) Core histones are glutaminyl substrates for tissue transglutaminase. J Biol Chem 271(31):18817–18824

Baranger K, Zani ML, Labas V, Dallet-Choisy S, Moreau T (2011) Secretory leukocyte protease inhibitor (SLPI) is, like its homologue trappin-2 (pre-elafin), a transglutaminase substrate. PLoS One 6(6):e20976. doi:10.1371/journal.pone.0020976

Baumgartner W, Weth A (2007) Transglutaminase 1 stabilizes beta-actin in endothelial cells correlating with a stabilization of intercellular junctions. J Vasc Res 44(3):234–240. doi:10.1159/000100422

Baumgartner W, Golenhofen N, Weth A, Hiiragi T, Saint R, Griffin M, Drenckhahn D (2004) Role of transglutaminase 1 in stabilisation of intercellular junctions of the vascular endothelium. Histochem Cell Biol 122(1):17–25. doi:10.1007/s00418-004-0668-y

Behrendt N, Ronne E, Dano K (1993) A novel, specific pro-urokinase complex on monocyte-like cells, detected by transglutaminase-catalyzed cross-linking. FEBS Lett 336(3):394–396

Belkin AM, Zemskov EA, Hang J, Akimov SS, Sikora S, Strongin AY (2004) Cell-surface-associated tissue transglutaminase is a target of MMP-2 proteolysis. Biochemistry 43 (37):11760–11769. doi:10.1021/bi049266z

Belkin AM, Tsurupa G, Zemskov E, Veklich Y, Weisel JW, Medved L (2005) Transglutaminase-mediated oligomerization of the fibrin(ogen) alphaC domains promotes integrin-dependent cell adhesion and signaling. Blood 105(9):3561–3568. doi:10.1182/blood-2004-10-4089

Bendixen E, Harpel PC, Sottrup-Jensen L (1995) Location of the major epsilon-(gamma-glutamyl) lysyl cross-linking site in transglutaminase-modified human plasminogen. J Biol Chem 270 (30):17929–17933

Boroughs LK, Antonyak MA, Johnson JL, Cerione RA (2011) A unique role for heat shock protein 70 and its binding partner tissue transglutaminase in cancer cell migration. J Biol Chem 286 (43):37094–37107. doi:10.1074/jbc.M111.242438

Bruce SE, Peters TJ (1983) The subcellular localization of transglutaminase in normal liver and in glucagon-treated and partial hepatectomized rats. Biosci Rep 3(12):1085–1090

Candi E, Oddi S, Terrinoni A, Paradisi A, Ranalli M, Finazzi-Agro A, Melino G (2001) Transglutaminase 5 cross-links loricrin, involucrin, and small proline-rich proteins in vitro. J Biol Chem 276(37):35014–35023. doi:10.1074/jbc.M010157200

Cho SY, Lee JH, Bae HD, Jeong EM, Jang GY, Kim CW, Shin DM, Jeon JH, Kim IG (2010) Transglutaminase 2 inhibits apoptosis induced by calcium- overload through down-regulation of BAX. Exp Mol Med 42(9):639–650. doi:10.3858/emm.2010.42.9.063

Chou CY, Streets AJ, Watson PF, Huang L, Verderio EA, Johnson TS (2011) A crucial sequence for transglutaminase type 2 extracellular trafficking in renal tubular epithelial cells lies in its N-terminal beta-sandwich domain. J Biol Chem 286(31):27825–27835. doi:10.1074/jbc.M111.226340

Clegg JS (1984) Properties and metabolism of the aqueous cytoplasm and its boundaries. Am J Physiol 246(2 Pt 2):R133–R151

D'Eletto M, Farrace MG, Falasca L, Reali V, Oliverio S, Melino G, Griffin M, Fimia GM, Piacentini M (2009) Transglutaminase 2 is involved in autophagosome maturation. Autophagy 5(8):1145–1154

Dardik R, Inbal A (2006) Complex formation between tissue transglutaminase II (tTG) and vascular endothelial growth factor receptor 2 (VEGFR-2): proposed mechanism for modulation of endothelial cell response to VEGF. Exp Cell Res 312(16):2973–2982. doi:10.1016/j.yexcr.2006.05.019

Davies PJ, Davies DR, Levitzki A, Maxfield FR, Milhaud P, Willingham MC, Pastan IH (1980) Transglutaminase is essential in receptor-mediated endocytosis of alpha 2-macroglobulin and polypeptide hormones. Nature 283(5743):162–167

Deasey S, Nurminsky D, Shanmugasundaram S, Lima F, Nurminskaya M (2013) Transglutaminase 2 as a novel activator of LRP6/beta-catenin signaling. Cell Signal 25 (12):2646–2651. doi:10.1016/j.cellsig.2013.08.016

Dedeoglu A, Kubilus JK, Jeitner TM, Matson SA, Bogdanov M, Kowall NW, Matson WR, Cooper AJ, Ratan RR, Beal MF, Hersch SM, Ferrante RJ (2002) Therapeutic effects of cystamine in a murine model of Huntington's disease. J Neurosci Off J Soc Neurosci 22(20):8942–8950

Dolge L, Aufenvenne K, Traupe H, Baumgartner W (2012) Beta-actin is a target for transglutaminase activity at synaptic endings in chicken telencephalic cell cultures. J Mol Neurosci MN 46(2):410–419. doi:10.1007/s12031-011-9601-8

Dudek SM, Johnson GV (1994) Transglutaminase facilitates the formation of polymers of the beta-amyloid peptide. Brain Res 651(1–2):129–133

Faverman L, Mikhaylova L, Malmquist J, Nurminskaya M (2008) Extracellular transglutaminase 2 activates beta-catenin signaling in calcifying vascular smooth muscle cells. FEBS Lett 582 (10):1552–1557. doi:10.1016/j.febslet.2008.03.053

Faye C, Inforzato A, Bignon M, Hartmann DJ, Muller L, Ballut L, Olsen BR, Day AJ, Ricard-Blum S (2010) Transglutaminase-2: a new endostatin partner in the extracellular matrix of endothelial cells. Biochem J 427(3):467–475. doi:10.1042/bj20091594

Festoff BW, SantaCruz K, Arnold PM, Sebastian CT, Davies PJ, Citron BA (2002) Injury-induced "switch" from GTP-regulated to novel GTP-independent isoform of tissue transglutaminase in the rat spinal cord. J Neurochem 81(4):708–718

Filiano AJ, Tucholski J, Dolan PJ, Colak G, Johnson GV (2010) Transglutaminase 2 protects against ischemic stroke. Neurobiol Dis 39(3):334–343. doi:10.1016/j.nbd.2010.04.018

Fischer J, Koblyakova Y, Latendorf T, Wu Z, Meyer-Hoffert U (2013) Cross-linking of SPINK6 by transglutaminases protects from epidermal proteases. J Invest Dermatol 133(5):1170–1177. doi:10.1038/jid.2012.482

Fisher M, Jones RA, Huang L, Haylor JL, El Nahas M, Griffin M, Johnson TS (2009) Modulation of tissue transglutaminase in tubular epithelial cells alters extracellular matrix levels: a potential mechanism of tissue scarring. Matrix Biol J Int Soc Matrix Biol 28(1):20–31. doi:10.1016/j.matbio.2008.10.003

Folk JE, Park MH, Chung SI, Schrode J, Lester EP, Cooper HL (1980) Polyamines as physiological substrates for transglutaminases. J Biol Chem 255(8):3695–3700

Furutani Y, Kato A, Fibriani A, Hirata T, Kawai R, Jeon JH, Fujii Y, Kim IG, Kojima S, Hirose S (2005) Identification, evolution, and regulation of expression of Guinea pig trappin with an unusually long transglutaminase substrate domain. J Biol Chem 280(21):20204–20215. doi:10.1074/jbc.M501678200

Grenard P, Bresson-Hadni S, El Alaoui S, Chevallier M, Vuitton DA, Ricard-Blum S (2001) Transglutaminase-mediated cross-linking is involved in the stabilization of extracellular matrix in human liver fibrosis. J Hepatol 35(3):367–375

Haddox MK, Russell DH (1981) Increased nuclear conjugated polyamines and transglutaminase during liver regeneration. Proc Natl Acad Sci U S A 78(3):1712–1716

Hamada K, Terauchi A, Nakamura K, Higo T, Nukina N, Matsumoto N, Hisatsune C, Nakamura T, Mikoshiba K (2014) Aberrant calcium signaling by transglutaminase-mediated posttranslational modification of inositol 1,4,5-trisphosphate receptors. Proc Natl Acad Sci U S A 111(38):E3966–E3975. doi:10.1073/pnas.1409730111

Hasegawa G, Suwa M, Ichikawa Y, Ohtsuka T, Kumagai S, Kikuchi M, Sato Y, Saito Y (2003) A novel function of tissue-type transglutaminase: protein disulphide isomerase. Biochem J 373 (Pt 3):793–803. doi:10.1042/bj20021084

Hebert SS, Daviau A, Grondin G, Latreille M, Aubin RA, Blouin R (2000) The mixed lineage kinase DLK is oligomerized by tissue transglutaminase during apoptosis. J Biol Chem 275 (42):32482–32490. doi:10.1074/jbc.M006528200

Hiiragi T, Sasaki H, Nagafuchi A, Sabe H, Shen SC, Matsuki M, Yamanishi K, Tsukita S (1999) Transglutaminase type 1 and its cross-linking activity are concentrated at adherens junctions in simple epithelial cells. J Biol Chem 274(48):34148–34154

Hitomi K, Presland RB, Nakayama T, Fleckman P, Dale BA, Maki M (2003) Analysis of epidermal-type transglutaminase (transglutaminase 3) in human stratified epithelia and cultured keratinocytes using monoclonal antibodies. J Dermatol Sci 32(2):95–103

Ichinose A (2005) Extracellular transglutaminase: factor XIII. Prog Exp Tumor Res 38:192–208. doi:10.1159/000084241

Jaovisidha K, Etim A, Yamakawa K, Masuda I, Gohr CM, Wakim BT, Boonapatcharoen N, Ninomiya J, Rosenthal AK (2006) The serine protease inhibitor trappin-2 is present in cartilage and synovial fluid in osteoarthritis. J Rheumatol 33(2):318–325

Jensen PH, Lorand L, Ebbesen P, Gliemann J (1993) Type-2 plasminogen-activator inhibitor is a substrate for trophoblast transglutaminase and factor XIIIa. Transglutaminase-catalyzed cross-linking to cellular and extracellular structures. Eur J Biochem/FEBS 214(1):141–146

Jiang WG, Ye L, Sanders AJ, Ruge F, Kynaston HG, Ablin RJ, Mason MD (2013) Prostate transglutaminase (TGase-4, TGaseP) enhances the adhesion of prostate cancer cells to extracellular matrix, the potential role of TGase-core domain. J Transl Med 11:269. doi:10.1186/1479-5876-11-269

Johnson KA, Terkeltaub RA (2005) External GTP-bound transglutaminase 2 is a molecular switch for chondrocyte hypertrophic differentiation and calcification. J Biol Chem 280 (15):15004–15012. doi:10.1074/jbc.M500962200

Kaartinen MT, El-Maadawy S, Rasanen NH, McKee MD (2002) Tissue transglutaminase and its substrates in bone. J Bone Miner Res Off J Am Soc Bone Miner Res 17(12):2161–2173. doi:10.1359/jbmr.2002.17.12.2161

Karpuj MV, Garren H, Slunt H, Price DL, Gusella J, Becher MW, Steinman L (1999) Transglutaminase aggregates huntingtin into nonamyloidogenic polymers, and its enzymatic activity increases in Huntington's disease brain nuclei. Proc Natl Acad Sci U S A 96(13):7388–7393

Kim SY, Grant P, Lee JH, Pant HC, Steinert PM (1999) Differential expression of multiple transglutaminases in human brain. Increased expression and cross-linking by transglutaminases 1 and 2 in Alzheimer's disease. J Biol Chem 274(43):30715–30721

Kim DS, Park SS, Nam BH, Kim IH, Kim SY (2006) Reversal of drug resistance in breast cancer cells by transglutaminase 2 inhibition and nuclear factor-kappaB inactivation. Cancer Res 66 (22):10936–10943. doi:10.1158/0008-5472.can-06-1521

Kojima S, Muramatsu H, Amanuma H, Muramatsu T (1995) Midkine enhances fibrinolytic activity of bovine endothelial cells. J Biol Chem 270(16):9590–9596

Kuo TF, Tatsukawa H, Kojima S (2011) New insights into the functions and localization of nuclear transglutaminase 2. FEBS J 278(24):4756–4767. doi:10.1111/j.1742-4658.2011.08409.x

Kuo TF, Tatsukawa H, Matsuura T, Nagatsuma K, Hirose S, Kojima S (2012) Free fatty acids induce transglutaminase 2-dependent apoptosis in hepatocytes via ER stress-stimulated PERK pathways. J Cell Physiol 227(3):1130–1137. doi:10.1002/jcp.22833

Le M, Gohr CM, Rosenthal AK (2001) Transglutaminase participates in the incorporation of latent TGFbeta into the extracellular matrix of aging articular chondrocytes. Connect Tissue Res 42 (4):245–253

Lee J, Kim YS, Choi DH, Bang MS, Han TR, Joh TH, Kim SY (2004) Transglutaminase 2 induces nuclear factor-kappaB activation via a novel pathway in BV-2 microglia. J Biol Chem 279 (51):53725–53735. doi:10.1074/jbc.M407627200

Mahoney SA, Wilkinson M, Smith S, Haynes LW (2000) Stabilization of neurites in cerebellar granule cells by transglutaminase activity: identification of midkine and galectin-3 as substrates. Neuroscience 101(1):141–155

Milakovic T, Tucholski J, McCoy E, Johnson GV (2004) Intracellular localization and activity state of tissue transglutaminase differentially impacts cell death. J Biol Chem 279 (10):8715–8722. doi:10.1074/jbc.M308479200

Mishra S, Murphy LJ (2004) Tissue transglutaminase has intrinsic kinase activity: identification of transglutaminase 2 as an insulin-like growth factor-binding protein-3 kinase. J Biol Chem 279 (23):23863–23868. doi:10.1074/jbc.M311919200

Mishra S, Saleh A, Espino PS, Davie JR, Murphy LJ (2006) Phosphorylation of histones by tissue transglutaminase. J Biol Chem 281(9):5532–5538. doi:10.1074/jbc.M506864200

Nadella V, Wang Z, Johnson TS, Griffin M, Devitt A (2015) Transglutaminase 2 interacts with syndecan-4 and CD44 at the surface of human macrophages to promote removal of apoptotic cells. Biochim Biophys Acta 1853(1):201–212. doi:10.1016/j.bbamcr.2014.09.020

Nakaoka H, Perez DM, Baek KJ, Das T, Husain A, Misono K, Im MJ, Graham RM (1994) Gh: a GTP-binding protein with transglutaminase activity and receptor signaling function. Science (New York, NY) 264(5165):1593–1596

Nara K, Aoyama Y, Iwata T, Hagiwara H, Hirose S (1992) Cell cycle-dependent changes in tissue transglutaminase mRNA levels in bovine endothelial cells. Biochem Biophys Res Commun 187(1):14–17

Nemes Z, Marekov LN, Fesus L, Steinert PM (1999a) A novel function for transglutaminase 1: attachment of long-chain omega-hydroxyceramides to involucrin by ester bond formation. Proc Natl Acad Sci U S A 96(15):8402–8407

Nemes Z, Marekov LN, Steinert PM (1999b) Involucrin cross-linking by transglutaminase 1. Binding to membranes directs residue specificity. J Biol Chem 274(16):11013–11021

Nemes Z, Petrovski G, Aerts M, Sergeant K, Devreese B, Fesus L (2009) Transglutaminase-mediated intramolecular cross-linking of membrane-bound alpha-synuclein promotes amyloid formation in Lewy bodies. J Biol Chem 284(40):27252–27264. doi:10.1074/jbc.M109.033969

Nunes I, Gleizes PE, Metz CN, Rifkin DB (1997) Latent transforming growth factor-beta binding protein domains involved in activation and transglutaminase-dependent cross-linking of latent transforming growth factor-beta. J Cell Biol 136(5):1151–1163

Nunomura K, Kawakami S, Shimizu T, Hara T, Nakamura K, Terakawa Y, Yamasaki A, Ikegami S (2003) In vivo cross-linking of nucleosomal histones catalyzed by nuclear transglutaminase in starfish sperm and its induction by egg jelly triggering the acrosome reaction. Eur J Biochem/FEBS 270(18):3750–3759

Ohba H, Harano T, Omura T (1981) Intracellular and intramembranous localization of a protein disulfide isomerase in rat liver. J Biochem 89(3):889–900

Pavlyukov MS, Antipova NV, Balashova MV, Shakhparonov MI (2012) Detection of Transglutaminase 2 conformational changes in living cell. Biochem Biophys Res Commun 421 (4):773–779. doi:10.1016/j.bbrc.2012.04.082

Pfundt R, van Ruissen F, van Vlijmen-Willems IM, Alkemade HA, Zeeuwen PL, Jap PH, Dijkman H, Fransen J, Croes H, van Erp PE, Schalkwijk J (1996) Constitutive and inducible expression of SKALP/elafin provides anti-elastase defense in human epithelia. J Clin Invest 98 (6):1389–1399. doi:10.1172/jci118926

Piacentini M, Farrace MG, Piredda L, Matarrese P, Ciccosanti F, Falasca L, Rodolfo C, Giammarioli AM, Verderio E, Griffin M, Malorni W (2002) Transglutaminase overexpression sensitizes neuronal cell lines to apoptosis by increasing mitochondrial membrane potential and cellular oxidative stress. J Neurochem 81(5):1061–1072

Pinkas DM, Strop P, Brunger AT, Khosla C (2007) Transglutaminase 2 undergoes a large conformational change upon activation. PLoS Biol 5(12):e327. doi:10.1371/journal.pbio.0050327

Ponnusamy M, Pang M, Annamaraju PK, Zhang Z, Gong R, Chin YE, Zhuang S (2009) Transglutaminase-1 protects renal epithelial cells from hydrogen peroxide-induced apoptosis through activation of STAT3 and AKT signaling pathways. Am J Physiol Renal Physiol 297 (5):F1361–F1370. doi:10.1152/ajprenal.00251.2009

Popov Y, Sverdlov DY, Sharma AK, Bhaskar KR, Li S, Freitag TL, Lee J, Dieterich W, Melino G, Schuppan D (2011) Tissue transglutaminase does not affect fibrotic matrix stability or regression of liver fibrosis in mice. Gastroenterology 140(5):1642–1652. doi:10.1053/j.gastro.2011.01.040

Priglinger SG, Alge CS, Neubauer AS, Kristin N, Hirneiss C, Eibl K, Kampik A, Welge-Lussen U (2004) TGF-beta2-induced cell surface tissue transglutaminase increases adhesion and migration of RPE cells on fibronectin through the gelatin-binding domain. Invest Ophthalmol Vis Sci 45(3):955–963

Robinson NA, Eckert RL (1998) Identification of transglutaminase-reactive residues in S100A11. J Biol Chem 273(5):2721–2728

Robinson NJ, Baker PN, Jones CJ, Aplin JD (2007) A role for tissue transglutaminase in stabilization of membrane-cytoskeletal particles shed from the human placenta. Biol Reprod 77(4):648–657. doi:10.1095/biolreprod.107.061747

Rodolfo C, Mormone E, Matarrese P, Ciccosanti F, Farrace MG, Garofano E, Piredda L, Fimia GM, Malorni W, Piacentini M (2004) Tissue transglutaminase is a multifunctional BH3-only protein. J Biol Chem 279(52):54783–54792. doi:10.1074/jbc.M410938200

Rossin F, D'Eletto M, Falasca L, Sepe S, Cocco S, Fimia GM, Campanella M, Mastroberardino PG, Farrace MG, Piacentini M (2015) Transglutaminase 2 ablation leads to mitophagy impairment associated with a metabolic shift towards aerobic glycolysis. Cell Death Differ 22(3):408–418. doi:10.1038/cdd.2014.106

Saber-Lichtenberg Y, Brix K, Schmitz A, Heuser JE, Wilson JH, Lorand L, Herzog V (2000) Covalent cross-linking of secreted bovine thyroglobulin by transglutaminase. FASEB J Off Publ Fed Am Soc Exp Biol 14(7):1005–1014

Sane DC, Moser TL, Pippen AM, Parker CJ, Achyuthan KE, Greenberg CS (1988) Vitronectin is a substrate for transglutaminases. Biochem Biophys Res Commun 157(1):115–120

Serrano K, Devine DV (2002) Intracellular factor XIII crosslinks platelet cytoskeletal elements upon platelet activation. Thromb Haemost 88(2):315–320

Shibata T, Sekihara S, Fujikawa T, Miyaji R, Maki K, Ishihara T, Koshiba T, Kawabata S (2013) Transglutaminase-catalyzed protein-protein cross-linking suppresses the activity of the NF-kappaB-like transcription factor relish. Sci Signal 6(285):ra61. doi:10.1126/scisignal.2003970

Shrestha R, Tatsukawa H, Shrestha R, Ishibashi N, Matsuura T, Kagechika H, Kose S, Hitomi K, Imamoto N, Kojima S (2015) Molecular mechanism by which acyclic retinoid induces nuclear localization of transglutaminase 2 in human hepatocellular carcinoma cells. Cell Death Dis. doi:10.1038/cddis.2015.339 (in press)

Shweke N, Boulos N, Jouanneau C, Vandermeersch S, Melino G, Dussaule JC, Chatziantoniou C, Ronco P, Boffa JJ (2008) Tissue transglutaminase contributes to interstitial renal fibrosis by favoring accumulation of fibrillar collagen through TGF-beta activation and cell infiltration. Am J Pathol 173(3):631–642. doi:10.2353/ajpath.2008.080025

Siegel M, Strnad P, Watts RE, Choi K, Jabri B, Omary MB, Khosla C (2008) Extracellular transglutaminase 2 is catalytically inactive, but is transiently activated upon tissue injury. PLoS One 3(3):e1861. doi:10.1371/journal.pone.0001861

Singh US, Erickson JW, Cerione RA (1995) Identification and biochemical characterization of an 80 kilodalton GTP-binding/transglutaminase from rabbit liver nuclei. Biochemistry 34 (48):15863–15871

Song Y, Kirkpatrick LL, Schilling AB, Helseth DL, Chabot N, Keillor JW, Johnson GV, Brady ST (2013) Transglutaminase and polyamination of tubulin: posttranslational modification for stabilizing axonal microtubules. Neuron 78(1):109–123. doi:10.1016/j.neuron.2013.01.036

Staab JF, Bradway SD, Fidel PL, Sundstrom P (1999) Adhesive and mammalian transglutaminase substrate properties of Candida albicans Hwp1. Science 283(5407):1535–1538

Steinert PM, Marekov LN (1995) The proteins elafin, filaggrin, keratin intermediate filaments, loricrin, and small proline-rich proteins 1 and 2 are isodipeptide cross-linked components of the human epidermal cornified cell envelope. J Biol Chem 270(30):17702–17711

Tatsukawa H, Fukaya Y, Frampton G, Martinez-Fuentes A, Suzuki K, Kuo TF, Nagatsuma K, Shimokado K, Okuno M, Wu J, Iismaa S, Matsuura T, Tsukamoto H, Zern MA, Graham RM, Kojima S (2009) Role of transglutaminase 2 in liver injury via cross-linking and silencing of transcription factor Sp1. Gastroenterology 136(5):1783–1795, e1710. doi:10.1053/j.gastro.2009.01.007

Tseng HC, Tang JB, Gandhi PS, Luo CW, Ou CM, Tseng CJ, Lin HJ, Chen YH (2012) Mutual adaptation between mouse transglutaminase 4 and its native substrates in the formation of copulatory plug. Amino Acids 42(2–3):951–960. doi:10.1007/s00726-011-1009-9

van den Akker J, VanBavel E, van Geel R, Matlung HL, Guvenc Tuna B, Janssen GM, van Veelen PA, Boelens WC, De Mey JG, Bakker EN (2011) The redox state of transglutaminase 2 controls arterial remodeling. PLoS One 6(8):e23067. doi:10.1371/journal.pone.0023067
Verderio E, Gaudry C, Gross S, Smith C, Downes S, Griffin M (1999) Regulation of cell surface tissue transglutaminase: effects on matrix storage of latent transforming growth factor-beta binding protein-1. J Histochem Cytochem Off J Histochem Soc 47(11):1417–1432
Verderio EA, Telci D, Okoye A, Melino G, Griffin M (2003) A novel RGD-independent cel adhesion pathway mediated by fibronectin-bound tissue transglutaminase rescues cells from anoikis. J Biol Chem 278(43):42604–42614. doi:10.1074/jbc.M303303200
Vig PJ, Wei J, Shao Q, Hebert MD, Subramony SH, Sutton LT (2007) Role of tissue transglutaminase type 2 in calbindin-D28k interaction with ataxin-1. Neurosci Lett 420(1):53–57. doi:10.1016/j.neulet.2007.04.005
Wang Z, Telci D, Griffin M (2011) Importance of syndecan-4 and syndecan −2 in osteoblast cell adhesion and survival mediated by a tissue transglutaminase-fibronectin complex. Exp Cell Res 317(3):367–381. doi:10.1016/j.yexcr.2010.10.015
Wang Z, Collighan RJ, Pytel K, Rathbone DL, Li X, Griffin M (2012) Characterization of heparin-binding site of tissue transglutaminase: its importance in cell surface targeting, matrix deposition, and cell signaling. J Biol Chem 287(16):13063–13083. doi:10.1074/jbc.M111.294819
Wang Z, Perez M, Caja S, Melino G, Johnson TS, Lindfors K, Griffin M (2013) A novel extracellular role for tissue transglutaminase in matrix-bound VEGF-mediated angiogenesis. Cell Death Dis 4:e808. doi:10.1038/cddis.2013.318
Wang L, Uhlig PC, Eikenberry EF, Robenek H, Bruckner P, Hansen U (2014) Lateral growth limitation of corneal fibrils and their lamellar stacking depend on covalent collagen cross-linking by transglutaminase-2 and lysyl oxidases, respectively. J Biol Chem 289(2):921–929. doi:10.1074/jbc.M113.496364
Yoo JO, Lim YC, Kim YM, Ha KS (2012) Transglutaminase 2 promotes both caspase-dependent and caspase-independent apoptotic cell death via the calpain/BAX protein signaling pathway. J Biol Chem 287(18):14377–14388. doi:10.1074/jbc.M111.326074
Zemskov EA, Mikhailenko I, Hsia RC, Zaritskaya L, Belkin AM (2011) Unconventional secretion of tissue transglutaminase involves phospholipid-dependent delivery into recycling endosomes. PLoS One 6(4):e19414. doi:10.1371/journal.pone.0019414

Chapter 3
Preferred Substrate Structure of Transglutaminases

Kiyotaka Hitomi and Hideki Tatsukawa

Abstract By catalytic reactions, transglutaminases (TGase) modify structure, functions and localizations of proteins. Identification of specific substrate proteins and analyses on the reaction products is the first step in clarifying the physiological significance of cross-linking and transamidation reactions. To date, in in vitro and in vivo analyses, a number of proteins have been identified as possible substrates by several methods. Additionally, each isozyme in the enzyme family has its specific preference for its recognition of the substrate proteins. During these investigations, structural properties and interaction with the enzyme appeared specific and the elegant mechanisms have been characterized.

In the first part of this chapter, we will describe the basic knowledge regarding the mechanism of substrate enzyme reactions, substrates for major isozymes, and substrate sequences available for several experiments. In the second part of this chapter, our recent studies will be introduced, discussing identification of highly reactive substrate peptides on each isozyme from a random peptide library, applications for detection of activity, and identification of substrate candidates.

Keywords Substrate • Intermediate • Cornified envelop • Assay • Detection

3.1 Introduction

In catalytic reactions, three types of transglutaminase modifications (TGase) have been mainly characterized: cross-linking, transamidation, and deamidation. For cross-linking, between peptide-bound glutamine and lysine residues in identical or different substrate proteins, N^{ε} (γ -glutamyl) lysine isopeptide bond is formed. This bond is covalent, which allows stable binding, resistance to proteolysis, and the possibility to change the original secondary structures in the substrates, while transamidation and deamidation execute the attachment of the primary amine to the glutamine residues and replacement from glutamine to glutamic acid, respectively.

K. Hitomi (✉) • H. Tatsukawa
Graduate School of Pharmaceutical Sciences, Nagoya University, Nagoya, Chikusa, Nagoya 464-8601, Japan
e-mail: hitomi@ps.nagoya-u.ac.jp

K. Hitomi et al. (eds.), *Transglutaminases*, DOI 10.1007/978-4-431-55825-5_3

Though directional regulation of the two reactions still remained unknown, this protein modification is also possible in changing the function of several proteins.

In order to ascertain physiological significance of the enzymes, investigation of specific substrates and their reaction products are inevitable. By characterization of several substrates and their reaction products, specific roles have been clarified for each isozyme so far. In these investigations, substrate proteins have been characterized either in vitro or in vivo. However, we have to recognize that an excellent substrate in vitro does not necessarily translate to the in vivo reaction. Hence, structural and functional change in the reaction product should be demonstrated at the cellular level. From this consideration, well-designed research will be required to demonstrate clear evidence that in vivo substrates are comparable in vitro substrates.

Several procedures to identify the reactive residues (glutamine and lysine) have been developed. As initial experiments, the enzymatic cross-linked products are proteolyzed and subjected to amino acid residue sequencing. Because isopeptide bonds cannot be proteolyzed using Edman reaction, the cross-linking sites are identifiable. In recent years, with the appearance of mass spectrometry, identification of the reactive residues has been accelerated because the peptides containing the modified glutamine or lysine residue can be detected.

In parallel to such investigations, interactions between the enzymes (each isozyme) and their substrates have been studied for several purposes. Recognition of substrates by each isozyme was quite specific; therefore this property would produce substrate preferences. In other words, not all the glutamine residues react with any enzyme but are selectively recognized. From this aspect, it is important not only to identify substrates but also to clarify the glutamine and the lysine residues in the substrate proteins. These results will provide information toward the search for counterpart substrates and also in the development of specific regulators and inhibitors of TGs.

In this chapter, we first describe the basic knowledge of substrate recognition and the catalytic reaction mechanism. Furthermore, the major profile of the substrate proteins and responsible sequences on each isozyme that have been reported so far will be described. In the latter part of the chapter, our studies will be introduced on the identification of "preferred substrate sequence" and application of detection tools for the enzymatic activity and identification of substrates.

3.2 Reaction Mechanism

3.2.1 Intermediate Formation and the Subsequent Reactions (Fig. 3.1)

In the catalytic reaction of the enzyme, generally, the intermediate between enzyme and substrate forms an unstable state, which quickly changes to the substrate-

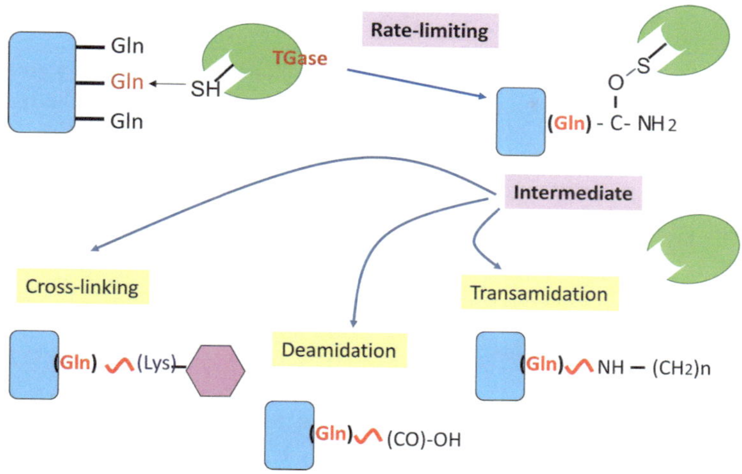

Fig. 3.1 Catalytic reactions by TGase. The γ-carboxamide group of the selected glutamine residue forms an intermediate with the thiol group of the cysteine residue in the activated TGase. Then the intermediate reacts with either the lysine residue of the substrate protein (cross-linking) or primary amine (transamidation) or water molecule (deamidation). In the reaction products, a new covalent isopeptide bond is formed

product complex. Then the product is released to allow the enzyme to interact with additional substrate molecules. In the reaction of TGase, as an initial step, the active site of cysteine of the enzyme binds γ-carboxiamide of the glutamine residue by a thioester bond. As a common mechanism in mammalian TGase, a calcium ion is necessary to expose the cysteine residue of the enzyme so that the substrate molecule will encounter the active site. Upon this intermediate formation, ammonia is released. Then, this acyl-enzyme intermediate goes forward to interact with nucleophilic molecules such as the ε-amino group of the lysine residue in proteins (cross-linking), primary amine (transamidation), and water molecules (deamidation). Cross-linking of proteins within identical proteins (oligomerization or polymerization) or distinct proteins will change their structural properties such as solubility and stability. Incorporation of polyamine and deamidation of glutamine side chains will add positive and negative charges, respectively. By these post-translational modifications, functional proteins will gain or lose their biochemical roles.

The competitive mechanisms between transamidation and deamidation have not been completely resolved. Fleckenstein et al. reported that the ratio of transamidation to deamidation decreased when pH was lowered under neutral pH, using gliadin-derived peptide (F-αI; QLQPFPQPQLPY) and biotin-pentylamine (Fleckenstein et al. 2002). At present, there is less information regarding the difference in the reactivity to the lysine residue and primary amine of the acyl-enzyme intermediate.

In these reactions, substrates that provide the glutamine residues for formation of the intermediate are designated as "glutamine-donor substrates". In the literature, the following synonyms are also used: acyl-donor, Q-substrate, and lysine-acceptor substrates. Conversely, the secondary substrate that provides the nucleophile is designated as "glutamine-acceptor" (or acyl-acceptor) substrate. In the case of protein, this molecule can be described as a lysine-donor substrate or "K-substrate".

3.2.2 Isozyme Specificity

The TG family comprises eight members of the isozymes, which exist in distinct areas in mammalian tissues. Particularly, FXIII, TG1, and TG2 have been exhaustively investigated regarding their tissue distribution and physiological functions. In such major isozymes, the substrates were also different regarding distribution and physiological roles in various tissues. Therefore, in spite of the same catalytic reaction, their interacting pattern of the glutamine residue of their substrates (also lysine residue of the glutamine-acceptor substrate) with the active site of the enzyme should be different. As the recognized glutamine residues differ among their substrate proteins, what determines the isozyme specificity upon binding between TGases and their substrates remains to be elucidated. Around the active site (Cys) in all the isozymes, eight residues are the same in the primary sequence (YGQCWVFA) resulting in formation of the interacting area for glutamine residues. However, the pocket area, which varies on each isozyme, may interact with the region of several amino acid residues surrounding the reactive glutamine residue in the substrate. Hence, the primary sequence around the glutamine residue should determine an isozyme-specific recognition.

3.3 Substrates and Characterization of Their Structure

In this section, we will summarize the investigations into the substrate structure and partial sequences that can act as prominent peptide.

3.3.1 Substrate Proteins and the Recognized Primary Sequences by the Enzymes

Here we summarize the favorable substrate sequences in mammalian proteins for each isozyme reaction. As described below, several substrates have been identified

Table 3.1 List of substrates on each isozyme: Factor XIII, Skin TG (TG1 and TG3), TG2, and TG4

Enzymes	Substrates	Identified substrate residues	References
FXIII	Fibrin (γ-γ chain)	Q398 and K406 (γ)	Lorand and Graham (2003)
	Fibrin (α-γ chain)	K556, K580 (α)	Sobel and Gawinowicz (1996)
	Fibrin (α-α chain)	Q223, Q237, Q366, Q563, K508, K539, K418, K556, K601	Wang (2011)
	α2-PI	NQEQVSPLTLLKLGN, K303 (to fibrin)	Cleary and Maurer (2006)
	Angiotensin receptor	Q315	AbdAlla et al. (2004)
	Fibronectin	Q3, Q7, Q9, Q246	Hoffmann et al. (2011)
TG1	SPR1	SSQQQKQP (1–8), QQKTKQK (82–88)	Candi et al. (1999)
TG3	SPR2	YQQQQCKQ (2–9), QPKCPPKSK (63–71)	Tarcsa et al. (1998)
	SPR3	QQQQVK (16–21), QQKTKQK (162–168)	Steinert et al. (1999)
	Involucrin	Q288, Q495, Q496	Steinert and Marekov (1997)
	Loricrin	Q3, Q6, Q10, Q156, Q215, Q216, Q305	Candi et al. (1995)
	S100A7	Q5	Ruse et al. (2001)
	S100A11	Q102	Robinson and Eckert (1998)
	Annexin I	Q19, Q23, K26	Robinson et al. (1997)
	SPINK6	Q4, Q11, Q30 (?)	Fischer et al. (2013)
TG2	Gluten (see Chap. 9)	PFPQPQLPYPQPQLPY	Sollid (2002)
	β-crystalline	TVQQEL	Berbers et al. (1984)
		EEKPAVTAAPKK	Groenen et al. (1992)
	Trappin family	GQDPVK	Schalkwijk et al. (1999)
	Papilloma virus protein E7	Q87,Q88	Jeon et al. (2003)
	IκBα	Q266, Q267, Q313, K177	Park et al. (2006)

(continued)

Table 3.1 (continued)

Enzymes	Substrates	Identified substrate residues	References
TG2	Fibronectin	Q3, Q7, Q9, Q246	Hoffmann et al. (2011)
	Glyceroaldehyde-3-phosphate dehydrogenase	Q78, Q204	Iwai et al. (2014)
	Osteopontin	Q34, Q42, Q193, Q248	Christensen et al. (2014)
TG4	SVL (seminal)	Q232 FYNSPEELA<u>Q</u>K Q254 TTEFP<u>Q</u>FSA	Tseng et al. (2009)

Substrates were selected in the reports which show the possible cross-linking sites. The glutamine and lysine residues identified as cross-linking product are described by either the amino acid number or the primary structure containing the neighboring sequences. The number of amino acids is referred to from the article (see references)

and investigation of responsible glutamine and lysine residues has been conducted by using a number of experiments. The first article, which discusses the influence of amino acid residue in the substrate sequence, is by Gorman and Folk (1984). They analyzed the preference of FXIII and guinea pig TG (TG2) for β-casein-derived peptide, LGLGQKVLG.

Here, we will mainly focus on the established substrates by each isozyme, particularly for the results on the primary structure (Table 3.1).

3.3.1.1 TG1 and TG3: Substrates in the Skin Epidermis

In the outermost area of the skin epidermis, barrier function is maintained by the supramolecule, cornified envelop (CE), which is constructed beneath the plasma membrane in the terminally differentiated keratinocytes. TGases in the keratinocytes are responsible for cross-linking structural proteins such as involucrin, loricrin, and SPRs (small proline-rich protein family). Most components of CE, as catalytically modified by transglutaminase, were purified and identified by protein sequence and mass spectrometry analysis.

Transglutaminase modifications of several substrate proteins have been characterized mainly in a series of studies by Steinert et al. According to their proposed model, briefly, loricrin cross-links SPRs at the cytoplasm, then resulting in movement close to the plasma membrane. Involucrin forms covalent binding with ceramide by TG1, where this product plays a role in providing a platform with other structural proteins such as envoplakin and periplakin. The SPRs-loricrin complex finally cross-links with involucrin, which is maturing into CE formation with the ceramide layer. However, a search for the uncharacterized members of CE components and also the precise mechanisms by which TGs cross-link is still under investigation.

In the sequences of the major substrates (involucrin and SPRs), although specific common sequences have not been observed, the lysine and glutamine residues are frequently observed in addition to the proline: involucrin, 7.7 % (Lys), 25.6 % (Gln) and 9.1 % (Pro); SPRs 12.4 % (Lys), 21.6 % (Gln) and 43.6 % (Pro) (Candi et al. 2005). Interestingly, as described below, in the preferred substrate sequences (as a glutamine-donor) for TG1 and TG3 that we identified from the random peptide library, lysine (or arginine; positive charged) residue are observed at the fixed position from the reactive glutamine residues (Sugimura et al. 2008a, b; Yamane et al. 2010).

3.3.1.2 TG2

Because substrates of TG2 are widely distributed and their functions are various, one may presume that there is less common structure in their structure. Apparent conformity in the primary sequence of TG2 substrates has not been shown although the amounts of the substrates have been identified. Hence, in this part, we briefly describe its preference from the studies on celiac disease influenced by TG2 activity.

Studies on celiac disease, a chronic inflammatory disorder, have contributed to the analyses of the preferred substrate sequences (Sollid 2002). In this disease, deamidated gliadin-derived peptides by TG2 stimulate the reactive T cells, which results in damage at the intestinal mucosa. By exhaustive investigation of the deamidation site in the peptide sequences (T cell epitope peptide: where underlined sequences are targeted), rules have been proposed for the recognition sequence where the spacing between the targeted glutamine and proline is significant; Q-x-P is the most favorable sequence, whereas Q-P, Q-G, Q-X-X-P, Q-X-X-G are not preferred (X denotes any amino acid). Another study of note is the amino acid substitution analysis using β-endorphin (Fleckenstein et al. 2002). This showed that the position −1, +1, +2, and +3 relative to the glutamine affects the reaction. These studies have contributed to the resolution of the mechanisms by which deamidation of the gluten-derived peptides cause unexpected inflammation and also may develop understanding in adding inhibitory drug production against the process. From our analysis using a phage-display peptide library (Sect. 4.1), this propensity is consistent with these investigations for the incorporation of the biotinylated primary amine.

3.3.1.3 Factor XIII

After the limited proteolysis by thrombin, the activated Factor XIII (FXIIIa) cross-links coagulating proteins, mainly fibrin monomer. In this final step in the coagulation pathway, fibrin γ-chains are cross-linked to form γ-γ dimers and further to react to the fibrin α monomer (Lorand and Graham 2003). In the carboxyl termini of the γ-chain as the only inter-molecular cross-links, the reactive glutamine and lysine residues were shown in the sequence <u>Q</u>QHHLGGA<u>K</u>QAGDV (Q398 and K406), where the underlined residues are involved in the reaction. The cross-linking of the α-chain contributes to link the γ-γ fibrils and also the plasma protein, α2PI. In these substrates, a couple of cross-linking sites have been identified. In a recent investigation by Wang, nine further cross-linking sites were newly identified by mass spectrometry (Wang 2011). Among them, there were no common motifs around the glutamine residues.

Activated FXIII also catalyzes cross-links of other proteins in plasma such as anti-plasmin, α2-macroglobulin, PAI-2, and vitronectin. Among them, in the sequence of α2 anti-plasmin, Maurer et al. reported that there was a tendency toward the reactive glutamine residue sites (Q2, Q4) and lysine residue (K12) on the peptide sequence, NQEQVSPLTLLKLGN (Cleary and Maurer 2006).

3.3.1.4 Other TGs: TG4, TG5, and TG6

On the substrates of these isozymes, less information has been obtained although the expression tissues and cells have been thoroughly investigated.

TG4 is prostate TGase and essential for plug formation in rodents. As its substrate, Tseng et al. characterized SVS I as being exclusively expressed in the seminal vesicle, and a substrate secretion protein (Tseng et al. 2009). In SVS I, 43 glutamine and 43 lysine residues exist in the total 796 amino acids. They concluded the most reactive residues as being Q232 (FYNSPEELAQK) and Q254 (TTEFPQFSA), though regulation for cross-linking and physiological significance has not been completely apparent.

TG6 is mainly expressed in the brain and the skin. In both tissues, however, there are no reports on the in vivo substrate proteins. Recently, regarding TG6, as with TG3, information on the recognition of glutamine residue has been obtained from studies on the extra-intestinal diseases caused by gluten-derived peptides such as dermatitis herpetiformis in the skin and gluten ataxia as neuropathy induced by TG3 and TG6, respectively (Stamnaes et al. 2010). In both cases, unexpected modification of gluten-derived peptides renders them toxic by stimulating T cells. In this analysis, even on the same peptide sequence, TG2 and TG6 prefer distinct glutamine residues as a target. In the sequence of gluten-derived peptide, PGQQQPFPQQPY, TG2 and TG6 favor Q34 and Q40, respectively; TG3 can target both residues.

TG5 is a novel member expressed in keratinocytes and was difficult to express as a soluble recombinant protein. Hence, there is less information on the biochemical analysis of the enzymatic reaction and the search for a substrate.

3.3.2 Substrate Peptides

In this section, two well-known peptides that have been characterized from natural substrate proteins and found to be efficient as peptide forms are described.

TVQQEL, a β-crystalline-derived peptide designated as A25, has been widely used as a glutamine-donor substrate peptide (Berbers et al. 1984). Crystallin is a family of structural proteins in the eye lens. Polymerization due to cross-linking reaction by TGase had been speculated as the cause of cataract formation. During the analyses of several crystalline family proteins, some molecules appeared to be good substrates in vitro. Hence, crystalline-derived peptides have been a target for glutamine and also lysine-donor substrate sequences, as describe later. Among them, A25 peptide has been used for an assay system that works as Q-donor substrate. However, there is sparse information regarding the isozyme specificity of this peptide.

Another substrate peptide is GQDPVK in the sequence of trappin (an acronym for transglutaminase substrate and whey acidic protein domain containing protein), which is a family of secreted proteins consisting of an N-terminal TGase substrate domain and a C-terminal four-disulfide bond harboring protease-binding domain (Schalkwijk et al. 1999). The repeating sequence in the substrate domains are composed of GQDPVK, which works as both a glutamine- and lysine-donor substrate. Not only repeating peptide, but also the single hexapeptide can act as a substrate, which was experimentally shown using biotin-labeled peptide (Zeeuwen et al. 1997).

3.3.3 *In Vivo and In Vitro Substrates: How Have TGase Substrates and Cross-Linking Sites Been Investigated?*

For TGase isozymes (FXIII, TG1-TG7), numerous substrates have been reported in various tissues. To search for substrates, in the cellular extracts from tissues or cultured cells, biotin-labeled primary amine was added and the incorporating proteins were affinity purified for identification. By this procedure, in vitro substrates have been identified by subjection to mass spectrometry and/or an amino acid sequencer.

This procedure also enables the identification of the cross-linking sites in the proteins. Ruppolo et al. investigated the incorporated pattern of biotin-labeled primary amine (biotin-pentylamine) in several proteins and peptides, by analyzing the abbreviated proteolyzed peptide fragments using LC-MS/MS (liquid chromatography with tandem mass spectrometry) (Ruoppolo et al. 2003). By this method they identified reactive glutamine residues in the well-known lysine-acceptor substrates. For analysis of glutamine-accepting sites, Ruppolo and colleagues used the A25 peptide (TVQQEL). Using a similar approach, Emanuelsson et al. demonstrated the system to know the susceptible sites in the known substrate (Emanuelsson et al. 2005). By the catalytic activity of guinea pig liver TG (an ortholog for human TG2), the cross-linking product between the labeled hexapeptides (GQDPVR and GNDPVK) as a probe and the small heat shock proteins were analyzed.

Identified substrates by incorporation of "substrate probe" are in vitro substrates, which do not necessarily reflect physiological events. Hence, as a next step, the proteins have to be confirmed whether the identified protein (either purified or produced as a recombinant protein) can be modified upon catalytic reaction by TGase. Alternatively, if a specific antibody is available, dimer or polymer products can be observed by western blotting. By the evidence of inclusion of the proteins in the cross-linking products at cellular and/or tissue level, they are then recognized as in vivo substrate.

3.3.4 Characterization of the Lysine-Donor Substrate Sequence

Initially, when compared to the glutamine-donor substrate, it is considered that there is less tendency in the favorable sequence and structure for the lysine-donor substrate. However, there have been reports of the peptide substrate sequence for the lysine-donor substrate mainly from two protein sequences: crystalline and fibrinogen.

As described, since crystalline is a good substrate, Groenen et al. found that carboxyl-terminal lysine αB-crystalline in calf lens is an amine-donor substrate: EEKPAVTAAPKK (Groenen et al. 1992). Other possible peptides were also found in βH-crystalline of which reactivity to TGase is enhanced by exposure to hydroxyl radical (Groenen et al. 1993). In fibrin(ogen), Sobel et al. biochemically characterized cross-linking sites for lysine residues as glutamine-acceptor sites: K406 in γ-chain and K556 and K580 in α-chain (Sobel and Gawinowicz 1996).

As non-protein type glutamine-acceptor substrate in tissue and cellular component, polyamine is a major substrate, as a primary amine. This molecule exists in most cells in significant amounts and is widely distributed from microorganism to higher animals. Other than polyamine, serotonin appeared as a functional substrate that modulates several proteins such as small GTPases (Rho and Rab family) and fibronectin (Walther et al. 2003; Wei et al. 2012).

3.3.5 Database for Substrates

Additionally, we have to shed light on the Transdab database: (http://genomics.dote.hu/wiki/index.php?title=Main_Page&oldid=1673). (Csosz et al. 2009). This web site for this database has been used to identify substrates for each isozyme, by listing on the site. From the database, we can obtain information on the substrates including the reactive residues (Gln, Lys) for the 363 substrates presently included (2015, Jan).

3.4 Screening and Characterization of Substrate Proteins and Preferable Sequence-Preference on the Recognized Primary Structure

In this section, we describe our established screening procedure to identify preferred substrate sequences using a random peptide library.

Upon formation of the enzyme-substrate intermediate, the glutamine residue is selected for interaction with the active cysteine residue in the enzyme. Therefore, feasibility of interaction for glutamine is quite important to decide how fast and efficiently the catalytic reaction proceeds. In this case, all the glutamine does not participate in the reaction. Surrounding amino acid sequences affect the interaction for the active thiol group of cysteine. Therefore, we attempted to clarify the primary structure surrounding the active cysteine using the peptide library.

In recent years, several kinds of peptide library have been developed which provide random amino acid sequences with various lengths. In our case, we used a M13 phage-displayed peptide library, which was originally developed to obtain a mimicking antibody molecule or to identify epitopes.

In this library, M13 phage is genetically modified on which a 12-mer random peptide is expressed as a fusion protein with the original coat protein. Each phage displays a unique peptide sequence, which can be identified by sequencing the encoded gene. Hence, the phage particle on which the displaying peptide has affinity to the target molecule can be selected among numerous phage solutions.

In our established system, the displayed peptide is used as a glutamine-donor substrate while biotin-labeled primary amine (cadaverine) is used as a glutamine-acceptor substrate in the presence of the active enzyme. After the enzymatic reaction, phage particles that display the peptide containing highly reactive glutamine residue(s), and also incorporate biotin-cadaverine, can be purified using avidin affinity chromatography. The purified particle can be amplified by infection of bacteria and then subjected to additional enzymatic reactions. This "panning" procedure is repeated four to five times and obtained phage particles are analyzed by their coding peptide sequences (Fig. 3.2).

The peptide sequences displayed on the selected phages appeared to have common amino acid sequences. These candidate sequences were further evaluated by analyses on higher reactivity and specificity for the isozymes. Finally, the sequence was examined to confirm the reactivity as a peptide form, which would be available for several application tools as described later.

So far, among the isozymes, we have obtained "minimum" and "preferred" substrate peptides on FXIII, TG1, TG2, TG3, TG6, and TG7 (Sugimura et al. 2006, 2008a; Yamane et al. 2010; Fukui et al. 2013; Kuramoto et al. 2013). As shown in Fig. 3.3, each preferred sequence had specificity and high reactivity when the peptide form was used for the reaction. For microbial transglutaminase (MTG) from *Streptomyces mobaraensis* that have been used in the food industry, we also obtained the specific preferred substrate sequence (YELQRPYHSELP) (Sugimura et al. 2008b).

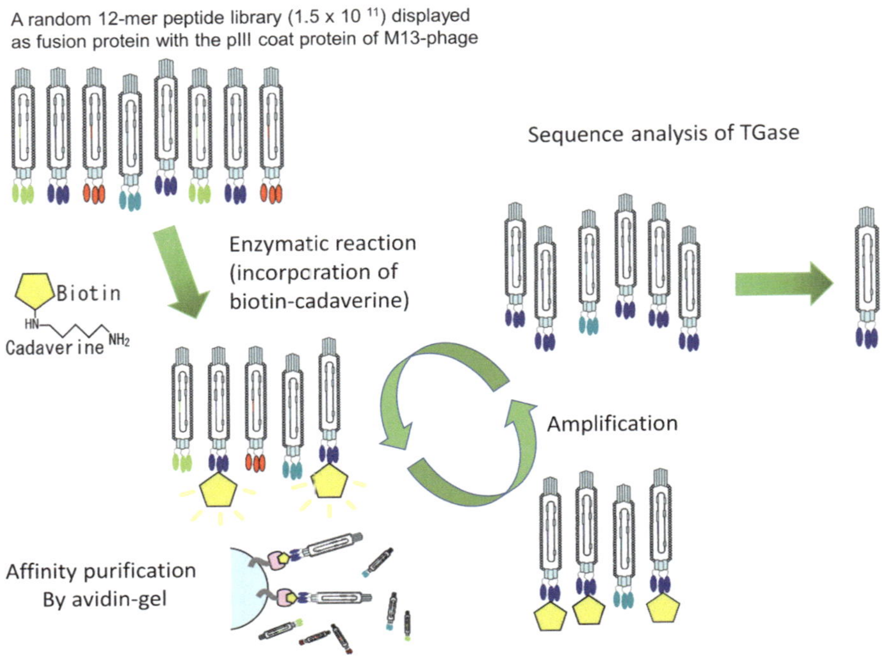

Fig. 3.2 Screening procedure using an M13 phage-displayed random peptide library. Phage particles displaying 12-mer peptide that worked as the glutamine-donor substrate were selected by the repeating procedure, i.e., the enzyme reaction, panning, and amplification. Upon panning, the phage particles that incorporated preferentially biotin-labeled primary amine (cadaverine) were purified. Selected phages were analyzed for their displaying peptide sequence

Regarding the lysine-donor substrate, there appeared to be a preferable sequence, which was less specific than the case of the glutamine-donor substrate. Our method may be applicable; however, at present we have failed to obtain specific sequences (using biotin-F11 peptide as a probe). We suggest that the condition of screening may be difficult probably due to lower specificity of the substrate.

Before we established this system, Fesus et al. demonstrated a distinct screening procedure using a phage-displayed peptide library. From the random hepta-peptide library, they conducted directly binding sequences on to the active form of TG2 (Keresztessy et al. 2006). This screening system also identified several glutamine-donor peptides such as GQQQTPY, GLQQASV, and WQTPMNS. Using an mRNA display system, recently, predicting peptide sequence (RLQQP) as a favorable sequence has been found (Lee et al. 2013). Furthermore, toward the MTG, a novel preferred tetra-peptide (TQGA) substrate was identified from a small focused synthetic peptide library (Caporale et al. 2015).

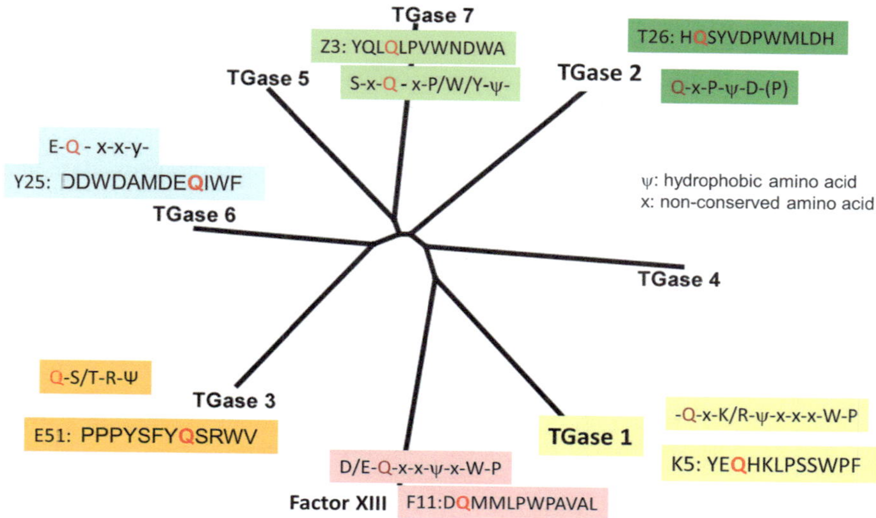

Fig. 3.3 TGase family members and the preferred substrate sequences. The location of the isozymes is expressed based on the phylogenic tree constructed by clustal X analysis of the primary sequences of TGase. Peptide sequences for F11 (FXIII), K5 (TG1), T26 (TG2), E51 (TG3), Y25 (TG6), and Z3 (TG7) are the most favorable substrate sequences screened from the random peptide library. The commonly observed amino acid residues (as motif-like sequences) in the selected peptide sequences are also described: ψ and x indicate the hydrophobic amino acids and any amino acid, respectively

3.5 Application of Highly Reactive Substrate Peptides

In this section, we describe applications using our identified preferred substrate peptides: sensitive detection systems for the enzymatic activity and identification of substrate candidates.

3.5.1 *Assay System Using Labeled Substrate*

So far, several assay systems for the enzyme have been established. As a principle, the labeled primary amine coupled with chemical compounds such as fluorescence or biotin is incorporated into the glutamine residues of the target substrate protein (dimethyl-casein that has no reactive lysine residue). Measurement of incorporated compound allows evaluation of the enzymatic activity. Regarding the compound coupled with the primary amine, an attempt to improve the labeling method has been made in enhancing both sensitivity and speed.

As far as labeled molecules are concerned, such as glutamine-acceptor substrate, it was impossible to discriminate the isozyme-specific activity. Although there is not such an assay system to resolve this problem, the system using our developed substrate peptide enabled rapid and isozyme-specific detection of the enzymatic activity. As a brief protocol, the lysine-donor substrate such as casein or primary amine is immobilized on the microtiter plate and then the labeled isozyme-specific peptide can be added with the enzyme-containing solution. The amounts of the incorporated peptides are measured by taking advantage of the labeling molecule of the peptide. In the case of biotin-peptide, addition of streptavidin-peroxidase and color development are conducted. In such an assay system, isozyme-specific detection is possible even in the cellular extract that contains different isozymes. This assay system is commercially available from Covalab S.A.R.L. (Lyon, France) (Perez-Alea et al. 2009; Hitomi et al. 2009).

3.5.2 *Visualization of In Situ Enzymatic Activity Using Highly Reactive Substrate Peptide*

3.5.2.1 Principle and Method

By taking advantage of the specificity and sensitivity, fluorescent-labeled peptides have been used for detection of in situ activity, which enables visual localization of the active enzyme (Fig. 3.4A). A non-fixed tissue section should be applied because the enzymes in the tissues have to maintain their catalytic activity. On the tissue section, FITC-labeled peptide cross-reacts with the lysine-donor substrates, visualized as a fluorescent signal. Because the peptide concentration of the reaction solution is adequate at less than 5–10 μM, it is not expensive. Instead, biotin-labeled peptide is also available for visualization by combination of peroxidase-conjugated streptavidin and appropriate no-water soluble dye (e.g., DAB; diaminobenzidine). These methods with this principle are applicable for most tissues.

The advantage of this method is that it can detect only the activated form of TGs. By a tentative immunodetection system, both the activated and inactivated TGs were unable to be discriminated. For example, during keratinocyte differentiation, TG1 cross-links several structural proteins to form a cornified envelop, as barrier functional components. Upon terminal differentiation, TG1 becomes inactivated probably because TG1 itself cross-links as substrate into the cornified envelop and/or results in proteolytic degradation by lysosomal proteases. The distribution pattern of immunostaining and in situ activity in several tissues is sometimes different. In parallel to the immunostaining, our method is specific to distinct localization of the active enzyme.

3.5.2.2 Detection of In Situ Activity in the Skin Epidermis

In our studies, we successfully visualized the localization of skin epidermal TGase, TG1, TG3, and TG6 (Sugimura et al. 2008a, b; Yamane et al. 2010;

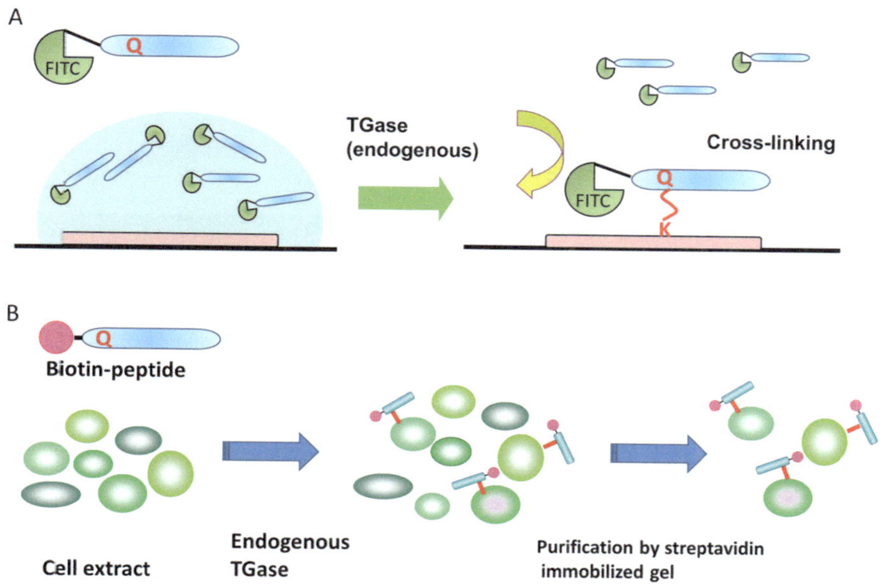

Fig. 3.4 (**a**) Detection procedure for in situ activity using a fluorescence-labeled substrate peptide. On the unfixed tissue section, FITC-labeled peptide is added to the reaction in an appropriate buffer solution. By the enzymatic activity of endogenous TG, the peptide is incorporated into the lysine-donor substrate (via covalent cross-linking). Detection of fluorescence is possible from the area depending on in situ enzymatic activity in the tissue section. (**b**) Purification of substrate proteins using affinity chromatography. Biotin-labeled peptide is incorporated into possible substrate proteins by endogenous enzymatic activity in cellular or tissue extracts. Using avidin-immobilized gel, the biotin-bound proteins are purified as possible substrates. Detection in the cellular extract using streptavidin-peroxidase is possible before purification

Fukui et al. 2013). Using each peptide conjugated to FITC, areas of active skin where TGase is present can be identified. The concentration of the peptide is less than 10 μM; therefore, there is no cross-reactivity for other isozymes among the peptide.

3.5.2.3 Whole Mouse Section

This method is applicable for any tissue section. We have succeeded in obtaining "Atlas" of in situ enzymatic activity using a whole mouse section. As shown in Fig. 3.5, fluorescent-labeled substrate peptides can cross-react with the lysine-donor substrates in the most tissues. In the case of both K5 (for TG1) and T26 (for TG2), in situ activity of the activated form of the TGs was clearly shown (Itoh et al. 2011).

A specific film and technique called the "Kawamoto method" enables us to prepare thin sections on the sheet, upon cryostat preparation (Kawamoto 2003). Originally, this method was for immune detection but it is possible to assay the enzymatic activity using labeled substrate. Double staining by in situ activity

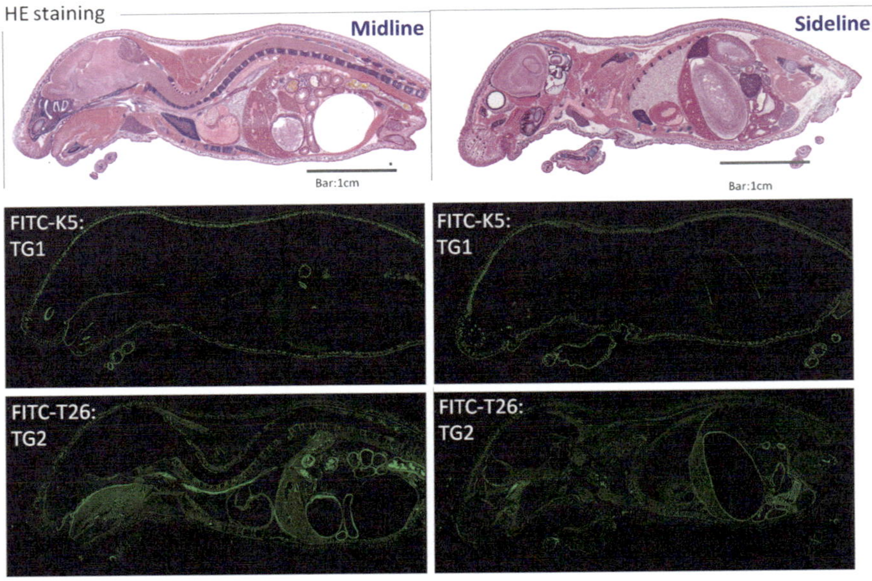

Fig. 3.5 Reaction patterns of in situ activity for whole mouse body section. Pictures for the midline and the sideline of the body sections are shown with hematoxylin and eosin staining (*upper*). Fluorescence signals of in situ activities for TG1 and TG2 are shown by using peptides for K5 (*middle*) and T26 (*lower*). Bar indicates 1 cm

followed by an immunostaining method is possible, which clearly shows the TG distributing pattern in activation and inactivation forms (Itoh et al. 2013).

3.5.3 Detection, Purification and Identification of the Lysine-Donor Substrate Proteins in an Isozyme-Specific Manner

Substrate glutamine-donor peptides can be used as an efficient method to detect and purify the possible in vitro substrates (Fig. 3.4B). Addition of the biotin-labeled peptide into cellular extracts, containing soluble proteins from any tissues, will produce the cross-linked products between the lysine-donor substrates and the biotin-labeled peptide. These products can be detected by avidin-conjugated peroxidase or a fluorescence molecule in either the cell extract or cultured cells (Watanabe et al. 2013; Fukui et al. 2013; Wang et al. 2014). From the total reaction products, possible lysine-donor substrates can be affinity purified using streptavidin-immobilized gel chromatography. The possible substrate candidate proteins in the fractions purified by the chromatography can be analyzed by mass spectrometry for identification. By this method, so far, isozyme-specific substrates for each TGase have been characterized in cellular extracts (Watanabe et al. 2013).

3.6 Concluding Remarks

The transglutaminase family comprises eight isozymes that are distributed in various tissues. However, unknown physiological roles for all the isozymes are still under investigation. Searching for substrates is the first step in investigating the physiological functions of the enzyme. Additional information on the substrate, either in vivo or in vitro, will help to clarify the roles of all the isozymes.

We have successfully identified several kinds of preferred substrate sequences that specifically react with each isozyme. Although this preferred tendency was not possible to examine via the database, several applications of "minimum substrate peptides" enabled detection of the enzymatic activity both in vitro and in situ. As a future analysis, with higher stability, incorporation into the animal body may be possible to detect activity.

References

AbdAlla S, Lother H, Langer A, el-Faramawy Y, Quitterer U (2004) Factor XIIIA crosslinks AT1 receptor dimers of monocytes at the onset of atherosclerosis. Cell 119:343–354

Berbers GA, Feenstra RW, van den Bos R, Hoekman WA, Bloemendal H, de Jong WW (1984) Lens transglutaminase selects specific beta-crystalline sequences as substrate. Proc Natl Acad Sci U S A 81:7017–7020

Candi E, Melino G, Mei G, Tarcsa E, Chung SI, Marekov LN, Steinert PM (1995) Biochemical, structural, and transglutaminase substrate properties of human loricrin, the major epidermal cornified cell envelope protein. J Biol Chem 270:26382–26390

Candi E, Tarcsa E, Idler WW, Kartasova T, Marekov LN, Steinert PM (1999) Transglutaminase cross-linking properties of the small proline-rich 1 family of cornified cell envelope proteins. Integration with loricrin. J Biol Chem 274:7226–7237

Candi E, Schmidt R, Melino G (2005) The cornified envelop: a model of cell death in the skin. Nat Rev Mol Cell Biol 6:328–340

Caporale A, Selis F, Sandomenico A, Jotti GS, Tonon G, Ruvo M (2015) The LQSP tetrapeptide is a new highly efficient substrate of microbial transglutaminase for the site-specific derivatization of peptides and proteins. Biotechnol J 10:154–161

Christensen B, Zachariae ED, Scavenius C, Thybo M, Callesen MM, Kløverpris S, Oxvig C, Enghild JJ, Sørensen ES (2014) Identification of transglutaminase reactive residues in human osteopontin and their role in polymerization. PLoS One 9:e113650

Cleary DB, Maurer MC (2006) Characterizing the specificity of activated Factor XIII for glutamine-containing substrate peptides. Biochim Biophys Acta 1764:1207–1217

Csosz E, Mesko B, Fesus L (2009) Transdabwiki: the interactive transglutaminase substrate database on web 2.0 surface. Amino Acids 36:615–617

Emanuelsson CS, Boros S, Hjernoe K, Boelens WC, Hojrup P (2005) Screening for transglutaminase-catalyzed modifications by peptide mass finger printing using multipoint recalibration on recognized peaks for high mass accuracy. J Biomol Tech 16:197–208

Fischer J, Koblyakova Y, Latendorf T, Wu Z, Meyer-Hoffert U (2013) Cross-linking of SPINK6 by transglutaminases protects from epidermal proteases. J Invest Dermatol 133:1170–1177
Fleckenstein B, Molberg Ø, Qiao SW, Schmid DG, von der Mulbe F, Elgstoen K, Jung G, Sollid LM (2002) Gliadin T cell epitope selection by tissue transglutaminase in celiac disease. J Biol Chem 277:34109–34116
Fukui M, Kuramoto K, Yamasaki R, Shimizu Y, Itoh M, Kawamoto T, Hitomi K (2013) Identification of a highly reactive peptide for TG6: detection of its transglutaminase activity in the skin epidermis using the peptide. FEBS J 280:1420–1429
Gorman JJ, Folk JE (1984) Structural features of glutamine substrates for transglutaminases. Role of extended interactions in the specificity of human plasma Factor XIIIa and of the guinea pig liver enzyme. J Biol Chem 259:9007–9010
Groenen PJ, Bloemendal H, de Jong WW (1992) The carboxy-terminal lysine αB-crystalline is an amine donor substrate for tissue transglutaminase. Eur J Biochem 205:671–674
Groenen PJ, Seccia M, Smulders RH, Gravela E, Cheeseman KH, Bloemendal H, de Jong WW (1993) Exposure of beta H-crystallin to hydroxyl radicals enhances the transglutaminase-susceptibility of its existing amine-donor and amine-acceptor sites. Biochem J 295:399–404
Hitomi K, Kitamura M, Perez-Alea M, Ismail C, Thomas V, El-Alaoui S (2009) A specific colorimetric assay for measuring of transglutaminase 1 and Factor XIII activities. Anal Biochem 394:281–283
Hoffmann BR, Annis D, Mosher DF (2011) Reactivity of the N-terminal region of fibronectin protein to transglutaminase 2 and FactorXIIIA. J Biol Chem 286:32220–32230
Itoh M, Kawamoto T, Tatsukawa H, Kojima S, Yamanishi K, Hitomi K (2011) In situ detection of active transglutaminases for keratinocyte-type (TGase 1) and tissue-type (TGase 2) using fluorescence-labeled highly reactive substrate peptides. J Histochem Cytochem 59:180–187
Itoh M, Tatsukawa H, Lee E-S, Yamanishi K, Kojima S, Hitomi K (2013) Variations in both TG1 and TG2 isozyme-specific in situ activities and protein expressions during mouse embryonic development. J Histochem Cytochem 61:793–801
Iwai K, Shibukawa Y, Yamazaki K, Wada Y (2014) Transglutaminase 2-dependent deamidation of glyceraldehyde-3-phosphate dehydrogenase promotes trophoblastic cell fusion. J Biol Chem 289:4989–4999
Jeon J-H, Choi K-H, Cho S-Y, Kim C-W, Shin D-M, Kwon J-C, Song K-Y, Park S-C, Kim I-G (2003) Transglutaminase 2 inhibits Rb binding of human papillomavirus E7 by incorporating polyamine. EMBO J 22:5273–5282
Kawamoto T (2003) Use of a new adhesive film for the preparation of multi-purpose fresh-frozen sections from hard tissues, whole-animals, insects and plants. Arch Histol Cytol 66:123–143
Keresztessy Z, Csosz E, Harsfalvi J, Csomos K, Gray J, Lightowlers RN, Lakey JH, Balajthy Z, Fesus L (2006) Phage display selection of efficient glutamine-donor substrate peptides for transglutaminase 2. Protein Sci 15:2466–2480
Kuramoto K, Yamasaki R, Shimizu Y, Tatsukawa H, Hitomi K (2013) Phage-displayed peptide library screening for preferred human substrate peptide sequences for transglutaminase 7. Arch Biophys Biochem 537:138–143
Lee JH, Song C, Kim DH, Park IH, Lee SG, Lee YS, Kim BG (2013) Glutamine (Q)-peptide screening for transglutaminase reaction using mRNA display. Biotechnol Bioeng 110:353–362
Lorand L, Graham RM (2003) Transglutaminases: crosslinking enzymes with pleiotropic functions. Nat Rev Mol Cell Biol 4:140–156
Park SS, Kim JM, Kim DS, Kim IH, Kim SY (2006) Transglutaminase 2 mediates polymer formation of I-kappaBalpha through C-terminal glutamine cluster. J Biol Chem 281:34965–34972
Perez-Alea M, Kitamura M, Martin G, Thomas V, Hitomi K, El-Alaoui S (2009) Development of an isoenzyme-specific colorimetric assay for tissue transglutaminase 2 cross-linking activity. Anal Biochem 389:150–156
Robinson NA, Eckert RL (1998) Identification of transglutaminase reactive residues in S100A11. J Biol Chem 273:2721–2728

Robinson NA, Lapic S, Welter JF, Eckert RL (1997) S100A11, S10A10, annexin 1, desmosomal proteins, small proline-rich protein, plasminogen activator inhibitor-2, and involucrin are components of the cornified envelop of cultured human epidermal keratinocytes. J Biol Chem 272:12035–12046

Ruoppolo M, Orrù S, D'Amato A, Francese S, Rovero P, Marino G, Esposito C (2003) Analysis of transglutaminase protein substrates by functional proteomics. Protein Sci 12:1290–1297

Ruse M, Lambert A, Robinson N, Ryan D, Shon K-J, Eckert RL (2001) S100A7, S100A10, and S100A11 are transglutaminase substrates. Biochemistry 40:3167–3173

Schalkwijk J, Wiedow O, Hirose S (1999) The trappin gene family: proteins defined by an N-terminal transglutaminase substrate domain and a C-terminal four-disulphide core. Biochem J 340:569–577

Sobel JH, Gawinowicz MA (1996) Identification of the α chain lysine donor sites involved in Factor XIIIa fibrin cross-linking. J Biol Chem 271:19288–19297

Sollid LM (2002) Coeliac disease: dissecting a complex inflammatory disorder. Nat Rev Immunol 2:647–655

Stamnaes J, Dorum S, Fleckenstein B, Aeschlimann D, Sollid LM (2010) Gluten T cell epitope targeting by TG3 and TG6; implications for dermatitis herpetiformis and gluten ataxia. Amino Acids 39:1183–1191

Steinert PM, Marekov LN (1997) Direct evidence that involucrin is a major early isopeptide cross-linked component of the keratinocyte cornified cell envelope. J Biol Chem 272:2021–2030

Steinert PM, Candi E, Tarcsa E, Marekov LN, Sette M, Paci M, Ciani B, Guerrieri P, Melino G (1999) Transglutaminase crosslinking and structural studies of the human small proline rich 3 protein. Cell Death Differ 6:916–930

Sugimura Y, Hosono M, Wada F, Yoshimura T, Maki M, Hitomi K (2006) Screening for the preferred substrate sequence of transglutaminase using a phage-displayed peptide library: Identification of peptide substrates for TGase 2 and Factor XIIIa. J Biol Chem 281:17699–17706

Sugimura Y, Hosono M, Kitamura M, Tsuda T, Yamanishi K, Maki M, Hitomi K (2008a) Identification of preferred substrate sequences for transglutaminase 1: development of a novel peptide that can efficiently detect cross-linking enzyme activity in the skin. FEBS J 275:5667–5677

Sugimura Y, Yokoyama K, Nio N, Maki M, Hitomi K (2008b) Identification of the preferred substrate sequences of microbial transglutaminase from *Streptomyces mobaraensis* using a phage-displayed peptide library. Arch Biochem Biophys 477:379–383

Tarcsa E, Candi E, Kartasova T, Idler WW, Marekov LN, Steinert PM (1998) Structural and transglutaminase substrate properties of the small proline-rich 2 family of cornified cell envelope proteins. J Biol Chem 273:23297–23303

Tseng HC, Lin HJ, Tang JB, Gandhi PS, Chang WC, Chen YH (2009) Identification of the major TG4 cross-linking sites in the androgen-dependent SVS I exclusively expressed in mouse seminal vesicle. J Cell Biochem 107:899–907

Walther DJ, Peter JU, Winter S, Höltje M, Paulmann N, Grohmann M, Vowinckel J, Alamo-Bethencourt V, Wilhelm CS, Ahnert-Hilger G, Bader M (2003) Serotonylation of small GTPases is a signal transduction pathway that triggers platelet alpha-granule release. Cell 115:851–862

Wang W (2011) Identification of respective lysine donor and glutamine acceptor sites involved in Factor XIIIa-catalyzed fibrin α chain cross-linking. J Biol Chem 286:44952–44964

Wang S, Cui C, Hitomi K, Kaartinen MT (2014) Detyrosinated Glu-tubulin is a substrate for cellular Factor XIIIA transglutaminase in differentiating osteoblasts. Amino Acids 46:1513–1526

Watanabe K, Tsunoda K, Itoh M, Fukui M, Mori H, Hitomi K (2013) Transglutaminase 2 and Factor XIII catalyze distinct substrates in differentiating osteoblastic cell line: utility of highly reactive substrate peptides: utility of highly reactive substrate peptides. Amino Acids 44:209–214

Wei L, Warburton RR, Preston IR, Roberts KE, Comhair SA, Erzurum SC, Hill NS, Fanburg BL (2012) Serotonylated fibronectin is elevated in pulmonary hypertension. Am J Physiol Lung Cell Mol Physiol 302:1273–1279
Yamane A, Fukui M, Sugimura Y, Itoh M, Alea MP, El Alaoui S, Akiyama M, Hitomi K (2010) Identification of a preferred substrate peptide for transglutaminase 3 and detection of in situ activity in skin and hair follicles. FEBS J 277:3564–3574
Zeeuwen PL, Hendriks W, de Jong WW, Schalkwijk J (1997) Identification and sequence analysis of two new members of the SKALP/elafin and SPAI-2 gene family. Biochemical properties of the transglutaminase substrate motif and suggestions for a new nomenclature. J Biol Chem 272:20471–20478

Chapter 4
Insights into Transglutaminase 2 Function Gained from Genetically Modified Animal Models

Siiri E. Iismaa

Abstract Transglutaminase 2 (TG2) has been implicated in a number of physiological processes and in disease. A number of different types of genetically engineered mice have been generated to help understand the role and regulation of TG2 in physiology and pathology and to help delineate the molecular basis of its actions. This chapter will briefly review guidelines for animal husbandry and animal study design, TG2 mouse models and recent insights into the physiology and pathophysiology of TG2 that have come from the study of these genetically engineered animal models.

Keywords Transglutaminase 2 • TG2 Mice • TG2 Transgenic • TG2 Knockout • TG2 Knock-in • Genetically modified mice • Physiology • Disease

4.1 Introduction

Proteins perform most of the work of living cells yet the questions of what a protein does inside a living cell, where, when, how and why, are not simple ones to answer. Many clues to the function(s) of a protein can be gained from in vitro biochemical experiments using purified protein or from cell culture or genetically modified mouse models, in which the protein of interest is over-expressed or knocked-down and its activity or activities is activated or inhibited.

Genetically modified mouse models play a vital role in evaluating the integrated physiological or pathological action(s) of a protein in diverse cellular contexts. The two most common types of genetically modified mice are transgenic overexpressing mice, which carry a segment of foreign DNA incorporated into the genome, and targeted mutant mice, in which a gene has been inactivated by deletion or disruption (knockout) or mice have a mutant, duplicated or new gene (knock-in). Gene inactivation in genetically modified mice may be permanent or

S.E. Iismaa (✉)
Victor Chang Cardiac Research Institute and University of New South Wales, Sydney, NSW, Australia
e-mail: s.iismaa@victorchang.edu.au

K. Hitomi et al. (eds.), *Transglutaminases*, DOI 10.1007/978-4-431-55825-5_4

conditional. Conditional inactivation of genes is extremely powerful to determine the complex roles of genes in tissue-specific contexts (in which a target gene is specifically inactivated in specific tissue(s)) and/or in developmental contexts (in which the target gene is specifically inactivated at a given time point during development or in adult life) [see (Pease and Saunders 2011) for a detailed review of animal transgenesis protocols].

Animal models can often recapitulate many aspects of human disease and as such are powerful tools for confirming disease causation, for examining disease pathogenesis and for assessing the efficacy of drugs or treatment strategies. There are limitations, however. Pharmacological inhibitors are commonly used in animal studies yet "in cell biology we are forced to realize that the specificity of an inhibitor is inversely proportional to our knowledge regarding its mode of action" (Lorand and Conrad 1984). The selectivity profile across proteins is, therefore, an important factor in the development of specific inhibitors for target validation and for therapeutic applications. If common biological effects are observed with diverse classes of compounds, it is more likely that the effects can be attributed to the inhibition of a specific protein or to the inhibition of a specific activity of the protein rather than to an off-target effect. It is also important to bear in mind, particularly when dealing with the overexpression or knockout of a protein for which there is more than one family member, that there may be tissue-specific compensatory changes, either at baseline or following experimental treatment(s), that may compensate for the gain or loss of the protein of interest. Moreover, the reliability and reproducibility of animal studies is dependent on adequate reporting on the strain background, husbandry, study design, conduct and analysis of the experiments, the published details of which are often severely lacking.

TG2 has been implicated in a number of physiological processes and in disease. This chapter will briefly review guidelines for animal husbandry and animal study design, TG2 mouse models and recent insights into the physiology and pathophysiology of TG2 that have come from the study of these genetically engineered animal models. The reader is also referred to a previous extensive review on this subject (Iismaa et al. 2009).

4.2 Guidelines for Animal Husbandry and Animal Study Design

A number of factors influence the reliability and replicability of animal studies, in particular the mice used (strain background and husbandry), study design and analysis (protocols, randomization, blinding, power, data handling), and study reporting. Deficiencies in any of these factors severely limit the interpretability and reproducibility of the findings. This section will only discuss the importance of genetic background and animal husbandry. A summary of universal guidelines that reflect rigorous study design is presented in Table 4.1 and the reader is referred to several excellent recent perspectives and reviews detailing common pitfalls and

Table 4.1 Universal guidelines for animal studies

Animals and husbandry	
Genetic background	Report full details of strain and source, including whether true littermates are used
	Where possible, use congenic lines and/or F1 background lines
Sex and age	Report sex and age
Environmental enrichment and husbandry	Utilize environmental enrichment such as nesting material, shredded paper, play tunnels. Report enrichment used
	Keep experimental animals group-housed in balanced, mixed-genotype groups unless prevented by male aggression or testing protocols. Report housing conditions, including separation of animals into single-housing cages as well as lighting conditions and phase of the diurnal cycle utilized for testing
	Tailor husbandry to needs of mice (e.g. providing wet feed to debilitated mice). Report special needs
Breeding scheme	Where possible, genotypes to be studied should be generated in the same litter to ensure comparable uterine environment and maternal influences. Report breeding scheme
Study design	
Randomization	Assign animals randomly to the various experimental groups. Report the allocation method followed
Blinding	Experimenters should not be aware of genotype or treatment when conducting testing or analyzing the data. Report on blinding
Sample size	Justify sample size by power analysis for the measure(s) of interest
Data handling	If survival is included as an endpoint include a description of how it is assessed (e.g. surrogate markers vs actual death)
	Rules for inclusion and exclusion of animals or data should be defined before the start of the study and reported
	Any data removed before analysis should be reported
	Report whether samples were processed randomly or in batches
	Report how often an experiment was repeated

Based on published guidelines (Landis et al. 2012; Menalled et al. 2014a)

providing clear guidance and recommended practices (Brennan 2011; Landis et al. 2012; Menalled et al. 2014a).

4.2.1 Strain Background and Animal Husbandry

Targeted mutations can have different phenotypes in different backgrounds (Silva et al. 1997; Iismaa et al. 2009). Differences between observed phenotypes can only be ascribed to a particular targeted mutation if the same genetic background is used across experiments.

Targeted transgenesis historically employed ES cells derived from 129 substrains and chimeras were mated with C57BL/6 mice (B6) to generate mutant mice on a mixed-strain F1 background (all TG2 knockout mice have been generated on a mixed-strain 129-B6 background). Colony maintenance of genetically modified mice on a mixed-strain background is strongly discouraged because over successive generations random segregation events result in greater genetic variability between individual mice (Fig. 4.1) (Silva et al. 1997). With all of the polymorphic alleles randomly segregating, there is no appropriate control for the mutant mice. Generations of inbreeding of mutant mice and wild-type mice results in homozygous and wild-type mice of very different genetic backgrounds (Fig. 4.1).

It is recommended that targeted mutant mice be backcrossed to an inbred line until they are fully congenic (genetically identical to the recipient except for the donor locus of interest, some genes linked to that locus, and random genetic elements from the donor genome). Genetic uniformity between mice thus enables direct comparison of all mice in a study without confounding complications from differing genetic backgrounds and reduces experimental variability. A congenic line is generated by backcrossing mice carrying the locus of interest to a "target strain", identifying the offspring with the locus of interest, backcrossing them to the "target strain", and repeating this procedure for a minimum of ten generations. Each successive generation retains the locus of interest but has a reduction of 50 % at each generation of the genomic material of the starting mouse. A strain is an incipient congenic after five to nine backcross cycles (N5 to N9) and a full congenic (99.9 % similar at all unlinked loci) after ten backcross cycles (N10). Historically, this procedure takes 2–3 years to produce a full congenic; however, marker-assisted congenic screening or speed congenics can now hasten this process.

It is tempting to maintain and breed congenic mice as separate homozygote and wild-type lines to save on genotyping. However, if only a few mating pairs are used, lines can become homozygous or fixed for random alleles (genetic drift). These may be phenotypically quiet and only be discovered by accident or under specific experimental conditions, or they may cause phenotypic abnormalities such as loss of spatial learning or high seizure susceptibility (Silva et al. 1997; Taft et al. 2006; Stevens et al. 2007). It is therefore recommended that full congenic targeted mutant mice be maintained with continuous backcrossing to reduce the chance of genetic drift. Figure 4.2 details a breeding strategy that allows maintenance of a stable congenic strain as a stem (or nucleus) colony and generation of animals for experimental use as an expansion colony.

4.3 TG2 Mouse Models

A number of different types of genetically engineered mice have been generated to help understand the role and regulation of TG2 in physiological and pathological processes and to help delineate the molecular basis of its actions.

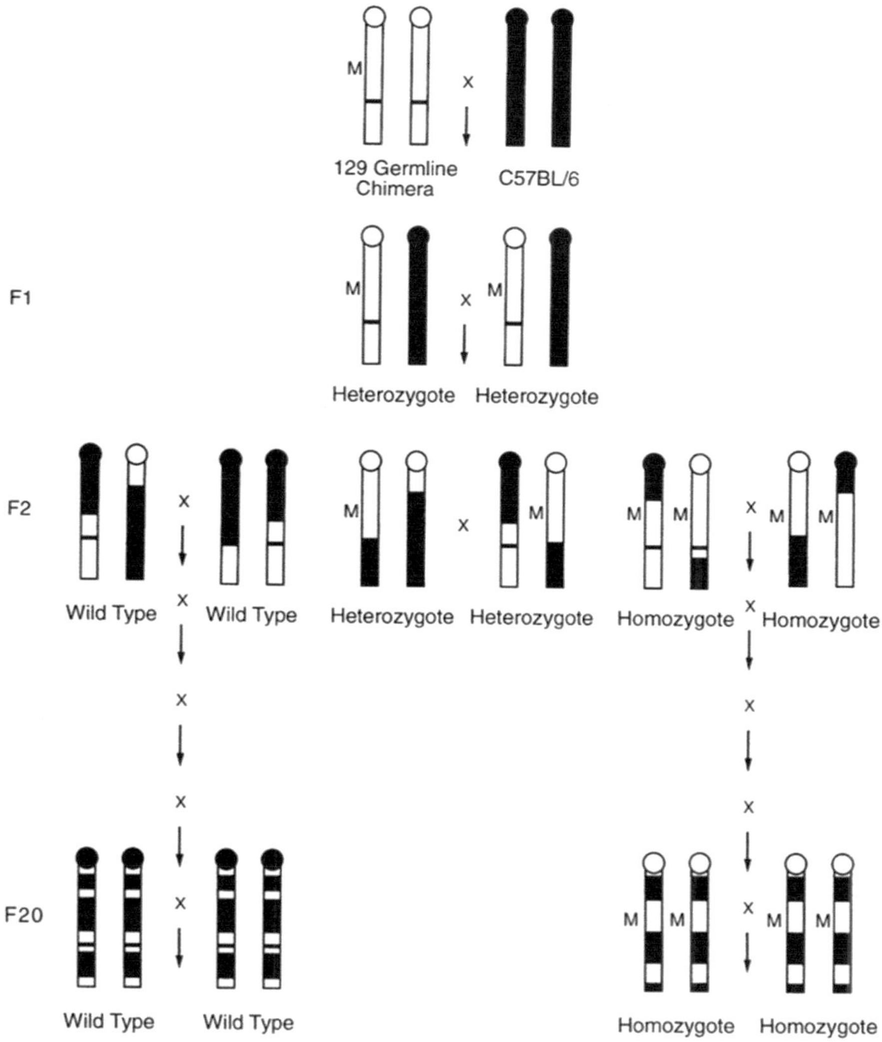

Fig. 4.1 Maintaining mixed-strain lines by inbreeding should be avoided. Starting from F2 mice, averaging 50:50 129 (white chromosome regions) and B6 (black chromosome regions), generations of inbreeding will inevitably result in homozygous and wild-type control lines of very different genetic backgrounds, the exact genotypes of which are not known and are not easily reproducible [Reprinted from Silva et al. 1997 with kind permission from Elsevier]

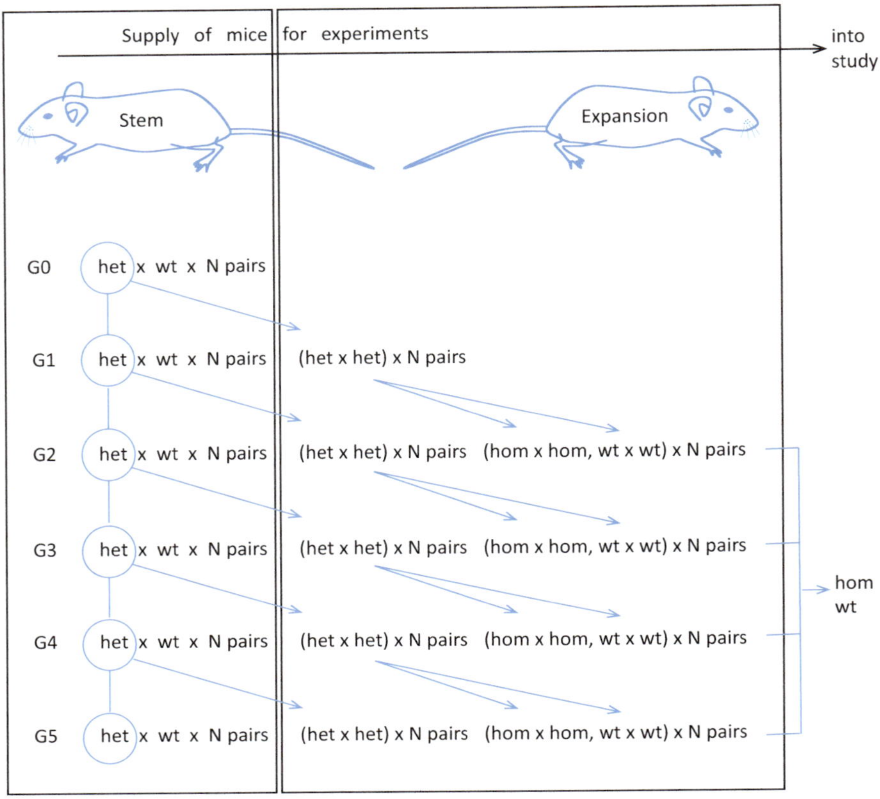

Fig. 4.2 Preferred breeding strategy for maintenance of stable congenic lines. The use of a stem and expansion colony ensures that breeding continues to propagate the line (stem colony) while generating animals for experimental use (expansion colony). An example is given here of a colony depicted over 5 generations (G0-G5), where the stem colony is populated by backcrossing targeted mutant mice to wild-type mice. Note that one pair only contributes future stem breeders. Heterozygous animals supply the expansion colony, where a sufficient number of mated pairs (N) are set up to produce the required animals for study. Expansion colony breeding animals should be refreshed from the stem colony at each generation to prevent disparity developing between the two colonies due to genetic drift [Adapted from Brennan 2011]

4.3.1 Transgenic Overexpressing Mice

In an effort to gain insight into the regulation of TG2 expression, the first genetically engineered TG2 mouse model expressed the *Escherichia coli* β-galactosidase (*lacZ*) reporter gene under the control of a 3.8 kb 5'-flanking region of the *Tgm2* promoter [founder mouse strain not reported but backcrossed to B6 for three generations] (Nagy et al. 1997). LacZ expression patterns were observed to correlate with some, but not all, patterns of endogenous TG2 expression (Nagy et al. 1997), indicating that this fragment does not contain all of the elements that are required for TG2 expression.

To investigate the effect of site-specific TG2 overexpression, transgenic mice have been generated that express TG2 under the control of different promoters. Two independent groups generated transgenic mice on an FVB/N (Small et al. 1999) or B6SJLF1 (Zhang et al. 2003) background that specifically overexpress rat TG2 cDNA in the heart under the control of the murine α-myosin heavy chain promoter. Transgenic mice overexpressing human TG2 cDNA in neurons under the control of the murine prion promoter have been generated on a B6 background (Tucholski et al. 2006).

4.3.2 Targeted Mutant Mice

To investigate the effect of TG2 deletion, three independent TG2 knockout ($TG2^{-/-}$) mouse lines have been generated by three different groups (De Laurenzi and Melino 2001; Nanda et al. 2001; Kim et al. 2010). All have used 129 substrain genomic DNA to target the region of the *Tgm2* gene that encodes the catalytic site: De Laurenzi and Melino (2001) used homologous recombination to replace part of exon 5 and all of exon 6 with the neomycin resistance gene, Nanda et al. (2001) utilized the Cre/*loxP* site-specific recombination system to delete exons 6–8, and Kim et al. (2010) used homologous recombination to replace exons 5 and 6 with the neomycin resistance gene. The targeting vectors were recombined in 129 substrain embryonic stem cells and injected into B6 blastocysts to generate chimeric mice, which were crossed with B6 mice to generate germ line-transmitting $TG2^{+/-}$ mice with a mixed-strain (129-B6) background. Two of these lines (De Laurenzi and Melino 2001; Nanda et al. 2001) are now available as congenic B6, and one of these (Nanda et al. 2001) is available as congenic 129. The Cre/*loxP* site-specific recombination system (Nanda et al. 2001) has the added advantage that TG2 can be ubiquitously deleted in all cells and tissues (by crossing "*Tgm2*-floxed [flanked by *loxP* sites]" mice with transgenic mice expressing Cre ubiquitously under the control of the human cytomegalovirus promoter) or TG2 can be selectively deleted in specific cells (by crossing *Tgm2*-floxed mice with transgenic mice expressing Cre under the control of a cell type-specific promoter). The congenic B6 *Tgm2*-floxed line is available from The Jackson Laboratory (Strain Name: B6.129S1-$Tgm2^{tm1Rmgr}$/J; Stock Number: 024694; URL: http://jaxmice.jax.org/strain/024694.html).

To investigate the effect of constitutively active transamidase activity, $TG2^{R579A}$ knock-in mice that have Ala substituted for Arg at position 579 of TG2 were generated by homologous recombination using B6 ES cells, injection into BALB/c blastocysts and mating of chimeras with B6 mice to generate germ line-transmitting $TG2^{R579A}$ mice on a B6 background (Iismaa et al. 2013). These mice are co-isogenic (genetically identical with their wild-type littermates except for the genetically modified allele).

4.3.3 Phenotype of $TG2^{-/-}$ Mice and Compensation by Other TGs in Targeted Mutant Mice

All three independently generated mixed-strain 129-B6 $TG2^{-/-}$ colonies were reported to be viable, of normal size and weight, born with the expected Mendelian frequency and with no obvious phenotypic or developmental abnormalities (De Laurenzi and Melino 2001; Nanda et al. 2001; Kim et al. 2010). Cardiac function, in terms of heart rate, systolic and diastolic blood pressures, and maximum rates of pressure development (dP/dtmax) and relaxation (dP/dtmin), was identical between wild-type and $TG2^{-/-}$ mice (Nanda et al. 2001).

TG2 mouse models have started to be characterized for tissue-specific compensatory changes in other TGs, with contradictory results. For example, one study compared male mice from two different colonies of $TG2^{-/-}$ and wild-type mice [$n = 6$ male mixed-strain 129-B6 $TG2^{-/-}$ mice bred separately from mixed-strain 129-B6 wild-type mice (De Laurenzi and Melino 2001); and $n = 5$ male B6 non-littermates compared to $TG2^{-/-}$ mice (Nanda et al. 2001) that were backcrossed >10 generations to B6 and maintained as homozygous knockouts] (Johnson et al. 2012). Transamidase activity in the aorta was similar between $TG2^{-/-}$ and wild-type mice, with Western analysis detecting equivalent levels of TG1 and TG4 in both genotypes. Another study used B6 and congenic B6 $TG2^{-/-}$ mice [sourced from Nanda et al. (2001) and backcrossed >10 generations to B6] (Deasey et al. 2013). Tissue was pooled together from two to three mice (but there was no reporting of the number of independent samples (n) used or replicates performed) for real-time PCR analysis and transamidase assays. There was transcriptional upregulation of some TGs in some tissues and upregulation of transamidase activity in other tissues (Deasey et al. 2013). In aorta, for example, transamidase activity was upregulated (in contrast to Johnson et al. 2012), without transcriptional upregulation of other TGs. Two other studies have examined only TG mRNA expression. No compensatory enhanced expression of other TGs was observed in a range of tissues and cells from $n = 3$ congenic 129 $TG2^{-/-}$, congenic B6 $TG2^{-/-}$ and co-isogenic $TG2^{R579A}$ knock-in mice relative to their respective wild-type littermates (Iismaa et al. 2013). Similarly, no compensatory induction of other TGs was observed relative to wild-type in a range of tissues from $n = 7$-10 $TG2^{-/-}$, $FXIIIA^{-/-}$ or $TG2^{-/-}/FXIIIA^{-/-}$ mice [these mice were all $ApoE^{-/-}$; no direct reference to strain background of $TG2^{-/-}$ (Nanda et al. 2001), $FXIIIA^{-/-}$ (Lauer et al. 2002) or $ApoE^{-/-}$ (Williams et al. 2010) mice, although likely all mixed-strain] (Cordell et al. 2015). $TG2^{-/-}$ mice [(De Laurenzi and Melino 2001) backcrossed 12 generations to B6] had no apparent compensation of transamidase activity by other TGs either at baseline or after carbon tetrachloride-induced fibrosis, relative to wild-type littermates (Popov et al. 2011).

4.4 Insights into TG2 Physiology and Pathophysiology from Genetically Engineered Animal Models

Studies directly related to the use of genetically engineered animal models to investigate TG2 function will be discussed here. The reader is asked to bear in mind the universal guidelines for animal husbandry and animal study design (Sect. 4.2), as shortcomings in these aspects have likely contributed to inconsistencies between studies.

4.4.1 Celiac Disease (See Chap. 9 for More Detail)

The role of TG2 in human disease has been best established for celiac disease, where the ingestion of gluten (present in wheat, barley and rye) induces the development of IgA antibodies against gluten epitopes and predominantly IgA autoantibodies against the open Ca^{2+}-activated conformation of TG2 (Iversen et al. 2013) that deposit in the small intestine of genetically susceptible individuals. Celiac disease manifests as an enteropathy driven by $CD4^{+}$ T cells that react with gluten-derived peptides when presented on the disease-associated HLA molecules, HLA-DQ2 and -DQ8 [see Iismaa et al. (2009) for review]. This causes flattening of the small intestinal mucosa resulting in malabsorption and chronic diarrhea. Sensitivity to gluten can also induce autoantibodies against TG3 in dermatitis herpetiformis, manifesting as a dermatopathy with blistering of the skin at extensor surfaces of the major joints and dermal IgA-TG3 deposits, or against TG6 in gluten neuropathy, characterized by cerebellar IgA-TG6 deposits and cerebellar ataxia due to the loss of Purkinje cells, as well as axonal neuropathy. Incubation of IgA-positive serum from untreated celiac and dermatitis herpetiformis patients with tissues from wild-type and $TG2^{-/-}$ mice [(De Laurenzi and Melino 2001) backcrossed to B6, number of generations not stated] confirmed that antibody binding was TG2 dependent, thereby verifying that TG2 is an autoantigen in celiac disease and dermatitis herpetiformis (Korponay-Szabo et al. 2003). Attempts to establish models of gluten sensitivity in rabbits, dogs and mice have been largely unsuccessful (Bethune and Khosla 2008); however, injection into nude (immune-compromised) mice of IgA-deficient celiac disease patient serum or total IgG has been reported to induce enteropathy with characteristics of early stage celiac disease in humans (Kalliokoski et al. 2015).

4.4.2 The Immune System

Immunity is the capability of the body to resist harmful microorganisms or viruses from entering it and involves both the innate and adaptive immune systems. The innate immune system provides resistance during the first critical hours and days of exposure to a new pathogen through physical, chemical and cellular (CD4 or helper T cells) approaches. It does not confer long-lasting or protective immunity to the host. The adaptive immune system takes a week or so to create immunological memory after an initial response to a specific pathogen, leads to an enhanced response to subsequent encounters with that pathogen, and provides long-lasting protection. The adaptive system includes both humoral immunity (secreted antibodies, complement proteins and certain antimicrobial peptide components) and cell-mediated immunity (activation of phagocytes, antigen-specific cytotoxic T-lymphocytes, and the release of various cytokines in response to an antigen).

4.4.2.1 Dendritic Cells

Dendritic cells (DCs) are antigen-presenting cells found in tissues that are in contact with the external environment, such as the skin and the inner lining of the nose, lungs, stomach and intestines. After activation by antigen, they migrate to the peripheral lymph nodes where, upon maturation, they present antigen to T cells and B cells to initiate and regulate the adaptive immune response. Dendritic cells are profoundly depleted in patients and mice suffering from septic shock. $TG2^{-/-}$ mice [(De Laurenzi and Melino 2001) backcrossed to B6, number of generations not stated] were partially resistant to lipopolysaccharide (LPS)-induced septic shock relative to wild-type mice (Falasca et al. 2008). Unlike B6 mice, LPS stimulation did not induce splenic DC depletion in B6 $TG2^{-/-}$ mice [(De Laurenzi and Melino 2001) backcrossed to B6 for ten generations] (Matic et al. 2010). $TG2^{-/-}$ splenic DCs were reported to express higher levels of the major histocompatibility class II (MHCII) molecule than wild-type DCs, but were unable to upregulate MHCII in response to LPS (Matic et al. 2010) and were impaired in their maturation. A later study found no difference in the expression of MHCII either before or after LPS activation of bone-marrow-derived DCs from B6 (Jackson Laboratory) and $TG2^{-/-}$ mice [(De Laurenzi and Melino 2001), no reference to strain background] (Kim et al. 2014b). Rather, cell-surface TG2 appeared to modulate DC-T cell interaction.

4.4.2.2 T Cells

TG2 expression is low in isolated naïve T cells and increases following activation (Kim et al. 2014a). The intensity of contact hypersensitivity reaction to dinitrofluorobenzene (DNFB) was attenuated in $TG2^{-/-}$ mice [(De Laurenzi and Melino 2001), no reference to strain background] relative to B6 mice (Jackson Laboratory) and this correlated with reduced splenic T cell proliferation and activation (Kim et al. 2014a), suggesting a possible role for TG2 in T cell activation.

Multiple sclerosis is a chronic, inflammatory and demyelinating disease of the central nervous system in which the fatty myelin sheath that insulates neurons and helps relay action potentials becomes thin or lost, resulting in neuronal damage. In a relevant animal model of multiple sclerosis, experimental autoimmune encephalomyelitis (EAE), disease severity was reduced in $TG2^{-/-}$ mice [(De Laurenzi and Melino 2001) backcrossed to B6 for 12 generations] relative to B6 mice (Jackson Laboratory). This was characterized by reduced $CD4^{+}$ T cell differentiation and infiltration into the central nervous system as well as reduced local inflammatory cytokines in the lesion, relative to B6 mice (Oh et al. 2012). This work supports a pathogenic role for TG2 in EAE. (See also Sect. 4.4.9.3, which discusses evidence that TG2 is involved in remyelination).

4.4.2.3 The Humoral Response

B cell development profiles in the spleen and bone marrow of wild-type and $TG2^{-/-}$ mice [(De Laurenzi and Melino 2001) backcrossed to B6 mice for 12 generations] were no different; however, in response to primary and secondary immunization with a protein antigen, TG2 was found to negatively regulate the humoral immune response (Kim et al. 2012).

Further studies following the recommended guidelines for animal husbandry and study design are required to confirm the role of TG2 in the immune system.

4.4.3 Inflammation and Fibrosis

Inflammation is one of the first host defense responses to tissue injury that occurs as a result of infection, wounding or chemical exposure and results in redness, heat, pain and swelling in the affected area. Inflammation involves activation of the master transcription factor, NFκB (nuclear factor kappa-light-chain-enhancer of activated B cells), resulting in the expression of adhesion molecules, cytokines and chemokines that together promote neutrophil (polymorphonuclear leukocyte, PMN) adhesion to vascular endothelial cells (ECs) and subsequent infiltration into tissue (Rahman and Fazal 2011). The body responds to chronic injury by fibrosis. Fibrosis is characterized by the activation of fibroblasts (or hepatic stellate cells in the liver) to myofibroblasts (contractile cells expressing both fibroblast and smooth muscle cell markers such as α-smooth muscle actin) that synthesize excess fibrillar collagens (largely collagens I and III), and the accumulation of extracellular matrix due to changes in matrix degradation and stabilization, leading to the destruction of normal tissue architecture and function.

4.4.3.1 Liver

Following in vivo treatment of mice with the hepatotoxin carbon tetrachloride, fibrosis was observed to be significantly increased in $TG2^{-/-}$ mice [(De Laurenzi and Melino 2001) backcrossed to B6, number of generations not stated] relative to B6 mice (Nardacci et al. 2003) and this was associated with an increased inflammatory response. In contrast, a later study showed that in two in vivo models of advanced fibrosis (hepatotoxicity induced by carbon tetrachloride or by thioacetamide) in TG2$^{-/-}$ mice [(De Laurenzi and Melino 2001) backcrossed 12 generations to B6] and their wild-type littermates, TG2 deletion had no effect on the degree of fibrosis, the stability of the fibrotic liver extracellular matrix or the reversibility of hepatic fibrosis following removal of the pro-fibrogenic stimuli (Popov et al. 2011). There was no apparent compensation by other TG isoforms as the reduced transamidase activity in normal TG2$^{-/-}$ livers did not increase in fibrotic TG2$^{-/-}$ livers (Popov et al. 2011). The contribution of TG2 to fibrogenesis and stabilization of the collagen matrix in the fibrotic liver may thus be dispensable.

4.4.3.2 Kidney (See Chap. 11 for More Detail)

Using TG2$^{-/-}$ tubular epithelial cells [(De Laurenzi and Melino 2001) backcrossed to B6, number of generations not stated], TG2 has been shown to contribute to renal extracellular matrix accumulation primarily by accelerating deposition of soluble collagens III and IV into the matrix (Fisher et al. 2009). Consistent with this, kidneys from TG2$^{-/-}$ mice [(De Laurenzi and Melino 2001) backcrossed to B6, number of generations not stated] subjected to unilateral ureteral obstruction (UUO) displayed reduced macrophage and myofibroblast infiltration, decreased collagen I synthesis, reduced fibrosis and similar levels of apoptosis, compared with wild-type mice (Shweke et al. 2008). Similarly, in a kidney injury model employing UUO and TNF-α-induced inflammation, TG2$^{-/-}$ kidneys [mixed strain 129-B6 (Kim et al. 2010)] exhibited less inflammation and reduced NFκB activity relative to wild-type controls (Kim et al. 2010). Two kidney fibrosis models (UOO and aristolochic acid nephropathy) in mixed-strain 129-B6 syndecan-4$^{-/-}$ mice (Ishiguro et al. 2000) and B6 mice indicated that the binding of TG2 to the heparan sulphate chains of the heparan sulphate proteoglycan, syndecan-4, allows extracellular trafficking of TG2 to the cell surface and matrix, where its transamidase activity is activated for the crosslinking of extracellular matrix substrate proteins (Scarpellini et al. 2014). Animal studies, thus, consistently support a role for TG2 in renal fibrosis and in renal ischemic injury.

4.4.3.3 Lung

TG2$^{-/-}$ mice [(De Laurenzi and Melino 2001), no reference to strain background] challenged with intraperitoneal injection of LPS to induce lung inflammation showed a marked reduction in NF-κB activation, vascular adhesion molecule expression (I-CAM-1, VCAM-1) and lung PMN sequestration relative to LPS-treated B6 mice (Jackson Laboratory) (Bijli et al. 2014). Thus TG2 may mediate EC inflammation and lung PMN infiltration associated with intravascular coagulation and sepsis.

TG2$^{-/-}$ mice [sourced from Nanda et al. (2001) and backcrossed 10 generations to B6 (Olsen et al. 2011); sourced from De Laurenzi and Melino (2001) and backcrossed 12 generations to B6 (Oh et al. 2011)] developed significantly reduced inflammation (Oh et al. 2011) and reduced fibrosis relative to B6 mice (Jackson Laboratory) in a mouse model of pulmonary fibrosis (Olsen et al. 2011; Oh et al. 2011). Lung epithelial TG2 was shown to trigger a distinct inflammatory subset of T helper cells, which in turn led to increases in fibroblast-derived TG2 involved in the subsequent fibrosis phase (Oh et al. 2011). In a mouse model of allergic asthma, airway hypersensitivity was attenuated in TG2$^{-/-}$ mice (source not reported, backcrossed to BALB/c for 12 generations) compared to BALB/c mice (Oh et al. 2013). This was associated with reduced expression of IL-33 in airway epithelial cells, which in turn led to reduced differentiation of a subset of T helper cells and reduced recruitment of eosinophils (important mediators of allergy and asthma). Further work is required to confirm and extend these findings.

4.4.4 Cell Death and Survival

4.4.4.1 Apoptosis

Apoptosis is a highly regulated, programmed cell death process that involves breakdown of cytoplasmic and nuclear proteins by caspases, DNA degradation by DNases, cell shrinkage, and cell-surface changes that lead to cell engulfment by macrophages or neighboring cells, without the release of cellular components into the extracellular milieu. Thus, there is essentially no inflammatory reaction. Failed clearance of apoptotic cells may lead to the development of autoimmune diseases (Michlewska et al. 2007). Apoptotic cells also release factors that can stimulate cell proliferation, differentiation, wound healing and tissue regeneration (Fuchs and Steller 2011). Apoptosis is triggered through two main signaling pathways: the intrinsic or mitochondrial pathway, which is initiated in response to signals from within the cell that result in mitochondrial release of pro-apoptotic proteins that activate caspases, and the extrinsic or death receptor pathway, which is initiated through activation of cell-surface pro-apoptotic receptors, such as Fas or TNF receptor, that activate caspases.

Despite a large body of data suggesting an involvement of TG2 in apoptosis, initial characterization of wild-type and TG2$^{-/-}$ mice [mixed-strain 129-B6 (De Laurenzi and Melino 2001; Nanda et al. 2001)] indicated that induced apoptosis of thymocytes and fibroblasts was identical, although TG2$^{-/-}$ thymocytes appeared to be more susceptible to late-stage apoptosis (Nanda et al. 2001). Aged TG2$^{-/-}$ erythrocytes [mixed-strain 129-B6 (De Laurenzi and Melino 2001)] appeared more resistant to osmotic stress-induced hemolysis than wild-type erythrocytes, but little else was different (Bernassola et al. 2002a).

Reports of hepatic injury in TG2$^{-/-}$ mice are conflicting. Supportive of a protective or anti-apoptotic role, TG2$^{-/-}$ mice [mixed strain 129-B6 (De Laurenzi and Melino 2001)] were more susceptible to Fas-mediated cell death than FVB mice (Sarang et al. 2005). In direct contrast, apoptosis was markedly reduced in TG2$^{-/-}$ hepatocytes [mixed strain 129-B6 (Nanda et al. 2001)] relative to wild-type mixed-strain hepatocytes following treatment of isolated hepatocytes with ethanol in vitro or induction of Fas-mediated cell death as an in vivo mouse model of alcoholic steatohepatitis (Tatsukawa et al. 2009). The mechanism accounting for this latter pro-apoptotic function of TG2 was shown to involve TG2 translocation into the nucleus followed by crosslinking and inactivation of the transcription factor Sp1, leading to down-regulation of the hepatocyte growth factor receptor, c-Met, and hepatocyte cell death (Tatsukawa et al. 2009). In another study, TG2$^{-/-}$ mice [(Nanda et al. 2001) backcrossed 5 generations to B6] were more resistant than wild-type mice to injury-mediated liver hypertrophy and the formation of Mallory bodies, which are protein deposits characteristic of liver disorders such as alcoholic and non-alcoholic steatohepatitis (Strnad et al. 2007). Hepatocellular damage was similar; however, TG2$^{-/-}$ mice had increased hepatic cholestasis, as evidenced by more gallstones, jaundice, and ductal proliferation, than wild-type mice (Strnad et al. 2007). In support of an anti-apoptotic role for TG2, TG2$^{-/-}$ mice [mixed strain 129-B6 (Kim et al. 2010), backcrossed more than 8 generations to B6] were more susceptible to hepatocyte apoptosis than WT controls in two models of TNF-dependent acute liver injury (Delhase et al. 2012). In an attempt to reconcile these seemingly conflicting findings it has been suggested that the role of TG2 as a pro-apoptotic or cell-survival factor is determined by multiple factors including the cell/tissue type, the type of death stimuli, the cellular localization and which activity of TG2 is activated (Fésüs and Szondy 2005). Clearly much more work following the recommended guidelines for animal husbandry and study design is required to determine the role of TG2 in cell death or survival.

4.4.4.2 Autophagy (See Chap. 4 for More Detail)

Autophagy is a normal physiological process that involves the degradation and recycling of cellular components to maintain energy levels and promote cell survival. The transamidase activity of TG2 seems to play a role in the autophagic clearance of damaged proteins and organelles (Rossin et al. 2012). Cardiac muscle from TG2$^{-/-}$ mice [(De Laurenzi and Melino 2001), no reference to strain background] that were starved for 48 h showed more damaged mitochondria and an earlier and more pronounced induction of autophagy than wild-type mice (D'Eletto et al. 2009). Livers of TG2$^{-/-}$ mice [(De Laurenzi and Melino 2001), no reference to strain background] accumulated more ubiquitinated protein aggregates into autophagosomes than wild-type mice, both in the fed state and following 48 h starvation, and hepatocytes were impaired in pexophagy (clearance of hyper-accumulated peroxisomes) (D`Eletto et al. 2012).

4.4.4.3 Mitochondrial Function

Conflicting results have been reported for heart and mitochondrial function in $TG2^{-/-}$ mice. Mixed-strain $TG2^{-/-}$ mice (Nanda et al. 2001) were no different from wild-type littermates in terms of baseline heart rate, systolic and diastolic blood pressure, heart contractility and relaxation rates or motility. No significant difference was noted in resting ATP levels in heart, liver, skeletal muscle (Mastroberardino et al. 2006) or brain (Battaglia et al. 2007) of $TG2^{-/-}$ mice [(De Laurenzi and Melino 2001) backcrossed 10 generations to B6] compared with wild-type controls, despite changes in mitochondrial respiratory complex activity in $TG2^{-/-}$ relative to wild-type mice (complex I activity reduced; complex II activity increased) (Battaglia et al. 2007). Following exercise, ATP levels in $TG2^{-/-}$ skeletal and cardiac muscle were more severely depleted than in wild-type mice and this was associated with a dramatic decrease in the general motility of $TG2^{-/-}$ mice (Mastroberardino et al. 2006). Upon challenge with a mitochondrial toxin that inhibits complex I, however, ATP levels in $TG2^{-/-}$ brain homogenates were less depleted than in wild-type mice (Battaglia et al. 2007). This correlated with a greater reduction in complex I activity and a greater increase in complex II activity in the treated brains of $TG2^{-/-}$ relative to wild-type mice (Battaglia et al. 2007). Consistent with this, $TG2^{-/-}$ mice were more sensitive than wild-type mice to a selective inhibitor of mitochondrial complex II (Battaglia et al. 2007). Surprisingly, a later study from the same group reported that heart rate, coronary flow, aortic flow, aortic pressure and myocardial ATP levels were reduced at rest in $TG2^{-/-}$ mice (background strain information and husbandry not defined) relative to B6 controls (Szondy et al. 2006; Sarang et al. 2009). Following ischemia/reperfusion injury, these $TG2^{-/-}$ mice had smaller infarct sizes, a lower incidence of reperfusion-induced ventricular fibrillation and a further reduction in ATP levels (Szondy et al. 2006). Yet another study from another group using B6 $TG2^{-/-}$ mice [(De Laurenzi and Melino 2001) backcrossed to B6, number of generations not stated] and littermate wild-type mice showed no significant difference in mitochondrial function (complex II and IV activity) between wild-type and $TG2^{-/-}$ mice (Van Strien et al. 2011).

An essential feature of normal mitochondrial function is the impermeability of the mitochondrial membrane to metabolites and ions due to the closed conformation of the permeability transition pore (PTP). An important component of the PTP is the adenine nucleotide transporter 1 (ANT1), which is involved in ADP/ATP exchange. Although a detailed study failed to find TG2 in mitochondria (Krasnikov et al. 2005), a small amount of TG2 has been localized to the inner mitochondrial membrane, where it interacts physically with ANT1 (Malorni et al. 2009). During apoptosis, ANT1 interaction with Bax results in dissipation of the mitochondrial membrane potential. The protein disulphide isomerase activity of TG2 appears to regulate the correct assembly and function of ANT1, with mitochondria from hearts, but not brains or livers, of $TG2^{-/-}$ mice [(De Laurenzi and Melino 2001) backcrossed 10 generations to B6] exhibiting markedly increased polymerization of

ANT1 and enhanced ADP/ATP carrier activity relative to wild-type mice (Malorni et al. 2009). Immortalized TG2$^{-/-}$ MEF cell lines were characterized by a lower mitochondrial membrane potential than immortalized wild-type MEF cell lines (Malorni et al. 2009). Immortalized TG2$^{-/-}$ MEFs also exhibited less "spontaneous" apoptosis than immortalized wild-type MEFs as well as reduced sensitivity to various inducers of apoptosis such as the glycolytic inhibitor 2-deoxy-D-glucose. This was proposed to be the result of impaired recruitment of Bax to TG2$^{-/-}$ mitochondria, resulting in decreased Bax/ANT1 interaction relative to wild-type mitochondria (Malorni et al. 2009). A later study from the same group reported that under basal conditions TG2$^{-/-}$ MEFs (source nor strain background defined) had a large number of fragmented, dysfunctional mitochondria due to an impairment in mitophagy (clearance of defective mitochondria) compared with wild-type MEFs (Rossin et al. 2014). TG2$^{-/-}$ MEFs were also increased in glycolytic activity relative to wild-type MEFs and in contrast to earlier studies (Malorni et al. 2009) were now more sensitive to inhibition of glycolysis by 2-deoxy-D-glucose than wild-type MEFs (Rossin et al. 2014). Clearly more work following the recommended guidelines for animal husbandry and study design is required to reach a consensus on these conflicting results.

4.4.5 *Phagocytosis*

The rate of in vitro phagocytosis by wild-type macrophages of TG2$^{-/-}$ red blood cells [(De Laurenzi and Melino 2001), no reference to strain background] was observed to be delayed relative to wild-type red blood cells and this correlated with delayed cell-surface phosphatidylserine exposure in TG2$^{-/-}$ red blood cells relative to wild-type red blood cells, suggesting that TG2 may promote cell-surface migration of phosphatidylserine (Sarang et al. 2007).

TG2$^{-/-}$ mice have been reported to be defective in macrophage clearance of necrotic or apoptotic cells in the following animal models: clearance of necrotic (Nardacci et al. 2003) or apoptotic (Falasca et al. 2005) cells following exposure to hepatotoxin [TG2$^{-/-}$: (De Laurenzi and Melino 2001) backcrossed to B6, number of generations not stated]; clearance of apoptotic thymic cells during thymic-induced involution [TG2$^{-/-}$: (De Laurenzi and Melino 2001), no reference to strain background] (Szondy et al. 2003); and clearance of apoptotic neutrophils during gout-like peritoneal inflammation [TG2$^{-/-}$: (Nanda et al. 2001) backcrossed five generations to B6] (Rose et al. 2006). The impaired phagocytosis by TG2$^{-/-}$ macrophages has been attributed in part to a lack of activation of latent transforming growth factor (TGF)-β1 (Szondy et al. 2003; Rose et al. 2006) by nucleotide-bound TG2 (Rose et al. 2006) and in part to a lack of stabilization of integrin β3-containing phagocytic portals on the macrophage by nucleotide-bound TG2 (Tóth et al. 2009a). TG2 on thioglycollate-elicited macrophages binds MFG-E8 (milk fat globulin EGF factor 8), which bridges integrin β3 to phosphatidylserine exposed on the surface of apoptotic cells (Tóth et al. 2009a). In TG2$^{-/-}$

mice [(De Laurenzi and Melino 2001), no reference to strain background (Tóth et al. 2009a); (De Laurenzi and Melino 2001) crossed with each other for several generations (Tóth et al. 2009b)], integrin β3 does not accumulate in the phagocytic cup, resulting in impaired integrin β3 signaling and impaired apoptosis (Tóth et al. 2009a, b). The long-term consequence of defects in the clearance of dead cells in TG2$^{-/-}$ mice [(De Laurenzi and Melino 2001), no reference to strain background] relative to B6 mice, appears to be autoimmunity (Szondy et al. 2003), with increased levels of NFκB potentially contributing to inflammation and autoimmunity development (Sarang et al. 2011).

Recent work indicates the removal of apoptotic cells may also be facilitated by TG2 interaction with syndecan-4 on the surface of human macrophages, whereby the transamidase activity of cell-surface TG2 crosslinks CD44, a phagocytic receptor on human macrophages, to promote apoptotic cell clearance (Nadella et al. 2015).

Interestingly, genetic deletion of TG2 did not compromise phagocytosis by yeast extract-elicited neutrophils from TG2$^{-/-}$ mice relative to those from wild-type mice [(De Laurenzi and Melino 2001), no reference to strain background] (Balajthy et al. 2006).

4.4.6 Cardiac Hypertrophy

Cardiac hypertrophy is an adaptive response to increased workload induced by pressure- or volume-overload; however, cardiac function deteriorates in response to chronic overload, resulting in decompensated heart failure. Transgenic cardiac-specific overexpression of TG2 resulted in a unique type of cardiac hypertrophy in FVB/N mice aged 3–4 months with mild elevation of some markers of hypertrophy (β-myosin heavy chain and α-skeletal actin) but not others (atrial natriuretic factor) (Small et al. 1999). Transgenic cardiac-specific overexpression of TG2 in B6SJLF1 mice resulted in increased thromboxane A2/TG2-dependent signaling leading to cardiac decompensation by age 7–10 months, characterized by cardiomyocyte apoptosis and fibrosis, and an increase in mortality by 15 months of age (Zhang et al. 2003).

4.4.7 Vascular Remodeling

4.4.7.1 Atherosclerosis and Arterial Calcification

Vascular remodeling can involve the lumen diameter becoming larger (outward remodeling) or smaller (inward remodeling), or the wall cross-sectional area becoming larger (hypertrophic remodeling), remaining constant (eutrophic remodeling) or becoming smaller (hypotrophic remodeling). Chronic vasoconstriction results in inward remodeling of small arteries whereas chronic exposure to vasodilators leads to outward remodeling. Atherosclerosis is characterized by progressive invasion of the artery wall by white blood cells and vascular smooth muscle cells to form a plaque that results in narrowing of the lumen and impairment of blood flow. In early atherosclerotic lesions, apoptotic macrophages are efficiently cleared by neighboring macrophages (efferocytosed), thereby limiting lesion size and progression; however, efferocytosis in advanced lesions is defective, resulting in a necrotic core that may rupture resulting in the formation of an occlusive thrombus (Thorp and Tabas 2009).

Studies on atherosclerosis progression and vascular remodeling in TG2$^{-/-}$ mice have yielded inconsistent results. Studies using lethally irradiated B6 LDLR$^{-/-}$ (low-density lipoprotein receptor knockout) mice that had been transplanted with bone marrow from wild-type or TG2$^{-/-}$ mice [(Nanda et al. 2001) backcrossed five generations to B6] and fed a high-fat diet for 16 weeks to induce atherosclerosis showed that recipients of TG2$^{-/-}$ bone marrow had larger atherosclerotic aortic valve lesions with expanded necrotic cores and deeper penetration of macrophages into the intima of the lesion than did recipients of wild-type bone marrow (Boisvert et al. 2006). PMA (phorbol myristate acetate) stimulation of peritoneal macrophages isolated from wild-type and TG2$^{-/-}$ mice indicated decreased phagocytosis of apoptotic leukocytes by TG2$^{-/-}$ macrophages relative to WT macrophages (Boisvert et al. 2006), consistent with previous work using unstimulated peritoneal macrophages from wild-type and TG2$^{-/-}$ mice [(De Laurenzi and Melino 2001), no reference to strain background] (Szondy et al. 2003). This work suggests macrophage-expressed TG2 promotes apoptotic cell clearance and limits atherosclerotic lesion size.

In another study, TG2$^{-/-}$ mice [mixed strain 129-B6 (De Laurenzi and Melino 2001)] were crossed with ApoE$^{-/-}$ mice (B6, Charles River Laboratories) to generate ApoE$^{-/-}$ and ApoE$^{-/-}$TG2$^{-/-}$ mice (both mixed strain 129-B6), and were fed a high-fat diet for 16 or 30 weeks (Van Herck et al. 2010). Consistent with previous studies (Szondy et al. 2003; Boisvert et al. 2006), peritoneal macrophages from ApoE$^{-/-}$TG2$^{-/-}$ mice were impaired in phagocytosis relative to those from ApoE$^{-/-}$ mice (Van Herck et al. 2010). In contrast to TG2 deficiency being associated with larger atherosclerotic lesions (Boisvert et al. 2006), atherosclerotic plaques in ApoE$^{-/-}$TG2$^{-/-}$ mice were about half the size of those of ApoE$^{-/-}$ mice after 16 and 30 weeks on a high-fat diet (Van Herck et al. 2010). Characterization of the plaques themselves showed that after 30 weeks on a high-fat diet, plaques from ApoE$^{-/-}$TG2$^{-/-}$ mice had fewer crosslinks and more macrophages than those from ApoE$^{-/-}$ mice, suggesting TG2 deficiency contributes to a more unstable plaque (Van Herck et al. 2010).

A second study that crossed ApoE$^{-/-}$ and TG2$^{-/-}$ mice used mixed-strain 129-B6 ApoE$^{-/-}$ mice (Plump et al. 1992) and mixed-strain 129-B6 TG2$^{-/-}$ mice (Nanda et al. 2001) to generate ApoE$^{-/-}$ and ApoE$^{-/-}$TG2$^{-/-}$ mice and fed mice a high-fat diet for 24 weeks (Williams et al. 2010). In contrast to other studies (Boisvert et al. 2006; Van Herck et al. 2010), TG2 deficiency in this study had no effect on lesion size or stability in brachiocephalic arteries or aortic sinuses

(Williams et al. 2010). Atherosclerotic lesions calcify over time due to osteogenic and chondrogenic differentiation of vascular smooth muscle cells in a process similar to bone formation (Abedin et al. 2004). Although previous work using cultured wild-type and $TG2^{-/-}$ arterial smooth muscle cells [(Nanda et al. 2001) backcrossed more than nine generations to B6] has indicated TG2 is required for ectopic calcification of arteries (Johnson et al. 2008), the extent of arterial calcification in the 129-B6 $ApoE^{-/-}TG2^{-/-}$ mice was not impaired relative to 129-B6 $ApoE^{-/-}$ mice (Williams et al. 2010). In support of the previous work (Johnson et al. 2008), ex vivo cultures of aortic rings dissected from B6 and $TG2^{-/-}$ mice [(Nanda et al. 2001) backcrossed more than ten generations to B6] demonstrated that vascular calcification induced by warfarin required TG2-mediated activation of β-catenin signaling in vascular smooth muscle cells via warfarin-dependent upregulation of TG2 protein expression and transamidating activity (Beazley et al. 2012). In another study, primary vascular smooth muscle cells from $TG2^{-/-}$ mice [(Nanda et al. 2001) backcrossed more than 10 generations to B6] were shown to be attenuated in chondrogenic differentiation relative to those from B6 mice (Beazley et al. 2013b). Moreover, the flavonoid quercetin, an antioxidant and anti-inflammatory agent found in many plants and foods, inhibited TG2 transamidase activity as well as warfarin-induced calcification in wild-type mouse aortic rings but had no effect on the already attenuated warfarin-induced calcification in $TG2^{-/-}$ aortic tissue (source nor strain background defined), implicating quercetin as an inhibitor of β-catenin signaling and TG2 as a primary target of quercetin (Beazley et al. 2013a). Further investigations following the recommended guidelines for animal husbandry and study design are necessary to clarify the role of TG2 in the formation of calcifying chondrogenic lesions in vascular smooth muscle cells and if it is a potential target for therapy.

Quercetin, a dietary flavonoid already widely used as a food supplement in the general population, has recently been implicated in female fertility impairment. Female B6 mice receiving dietary quercetin were reported to have increased spacing between litters born and more rapid maturation of follicles at the expense of reducing the numbers of earlier stage follicles (Beazley and Nurminskaya 2015). $TG2^{-/-}$ mice [(Nanda et al. 2001) backcrossed more than ten generations to B6] mirrored the effects of quercetin on female B6 mice with respect to follicular development (Beazley and Nurminskaya 2015). However, the sample sizes were small for a breeding study ($n=4$). In addition, our unpublished observations spanning 4 years of breeding the same strain of mice [(Nanda et al. 2001) backcrossed more than 10 generations to B6] indicated no significant difference between wild-type ($n = 13$) and $TG2^{-/-}$ ($n = 11$) breeding pairs in terms of the litter size of each of the first four litters (wild-type: 5.5 ± 1.4SD, 7.5 ± 1.9SD, 7.3 ± 2.5SD, 5.8 ± 2.1SD; $TG2^{-/-}$: 6.6 ± 1.6SD, 7 ± 3.6SD, 6 ± 2.2SD, 7.4 ± 1.9SD) or birth spacing between these litters (wild-type: 4.5 ± 1.4SD weeks; $TG2^{-/-}$: 4.2 ± 1.2SD weeks).

4.4.7.2 Arterial Remodeling

The role of TG2 in small artery remodeling has been investigated with mostly consistent results. In response to surgical reduction of blood flow (Bakker et al. 2006) or nitric oxide inhibitor (L-NAME)-induced hypertension (Pistea et al. 2008), $TG2^{-/-}$ mice [(De Laurenzi and Melino 2001), mixed strain 129-B6] exhibited delayed inward remodeling (decrease in lumen diameter due to intimal hyperplasia) of small mesenteric arteries (that supply the intestine) relative to 129-B6 mice, whereas outward remodeling (increase in vessel size) was unaffected (Bakker et al. 2006). Chronic L-NAME treatment to induce hypertension led to enhanced blood pressure in $TG2^{-/-}$ mice [(De Laurenzi and Melino 2001) or (Nanda et al. 2001), mixed strain 129-B6] relative to 129-B6 mice (Pistea et al. 2008; Santhanam et al. 2010). Unlike wild-type mice, where carotid artery compliance decreased with L-NAME treatment, compliance in $TG2^{-/-}$ mice [(Nanda et al. 2001), mixed strain 129-B6] was unchanged (Santhanam et al. 2010), indicating that endothelium-derived nitric oxide mediates artery stiffness primarily through TG2. Indeed, there is a clear link between increased TG2 transamidating activity and increased arterial stiffness in aging rats (Santhanam et al. 2010). Exercise-mediated augmentation of nitric oxide production in rats attenuated TG2 transamidating activity through increased S-nitrosylation of TG2, and led to improved vascular elasticity ex vivo; however, there was no direct effect of exercise on vascular stiffness in vivo (Steppan et al. 2014). Collagen matrix remodeling by explant cultures of mesenteric small artery smooth muscle cells from $TG2^{-/-}$ mice [(De Laurenzi and Melino 2001), mixed strain 129-B6] was slower and gel compaction following cytoskeletal disruption was more easily reversed relative to cultures from 129-B6 mice (van den Akker et al. 2012).

Growth factors such as PDGF and β-catenin signaling pathways contribute to neointima formation via increased cell proliferation of vascular smooth muscle cells (Quasnichka et al. 2006). Studies using $TG2^{-/-}$ mice [(Nanda et al. 2001) backcrossed more than 10 generations to B6] showed that TG2 appears to promote PDGF-induced neointima formation by mediating the activation of both PDGFR/Akt1 and β-catenin signaling cascades in vascular smooth muscle cells (Nurminskaya et al. 2014). In contrast, a recent study found no significant difference in intimal hyperplasia between B6 and $TG2^{-/-}$ mice [(De Laurenzi and Melino 2001), no reference to strain background] following 2 or 4 weeks of carotid artery ligation (Min et al. 2014). However, this study found that 2 or 4 weeks after endothelial denudation of the external carotid artery, intimal thickness was reduced in $TG2^{-/-}$ mice compared with wild-type mice (Min et al. 2014), again indicating a role for TG2 in the development of intimal hyperplasia.

Interestingly, unlike the remodeling of small arteries described above, remodeling of large arteries and veins in response to high blood pressure does not appear to be TG2-dependent (Petersen-Jones et al. 2015). Indeed TG2 activity, expression and functionality were decreased in the aorta (which undergoes remodeling) but not in the vena cava (which does not undergo remodeling) in two models of hypertension in the rat (Petersen-Jones et al. 2015).

4.4.8 *Cell Adhesion and Spreading*

In the initial characterization of $TG2^{-/-}$ mice (Nanda et al. 2001) primary MEFs isolated from mixed-strain 129-B6 $TG2^{-/-}$ mice were reported to exhibit decreased adhesion relative to those from wild-type mice (Nanda et al. 2001). Backcrossing these $TG2^{+/-}$ mice to 129 or B6 to generate congenic lines revealed that MEFs isolated from congenic 129 $TG2^{-/-}$ mice are defective in both cell adhesion and spreading relative to MEFs isolated from wild-type 129 littermates, whereas MEFs isolated from congenic B6 $TG2^{-/-}$ mice are identical to wild-type B6 littermate MEFs in adhesion and spreading (TW Yiu, RM Graham, SE Iismaa, unpublished observations). This is reflected in delayed wound healing in congenic 129, but not in congenic B6, $TG2^{-/-}$ mice relative to their respective wild-type littermate mice (TW Yiu, RM Graham, SE Iismaa, unpublished observations). This provides a good example of a targeted mutation having a different phenotype in different backgrounds, and the influence of strain background on phenotype in mixed-strain backgrounds.

4.4.9 *Neurodegenerative Diseases (See Chap. 14 for More Detail)*

4.4.9.1 Huntington's Disease (HD)

Genetic deletion of TG2 was shown to extend life span and improve motor function in two mixed-strain mouse models of HD: R6/1 [heterozygous CBA-B6 R6/1 mice from Jackson Laboratory were backcrossed eight generations to B6 before breeding with $TG2^{+/-}$ mice (De Laurenzi and Melino 2001) that were backcrossed five generations to B6) (Mastroberardino et al. 2002) and R6/2 (CBA-B6 R6/2 mice (Mangiarini et al. 1996) were bred with 129-B6 $TG2^{-/-}$ mice (Nanda et al. 2001)] (Bailey et al. 2004; Bailey and Johnson 2006). In contrast, constitutive TG2 overexpression in B6 mouse neurons (Tucholski et al. 2006) did not modify disease onset or progression in R6/2 on a B6 background (R6/2B) (Kumar et al. 2012). Nor did genetic deletion of TG2 [(Nanda et al. 2001) backcrossed 12 generations to B6] in two different HD models, where experiments were performed according to the recommended guidelines for animal husbandry and study design (Menalled et al. 2014b), calling into question the role of TG2 in HD.

4.4.9.2 Stroke

Transgenic B6 mice overexpressing TG2 in neurons were phenotypically normal, but when challenged with oxidative stress, developed more neuronal damage, with increased duration and severity of seizures, than their wild-type B6 controls (Tucholski et al. 2006). After permanent ligation of the middle cerebral artery, the same transgenic mice had significantly smaller infarcts than wild-type B6 controls (Filiano et al. 2010). Surprisingly, so too did $TG2^{-/-}$ mice [(Nanda et al. 2001) backcrossed 12 generations to B6] relative to B6 mice, where it was shown that TG2 supports neuronal survival yet decreases astrocyte survival and astrocytic support of neurons following ischemia (Colak and Johnson 2012). Cortical neurons from $TG2^{-/-}$ mice [(Nanda et al. 2001) backcrossed 12 generations to B6], like wild-type littermates, upregulate TG1 in response to oxidative stress (Basso et al. 2012) suggesting TG1 may also participate in oxidative stress-induced cell death signaling.

4.4.9.3 Remyelination

$TG2^{-/-}$ [(De Laurenzi and Melino 2001) backcrossed to B6, number of generations not stated] and littermate wild-type mice showed no difference in myelination; however, following induction of demyelination, subsequent remyelination was delayed in $TG2^{-/-}$ mice (Van Strien et al. 2011). There was no difference in inflammatory mediator production (in contrast to the EAE model, Sect. 4.4.2.2), $TG2^{-/-}$ mice had more oligodendrocyte precursor cells and were delayed in their recovery of motor function (Van Strien et al. 2011). Contrary to previous reports (Mastroberardino et al. 2006; Szondy et al. 2006), there was no impairment of mitochondrial function (assessed by the activities of mitochondrial complex II and IV enzymes) in the hearts of these mice (Van Strien et al. 2011). Thus, TG2 appears to be involved in neuronal remyelination.

4.4.9.4 Parkinson's Disease (PD)

$TG2^{-/-}$ mice [(De Laurenzi and Melino 2001) backcrossed ten generations to B6] treated with MPTP (1-methyl-4-phenyl-1,2,3,6-tetrahydropyridine) to cause experimental PD were more resistant to neuronal damage than wild-type mice (Battaglia et al. 2007). Double transgenic mice overexpressing TG2 [B6 (Tucholski et al. 2006)] and α-synuclein [B6xDBA (Rockenstein et al. 2002)] in neurons exhibited enhanced high-molecular-weight insoluble aggregates of α-synuclein, greater neuroinflammation, exacerbated nerve fiber damage and impaired motor behavioral performance on the balance beam test and in nest-building (Grosso et al. 2014), suggesting TG2 may contribute to the pathogenesis of α-synucleinopathies such as PD.

4.4.10 *Diabetes*

Type 2 diabetes mellitus (T2DM) is characterized by defects in end-organ responsiveness to insulin (insulin resistance) and in the regulation of insulin release by pancreatic β-cells. Maturity-onset diabetes (MODY) is a monogenic heritable variant of T2DM that occurs in non-obese individuals and presents in adolescence or when subjects are in their early 20s. TG2 has been reported to be a candidate gene for MODY, with heterozygous mutations that impair transamidase activity (M330R, I331N, or N333S) having been identified in one or more affected individuals from three different families (Bernassola et al. 2002b; Porzio et al. 2007). In vivo (Bernassola et al. 2002b) and less convincing in vitro (Salter et al. 2012) studies have reported impaired glucose-stimulated insulin secretion by pancreatic islets from $TG2^{-/-}$ mice [(De Laurenzi and Melino 2001), mixed-strain 129-B6 or B6, respectively] relative to wild-type mice. However, another study (performed according to the recommended guidelines for animal husbandry and study design), using congenic B6 and 129 $TG2^{-/-}$ lines [(Nanda et al. 2001) backcrossed 12 generations to the respective inbred strains], showed no impairment in vivo or in vitro of glucose-stimulated insulin secretion by pancreatic islets from $TG2^{-/-}$ mice relative to wild-type littermates (Iismaa et al. 2013). Insulin tolerance tests were also no different. Moreover, intraperitoneal glucose and insulin tolerance tests revealed no significant differences between wild-type littermates and mice expressing a mutant form of TG2 that is constitutively active for transamidase activity ($TG2^{R579A}$) in a B6 co-isogenic mouse line (Iismaa et al. 2013). This indicates glucose homeostasis is TG2 independent.

4.4.11 *Bone and Osteoarthritis*

4.4.11.1 Bone Development (See Chap. 12 for More Detail)

Neither TG2 knockout mice [(De Laurenzi and Melino 2001; Nanda et al. 2001), mixed 129-B6 background] nor FXIIIA knockout mice [(Lauer et al. 2002), mixed B6-129-Swiss albino-CBACa background] exhibited skeletal abnormalities, indicating neither TG alone is critical for bone development. One group documented no substantial overexpression of FXIIIA or other TGs in either mixed-strain knockout (Nakano et al. 2007) whereas another group reported marginally increased expression of FXIIIA in the mixed-strain TG2 knockout (Tarantino et al. 2009). Surprisingly, bone deposition was also normal in $FXIIIA^{-/-}TG2^{-/-}$ double knockout mice [$FXIIIA^{-/-}$: (Lauer et al. 2002), no reference to strain background; $TG2^{-/-}$: (Nanda et al. 2001), mixed strain 129-B6] (Cordell et al. 2015). Although low levels of TG1 were expressed in wild-type and double knockout bone, there was no upregulation of TG1 or other TGs relative to wild-type controls (Cordell et al. 2015). Further studies following the recommended guidelines for animal husbandry and study design are required to confirm the role of TG2 in bone development.

4.4.11.2 Osteoarthritis

Normal articular cartilage, which cushions the ends of bones in joints by secreting synovial fluid, contains resting chondrocytes that do not undergo terminal differentiation. In osteoarthritis, foci of articular chondrocyte hypertrophy develop with formation of osteophytes (bony outgrowths) and cartilage degeneration. Studies using wild-type and TG2$^{-/-}$ littermates [(Nanda et al. 2001), mixed-strain 129-B6] demonstrated that articular chondrocyte hypertrophy is mediated by TG2-dependent and -independent pathways (Johnson et al. 2003; Cecil and Terkeltaub 2008). In a surgically induced knee joint instability model of osteoarthritis, TG2$^{-/-}$ mice [(De Laurenzi and Melino 2001), mixed strain 129-B6] exhibited less cartilage destruction and greater osteophyte formation relative to wild-type littermates (Orlandi et al. 2009). Indeed chondrocyte expression of TG2 in situ and its release into the synovial fluid correlates with osteoarthritis severity (Huebner et al. 2009). These studies indicate that TG2 modulates cartilage repair and stability in response to certain pathological stressors and identify TG2 as a potential biomarker of osteoarthritis and a target for therapeutic intervention. Further studies are required to confirm and extend these findings.

4.4.12 The Eye

4.4.12.1 Lens

Lenses from TG2$^{-/-}$ mice [(De Laurenzi and Melino 2001), no reference to strain background], cultured with TGF-β2 for 10 days to mimic oxidative stress, were resistant to cortical opacification as compared to B6 mice (Charles River Laboratories), the lenses of which opacified within 5 days of TGF-β2 treatment and were completely opaque by day 10 (Shin et al. 2008). This work suggests a role for TG2 in the formation of protein aggregates.

4.4.12.2 Cornea

TG2 is expressed in murine corneal epithelial layers and in the basement membrane of corneas (Tong et al. 2013). Congenic B6 TG2$^{-/-}$ and wild-type mice [(Nanda et al. 2001) backcrossed more than 12 generations to B6] were subjected to a model of mechanical abrasion of the corneal surface and subsequent temporal observation of epithelial healing (Tong et al. 2013). Relative to wild-type mice, corneal wound closure was delayed in TG2$^{-/-}$ mice with the new migrating corneal epithelial sheet lacking strong adhesion at the leading edge of the wound. At the molecular level, this was associated with reduced cytoskeletal remodeling and implicated TG2-dependent activation of β3 integrin signaling in corneal epithelial cell adhesion and migration (Tong et al. 2013).

4.4.13 Cancer

TG2 is a major crosslinking enzyme in the extracellular matrix and has been extensively implicated in cancer, where increased density, elasticity and crosslinking of extracellular matrix is correlated with cancer malignancy. Subcutaneous implantation of B16F1 (Jones et al. 2006) or B16-F10 (Di Giacomo et al. 2009) melanoma cells into wild-type or $TG2^{-/-}$ mice [(De Laurenzi and Melino 2001) backcrossed to B6, number of generations not stated] resulted in increased tumor size in (Jones et al. 2006; Di Giacomo et al. 2009), and reduced survival of, $TG2^{-/-}$ mice compared to wild-type mice (Jones et al. 2006). Metastasis to lung and other organs was increased, although the individual metastasis size was smaller in $TG2^{-/-}$ relative to wild-type mice (Di Giacomo et al. 2009). TG2 was shown to interact in the extracellular matrix with the extracellular domain of the adhesion G protein-coupled receptor, GPR56 (Xu et al. 2006; Yang et al. 2014), which is a suppressor of melanoma tumor growth and metastasis, and which is markedly downregulated in highly metastatic cells. Xenograft studies in $TG2^{-/-}$ mice [(De Laurenzi and Melino 2001), no reference to strain background] bred into immunodeficient $Rag2^{-/-}$ mice (no reference to source or strain background) indicated that TG2 and its crosslinking activity promote melanoma growth (Yang et al. 2014). GPR56 inhibition of melanoma progression involves binding of the N-terminus of GPR56 to its ligand, TG2, in the melanoma extracellular matrix, followed by internalization and degradation of TG2 (Yang et al. 2014). TG2 is tightly associated with the major extracellular matrix protein, fibronectin, and downregulation of TG2 resulted in reduced fibronectin deposition in the extracellular matrix (Yang et al. 2014). This work supports a role for TG2 in promotion of tumor growth.

4.5 Conclusions

Genetically modified mouse models are an important tool to dissect the contribution of TG2 to various physiological and pathological processes. Advances continue to be made that allow better understanding of the mechanisms of TG2 action and help guide the development of better therapeutic and diagnostic strategies. Animal studies have consistently supported a role for TG2 in phagocytosis, autophagy, celiac disease, kidney and lung fibrosis, arterial remodeling, osteoarthritis and cancer. New potential roles for TG2 are emerging in the immune system, the eye, multiple sclerosis and Parkinson's disease. Recent rigorous animal studies have discounted a role for TG2 in liver fibrosis, Huntington's disease and diabetes. Some of the results from studies of TG2 that have been performed with genetically modified animals are contradictory, particularly in the areas of apoptosis, mitochondrial function and vascular remodeling. Shortcomings with respect to strain background, animal husbandry, sample size or reporting are evident in these studies. These areas would benefit from studies using congenic mouse lines to limit interaction of any confounding traits attributable to allele differences between different inbred strains. In addition, improving the rigor with which animal studies are designed, executed and reported will accelerate both scientific progress and the development of new therapies.

References

Abedin M, Tintut Y, Demer LL (2004) Vascular calcification: mechanisms and clinical ramifications. Arterioscler Thromb Vasc Biol 24(7):1161–1170

Bailey CD, Johnson GV (2006) The protective effects of cystamine in the R6/2 Huntington's disease mouse involve mechanisms other than the inhibition of tissue transglutaminase. Neurobiol Aging 27(6):871–879

Bailey CD, Graham RM, Nanda N, Davies PJ, Johnson GV (2004) Validity of mouse models for the study of tissue transglutaminase in neurodegenerative diseases. Mol Cell Neurosci 25 (3):493–503

Bakker EN, Pistea A, Spaan JA, Rolf T, de Vries CJ, van Rooijen N, Candi E, VanBavel E (2006) Flow-dependent remodeling of small arteries in mice deficient for tissue-type transglutaminase: possible compensation by macrophage-derived factor XIII. Circ Res 99(1):86–92

Balajthy Z, Csomós K, Vámosi G, Szántó A, Lanotte M, Fésüs L (2006) Tissue-transglutaminase contributes to neutrophil granulocyte differentiation and functions. Blood 108(6):2045–2054

Basso M, Berlin J, Xia L, Sleiman SF, Ko B, Haskew-Layton R, Kim E, Antonyak MA, Cerione RA, Iismaa SE, Willis D, Cho S, Ratan RR (2012) Transglutaminase inhibition protects against oxidative stress-induced neuronal death downstream of pathological ERK activation. J Neurosci 32(19):6561–6569

Battaglia G, Farrace MG, Mastroberardino PG, Viti I, Fimia GM, Van Beeumen J, Devreese B, Melino G, Molinaro G, Busceti CL, Biagioni F, Nicoletti F, Piacentini M (2007) Transglutaminase 2 ablation leads to defective function of mitochondrial respiratory complex I affecting neuronal vulnerability in experimental models of extrapyramidal disorders. J Neurochem 100(1):36–49

Beazley KE, Nurminskaya M (2015) Effects of dietary quercetin on female fertility in mice: implication of transglutaminase 2. Reprod Fertil Dev. doi:10.1071/RD14155 [Epub ahead of print]

Beazley KE, Deasey S, Lima F, Nurminskaya MV (2012) Transglutaminase 2-mediated activation of β-catenin signaling has a critical role in warfarin-induced vascular calcification. Arterioscler Thromb Vasc Biol 32(1):123–130

Beazley KE, Banyard D, Lima F, Deasey SC, Nurminsky DI, Konoplyannikov M, Nurminskaya MV (2013a) Transglutaminase inhibitors attenuate vascular calcification in a preclinical model. Arterioscler Thromb Vasc Biol 33(1):43–51

Beazley KE, Reckard S, Nurminsky D, Lima F, Nurminskaya M (2013b) Two sides of MGP null arterial disease: chondrogenic lesions dependent on transglutaminase 2 and elastin fragmentation associated with induction of adipsin. J Biol Chem 288(43):31400–31408

Bernassola F, Boumis G, Corazzari M, Bertini G, Citro G, Knight RA, Amiconi G, Melino G (2002a) Osmotic resistance of high-density erythrocytes in transglutaminase 2-deficient mice. Biochem Biophys Res Commun 291(5):1123–1127

Bernassola F, Federici M, Corazzari M, Terrinoni A, Hribal ML, De Laurenzi V, Ranalli M, Massa O, Sesti G, Mclean WHI, Citro G, Barbetti F, Melino G (2002b) Role of transglutaminase 2 in glucose tolerance: knockout mice studies and putative mutation in a MODY patient. FASEB J 16(11):1371–1378

Bethune MT, Khosla C (2008) Parallels between pathogens and gluten peptides in celiac sprue. PLoS Pathog 4(2):e34

Bijli KM, Kanter BG, Minhajuddin M, Leonard A, Xu L, Fazal F, Rahman A (2014) Regulation of endothelial cell inflammation and lung polymorphonuclear lymphocyte infiltration by transglutaminase 2. Shock 42(6):562–569

Boisvert WA, Rose DM, Boullier A, Quehenberger O, Sydlaske A, Johnson KA, Curtiss LK, Terkeltaub R (2006) Leukocyte transglutaminase 2 expression limits atherosclerotic lesion size. Arterioscler Thromb Vasc Biol 26(3):563–569

Brennan K (2011) Colony management. In: Pease S, Saunders TL (eds) Advanced protocols for animal transgenesis. Springer, Berlin/Heidelberg, pp 535–576

Cecil DL, Terkeltaub R (2008) Transamidation by transglutaminase 2 transforms S100A11 calgranulin into a procatabolic cytokine for chondrocytes. J Immunol 180(12):8378–8385

Colak G, Johnson GV (2012) Complete transglutaminase 2 ablation results in reduced stroke volumes and astrocytes that exhibit increased survival in response to ischemia. Neurobiol Dis 45(3):1042–1050

Cordell PA, Newell LM, Standeven KF, Adamson P, Simpson KR, Smith KA, Jackson CL, Grant PJ, Pease RJ (2015) Normal bone deposition occurs in mice deficient in factor XIII-A and transglutaminase 2. Matrix Biol 43(Apr):85–96

De Laurenzi V, Melino G (2001) Gene disruption of tissue transglutaminase. Mol Cell Biol 21 (1):148–155

Deasey S, Shanmugasundaram S, Nurminskaya M (2013) Tissue-specific responses to loss of transglutaminase 2. Amino Acids 44(1):179–187

D'Eletto M, Farrace MG, Falasca L, Reali V, Oliverio S, Melino G, Griffin M, Fimia GM, Piacentini M (2009) Transglutaminase 2 is involved in autophagosome maturation. Autophagy 5(8):1145–1154

D'Eletto M, Farrace MG, Rossin F, Strappazzon F, Giacomo GD, Cecconi F, Melino G, Sepe S, Moreno S, Fimia GM, Falasca L, Nardacci R, Piacentini M (2012) Type 2 transglutaminase is involved in the autophagy-dependent clearance of ubiquitinated proteins. Cell Death Differ 19 (7):1228–1238

Delhase M, Kim SY, Lee H, Naiki-Ito A, Chen Y, Ahn ER, Murata K, Kim SJ, Lautsch N, Kobayashi KS, Shirai T, Karin M, Nakanishi M (2012) TANK-binding kinase 1 (TBK1) controls cell survival through PAI-2/serpinB2 and transglutaminase 2. Proc Natl Acad Sci U S A 109(4):E177–E186

Di Giacomo G, Lentini A, Beninati S, Piacentini M, Rodolfo C (2009) In vivo evaluation of type 2 transglutaminase contribution to the metastasis formation in melanoma. Amino Acids 36 (4):717–724

Falasca L, Iadevaia V, Ciccosanti F, Melino G, Serafino A, Piacentini M (2005) Transglutaminase type II is a key element in the regulation of the anti-inflammatory response elicited by apoptotic cell engulfment. J Immunol 174(11):7330–7340

Falasca L, Farrace MG, Rinaldi A, Tuosto L, Melino G, Piacentini M (2008) Transglutaminase type II is involved in the pathogenesis of endotoxic shock. J Immunol 180(4):2616–2624

Fésüs L, Szondy Z (2005) Transglutaminase 2 in the balance of cell death and survival. FEBS Lett 579(15):3297–3302

Filiano AJ, Tucholski J, Dolan PJ, Colak G, Johnson GV (2010) Transglutaminase 2 protects against ischemic stroke. Neurobiol Dis 39(3):334–343

Fisher M, Jones RA, Huang L, Haylor JL, El Nahas M, Griffin M, Johnson TS (2009) Modulation of tissue transglutaminase in tubular epithelial cells alters extracellular matrix levels: a potential mechanism of tissue scarring. Matrix Biol 28(1):20–31

Fuchs Y, Steller H (2011) Programmed cell death in animal development and disease. Cell 147 (4):742–758

Grosso H, Woo J-M, Lee K-W, Im J-Y, Masliah E, Junn E, Mouradian MM (2014) Transglutaminase 2 exacerbates α-synuclein toxicity in mice and yeast. FASEB J 28(10):4280–4291

Huebner JL, Johnson KA, Kraus VB, Terkeltaub RA (2009) Transglutaminase 2 is a marker of chondrocyte hypertrophy and osteoarthritis severity in the Hartley guiniea pig model of knee OA. Osteoarthritis Cartilage 17(8):1056–1064

Iismaa SE, Mearns BM, Lorand L, Graham RM (2009) Transglutaminases and disease: Lessons from genetically engineered mouse models and inherited disorders. Physiol Rev 89 (3):991–1023

Iismaa SE, Aplin M, Holman S, Yiu TW, Jackson K, Burchfield JG, Mitchell CJ, O'Reilly L, Davenport A, Cantley J, Schmitz-Peiffer C, Biden TJ, Cooney GJ, Graham RM (2013) Glucose homeostasis in mice is transglutaminase 2 independent. PLoS One 8(5):e66346

Ishiguro K, Kadomatsu K, Kojima T, Muramatsu H, Tsuzuki S, Nakamura E, Kusugami K, Saito H, Muramatsu T (2000) Syndecan-4 deficiency impairs focal adhesion formation only under restricted conditions. J Biol Chem 275(8):5249–5252

Iversen R, Di Niro R, Stamnaes J, Lundin KE, Wilson PC, Sollid LM (2013) Transglutaminase 2-specific autoantibodies in celiac disease target clustered, N-terminal epitopes not displayed on the surface of cells. J Immunol 190(12):5981–5991

Johnson KA, van Etten D, Nanda N, Graham RM, Terkeltaub RA (2003) Distinct transglutaminase 2-independent and transglutaminase 2-dependent pathways mediate articular chondrocyte hypertrophy. J Biol Chem 278(21):18824–18832

Johnson KA, Polewski M, Terkeltaub RA (2008) Transglutaminase 2 is central to induction of the arterial calcification program by smooth muscle cells. Circ Res 102(5):529–537

Johnson KB, Petersen-Jones H, Thompson JM, Hitomi K, Itoh M, Bakker EN, Johnson GV, Colak G, Watts SW (2012) Vena cava and aortic smooth muscle cells express transglutaminases 1 and 4 in addition to transglutaminase 2. Am J Physiol Heart Circ Physiol 302 (7):H1355–H1366

Jones RA, Kotsakis P, Johnson TS, Chau DYS, Ali S, Melino G, Griffin M (2006) Matrix changes induced by transglutaminase 2 lead to inhibition of angiogenesis and tumor growth. Cell Death Differ 13(9):1442–1453

Kalliokoski S, Caja S, Frias R, Laurila K, Koskinen O, Niemelä O, Mäki M, Kaukinen K, Korponay-Szabó IR, Lindfors K (2015) Injection of celiac disease patient sera or immunoglobulins to mice reproduces a condition mimicking early developing celiac disease. J Mol Med 93(1):51–62

Kim DS, Kim B, Tahk H, Kim DH, Ahn ER, Choi C, Jeon Y, Park SY, Lee H, Oh SH, Kim SY (2010) Transglutaminase 2 gene ablation protects against renal ischemic injury by blocking constant NF-κB activation. Biochem Biophys Res Commun 403(3–4):479–484

Kim J-H, Jeong EM, Jeong Y-J, Lee WJ, Kang JS, Kim I-G, Hwang Y-I (2012) Transglutaminase 2 modulates antigen-specific antibody response by suppressing Blimp-1 and AID expression of B cells in mice. Immunol Lett 147(1–2):18–28

Kim J-H, Hong JM, Jeong EM, Lee WJ, Kim H-R, Kang JS, Kim I-G, Hwang Y-I (2014a) Lack of transglutaminase 2 diminished T-cell responses in mice. Immunology 142(3):506–515

Kim J-H, Jeong EM, Jeong Y-J, Lee WJ, Kang JS, Kim I-G, Hwang Y-I (2014b) Transglutaminase 2 on the surface of dendritic cells is proposed to be involved in dendritic cell-T cell interaction. Cell Immunol 289(1–2):55–62

Korponay-Szabo IR, Laurila K, Szondy Z, Halttunen T, Szalai Z, Dahlbom I, Rantala I, Kovacs JB, Fesus L, Maki M (2003) Missing endomysial and reticulin binding of coeliac antibodies in transglutaminase 2 knockout tissues. Gut 52(2):199–204

Krasnikov BF, Kim SY, McConoughey SJ, Ryu H, Xu H, Stavrovskaya I, Iismaa SE, Mearns BM, Ratan RR, Blass JP, Gibson GE, Cooper AJ (2005) Transglutaminase activity is present in highly purified nonsynaptosomal mouse brain and liver mitochondria. Biochemistry 44 (21):7830–7843

Kumar A, Kneynsberg A, Tucholski J, Perry G, van Groen T, Detloff PJ, Lesort M (2012) Tissue transglutaminase overexpression does not modify the disease phenotype of the R6/2 mouse model of Huntington's disease. Exp Neurol 237(1):78–89

Landis SC, Amara SG, Asadullah K, Austin CP, Blumenstein R, Bradley EW, Crystal RG, Darnell RB, Ferrante RJ, Fillit H, Finkelstein R, Fisher M, Gendelman HE, Golub RM, Goudreau JL, Gross RA, Gubitz AK, Hesterlee SE, Howells DW, Huguenard J, Kelner K, Koroshetz W, Krainc D, Lazic SE, Levine MS, Macleod MR, McCall JM, Moxley RT, Narasimhan K, Noble LJ, Perrin S, Porter JD, Steward O, Unger E, Utz U, Silberberg SD (2012) A call for transparent reporting to optimize the predictive value of preclinical research. Nature 490(7419):187–191

Lauer P, Metzner HJ, Zettlmeissl G, Li M, Smith AG, Lathe R, Dickneite G (2002) Targeted inactivation of the mouse locus encoding coagulation factor XIII-A: hemostatic abnormalities in mutant mice and characterization of the coagulation deficit. Thromb Haemost 88 (6):967–974

Lorand L, Conrad SM (1984) Transglutaminases. Mol Cell Biochem 58(1–2):9–35

Malorni W, Farrace MG, Matarrese P, Tinari A, Ciarlo L, Mousavi-Shafaei P, D'Eletto M, Di Giacomo G, Melino G, Palmieri L, Rodolfo C, Piacentini M (2009) The adenine nucleotide translocator 1 acts as a type 2 transglutaminase substrate: implications for mitochondrial-dependent apoptosis. Cell Death Differ 16(11):1480–1492

Mangiarini L, Sathasivam K, Seller M, Cozens B, Harper A, Hetherington C, Lawton M, Trottier Y, Lehrach H, Davies SW, Bates GP (1996) Exon 1 of the HD gene with an expanded CAG repeat is sufficient to cause a progressive neurological phenotype in transgenic mice. Cell 87(3):493–506

Mastroberardino PG, Iannicola C, Nardacci R, Bernassola F, De Laurenzi V, Melino G, Moreno S, Pavone F, Oliverio S, Fesüs L, Piacentini M (2002) 'Tissue' transglutaminase ablation reduces neuronal death and prolongs survival in a mouse model of Huntington's disease. Cell Death Differ 9(9):873–880

Mastroberardino PG, Farrace MG, Viti I, Pavone F, Fimia GM, Melino G, Rodolfo C, Piacentini M (2006) "Tissue" transglutaminase contributes to the formation of disulphide bridges in proteins of mitochondrial respiratory complexes. Biochim Biophys Acta 1757(9–10):1357–1365

Matic I, Sacchi A, Rinaldi A, Melino G, Khosla C, Falasca L, Piacentini M (2010) Characterization of transglutaminase type II role in dendritic cell differentiation and function. J Leukoc Biol 88(1):181–188

Menalled L, Lutz C, Ramboz S, Brunner D, Lager B, Noble S, Park L, Howland D (2014a) A field guide to working with mouse models of Huntington's disease. https://www.chdifoundation.org/wp-content/uploads/HD_Field_Guide_040414.pdf

Menalled LB, Kudwa AE, Oakeshott S, Farrar A, Paterson N, Filippov I, Miller S, Kwan M, Olsen M, Beltran J, Torello J, Fitzpatrick J, Mushlin R, Cox K, McConnell K, Mazzella M, He D, Osborne GF, Al-Nackkash R, Bates GP, Tuunanen P, Lehtimaki K, Brunner D, Ghavami A, Ramboz S, Park L, Macdonald D, Munoz-Sanjuan I, Howland D (2014b) Genetic deletion of transglutaminase 2 does not rescue the phenotypic deficits observed in R6/2 and zQ175 mouse models of Huntington's disease. PLoS One 9 (6):e99520

Michlewska S, McColl A, Rossi AG, Megson IL, Dransfield I (2007) Clearance of dying cells and autoimmunity. Autoimmunity 40(4):267–273

Min SK, Min SI, Jeong EM, Cho SY, Ha J, Kim SJ, Kim IG (2014) Intimal hyperplasia in loop-injured carotid arteries is attenuated in transglutaminase 2-null mice. J Korean Med Sci 29 (3):363–369

Nadella V, Wang Z, Johnson TS, Griffin M, Devitt A (2015) Transglutaminase 2 interacts with syndecan-4 and CD44 at the surface of human macrophages to promote removal of apoptotic cells. Biochim Biophys Acta 1853(1):201–212

Nagy L, Thomázy VA, Saydak MM, Stein JP, Davies PJ (1997) The promoter of the mouse tissue transglutaminase gene directs tissue-specific, retinoid-regulated and apoptosis-linked expression. Cell Death Differ 4(7):534–547

Nakano Y, Al-Jallad HF, Mousa A, Kaartinen MT (2007) Expression and localization of plasma transglutaminase factor XIIIA in bone. J Histochem Cytochem 55(7):675–685

Nanda N, Iismaa SE, Owens WA, Husain A, Mackay F, Graham RM (2001) Targeted inactivation of G_h/tissue transglutaminase II. J Biol Chem 276(23):20673–20678

Nardacci R, Lo Iacono O, Ciccosanti F, Falasca L, Addesso M, Amendola A, Antonucci G, Craxí A, Fimia GM, Iadevaia V, Melino G, Ruco L, Tocci G, Ippolito G, Piacentini M (2003) Transglutaminase type II plays a protective role in hepatic injury. Am J Pathol 162 (4):1293–1303

Nurminskaya M, Beazley KE, Smith EP, Belkin AM (2014) Transglutaminase 2 promotes PDGF-mediated activation of PDGFR/Akt1 and β-catenin signaling in vascular smooth muscle cells and supports neointima formation. J Vasc Res 51(6):418–428

Oh K, Park H-B, Byoun O-K, Shin D-M, Jeong EM, Kim YW, Kim YS, Melino G, Kim I-G, Lee D-S (2011) Epithelial transglutaminase 2 is needed for T cell interleukin-17 production and subsequent pulmonary inflammation and fibrosis in bleomycin-treated mice. J Exp Med 208 (8):1707–1719

Oh K, Park H-B, Seo MW, Byoun O-J, Lee D-S (2012) Tranglutaminase 2 exacerbates experimental autoimmune encephalomyelitis through positive regulation of encephalitogenic T cell differentiation and inflammation. Clin Immunol 145(2):122–132

Oh K, Seo MW, Lee GY, Byoun O-J, Kang H-R, Cho S-H, Lee D-S (2013) Airway epithelial cells initiate the allergen response through transglutaminase 2 by inducing IL-33 expression and a subsequent Th2 response. Respir Res 14:35

Olsen KC, Sapinoro RE, Kottmann RM, Kulkarni AA, Iismaa SE, Johnson GV, Thatcher TH, Phipps RP, Sime PJ (2011) Transglutaminase 2 and its role in pulmonary fibrosis. Am J Respir Crit Care Med 184(6):699–707

Orlandi A, Oliva F, Taurisano G, Candi E, Di Lascio A, Melino G, Spagnoli LG, Tarantino U (2009) Transglutaminase-2 differently regulates cartilage destruction and osteophyte formation in a surgical model of osteoarthritis. Amino Acids 36(4):755–763

Pease S, Saunders TL (2011) Advanced protocols for animal transgenesis. An ISTT manual, Springer protocols. Springer, Berlin/Heidelberg

Petersen-Jones HG, Johnson KB, Hitomi K, Tykocki NR, Thompson JM, Watts SW (2015) Transglutaminase activity is decreased in large arteries from hypertensive rats, compared to normotensive controls. Am J Physiol Heart Circ Physiol 308(6):H592–H602

Pistea A, Bakker EN, Spaan JA, Hardeman MR, van Rooijen N, VanBavel E (2008) Small artery remodeling and erythrocyte deformability in L-NAME-induced hypertension: role of transglutaminases. J Vasc Res 45(1):10–18

Plump AS, Smith JD, Hayek T, Aalto-Setälä K, Walsh A, Verstuyft JG, Rubin EM, Breslow JL (1992) Severe hypercholesterolemia and atherosclerosis in apolipoprotein E-deficient mice created by homologous recombination in ES cells. Cell 71(2):343–353

Popov Y, Sverdlov DY, Sharma AK, Bhaskar KR, Li S, Freitag TL, Lee J, Dieterich W, Melino G, Schuppan D (2011) Tissue transglutaminase does not affect fibrotic matrix stability or regression of liver fibrosis in mice. Gastroenterology 140(5):1642–1652

Porzio O, Massa O, Cunsolo V, Colombo C, Malaponti M, Bertuzzi F, Hansen T, Johansen A, Pedersen O, Meschi F, Terrinoni A, Melino G, Federici M, Decarlo N, Menicagli M, Campani D, Marchetti P, Ferdaoussi M, Froguel P, Federici G, Vaxillaire M, Barbetti F (2007) Missense mutations in the TGM2 gene encoding transglutaminase 2 are found in patients with early-onset type 2 diabetes. Hum Mutat 28(11):1150

Quasnichka H, Slater SC, Beeching CA, Boehm M, Sala-Newby GB, George SJ (2006) Regulation of smooth muscle cell proliferation by beta-catenin/T-cell factor signaling involves modulation of cyclin D1 and p21 expression. Circ Res 99(12):1329–1337

Rahman A, Fazal F (2011) Blocking NF-κB. An inflammatory issue. Proc Am Thorac Soc 8 (6):497–503

Rockenstein E, Mallory M, Hashimoto M, Song D, Shults CW, Lang I, Masliah E (2002) Differential neuropathological alterations in transgenic mice expressing alpha-synuclein from the platelet-derived growth factor and Thy-1 promoters. J Neurosci Res 68(5):568–578

Rose DM, Sydlaske AD, Agha-Babakhani A, Johnson K, Terkeltaub R (2006) Transglutaminase 2 limits murine peritoneal acute gout-like inflammation by regulating macrophage clearance of apoptotic neutrophils. Arthritis Rheum 54(10):3363–3371

Rossin F, D'Eletto M, Macdonald D, Farrace MG, Piacentini M (2012) TG2 transamidating activity acts as a reostat controlling the interplay between apoptosis and autophagy. Amino Acids 42(5):1793–1802

Rossin F, D'Eletto M, Falasca L, Sepe S, Cocco S, Fimia GM, Campanella M, Mastroberardino PG, Farrace MG, Piacentini M (2014) Transglutaminase 2 ablation leads to mitophagy impairment associated with a metabolic shift towards aerobic glycolysis. Cell Death Differ 22(3):408–418

Salter NW, Ande SR, Nguyen HK, Nyomba BL, Mishra S (2012) Functional characterization of naturally occurring transglutaminase 2 mutants implicated in early-onset type 2 diabetes. J Mol Endocrinol 48(3):203–216

Santhanam L, Tuday EC, Webb AK, Dowzicky P, Kim JH, Oh YJ, Sikka G, Kuo M, Halushka MK, Macgregor AM, Dunn J, Gutbrod S, Yin D, Shoukas A, Nyhan D, Flavahan NA, Belkin AM, Berkowitz DE (2010) Decreased S-nitrosylation of tissue transglutaminase contributes to age-related increases in vascular stiffness. Circ Res 107(1):117–125

Sarang Z, Molnar P, Nemeth T, Gomba S, Kardon T, Melino G, Cotecchia S, Fesus L, Szondy Z (2005) Tissue transglutaminase (TG2) acting as G protein protects hepatocytes against Fas-mediated cell death in mice. Hepatology 42(3):578–587

Sarang Z, Mádi A, Koy C, Varga S, Glocker MO, Ucker DS, Kuchay S, Chishti AH, Melino G, Fésüs L, Szondy Z (2007) Tissue transglutaminase (TG2) facilitates phosphatidylserine exposure and calpain activity in calcium-induced death of erythrocytes. Cell Death Differ 14 (10):1842–1844

Sarang Z, Tóth B, Balajthy Z, Köröskényi K, Garabuczi E, Fésüs L, Szondy Z (2009) Some lessons from the tissue transglutaminase knockout mouse. Amino Acids 36(4):625–631

Sarang Z, Köröskényi K, Pallai A, Duró E, Melino G, Griffin M, Fésüs L, Szondy Z (2011) Transglutaminase 2 null macrophages respond to lipopolysaccharide stimulation by elevated proinflammatory cytokine production due to an enhanced αvβ3 integrin-induced Src tyrosine kinase signaling. Immunol Lett 138(1):71–78

Scarpellini A, Huang L, Burhan I, Schroeder N, Funck M, Johnson TS, Verderio EA (2014) Syndecan-4 knockout leads to reduced extracellular transglutaminase-2 and protects against tubulointerstitial fibrosis. J Am Soc Nephrol 25(5):1013–1027

Shin DM, Jeon JH, Kim CW, Cho SY, Lee HJ, Jang GY, Jeong EM, Lee DS, Kang JH, Melino G, Park SC, Kim IG (2008) TGFbeta mediates activation of transglutaminase 2 in response to oxidative stress that leads to protein aggregation. FASEB J 22(7):2498–2507

Shweke N, Boulos N, Jouanneau C, Vandermeersch S, Melino G, Dussaule JC, Chatziantoniou C, Ronco P, Boffa JJ (2008) Tissue transglutaminase contributes to interstitial renal fibrosis by favoring accumulation of fibrillar collagen through TGF-beta activation and cell infiltration. Am J Pathol 173(3):631–642

Silva AJ, Simpson EM, Takahashi JS, Lipp H-P, Nakanishi S, Wehner JM, Giese KP, Tully T, Abel T, Chapman PF, Fox K, Grant S, Itohara S, Lathe R, Mayford M, McNamara JO, Moriis RJ, Picciotto M, Roder J, Shin H-S, Slesinger PA, Storm DR, Stryker MP, Tonegawa S, Wang Y, Wolfer DP (1997) Mutant mice and neuroscience: recommendations concerning genetic background. Banbury conference on genetic background in mice. Neuron 19 (4):755–759

Small K, Feng JF, Lorenz J, Donnelly ET, Yu A, Im MJ, Dorn GW 2nd, Liggett SB (1999) Cardiac specific overexpression of transglutaminase II (G_h) results in a unique hypertrophy phenotype independent of phospholipase C activation. J Biol Chem 274(30):21291–21296

Steppan J, Sikka G, Jandu S, Barodka V, Halushka MK, Flavahan NA, Belkin AM, Nyhan D, Butlin M, Avolio A, Berkowitz DE, Santhanam L (2014) Exercise, vascular stiffness, and tissue transglutaminase. J Am Heart Assoc 3(2):e000599

Stevens JC, Banks GT, Festing MF, Fisher EM (2007) Quiet mutations in inbred strains of mice. Trends Mol Med 13(12):512–519

Strnad P, Harada M, Siegel M, Terkeltaub RA, Graham RM, Khosla C, Omary MB (2007) Transglutaminase 2 regulates Mallory body inclusion formation and injury-associated liver enlargement. Gastroenterology 132(4):1515–1526

Szondy Z, Sarang Z, Molnár P, Németh T, Piacentini M, Mastroberardino PG, Falasca L, Aeschlimann D, Kovács J, Kiss I, Szegezdi E, Lakos G, Rajnavölgyi E, Birckbichler PJ, Melino G, Fésüs L (2003) Transglutaminase $2^{-/-}$ mice reveal a phagocytosis-associated crosstalk between macrophages and apoptotic cells. Proc Natl Acad Sci U S A 100 (13):7812–7817

Szondy Z, Mastroberardino PG, Váradi J, Farrace MG, Nagy N, Bak I, Viti I, Wieckowski MR, Melino G, Rizzuto R, Tósaki A, Fesus L, Piacentini M (2006) Tissue transglutaminase (TG2) protects cardiomyocytes against ischemia/reperfusion injury by regulating ATP synthesis. Cell Death Differ 13(10):1827–1829

Taft RA, Davisson M, Wiles MV (2006) Know thy mouse. Trends Genet 22(12):649–653

Tarantino U, Oliva F, Taurisano G, Orlandi A, Pietroni V, Candi E, Melino G, Maffulli N (2009) FXIIIA and TGF-β over-expression produces normal musculo-skeletal phenotype in $TG2^{-/-}$ mice. Amino Acids 36(4):679–684

Tatsukawa H, Fukaya Y, Frampton G, Martinez-Fuentes A, Suzuki K, Kuo TF, Nagatsuma K, Shimokado K, Okuno M, Wu J, Iismaa S, Matsuura T, Tsukamoto H, Zern MA, Graham RM, Kojima S (2009) Role of transglutaminase 2 in liver injury via cross-linking and silencing of transcription factor Sp1. Gastroenterology 135(5):1783–1795

Thorp E, Tabas I (2009) Mechanisms and consequences of efferocytosis in advanced atherosclerosis. J Leukoc Biol 86(5):1089–1095

Tong L, Png E, AiHua H, Yong SS, Yeo HL, Riau A, Mendoz E, Chaurasia SS, Lim CT, Yiu TW, Iismaa SE (2013) Molecular mechanism of transglutaminase-2 in corneal epithelial migration and adhesion. Biochim Biophys Acta 1833(6):1304–1315

Tóth B, Garabuczi E, Sarang Z, Vereb G, Vámosi G, Aeschlimann D, Blaskó B, Bécsi B, Erdõdi F, Lacy-Hulbert A, Zhang A, Falasca L, Birge RB, Balajthy Z, Melino G, Fésüs L, Szondy Z (2009a) Transglutaminase 2 is needed for the formation of an efficient phagocyte portal in macrophages engulfing apoptotic cells. J Immunol 182(4):2084–2092

Tóth B, Sarang Z, Vereb G, Zhang A, Tanaka S, Melino G, Fésüs L, Szondy Z (2009b) Over-expression of integrin beta3 can partially overcome the defect of integrin beta3 signaling in transglutaminase 2 null macrophages. Immunol Lett 126(1–2):22–28

Tucholski J, Roth KA, Johnson GV (2006) Tissue transglutaminase overexpression in the brain potentiates calcium-induced hippocampal damage. J Neurochem 97(2):582–594

van den Akker J, Tuna BG, Pistea A, Sleutel AJ, Bakker EN, van Bavel E (2012) Vascular smooth muscle cells remodel collagen matrices by long-distance action and anisotropic interaction. Med Biol Eng Comput 50(7):701–715

Van Herck JL, Schrijvers DM, De Meyer GR, Martinet W, Van Hove CE, Bult H, Vrints CJ, Herman AG (2010) Transglutaminase 2 deficiency decreases plaque fibrosis and increases plaque inflammation in apolipoprotein-E-deficient mice. J Vasc Res 47(3):231–240

Van Strien ME, Baron W, Bakker EN, Bauer J, Bol JG, Brevé JJ, Binnekade R, Van Der Laarse WJ, Drukarch B, Van Dam AM (2011) Tissue transglutaminase activity is involved in the differentiation of oligodendrocyte precursor cells into myelin-forming oligodendrocytes during CNS remyelination. Glia 59(11):1622–1634

Williams H, Pease RJ, Newell LM, Cordell PA, Graham RM, Kearney MT, Jackson CL, Grant PJ (2010) Effect of transglutaminase 2 (TG2) deficiency on atherosclerotic plaque stability in the apolipoprotein E deficient mouse. Atherosclerosis 210(1):94–99

Xu L, Begum S, Hearn JD, Hynes RO (2006) GPR56, an atypical G protein-coupled receptor, binds tissue transglutaminase, TG2, and inhibits melanoma tumor growth and metastasis. Proc Natl Acad Sci U S A 103(24):9023–9028

Yang L, Friedland S, Corson N, Xu L (2014) GPR56 inhibits melanoma growth by internalizing and degrading its ligand TG2. Cancer Res 74(4):1022–1031

Zhang Z, Vezza R, Plappert T, McNamara P, Lawson JA, Austin S, Pratico D, Sutton MS-J, FitzGerald GA (2003) COX-2 dependent cardiac failure in Gh/tTG transgenic mice. Circ Res 92(10):1153–1161

Chapter 5
Transglutaminase in Invertebrates

Toshio Shibata and Shun-ichiro Kawabata

Abstract In mammals, transglutaminases play roles in a variety of essential functions, including blood coagulation, skin formation, and signal transduction, by catalyzing the isopeptide bond formation between Lys and Gln residues to form ε-(γ-glutamyl) lysine bonds between appropriate substrates. Similarly, in invertebrates, such as the horseshoe crab *Tachypleus tridentatus*, the crayfish *Pacifastacus leniusculus*, and the fruit fly *Drosophila melanogaster*, transglutaminase is conserved and exhibits pleiotropic functions. In 1993, the first report of a nucleotide sequence of the intracellular transglutaminase in *T. tridentatus* was made. Today, the functions of invertebrate transglutaminases are widely investigated using biochemical and genetic techniques. Here we review the existing knowledge of invertebrate transglutaminases with an emphasis on the importance of various physiological properties in innate immune reactions.

Keywords Invertebrate • Horseshoe crab • Clotting • *Drosophila*

5.1 Roles of Transglutaminase-Dependent Protein Cross-Linking of a Clottable Protein in Horseshoe Crabs

The horseshoe crab *T. tridentatus* is one of the most investigated animals in the research field of invertebrate transglutaminases. Horseshoe crab transglutaminase is functionally and structurally similar to the mammalian transglutaminase 2. It contains 764 amino acid residues with a unique N-terminal extension sequence of 60 residues without a secretion signal sequence (Tokunaga et al. 1993). No transglutaminase activity has been found in plasma, but transglutaminase has been observed in various tissues, and is known to be mainly localized to the hemocyte cytosol (Tokunaga et al. 1993). Hemocytes release transglutaminase into the extracellular fluid in response to stimulation with lipopolysaccharide (LPS) on Gram-negative bacteria, but the molecular mechanism of the secretion remains unknown (Osaki et al. 2002). In horseshoe crabs, a hemolymph coagulation cascade consisting of four serine protease zymogens is triggered by LPS or β-1,3-D-

T. Shibata • S.-i. Kawabata (✉)
Faculty of Science, Department of Biology, Kyushu University, Fukuoka, 819-0395 Japan
e-mail: skawascb@kyudai.jp

K. Hitomi et al. (eds.), *Transglutaminases*, DOI 10.1007/978-4-431-55825-5_5

glucans. At the final stage of the coagulation cascade, a serine protease named the clotting enzyme cleaves a procoagulant protein, coagulogen, into a two-chain form of coagulin, with release of the fragment peptide C. Coagulin monomers associate non-covalently in a head-to-tail manner and the interaction between coagulin monomers does not require Ca^{2+} or Mg^{2+} (Bergner et al. 1996; Kawasaki et al. 2000). The three-dimensional structure of coagulogen suggests a possible polymerization mechanism, by which the release of the peptide C would expose a hydrophobic cove on the "head"; this cove would then interact with the hydrophobic "tail" of a second molecule, resulting in formation of a coagulin homopolymer. Not only the coagulin monomer but also coagulogen could be incorporated into a coagulin fiber, which would subsequently be converted to coagulin by the clotting enzyme, leading to an extension of the fiber (Fig. 5.1). However, horseshoe crab transglutaminase catalyzes neither homopolymerization of coagulin nor monodansylcadaverine incorporation into coagulin. On the other hand, clots from the American horseshoe crab *Limulus polyphemus* contain products cross-linked by transglutaminase (Wilson et al. 1992). Therefore, coagulin may be cross-linked by transglutaminase with other proteins. We have identified two hemocyte-derived proteins – the proline-rich protein proxin (Osaki et al. 2002) and the cysteine-rich protein stablin (Matsuda et al. 2007b) – that are localized in the large granules of hemocytes and are secreted by LPS-induced exocytosis. Proxin interacts non-covalently with coagulin, and then transglutaminase cross-links the two proteins to form a high molecular weight product. Moreover, proxin also interacts with stablin with $K_d = 4.0 \times 10^{-9}$ M and transglutaminase covalently cross-links proxin and stablin. In addition, stablin interacts with LPS and lipoteichoic acids on the surface of bacteria, and exhibits bacterial agglutinating activity. This transglutaminase-mediated formation of coagulin-proxin-stablin complexes promotes the clotting mesh and the immobilization of invading microbes at injury sites (Fig. 5.1). In mammals, proline-rich proteins such as cornifins and small proline-rich proteins are cross-linked by transglutaminase and involved in the formation of the cornified cell envelope, which is a highly insoluble structure in the outermost layer of the skin. This rigid structure of skin serves as a frontline physical barrier against invading pathogens in the manner of the exoskeleton in horseshoe crabs. Although cornifins have no significant sequence similarity to proxin, they have an N-terminal glutamine-rich portion and proline-containing tandem repeats. Cornifins function as amine acceptors through glutamine residues of the N-terminal portion. A similar glutamine cluster is also present at the N- and C-terminal regions of proxin.

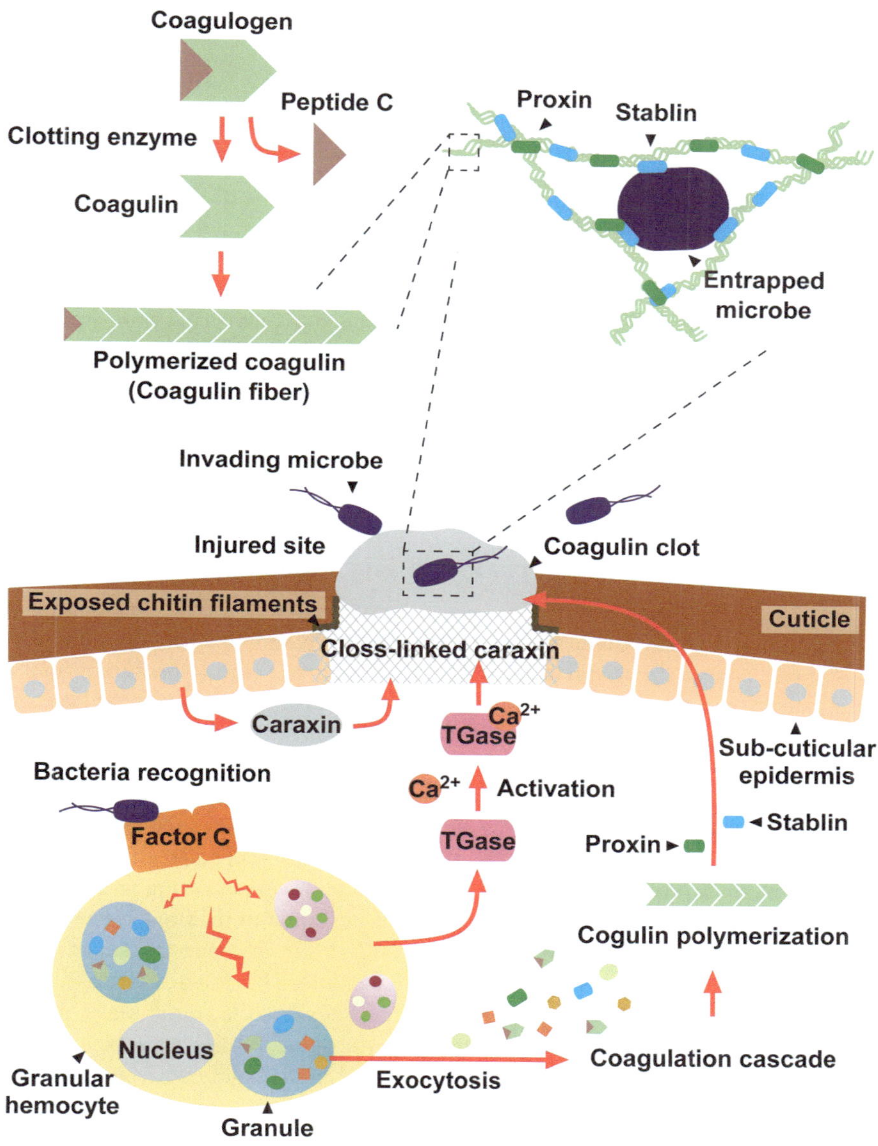

Fig. 5.1 A hypothetical scheme for transglutaminase-dependent hemolymph coagulation and wound healing at injured sites in horseshoe crabs. In response to stimulation with LPS, transglutaminase is secreted from hemocytes at injured sites and immediately activated by Ca^{2+} in plasma, which catalyzes cross-linking of caraxins localized in the sub-cuticular epithelial cells. Caraxin mesh may function to seal the wound, while also serving as a barrier to pathogen entry. In addition, transglutaminase-mediated formation of coagulin-proxin-stablin complexes promotes the clotting mesh and the immobilization of invading microbes at injury sites

5.2 Cuticular Formation and Wound Repair in Horseshoe Crabs

The arthropod cuticle of the exoskeleton mainly acts as a barrier against invading microbes, and is composed of chitin filaments, chitin-binding proteins, lipids, catecholamine derivatives, and minerals. Cuticular proteins of the horseshoe crab *T. tridentatus* were extracted and separated by two-dimensional SDS-PAGE. The N-terminal amino acid sequencing analysis revealed that there are many proteins containing a chitin-binding motif, the so-called Rebers and Riddiford (R & R) consensus (Iijima et al. 2005). Interestingly, monodansylcadaverine is incorporated into most of these cuticular R & R consensus-containing proteins, indicating that transglutaminase-dependent cross-linking is important for cuticular barrier formation in arthropods, and that mammals and arthropods share an evolutionarily conserved molecular system for skin/cuticular formation.

5.3 The Wound Repair System in Horseshoe Crabs

An epidermal barrier wound repair pathway has been shown to be evolutionarily conserved between *Drosophila* and mice. In *Drosophila*, the transcription factor grainy head regulates the production of enzymes such as dopa decarboxylase and tyrosine hydroxylase, which are involved in covalent cross-linking of cuticular structural components (Mace et al. 2005). Mice lacking a homolog of *Drosophila* grainy head (grainy head like-3) display defective skin barrier function and deficient wound repair. These phenotypes arise due to a reduction in the expression of transglutaminase-1 (Ting et al. 2005). Transglutaminase-dependent cross-linking plays an important role in host defense in the cuticle of horseshoe crabs, analogous to that observed in the epidermal cornified cell envelope in mammals. Horseshoe crab hemocytes are actively motile (Armstrong 1980), and one of the principal functions of the hemocyte is to seal scars in the cuticle. This function is fulfilled in part by the adherence of hemocytes to injured sites. At the initial stage of this wound repair process sufficient quantities of transglutaminase may be secreted from hemocytes recruited to the site of injury, and immediately activated by Ca^{2+} in hemolymph plasma. Horseshoe crab transglutaminase cross-links carapace-derived chitin-binding proteins, caraxins, which are specifically localized to the sub-cuticular epidermis (Matsuda et al. 2007a). One of these members, caraxin-1, comprises 135 amino acid residues and consists of N- and C-terminal heptad repeats that flank a central domain consisting of a pentapeptide tandem repeat structure. Wild-type caraxin-1 expressed in *Escherichia coli* exists as an oligomer of ~20-mer in solution, and a plurality of these oligomers are cross-linked by transglutaminase to form honeycomb-like mesh structures. The LPS-induced clotting fibrils and caraxin mesh may function to seal the wound to stop bleeding, while also serving as a barrier to pathogen entry, thereby facilitating wound healing

(Fig. 5.1). In addition, stablin also binds to chitin ($K_d = 1.5 \times 10^{-8}$ M), suggesting that stablin promotes not only the formation of stable clotting fibrils but also the immobilization of invading microbes at injured sites.

5.4 Transglutaminase in Crustaceans

Unlike the horseshoe crab *T. tridentatus*, in crustaceans, such as crayfish and shrimp, clotting is directly caused by the activity of transglutaminase against clottable proteins. Transglutaminase is released by hemocytes or muscle cells upon wounding and cross-links clottable hemolymph proteins such as the clotting protein (CP), which was found to form clots after injury (Kopácek et al. 1993; Hall et al. 1999). CPs have been isolated from a number of crustacean species, including a crayfish (Kopácek et al. 1993; Hall et al. 1999), shrimp (Yeh et al. 1998; Hall et al. 1999) and sand crayfish (Komatsu and Ando 1998). CPs are often large (~200 kDa) lipoproteins that are homologous to vitellogenins and exist in hemolymph in a soluble form. CPs are cross-linked by transglutaminase without cleaving. The proteinase inhibitor α_2-macroglobulin is also a substrate for crayfish transglutaminase (Hall and Söderhäll 1994). α_2-Macroglobulin-incorporated CP clotting may contribute to the resistance against proteolytic degradation caused by proteinases from invading pathogens trying to penetrate through the wound site into the body cavity.

5.5 Transglutaminase Is Involved in Cuticular Morphogenesis in *Drosophila*

Drosophila melanogaster is commonly used as a powerful animal model for studying innate immune responses. Recently, several groups have reported that *Drosophila* transglutaminase is involved in many physiological functions just as transglutaminase in mammals: these include hemolymph coagulation, cuticular morphogenesis, and antibacterial response. However, while the transglutaminase family in mammals comprises eight members, *Drosophila* have only a single transglutaminase gene that encodes a protein of 87 kDa (CG7356) (Theopold et al. 2002). Comparisons of the amino acid sequences of *Drosophila* transglutaminase with those of the horseshoe crab *T. tridentatus* and human transglutaminase have indicated that *Drosophila* TG shows about 30% identity and 50% similarity to these other enzymes, respectively. The authentic catalytic triad is conserved in *Drosophila* transglutaminase and Ca^{2+} is also required for its activation. The mechanisms underlying the expression of *Drosophila* transglutaminase have not been elucidated, but it is known that transglutaminase expression is upregulated upon injury (Shibata et al. 2010). Moreover, in cardiac cells, myocyte enhancer factor-2 (MEF2) directly regulates *transglutaminase* gene expression (Ikle et al. 2008). To investigate the physiological functions of transglutaminase *in vivo*, several genetic tools have been used, including GAL4/UAS-mediated overexpression and RNAi systems. The overexpression of transglutaminase in the eye imaginal disc results in a rough eye phenotype in adult flies. This phenotype is suppressed by the co-expression of the apoptosis inhibitor P35, suggesting that the rough eye phenotype is caused by apoptosis (Umehara et al. 2010). Furthermore, the overexpression of transglutaminase in the wing imaginal disc induces an extra wing crossvein phenotype in adult wings (Ichikawa et al. 2010). In contrast, following the RNAi of *transglutaminase*, approximately 80% of flies exhibited a pupal lethal phenotype (Shibata et al. 2010). Even after eclosion, *transglutaminase*-RNAi flies showed a wing blister phenotype and the disappearance of several melanized segments on the abdominal tergite. Also, *transglutaminase*-RNAi flies showed a significantly shorter life span than did their wild-type counterparts, with approximately 90% of the flies dying within 30 days of eclosion. However, RNAi after the late pupal stage did not cause morphological disorders, indicating that transglutaminase plays a key role at or before the early pupal stage. LC-MS/MS analysis using wing proteins of *transglutaminase*-RNAi flies revealed four proteins, two cuticular chitin-binding proteins (Cpr97Eb, Cpr76Bd), larval serum protein 2, and a putative C-type lectin (Clect27), that functioned as transglutaminase substrates. RNAi of their corresponding genes caused a lethal phenotype or cuticle abnormality.

5.6 Transglutaminase-Mediated Coagulation and Host Defense in *Drosophila*

Clotting in insects is critical to limiting hemolymph loss and initiating wound healing, as in mammals. *Drosophila* transglutaminase is also involved in hemolymph coagulation, and the inhibition of transglutaminase by monodansylcadaverine and RNAi of this gene affect the clotting properties (Lindgren et al. 2008). A proteomic analysis of the *Drosophila* clot showed the presence of transglutaminase substrates and the protein-protein cross-linking activity of transglutaminase in the hemolymph (Karlsson et al. 2004). A hemocyte-specific gene, *hml*, has been demonstrated to be required for efficient clot formation. Hemolymph from *hml*-RNAi flies fails to form clot fibers or promote wound healing (Goto et al. 2003). Another clotting factor of *Drosophila*, Fondue, has been shown to be a key protein in the clots (Lindgren et al. 2008). Its gene (*fon*) is induced following injury, and *fon*-RNAi causes defects in clotting and wound closure. Interestingly, *Drosophila* transglutaminase also catalyzes protein-protein cross-linking for microbial surface substances (Wang et al. 2010). In the presence of biotin cadaverine and hemolymph, small punctate signals by fluorescence-labeled streptavidin are detected on the surfaces of bacteria and the nematodes

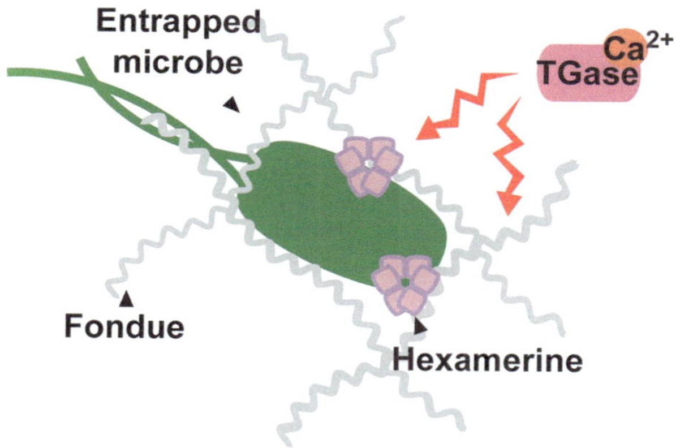

Fig. 5.2 A schematic model of transglutaminase-mediated entrapment of invading pathogens. Unknown transglutaminase substrates located on the surfaces of microbes could be subsequently cross-linked to humoral protein hexamerin and proteins in the clot, such as Fondue

Heterorhabditis bacteriophora. Analysis of bacterial lysates after incubation with biotin cadaverine and hemolymph revealed that a humoral procoagulant, hexamerin, is bound to bacteria through transglutaminase activity. Hexamerin may be assembled with bacterial surface proteins by transglutaminase-dependent cross-linking. Clots isolated from bacteria-injected *transglutaminase*-RNAi larvae contain fewer bacteria than those of control larvae. Transglutaminase substrates located on the surfaces of microbes could be subsequently cross-linked to proteins in the clot, such as Fondue (Wang et al. 2010). Thus, transglutaminase-mediated cross-linking is essential for the entrapment of invading pathogens (Fig. 5.2).

5.7 Maintenance of Commensal Bacteria in *Drosophila*

In the gut of *Drosophila*, antimicrobial peptides such as diptericin and cecropin play a pivotal role in sterilizing invading or commensal bacteria. The production of antimicrobial peptides is regulated by the immune deficiency (IMD) pathway, and this pathway is activated by peptidoglycans derived from Gram-negative bacteria. The signal is transduced through the receptor peptidoglycan recognition protein (PGRP)-LC, which is expressed on gut epithelial cells, and then through its adaptor protein IMD, and finally, one of the nuclear factor (NF)- κB-like transcription factors in *Drosophila*, Relish, is proteolytically cleaved by a caspase-8-homolog, Dredd, to translocate the N-terminal region of Relish into the nuclei (Royet and Dziarski 2007; Lemaitre and Hoffmann 2007). This signaling pathway is triggered by peptidoglycans derived from not only pathogenic bacteria but also commensal bacteria, which causes the overproduction of antimicrobial peptides and affects commensal homeostasis (Ryu et al. 2008). Through the investigation of this pathway, several systems that suppress the IMD pathway have been reported (Zaidman-Remy et al. 2006; Kleino et al. 2008; Basbous et al. 2011; Bosco-Drayon et al. 2012; Paredes et al. 2011; Shibata et al. 2013; Ryu et al. 2008). Recently, we have found that transglutaminase is involved in negative regulation of the IMD pathway (Shibata et al. 2013). RNAi of the *transglutaminase* gene in the gut reduces the life span of flies and enhances the expression of genes encoding antimicrobial peptides of the IMD pathway. Conversely, those phenotypes do not arise under germ-free conditions. Wild-type flies that ingested gut lysates prepared from conventionally reared *transglutaminase*-RNAi flies exhibit shorter life spans. The overproduction of antimicrobial peptides in conventionally reared *transglutaminase*-RNAi flies results in changes in the composition of the commensal community. *Drosophila* transglutaminase catalyzes polymerization of the active form of the N-terminal portion of Relish, but not the full-length Relish, by protein-protein cross-linking to inhibit nuclear translocation and thereby suppress the expression of genes encoding IMD-controlled antimicrobial peptides. Under non-pathogenic conditions in the gut, transglutaminase-mediated protein-protein cross-linking is important to suppress an excess antimicrobial reaction and to enable immune tolerance against commensal microbes (Fig. 5.3).

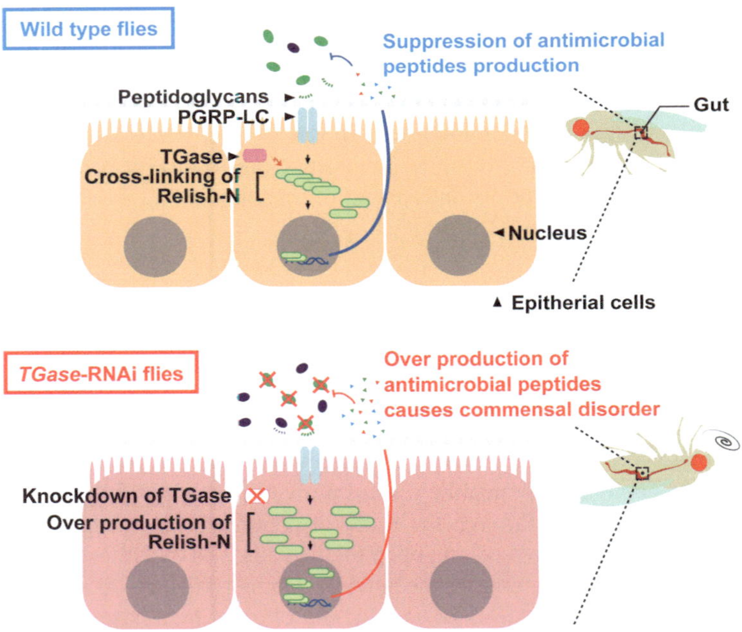

Fig. 5.3 A schematic model of transglutaminase-mediated gut homeostasis in *Drosophila*. Under conventionally reared conditions, transglutaminase suppresses an excess expression of antimicrobial peptides by inactivation of transcriptional factor Relish. This suppression enables immune tolerance against commensal microbes. In the case of transglutaminase-RNAi flies, overproduction of antimicrobial peptides would result in the disorganization of commensal microbes and a short life span of flies

5.8 Transglutaminase in *Anopheles gambiae*

Collectively, the genomes of mosquito species include three genes for transglutaminases. In the malaria vector mosquito, *Anopheles gambiae*, AGAP009100 or *AgTG1* is thought to be a homolog to *Drosophila* transglutaminase. *Culex* and *Anopheles* mosquitoes share a second *transglutaminase* gene, AGAP009098 or *AgTG2*, which is involved in the immune response to *Plasmodium falciparum*. *A. gambiae* has a third *transglutaminase*, *AgTG3*, which is the most characterized transglutaminase in *A. gambiae*. AgTG3 forms a dimer in solution and exhibits Ca^{2+}-dependent nonproteolytic activation, analogous to mammalian cytoplasmic transglutaminase, factor XIII (Le et al. 2013). Western blot analysis revealed that AgTG1 and AgTG2 are ubiquitously expressed at low levels, whereas AgTG3 is expressed exclusively in the male accessory glands. AgTG3 is a male-specific transglutaminase and coagulates seminal fluids to form the mating plug required for many aspects of the female physiology and behavior, including longevity, egg production, sperm storage, and remating. A substrate for AgTG3, plugin, is a key factor for this coagulation of seminal fluids (Le et al. 2013; Rogers et al. 2009). Formation of the mating plug by cross-linking plugin is necessary for efficient sperm storage in females. These data clearly indicate that AgTG3 and plugin play crucial roles in reproduction, and that the enzyme and substrate would be good targets for the extermination of malaria.

References

Armstrong PB (1980) Adhesion and spreading of *Limulus* blood cells on artificial surfaces. J Cell Sci 44:243–262

Basbous N, Coste F, Leone P, Vincentelli R, Royet J, Kellenberger C, Roussel A (2011) The *Drosophila* peptidoglycan-recognition protein LF interacts with peptidoglycan-recognition protein LC to down-regulate the Imd pathway. EMBO Rep 12(4):327–333

Bergner A, Oganessyan V, Muta T, Iwanaga S, Typke D, Huber R, Bode W (1996) Crystal structure of a coagulogen, the clotting protein from horseshoe crab: a structural homologue of nerve growth factor. EMBO J 15(24):6789–6797

Bosco-Drayon V, Poidevin M, Boneca IG, Narbonne-Reveau K, Royet J, Charroux B (2012) Peptidoglycan sensing by the receptor PGRP-LE in the *Drosophila* gut induces immune responses to infectious bacteria and tolerance to microbiota. Cell Host Microbe 12(2):153–165

Goto A, Kadowaki T, Kitagawa Y (2003) *Drosophila hemolectin* gene is expressed in embryonic and larval hemocytes and its knock down causes bleeding defects. Dev Biol 264(2):582–591

Hall M, Söderhäll K (1994) Crayfish α-macroglobulin as a substrate for transglutaminases. Comp Biochem Physiol Part B Comp Biochem 108(1):65–72

Hall M, Wang R, van Antwerpen R, Sottrup-Jensen L, Söderhäll K (1999) The crayfish plasma clotting protein: a vitellogenin-related protein responsible for clot formation in crustacean blood. Proc Natl Acad Sci U S A 96(5):1965–1970

Ichikawa A, Yamada A, Sakamoto H, Umehara M, Yoshioka Y, Yamaguchi M, Ikura K (2010) Overexpression of transglutaminase in the *Drosophila* wing imaginal disc induced an extra wing crossvein phenotype. Biosci Biotechnol Biochem 74(12):2494–2496

Iijima M, Hashimoto T, Matsuda Y, Nagai T, Yamano Y, Ichi T, Osaki T, Kawabata S (2005) Comprehensive sequence analysis of horseshoe crab cuticular proteins and their involvement in transglutaminase-dependent cross-linking. FEBS J 272(18):4774–4786

Ikle J, Elwell JA, Bryantsev AL, Cripps RM (2008) Cardiac expression of the *Drosophila Transglutaminase* (CG7356) gene is directly controlled by myocyte enhancer factor-2. Dev Dyn Off Publ Am Assoc Anatomists 237(8):2090–2099

Karlsson C, Korayem AM, Scherfer C, Loseva O, Dushay MS, Theopold U (2004) Proteomic analysis of the *Drosophila* larval hemolymph clot. J Biol Chem 279(50):52033–52041

Kawasaki H, Nose T, Muta T, Iwanaga S, Shimohigashi Y, Kawabata S (2000) Head-to-tail polymerization of coagulin, a clottable protein of the horseshoe crab. J Biol Chem 275 (45):35297–35301

Kleino A, Myllymaki H, Kallio J, Vanha-aho LM, Oksanen K, Ulvila J, Hultmark D, Valanne S, Ramet M (2008) Pirk is a negative regulator of the *Drosophila* IMD pathway J Immunol 180 (8):5413–5422 (Baltimore, Md.: 1950)

Komatsu M, Ando S (1998) A very-high-density lipoprotein with clotting ability from hemolymph of sand crayfish, *Ibacus ciliatus*. Biosci Biotechnol Biochem 62(3):459–463

Kopácek P, Hall M, Söderhäll K (1993) Characterization of a clotting protein, isolated from plasma of the freshwater crayfish *Pacifastacus leniusculus*. Eur J Biochem/FEBS 213 (1):591–597

Le BV, Nguyen JB, Logarajah S, Wang B, Marcus J, Williams HP, Catteruccia F, Baxter RH (2013) Characterization of *Anopheles gambiae* transglutaminase 3 (AgTG3) and its native substrate Plugin. J Biol Chem 288(7):4844–4853

Lemaitre B, Hoffmann J (2007) The host defense of *Drosophila melanogaster*. Annu Rev Immunol 25:697–743

Lindgren M, Riazi R, Lesch C, Wilhelmsson C, Theopold U, Dushay MS (2008) Fondue and transglutaminase in the *Drosophila* larval clot. J Insect Physiol 54(3):586–592

Mace KA, Pearson JC, McGinnis W (2005) An epidermal barrier wound repair pathway in *Drosophila* is mediated by grainy head. Science 308(5720):381–385 (New York, NY)

Matsuda Y, Koshiba T, Osaki T, Suyama H, Arisaka F, Toh Y, Kawabata S (2007a) An arthropod cuticular chitin-binding protein endows injured sites with transglutaminase-dependent mesh. J Biol Chem 282(52):37316–37324

Matsuda Y, Osaki T, Hashii T, Koshiba T, Kawabata S (2007b) A cysteine-rich protein from an arthropod stabilizes clotting mesh and immobilizes bacteria at injury sites. J Biol Chem 282 (46):33545–33552

Osaki T, Okino N, Tokunaga F, Iwanaga S, Kawabata S (2002) Proline-rich cell surface antigens of horseshoe crab hemocytes are substrates for protein cross-linking with a clotting protein coagulin. J Biol Chem 277(42):40084–40090

Paredes JC, Welchman DP, Poidevin M, Lemaitre B (2011) Negative regulation by amidase PGRPs shapes the *Drosophila* antibacterial response and protects the fly from innocuous infection. Immunity 35(5):770–779

Rogers DW, Baldini F, Battaglia F, Panico M, Dell A, Morris HR, Catteruccia F (2009) Transglutaminase-mediated semen coagulation controls sperm storage in the malaria mosquito. PLoS Biol 7(12):e1000272

Royet J, Dziarski R (2007) Peptidoglycan recognition proteins: pleiotropic sensors and effectors of antimicrobial defences, *Nature reviews*. Microbiology 5(4):264–277

Ryu JH, Kim SH, Lee HY, Bai JY, Nam YD, Bae JW, Lee DG, Shin SC, Ha EM, Lee WJ (2008) Innate immune homeostasis by the homeobox gene *caudal* and commensal-gut mutualism in *Drosophila*. Science 319(5864):777–782 (New York, NY)

Shibata T, Ariki S, Shinzawa N, Miyaji R, Suyama H, Sako M, Inomata N, Koshiba T, Kanuka H, Kawabata S (2010) Protein crosslinking by transglutaminase controls cuticle morphogenesis in *Drosophila*. PLoS One 5(10):e13477

Shibata T, Sekihara S, Fujikawa T, Miyaji R, Maki K, Ishihara T, Koshiba T, Kawabata S (2013) Transglutaminase-catalyzed protein-protein cross-linking suppresses the activity of the NF-kappaB-like transcription factor Relish. Sci Signal 6(285):ra61

Theopold U, Li D, Fabbri M, Scherfer C, Schmidt O (2002) The coagulation of insect hemolymph. Cell Mol Life Sci CMLS 59(2):363–372

Ting SB, Caddy J, Hislop N, Wilanowski T, Auden A, Zhao LL, Ellis S, Kaur P, Uchida Y, Holleran WM, Elias PM, Cunningham JM, Jane SM (2005) A homolog of *Drosophila* grainy head is essential for epidermal integrity in mice. Science 308(5720):411–413 (New York, NY)

Tokunaga F, Muta T, Iwanaga S, Ichinose A, Davie EW, Kuma K, Miyata T (1993) *Limulus* hemocyte transglutaminase. cDNA cloning, amino acid sequence, and tissue localization. J Biol Chem 268(1):262–268

Umehara M, Ichikawa A, Sakamoto H, Yamada A, Yoshioka Y, Yamaguchi M, Ikura K (2010) Over-expression of transglutaminase in the *Drosophila* eye imaginal disc induces a rough eye phenotype. Mol Cell Biochem 342(1–2):223–232

Wang Z, Wilhelmsson C, Hyrsl P, Loof TG, Dobes P, Klupp M, Loseva O, Morgelin M, Ikle J, Cripps RM, Herwald H, Theopold U (2010) Pathogen entrapment by transglutaminase – a conserved early innate immune mechanism. PLoS Pathog 6(2):e1000763

Wilson J, Rickles FR, Armstrong PB, Lorand L (1992) N epsilon(gamma-glutamyl)lysine crosslinks in the blood clot of the horseshoe crab, *Limulus polyphemus*. Biochem Biophys Res Commun 188(2):655–661

Yeh MS, Chen YL, Tsai IH (1998) The hemolymph clottable proteins of tiger shrimp, *Penaeus monodon*, and related species. Comp Biochem Physiol Part B Biochem Mol Biol 121 (2):169–176

Zaidman-Remy A, Herve M, Poidevin M, Pili-Floury S, Kim MS, Blanot D, Oh BH, Ueda R, Mengin-Lecreulx D, Lemaitre B (2006) The *Drosophila* amidase PGRP-LB modulates the immune response to bacterial infection. Immunity 24(4):463–473

Chapter 6
A New Integrin-Binding Site on a Transglutaminase-Catalyzed Polymer

Yasuyuki Yokosaki

Abstract Transglutaminase crosslinks Gln and Lys residues intra or inter proteins. Inter protein crosslinks result in anchoring of a monomeric protein to large molecules, such as the extracellular matrix, or in the generation of polymers of the monomeric substrates. Such enzymatically crosslinked polymers exert biological functions, including hemostasis, apoptosis and chemotaxis. However, many of the molecular mechanisms by which these polymers function are still not entirely clear. Although evidence concerning these functional polymers has been reported, some pieces to complete the whole picture may be missing. In this example, we describe the generation of a protein interaction site on the polymer. Binding of this site to an integrin receptor elicits a new function, chemoattractant for neutrophils, that has not previously been associated with the monomer. Formation of the specific conformational site is supported by the presence of isopeptide bonding-specific monoclonal antibodies and is further supported by descriptions of oligomer-specific antibodies for amyloid proteins. Here we describe the interaction of polymeric osteopontin with integrin α9β1 in comparison with other active polymers.

Keyword Integrin α9β1 • Transglutaminase • Osteopontin • Polymer • Neutrophil • Chemotaxis • Thrombin • Focal contact • Migration

6.1 Protein Polymerization by Transglutaminase

Transglutaminases, a family of enzymes that covalently link the side chains of Gln and Lys residues by forming isopeptide bonds, crosslink a variety of proteins in a substrate-specific fashion (Mycek et al. 1959; Kornguth and Waelsch 1963; Lorand and Graham 2003; Sugimura et al. 2006). The biochemistry of this acyl-transfer reaction has been characterized (Folk and Cole 1965; Folk 1969). The most well-known physiological transglutaminase reaction is the crosslinking of fibrinogen to form a clot. This reaction is catalyzed by factor XIII, a member of the transglutaminase family (Lorand et al. 1966; Matacic and Loewy 1966). Many other

Y. Yokosaki, M.D., Ph.D. (✉)
Cell-Matrix Frontier Laboratory, Health Administration Center, 216 Biomedical Research Building, Hiroshima University, 1-2-3 Kasumi, Minamiku, Hiroshima 734-8551, Japan
e-mail: yokosaki@hiroshima-u.ac.jp

K. Hitomi et al. (eds.), *Transglutaminases*, DOI 10.1007/978-4-431-55825-5_6

physiological and pathological roles of transglutaminase-induced polymers are still under investigation. The early investigations of protein crosslinking activity focused primarily on extracellular matrix proteins. By crosslinking macromolecular matrix proteins, such as fibronectin (Mosher 1977, 1978; Jones et al. 1997), collagen (Mosher and Schad 1979; Mosher et al. 1979; Jelenska et al. 1980), and laminin (Aeschlimann and Paulsson 1991), transglutaminases stabilize extracellular tissues, conferring rigidity, organization, and integration on these structures. One of the most critical regulators of extracellular matrix protein expression is transforming growth factor beta (TGF-β), which is stored within matrix proteins as a "large latent complex", in a non-active form. When initially secreted from cells, TGF-β is covered with its propeptide latency associated protein (LAP) through disulfide bonds, forming a "small latent complex". This complex is linked to latent TGF-β associated protein (LTBP-1) by disulfide bonding and is stored in extracellular matrix proteins. This elaborate storage, together with integrin contact to the small latent complex, allow the matrix environment to release TGF-β in a active form as soon as it is required. Integrin binding to the Arg-Gly-Asp (RGD) tripeptide sequence contained within LAP creates the tensile force and removes LAP from the small latent complex (Munger et al. 1999). Cell contraction or other tissue distortions initiate the tensile force, for which fixation of LTBP-1 in place is required. LTBP-1 serves as a substrate for transglutaminase and is crosslinked to the surrounding matrix proteins or polymerized (Nunes et al. 1997; Verderio et al. 1999). Thus, transglutaminase-induced polymerization plays a critical role in the regulation of TGF-β function.

The above polymers fibrin and LTBP-1 physically exert their functions straightforwardly. There are, however, many other polymers that look even more intelligent, bringing diverse biological consequences using versatile techniques. They elevate their activities, protect themselves from proteolysis, and bind to new receptors; or inversely, they stop binding to previous partners. In the following sections, I will illustrate several transformations that occur through polymerizations catalyzed by transglutaminase. Then, I will focus on osteopontin (Oldberg et al. 1986; Patarca et al. 1989), a particular substrate for polymerization, and describe how it forms a new binding site for an integrin and gains a new function as a chemoattractant for neutrophils (Nishimichi et al. 2009, 2011).

6.2 Functional Changes of the Polymers

Proteins with active functions, such as enzymes, ligands, receptors, channels, signaling molecules, transcription factors and antibodies, generally bind to specific binding sites on other biomolecules, including proteins, nucleic acids, carbohydrates, or lipids, to exert their functions. There are several such functional proteins that serve as substrates for transglutaminase-induced polymerization (Table 6.1). When these proteins are polymerized, their previously exposed monomer surfaces become partially buried within the interface of the polymer. Local conformations

Table 6.1 Functional transglutaminase-catalyzed polymers

Substrate	Physiological function	Changes in function	Role of TGase	Biological consequence	Mode of action	References
AT1	GPCR [a]	Up	Dimerize	Monocyte activation	Clustering of receptor	AbdAlla et al. (2004)
ATII	GPCR	Up	Dimerize	Associated with AD	$G_{\alpha q}$ suppression	AbdAlla et al. (2009)
Cystatin	Protease inhibitor	New	Polymerize	Stabilize amyloid	Unknown	von Horsten et al. (2007)
DLK	Kinase	Up	Polymerize	Apoptosis	JNK activation	Robitaille et al. (2004)
Fibrinogen	Coagulation factor	New	Polymerize	Hemostasis	Clot formation	Lorand et al. (1968)
Galectin-3	Lectin	Unknown	Polymerize	Melanoma cell spreading	Unknown	van den Brûle et al. (1998)
Hedgehog	Ligand for patched	Up	Polymerize	Chondrocyte differentiation	Unknown	Dierker et al. (2009)
Heat shock protein	Chaperon	Unknown	Polymerize	Associated with amyloid β	Unknown	Boros et al. (2004)
Keratin 8/18	Intermediate filament	Unknown	Polymerize	Mallory body	Unknown	Strnad et al. (2007)
LTBP-1[b]	TGFβ-associated	Enhanced	Anchoring	TGFβ storage	Enhance anchoring	Nunes et al. (1997)
Midkine	Growth factor	Up	Polymerize	Monocyte chemotaxis	Unknown	Kojima et al. (1997)
Osteopontin	ECM[c], cytokine	New	Polymerize	Neutrophil chemotaxis	Binding to integrin α9β1	Nishimichi et al. (2011)
Rb	Cell cycle	Prolong	Polymerize	Anti-apoptotic	Resistant to caspase degradation	Boehm et al. (2002)
S100A4	Ca^{2+}-binding	Up	Polymerize	Mammary tumor migration	Unknown	(Ruse et al. 2001)
S19	Ribosomal protein	Up	Polymerize	Monocyte chemotaxis	Unknown	Nishimura et al. (1996)
Sp1	Transcription factor	Down	Polymerize	Pro-apoptotic	Inhibit c-Met transcription	Tatsukawa et al. (2009)
TRPV5	Ion channel	Down	Polymerize	Inhibit Ca^{2+} transport	Decrease in pore size	Boros et al. (2012)

[a]G-protein coupled receptor
[b]Polymerization is seen in figures of the reference
[c]Extracellular matrix

on the remaining surface of the polymer may be affected by physical or chemical forces, in such a way that an interaction site with a biomolecule is no longer active. Changes in transglutaminase substrates upon polymerization would thus appear to be a loss of function, rather than a gain of function.

In fact, some proteins have been shown to be down-regulated by transglutaminase 2 (TG2)-induced polymerization. The polymerization of transcription factor Sp1 was demonstrated by SDS-PAGE; then the polymer was assessed for transcriptional activity in a reporter assay involving c-Met, which is a critical transcript of Sp1 in hepatocytes. The down-regulation of Sp1 by polymerization leads to apoptosis in hepatocytes, as a mechanism underlying alcoholic hepatitis (Tatsukawa et al. 2009). Another example involves two members of the transient receptor potential vanilloid (TRPV) family of cation channels (Kedei et al. 2001; Cheng et al. 2010), TRPV4 (Mamenko et al. 2015) and TRPV5 (Boros et al. 2012). TRPV5 plays a role in Ca^{2+} reabsorption in the distal tubule of the nephron; it is expressed on the tubular cell membrane in a quaternary structure. TG2 secreted into the pro-urine polymerizes TRPV5 by crosslinking several Gln and two Lys residues in the first extracellular loop (Huynh et al. 2014). This polymerization decreases the pore size of the channel, down-regulating TRPV5 activity. Sp1 and TRPV5 are both down-regulated as a consequence of TG2 crosslinks, however, these two mechanisms are apparently distinct from each other. The Sp1 multimer appears to lose a binding activity needed to form a transcriptional complex. On the other hand, functional down-regulation of TRPV5 as a Ca^{2+} channel involves narrowing of the pore due to physical forces driven by the crosslink.

In contrast, there are many monomers that exhibit functional up-regulation upon forming dimers or polymers. Angiotensin II type 1 receptor (AT1), a G-protein coupled receptor, was dimerized on the monocytes of atherosclerotic patients, while the receptor remained a monomer in healthy controls (AbdAlla et al. 2004). The homodimers were crosslinked by factor XIII, and enhanced signals on the monocytes, compared with those expressing monomers (Thomas 2005; Iismaa et al. 2006). This result was obtained by measuring AT1-activated, $G\alpha_{q/11}$-mediated levels of inositol phosphate. On the other hand, angiotensin II type 2 receptor (AT2) inhibits downstream signaling, suppressing $G\alpha_{q/11}$-mediated signals (AbdAlla et al. 2009). AT2 undergoes oligomerization through the following two-step mechanism: it is first dimerized by reactive oxygen species and then oligomerized by TG2. The oligomers sequester $G\alpha_{q/11}$, which disrupt $G\alpha_{q/11}$ signaling. The inhibitory activity of AT2 is thus enhanced by the TG2-induced polymerization. Functional enhancement by the polymerization is further reported for species of molecules other than receptors, such as kinases, cytokines, and ribosomal proteins. Dual leucine zipper-breaking kinase (DLK), a mitogen-activated protein kinase (MAPK) kinase kinase (MAPKKK), is oligomerized by TG2, which facilitates apoptosis by elevating its kinase activity (Hebert et al. 2000; Robitaille et al. 2004). The activation leads to the initiation of downstream c-Jun transcription through c-Jun N-terminal kinase (JNK) activation. c-Jun induces apoptosis through transcription of pro-apoptotic genes such as *Bim* (Shamas-Din et al. 2015). Midkine is a 13 kDa heparin-binding growth factor that activates plasminogen activator (Kojima et al. 1995; Iwasaki et al. 1997), which is enhanced by TG2-induced dimerization (Mahoney et al. 1996; Kojima et al. 1997). Ribosomal protein S19 in blood plasma is dimerized by factor XIIIa. The dimer, but not monomer, functions as a chemoattractant for monocyte/macrophages by interacting with the C5a receptor (Nishimura et al. 1996; Chen et al. 2014). Cell cycle G1/S checkpoint regulator protein Rb, also known as a tumor suppressor protein, exhibits anti-apoptotic function, of which caspase-induced degradation is protected by a TG2-induced polymer that acquires protease resistance (Oliverio et al. 1997; Ju-Hong Jeon et al. 2003; Boehm et al. 2002).

It should be noted that polymerization up-regulates the functions of many TG2 substrate proteins, and some substrates gain a new function that is not the monomer. It appears that the above expectation that polymerization would down-regulate functions by steric hindrance or destruction of active sites accounts for a minor population of the diverse polymers. Although some polymers change their activity through mechanisms that are consistent with known concepts, for example enhancement of the signals by receptor clustering, it is surprising that, in a majority of cases, the mechanism underlying the up-regulation or gain of function is unknown. However, it is conceivable that a novel conformation that fits with binding of other biomolecules could be formed along or within the interfaces of proteins; if transglutaminase-catalyzed binding occurred in a random, unregulated fashion, the chance to form such specific binding site would be infinitely low, however, transglutaminase-induced polymerization is a regulated reaction of specific Gln and Lys residues. In the following sections, a distinct case will be introduced in which a new function is acquired by a polymer associated with a new interaction site with a cell surface receptor integrin.

Incidentally, there is a class of polymerization substrate proteins that are involved in amyloid tissues, including causative proteins of neurodegenerative diseases (Grosso and Mouradian 2012) and non-causative proteins, such as cystatin (Cornwall et al. 2011).

6.3 Osteopontin – A Distinct Transglutaminase-Catalyzed Polymer

6.3.1 Structure and Functions

Osteopontin (OPN) is an acidic, phosphorylated, extracellular matrix glycoprotein. The mature protein consisting of 298 amino acids contains both RGD and non-RGD integrin-binding sites in the middle of the molecule (Denhardt et al. 2001) (Fig. 6.1). The theoretical molecular weight is 33.7 kDa, but its mobility in SDS-PAGE gels ranges from 50 to 70 kDa, depending on phosphorylation and glycosylation. Interestingly, this secreted protein is also classified as a Th1 cytokine. As this classification, OPN plays diverse roles in a variety of biological contexts, including tissue remodeling, inflammation, tissue mineralization,

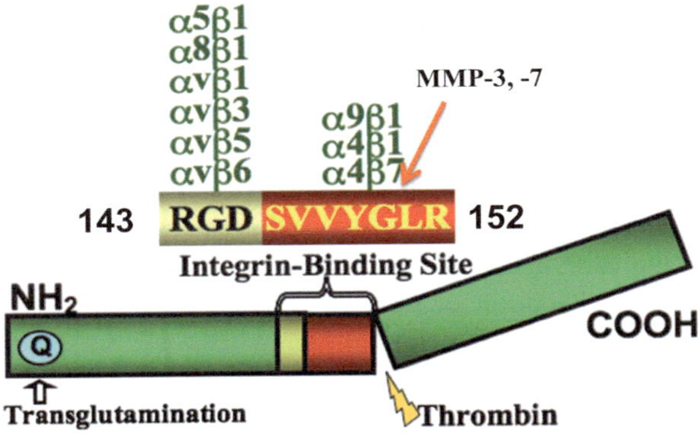

Fig. 6.1 Schematic diagram of osteopontin. There are three critical sites for osteopontin function. There are two integrin-binding sites in the middle of the molecule, the C-terminal end of which is cleaved by thrombin. The otherwise cryptic SVVYGLR sequence is exposed upon thrombin-catalyzed cleavage, allowing integrin α9β1 access. Canonical integrin-binding tripeptide RGD is immediately upstream of the SVVYGLR sequence. Six integrins bind to RGD, which does not require thrombin-catalyzed cleavage. There are a few Gln residues in the N-terminal position that serve as acyl donors for transglutaminase 2

immunomodulation, and tumor metastasis. These functions are exerted through its receptors, the integrins and CD44. At least ten integrins are known to function as OPN receptors, including α5β1 (Nasu et al. 1995), α8β1 (Denda et al. 1998), αvβ1 (Hu et al. 1995), αvβ3 (Miyauchi et al. 1991), αvβ5 (Hu et al. 1995), αvβ6 (Yokosaki et al. 2005), α9β1 (Smith et al. 1996), α4β1 (Bayless et al. 1998), α4β7 (Green et al. 2001), and αXβ2 (Schack et al. 2009). Interactions with these receptors are influenced by phosphorylation and glycosylation (Christensen et al. 2007). In addition, OPN serves as a substrate for many proteases, including thrombin, matrix metalloproteinases, cathepsin D, and plasmin, with at least 20 cleavage sites (Christensen et al. 2010). The thrombin cleavage site at 152R–153S was the first of these to be discovered; the resultant N-terminal fragment (nOPN) is the most characterized of the cleaved fragments (Senger et al. 1994). Cleavage of nOPN results in an up-regulated capacity for cell adhesion. nOPN also serves as a niche for glioma cells (Yamaguchi et al. 2013), is a marker of osteoarthritis (Hasegawa et al. 2011), and gains a new role as a neutrophil chemoattractant (Nishimichi et al. 2011). Interestingly, the two integrin-binding sites are contiguous and followed by the thrombin cleavage site immediately in the C-terminus (Yokosaki et al. 1999, 2005). Upon cleavage by thrombin the sequence SVVYGLR is exposed at the C-terminus of the fragment and serves as a binding site for integrin α9β1 that does not bind to full-length OPN (Yokasaki and Sheppard 2000). The other integrin-binding site is RGD of canonical integrin-binding tripeptide. Binding of α5β1 to the RGD site is up-regulated by thrombin cleavage, but binding of αvβ3, αvβ5, and αvβ6 to the RGD site is unaffected. The regulation of integrin-OPN interactions by enzymatic cleavage is further managed by another protease cleavage site. MMP-3 and MMP-7 share a cleavage site at 150GL151, which is within the integrin-binding sequence SVVYGLR (Agnihotri et al. 2001). The resulting C-terminus SVVYG no longer supports α9β1 binding (Yokosaki et al. 2005). The overlap of receptors for OPN does not necessarily mean that these integrins merely provide redundancy because each cell type has different integrin repertoires and the combinations of utilized integrins vary widely among cell types. In addition, interactions of different integrins with a single ligand can exert distinct effects on cell behavior in a single cell type. For example, we have previously reported that signals generated by binding of a single ligand, tenascin-C, to αvβ3, αvβ6, or α9β1 differently affected cell adhesion, spreading, and proliferation of the colon cancer cell line SW480 (Yokosaki et al. 1996). Furthermore, intact OPN or thrombin- or matrix metalloproteinase-cleaved OPN interact with distinct subsets of integrins and exhibit distinct effects on cell behavior. Collectively, some of the functional diversity of OPN could be attributed to this multiplicity of receptors and responses.

6.3.2 Polymerization

Besides the post-translational glycosylation, phosphorylation, and enzymatic digestions, OPN undergoes a post-translational modification, polymerization catalyzed by TG2 or factor XIII (Prince et al. 1991). In our previous mutational study, Gln34, 36, 42, and 55 were compatible with the binding residues (Nishimichi et al. 2009). A mass-spectrometric analysis revealed them to be Gln34 and 42 (Christensen et al. 2014). Matrix stabilization is a widely accepted role for the action of the transglutaminase on OPN. OPN is crosslinked to fibronectin to increase matrix integrity (Beninati et al. 1994) and is polymerized to enhance its collagen binding activity (Kaartinen et al. 1999). OPN is also covalently linked to bone matrix (Kaartinen et al. 2002) and urinary stones (Kohri et al. 1992; Hamamoto et al. 2011). In the mineralized compartment of intramembranous rat bone there are three different transglutaminase substrates: bone sialoprotein, α2 HS-glycoprotein, and OPN. Each of these substrates are polymerized to form homopolymers that do not interact with substrates different from themselves (Kaartinen et al. 2002). Experiments using two-dimensional gel electrophoresis and immunoblotting analyses have clearly demonstrated the specific homopolymer formation.

For OPN, there is evidence for in vivo polymerization. Western blotting with anti-OPN antibody showed a smeared and high molecular weight band from rat bone and the aorta of Matrix Gla protein-deficient mice (Kaartinen et al. 2007). We described the polymerization of exogenous recombinant OPN 3 h after injection into the peritoneal space of mice (Nishimichi et al. 2011). Functionally, polymeric OPN showed dramatically enhanced cell adhesion and migration as well as focal contact formation through interaction with RGD-binding integrins (Fig. 6.2) (Higashikawa et al. 2007). This enhancement supports the idea that the polymerization of ligands potentiates interactions with integrins by concentrating ligand-binding sites, thereby potentiating integrin clustering. At this point, we questioned

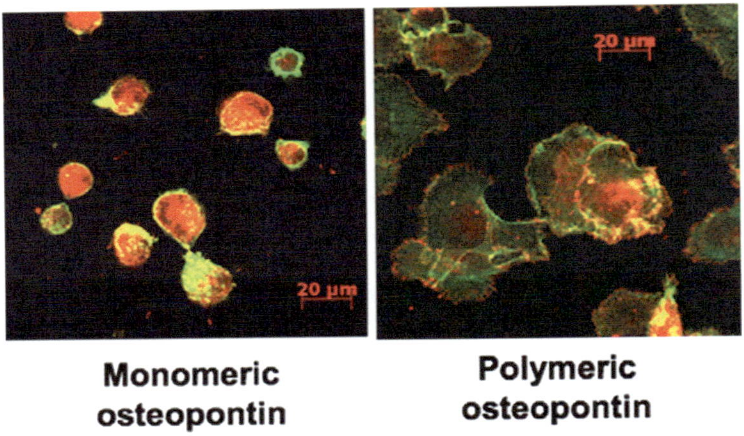

Fig. 6.2 Spreading of cells and focal contact formation on polymeric osteopontin. HUVE cells in serum free DMEM were plated on plates coated with 1.0 μg/ml of monomeric or polymeric osteopontin. In the course of 30 min, cells spread dramatically on the polymeric osteopontin, where apparent focal contacts were formed. This research was originally published in FEBS Letters (Higashikawa et al. 2007 © Federation of European Biochemical Societies)

whether polymeric OPN expresses SVVYGL on its surface, or is cryptic, as it is in full-length OPN. Polymeric OPN did support adhesion and migration of α9-transfected SW480 cells in the presence of anti-RGD-integrin antibodies. Further, the polymeric OPN bound to recombinant α9β1 in a dose-dependent manner, as shown by surface plasmon resonance analysis (Fig. 6.3). These interactions were all abrogated by an anti-α9β1 blocking mAb. Thus, it was convincing that, like the RGD site, the SVVYGLR site is exposed on the surface of the polymer and retains its capacity to bind α9β1. It was surprising, however, that mutations in the SVVYGLR sequence that disrupt interaction of nOPN with α9β1 (Yokosaki et al. 1999) had no effect at all on its binding to polymeric OPN. Moreover, anti-SVVYGLR antibody did not bind to polymeric OPN, although it clearly bound to nOPN (Nishimichi et al. 2011). These results indicated that polymeric OPN is a ligand for integrin α9β1, as is nOPN, but unlike nOPN, the binding site is not SVVYGLR. The binding site of α9β1 in polymeric OPN has not yet been identified.

6.3.3 *Gain of a New Function of Polymeric OPN Binding to Integrin α9β1*

Tissues expressing integrin α9β1 include the basal layer of epithelial cells (Palmer et al. 1993), the valves of lymphatic vessels (Bazigou et al. 2009), and neutrophils (Shang et al. 1999). The β2 integrins are a class of primary leukocyte integrins involved in tethering to and rolling on endothelial cells. Beside β2 integrins, α5, α6, and α9 subunits are also expressed on neutrophils as β1 partners, but in contrast to

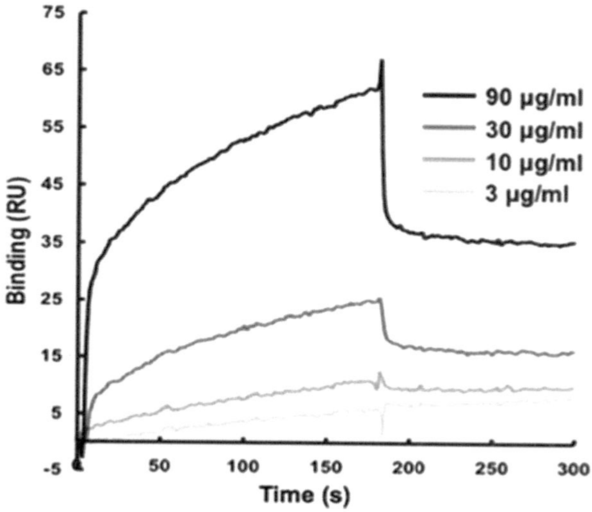

Fig. 6.3 Binding of osteopontin to integrin α9β1. Surface plasmon resonance (Biacore) analysis of recombinant α9β1 binding to polymeric osteopontin. The vertical axis shows the surface plasmon resonance intensity in resonance units (RU). The horizontal axis shows the duration of flow of buffer containing 1 mM Mn^{2+}. Recombinant α9β1 (10 μg/ml) was bound to anti-V5 that had been coupled to the sensor chip. Polymeric osteopontin, at concentrations of 3, 10, 30, or 90 μg/ml in buffer, was added over the sensor chip at a flow rate of 10 μl/min for 180 s (This research was originally published in The Journal of Biological Chemistry. Nishimichi et al. (2009) © the American Society for Biochemistry and Molecular Biology)

the unchanged expression levels of α5 and α6 upon neutrophil activation, expression of α9 is markedly up-regulated and plays a synergistic role with β2 (Mambole et al. 2010). To find if there is any biological significance to the interaction of α9β1 with polymeric OPN, we focused on the neutrophil response to polymeric OPN. In our previous study (Fig. 6.4), neutrophil migration was first observed with a horizontal migration apparatus, which has a 5-μm-deep, flat channel where chemoattractants form a gradient between one chamber and the other. Neutrophils collected from human blood migrated through the gradient of polymeric OPN. This activity was inhibited by anti-α9β1 antibody, but not by an anti-β2 blocking mAb (Nishimichi et al. 2009). Similar results were obtained by random migration analysis video-captured in the presence or absence of homogeneous concentrations of polymeric or monomeric full-length OPN (Nishimichi et al. 2011). nOPN also elicited random migration, which was anticipated because, unlike full-length OPN, nOPN interacts with α9β1 (Nishimichi et al. 2011). Next, we observed the chemotactic effect of polymeric OPN in vivo (Nishimichi et al. 2011). As expected, intraperitoneal injection of polymeric OPN-induced accumulation of neutrophils peaking at 3 h after injection. The control full-length monomeric OPN (intact OPN), however, unexpectedly induced the accumulation. Although the peak of the increase was delayed more than with polymeric OPN, the induction was in contrast to the effect in vitro. Therefore, we hypothesized that intact OPN would

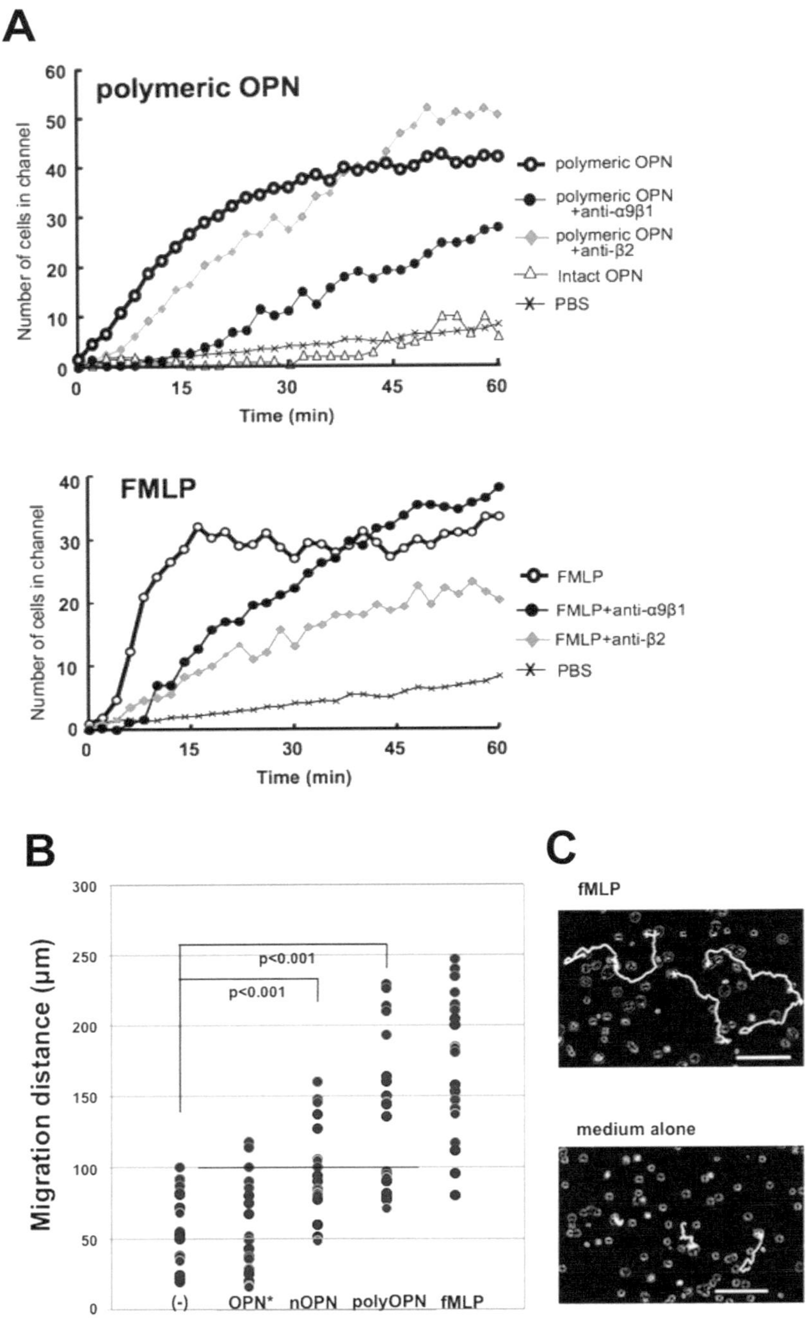

Fig. 6.4 Neutrophil migration induced by polymeric osteopontin. (**a**) the numbers of neutrophils that had migrated into and were present within the 5-μm horizontal channels containing gradients of polymeric osteopontin (*upper panel*) or fMLP (*lower panel*) at each time point during 60 min. Neutrophils were pre-incubated in the presence or absence of antibodies as indicated at the *right*.

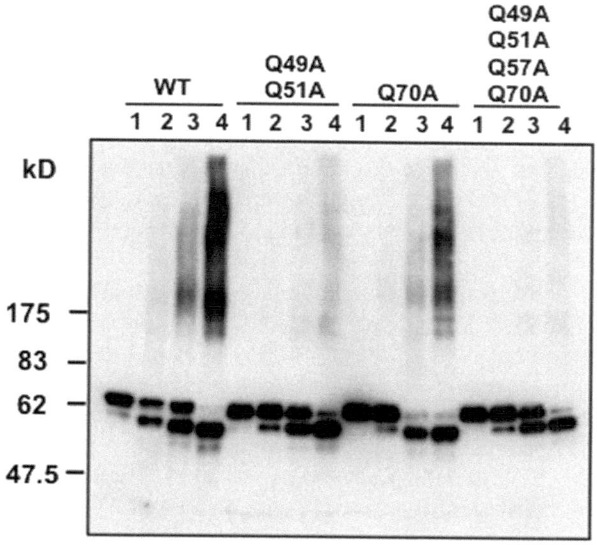

Fig. 6.5 Polymerization-incompetent mutant osteopontin. Western blotting of the transglutaminase 2-catalyzed polymerization reaction of human osteopontin. Lane 1 of each osteopontin variant shows a reaction with no transglutaminase 2; lane 2, 5 μg/ml; lane 3, 10 μg/ml; and lane 4, 20 μg/ml. Wild type and three mutant osteopontins (mouse recombinant) in which one, two, or four Gln residues, as indicated above, are replaced with Ala were polymerized and probed with an anti-OPN mAb that recognizes both monomeric and polymeric osteopontin (This research was originally published in The Journal of Biological Chemistry. Nishimichi et al. (2011) © the American Society for Biochemistry and Molecular Biology)

have been polymerized in vivo. To test this hypothesis of in vivo polymerization, we generated a mutant OPN that does not polymerize (Fig. 6.5) and injected it into mice peritoneal spaces. Neutrophil accumulation was of significantly lower magnitude than with intact OPN. To observe the in vivo polymerization directly, the peritoneal space was washed by PBS 3 h after injection of polymeric OPN, and analyzed by western blotting for immunoreactivity of OPN. The blot showed a high molecular weight smeared band characteristic of the appearance of polymerized OPN. Further ELISA measurement showed that the mutant did not decrease as

Fig. 6.4 (continued) (**b**) chemokinesis of neutrophils induced by three different forms of osteopontin. Each *dot* represents the distance of human neutrophil migration induced by OPN* (full-length monomeric osteopontin, $n = 20$), nOPN (N-terminal fragment of thrombin-cleaved osteopontin, $n = 22$), or polymeric osteopontin ($n = 20$) with negative (medium alone; $n = 20$) and positive (fMLP; $n = 21$) controls. Migrations of 20 or more randomly selected neutrophils were video-captured, and cell tracks were traced for 20 min. Cells that did not migrate were excluded as non-viable. Cells were kept at 37 °C in a humidified atmosphere of 5 % CO_2 during the migration. (**c**) representative computer-traced cell tracks for three cells induced by fMLP or medium alone. *Bars*, 100 μm (This research was originally published in The Journal of Biological Chemistry. Nishimichi et al. (2011) © the American Society for Biochemistry and Molecular Biology)

intact OPN did in the peritoneal space. These result indicate that intact OPN is polymerized in vivo and attracts neutrophils. But these result do not preclude the possibility that nOPN was also generated in the peritoneal space and contributed to the neutrophil accumulation. nOPN specific ELISA measurement indicated that nOPN was there, although it was at the level of ~1 ng/ml while intact OPN was at ~1 μg/ml. Then we generated thrombin cleavage-incompetent mutant OPN to determine whether nOPN attracts any neutrophils after full-length OPN injection. There was no difference in neutrophil accumulation between injections of wild-type and mutant OPN, excluding the possibility that the small amount of nOPN generated support the neutrophil accumulation. In these context, neutrophils were attracted by polymeric OPN but not by intact OPN or thrombin-cleaved OPN.

The requirement of polymerization for OPN to attract neutrophils suggests the recruitment was mediated by α9β1, as it is in vitro. However, in vivo there would be other factors for polymeric OPN than the interaction with α9β1. For example, polymeric OPN may stimulate cells to produce chemoattractants more avidly than monomeric OPN, including peritoneal mesothelial cells and monocyte/macrophages,. To assess the contribution of polymeric OPN, we generated mice lacking integrin subunit α9 expression on all leukocytes. We then compared neutrophil recruitment into the peritoneal space of leukocyte-α9-null mice induced by intact OPN, or polymerization-incompetent mutant OPN with that of wild-type mice (Fig. 6.6). Neutrophil recruitment induced by injection of intact OPN was

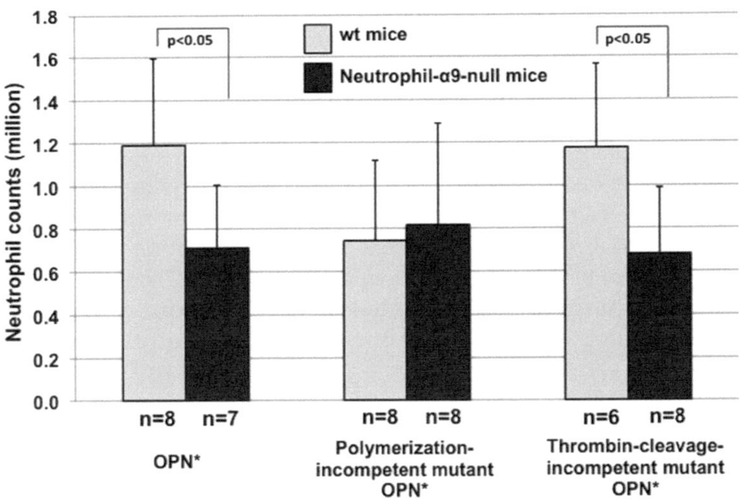

Fig. 6.6 Neutrophil accumulation in wild-type and neutrophil α9-null mice by injection of WT, polymerization-incompetent mutant, or thrombin cleavage-incompetent osteopontin. One of three species of osteopontin shown *below* was injected into the peritoneal space of mice that lack the integrin α9 subunit on their neutrophils. Neutrophils in the peritoneal space were counted 4 h after injection. The numbers of mice are indicated *below* each *bar*. Values represent the mean ± S.D. (*error bars*). OPN*, full-length monomeric osteopontin (This research was originally published in The Journal of Biological Chemistry. Nishimichi et al. (2011) © the American Society for Biochemistry and Molecular Biology)

significantly reduced in mice that lack the integrin α9 subunit on their leukocytes. In contrast, recruitment in response to the polymerization-incompetent mutant was reduced in wild-type mice, but was unaffected by loss of the integrin α9 subunit on leukocytes. These results indicate that intact OPN-induced neutrophil accumulation is mediated, at least in part, by α9β1 and that polymerization is critical for the recruitment.

Before the description of the requirement of polymerization, OPN was already regarded as a chemoattractant for neutrophils. Two independent groups had reported that neutrophil chemotaxis was a consequence of peritoneal OPN injection (Carrigan et al. 2007; Koh et al. 2007). It should be noted, however, that there are few data supporting the chemotactic activity of OPN in vitro. The only report that claimed to show OPN's in vivo neutrophil chemotactic activity used OPN purified from cells of a macrophage cell line, not recombinant OPN (Carrigan et al. 2007), where polymerized OPN could be involved. All of the previous publications are consistent with the finding that OPN stimulates chemotaxis after polymerization.

6.3.4 *The Specific Integrin α9β1 Binding Site on Polymeric OPN*

There must be a specific conformation that fits with the α9β1 ligand-binding pocket on the surface of the polymer. Two possibilities can explain how the conformation is provided upon polymerization; either an otherwise cryptic structure is exposed, or a new conformation arises as consequence of the atomic interactions, particularly at molecular interfaces. The relevance of SVVYGLR has been excluded, as described above. However, another cryptic binding sequence could be exposed by polymerization. The other possibility, i.e., generation of a new site, might be associated with the regulated enzymatic reaction as discussed in Sect. 6.2. In fact, there are few mAbs, including 81D4, that react with the isopeptide bond N^{ε}-(γ-L-glutamyl)-L-lysine (el Alaoui et al. 1991; Thomas et al. 2004), which indicates that the isopeptide bond has a specific conformation that serves an epitope of these mAbs. Further, a polymer-specific antiserum for amyloid β protein does not recognize the monomeric fibril (Kayed et al. 2003). The authors further developed monoclonal antibodies that recognize generic epitopes on amyloid β oligomers that do not depend on a specific linear amino acid sequence. These mAbs display distinct preferences for different subsets of prefibrillar oligomers (Kayed et al. 2010). The presence of such epitopes on the amyloid β polymer supports the hypothesis that polymeric OPN has a specific binding site for integrins. Note that amyloid β serves as a substrate for transglutaminase (Hartley et al. 2008; Schmid et al. 2011).

Although there are many Gln and Lys residues in the OPN sequence, only four candidate Gln residues and nine candidate Lys residues have been reported as potential TG2 substrates. These residues could theoretically form a maximum of

36 different isopeptide bonds between OPN molecules. However, not all of the combinations may occur because the preferences of the enzyme for substrate sequences are not the same among the Gln and Lys residues. If there are one or a few dominant combinations selected by the enzyme, the identical conformation would be scattered on the surface of the polymer.

In biological terms, this polymerization also provides us with an intriguing question. Native, full-length non-polymerized OPN does not bind to integrin $\alpha 9\beta 1$, but it does bind after a post-translational modification: thrombin-catalyzed cleavage. Further, another post-translational modification, polymerization, also confers the capacity to bind to $\alpha 9\beta 1$, even creating a binding site different from SVVYGLR. From these consequences, the interaction of OPN with $\alpha 9\beta 1$ is thought to be so critical that two ways of regulation are provided. As shown above, OPN acquires chemotactic activity when interacting with $\alpha 9\beta 1$ after either cleavage or polymerization. How are these two ways differentially utilized? One answer provided by our experimental work is that polymerization is overwhelmingly dominant over cleavage when recombinant, full-length monomeric OPN is injected into the peritoneal space of mice (Nishimichi et al. 2011). In this case, polymerization occurs systemically, rather than locally. Whether the activated thrombin that cleaves OPN is concentrated locally, such as on blood vessels as in thrombosis, may need to be determined.

6.4 Regulation of Polymerization by TG2

Recently, it was found from crystal structure studies that there are at least two different conformations of TG2: active open and inactive folded (Pinkas et al. 2007). As with many other proteins, Ca^{2+} is one of the most critical regulators of TG2 activation; it is also involved in the regulation of the conformational change of TG2 (Belkin 2011; Kiraly et al. 2011). TG2 works in both the extra- and intracellular environments. To exert catalytic function under both conditions, TG2 changes its own Ca^{2+} sensitivity. The Ca^{2+} concentration displays a huge gradient of $>10{,}000$ between the extra- and intracellular spaces; the cytoplasmic environment is highly reductive, with low Ca^{2+} and high GTP concentrations. GTP is another major regulator of TG2 that binds to TG2 between its core and β-barrel domain. The redox system also changes the properties of TG2, including its Ca^{2+} sensitivity. There are three Cys residues in the catalytic center of the core domain that reversibly form disulfide bonds in response to redox status (Stamnaes et al. 2010). In addition, other mechanisms are further involved in the regulation of extra- and intracellular activation of TG2, including S-nitrosylation of Cys residues, mechanical forces on the molecule, glycosaminoglycans, and the buffering effects of other Ca^{2+}-binding proteins. Regulation of the intracellular Ca^{2+} signal is also tightly controlled through chelation, sequestration, or removal (Clapham 2007).

Although regulatory compartments have been identified, there is little information about the systemic control of TG2 activation and polymerization. Many emerging questions still remain, particularly about regulation of the polymerization of a variety of substrates. For example, there are multiple transglutaminase-polymerized proteins that regulate apoptosis. When the multiple substrates for polymers that cause opposing effects on apoptosis are in place, how does transglutaminase manage these polymerizations? As mentioned in the first section, a transcription factor, Sp1, no longer initiates transcription of c-Met upon polymerization and initiate apoptosis in hepatocytes. Contrarily, the cell cycle regulator Rb enhances its protective effect (Boehm et al. 2002) against calphostin C-induced apoptosis upon polymerization (Oliverio et al. 1997). Is it possible for transglutaminase to specifically polymerize Sp1 while leaving Rb monomeric, thus inducing apoptosis in hepatocytes? Alternatively, DLK up-regulates its kinase activity upon polymerization and activates JNK in cells undergoing calphostin C-induced apoptosis (Robitaille et al. 2004), thus promoting apoptosis. DLK polymerization leads to an effect obviously opposed to the effect of Rb polymerization. In cells undergoing calphostin C-induced apoptosis, should transglutaminase be active or remain silent to regulate the apoptosis? These three kinds of apoptosis-related polymers illustrate the complexity of the regulation of transglutaminase-catalyzed polymerization.

One of the simplest explanations for the contradiction represented by the pro- and anti-apoptotic effects of polymeric Sp1 and Rb, respectively, may be cell type specificity. Because the effect of Sp1 polymerization is exhibited through loss of c-Met expression, such apoptosis would be exerted only in cells where the hepatocyte growth factor (HGF) signal is essential for survival. Another simple rationalization is the differential activation of TG2 so that it polymerizes one, but does not polymerize the other, i.e., an activation state of TG2 that specifically polymerizes one among multiple substrates. Because Sp1 is polymerized by ethanol-activated TG2, whereas Rb is polymerized by calphostin C-activated TG2, it is conceivable that these TG2-activated states are not the same. In this case, TG2 holds multiple activation states that can discriminate between the polymerization of Sp1 and Rb proteins. However, for dual polymerizations of DLK and Rb, the other specific example, since TG2 is activated through the same activator calphostin C it is less likely that TG2 could be differentially activated and exhibit specific polymerization of DLK or Rb. One possibility is that polymerization activity is dependent on each TG2 binding sequence of the substrate. This conflicting polymerization may be attributed to spatial or temporal activation of transglutaminase. If cells are in situations that promote apoptosis, activation of transglutaminase may occur locally within the region of the cell where pro-apoptotic DLK functions, but not in the region where anti-apoptotic Rb is concentrated. This could be the case because DLK phosphorylates MAPK kinase in the cytoplasm, whereas Rb supports cell survival by regulating cell cycles in the nucleus. Further speculation can be developed; transglutaminase may polymerize DLK, first, to activate the JNK pathway and initiate c-Jun to transcript pro-apoptotic molecules such as Bim, and then transglutaminase translocates into the nucleus to polymerize Rb to protect from apoptosis until Bim expression has been established. These differential activation states that allow specific polymerization among many substrate proteins, as well as spatial and temporal regulation of TG2 activation in total, conduct the polymerization orchestrating molecular instruments.

On comparing the regulation of TG2 polymerization with that of protein expression, although transcriptional regulation appears to be much more sophisticated, the principle of regulation may be shared. It is recognized that not all of the proteins expressed are required. The expression of superfluous proteins, in addition to the protein of interest, is allowed as far as they are non-deleterious and have no opposing effects on the protein of interest (Erickson 1993). This is probably because, in view of cellular economy, sharing the switch to turn on the expression of several proteins is preferable to preparing one switch for each protein. As far as it is not toxic and not opposing to the required function of the polymer being generated, superfluous polymerization could be allowed. For example, three substrates for TG2-polymerization were identified from bone extract, OPN, bone sialoprotein, and α2 HS-glycoprotein (Kaartinen et al. 2002). This list might include one or two biologically nonfunctional or needless polymers, but one or more of the polymers is required. Every transglutaminase-induced polymers found in vivo may not necessarily be functional or required at the time of polymerization; some of them may exist without a justified raison d'etre.

6.5 Summary and Future Direction

We have shown in vivo and in vitro experimental evidence for polymerization of OPN. Besides, human secreted OPN was reported to be highly polymerized in the airways but less polymerized in patients with asthma (Arjomandi et al. 2011). Since the magnitude of polymerization is associated with the disease state, the polymerization has to be determined in various diseases in the view of magnitude and disease state. The elucidation of the mechanisms underlying the polymerization and roles in pathological states help understand and manage the diseases.

Increasing numbers of substrates for transglutaminases have been identified. Although many of them are polymerized by the same transglutaminase enzyme, functional alterations resulting from the reaction are diverse, including up-regulation, down-regulation, loss of function, prolonged half-life, and even a gain of new function. Although the in vivo polymerization of some of the polymers has been confirmed, there are many polymers that are reportedly functional in vitro. This may arise because in vitro conditions where high concentrations of transglutaminase are present could produce many polymers that are not polymerized under physiological conditions. Nevertheless, collecting evidence for polymerization in vitro would be of importance because in pathological conditions transglutaminase could be highly activated locally. The identification of the conditions that are specifically seen in any disease would be intriguing.

OPN is one of the substrates for which there is evidence of in vivo polymerization (Kaartinen et al. 2002). Further, OPN gains a new function, becoming a chemoattractant for neutrophils upon polymerization and serving as a ligand for integrin $\alpha9\beta1$, proved using two different in vitro chemotaxis assays and in vivo injection (Higashikawa et al. 2007; Nishimichi et al. 2009, 2011). It should be noted that we have demonstrated that native OPN is not a chemotactic for neutrophils in vitro at all, but transforms into a chemoattractant by undergoing polymerization. Although chemotactic activity for neutrophils has been accepted as a function of OPN, the post-translational modifications have not been a point of discussion. This illustrates that a function that is exerted only by a polymer may not be discerned.

One should be careful to detect transglutaminase-mediated polymers by electrophoresis because the polymers are prone to be trapped within the stacking gel and do not migrate into the running gel. These polymers have been detected, so far, by immunohistochemistry with isopeptide-bond-specific antibodies or western blotting with antibodies reactive for both the monomer and the polymer. Development of mAbs that are specific for each polymer would greatly facilitate the study of these structures and the functions of the polymers. Such mAbs could be applied not only in qualitative immunoblotting, but also in quantitative assays, particularly in pathological aspects, with much expectation.

References

AbdAlla S, Lother H, Langer A, el Faramawy Y, Quitterer U (2004) Factor XIIIA transglutaminase crosslinks AT1 receptor dimers of monocytes at the onset of atherosclerosis. Cell 119 (3):343–354. doi:10.1016/j.cell.2004.10.006

AbdAlla S, Lother H, el Missiry A, Langer A, Sergeev P, el Faramawy Y, Quitterer U (2009) Angiotensin II AT2 receptor oligomers mediate G-protein dysfunction in an animal model of Alzheimer disease. J Biol Chem 284(10):6554–6565. doi:10.1074/jbc.M807746200

Aeschlimann D, Paulsson M (1991) Cross-linking of laminin-nidogen complexes by tissue transglutaminase: a novel mechanism for basement membrane stabilization. J Biol Chem 266(23):15308–15317

Agnihotri R, Crawford HC, Haro H, Matrisian LM, Havrda MC, Liaw L (2001) Osteopontin, a novel substrate for matrix metalloproteinase-3 (stromelysin-1) and matrix metalloproteinase-7 (matrilysin). J Biol Chem 276(30):28261–28267

Arjomandi M, Frelinger J, Donde A, Wong H, Yellamilli A, Raymond W (2011) Secreted osteopontin is highly polymerized in human airways and fragmented in asthmatic airway secretions. PLoS One 6(10):e25678. doi:10.1371/journal.pone.0025678

Bayless KJ, Meininger GA, Scholtz JM, Davis GE (1998) Osteopontin is a ligand for the $\alpha4\beta1$ integrin. J Cell Sci 111:1165–1174

Bazigou E, Xie S, Chen C, Weston A, Miura N, Sorokin L, Adams R, Muro AF, Sheppard D, Makinen T (2009) Integrin-$\alpha9$ is required for fibronectin matrix assembly during lymphatic valve morphogenesis. Dev Cell 17(2):175–186. doi:10.1016/j.devcel.2009.06.017

Belkin AM (2011) Extracellular TG2: emerging functions and regulation. FEBS J 278 (24):4704–4716. doi:10.1111/j.1742-4658.2011.08346.x

Beninati S, Senger DR, Cordella-Miele E, Mukherjee AB, Chackalaparampil I, Shanmugam V, Singh K, Mukherjee BB (1994) Osteopontin: its transglutaminase catalyzed posttranslational modifications and cross linking to fibronectin. J Biochem 115(4):675–682

Boehm JE, Singh U, Combs C, Antonyak MA, Cerione RA (2002) Tissue transglutaminase protects against apoptosis by modifying the tumor suppressor protein p110 Rb. J Biol Chem 277(23):20127–20130. doi:10.1074/jbc.C200147200

Boros S, Kamps B, Wunderink L, de Bruijn W, de Jong WW, Boelens WC (2004) Transglutaminase catalyzes differential crosslinking of small heat shock proteins and amyloid β. FEBS Lett 576(1–2):57–62. doi:10.1016/j.febslet.2004.08.062

Boros S, Xi Q, Dimke H, van der Kemp AW, Tudpor K, Verkaart S, Lee KP, Bindels RJ, Hoenderop JG (2012) Tissue transglutaminase inhibits the TRPV5-dependent calcium transport in an N-glycosylation-dependent manner. Cell Mol Life Sci 69(6):981–992. doi:10.1007/s00018-011-0818-z

Carrigan S, Pink D, Stadnyk A (2007) Neutrophil transepithelial migration in response to the chemoattractant fMLP but not C5a is phospholipase D-dependent and related to the use of CD11b/CD18. J Leukoc Biol 82(6):1575–1584

Chen J, Fujino R, Zhao R, Semba U, Araki K, Yamamoto T (2014) Role of blood ribosomal protein S19 in coagulum resorption: a study using Gln137Glu-ribosomal protein S19 gene knock-in mouse. Pathol Int 64(11):543–550. doi:10.1111/pin.12208

Cheng X, Jin J, Hu L, Shen D, Dong XP, Samie MA, Knoff J, Eisinger B, Liu ML, Huang SM, Caterina MJ, Dempsey P, Michael LE, Dlugosz AA, Andrews NC, Clapham DE, Xu H (2010) TRP channel regulates EGFR signaling in hair morphogenesis and skin barrier formation. Cell 141(2):331–343. doi:10.1016/j.cell.2010.03.013

Christensen B, Kazanecki C, Petersen T, Rittling S, Denhardt D, Sorensen E (2007) Cell type-specific post-translational modifications of mouse osteopontin are associated with different adhesive properties. J Biol Chem 282(27):19463–19472

Christensen B, Schack L, Klaning E, Sorensen ES (2010) Osteopontin is cleaved at multiple sites close to its integrin-binding motifs in milk and is a novel substrate for plasmin and cathepsin D. J Biol Chem 285(11):7929–7937. doi:M109.075010 [pii] 10.1074/jbc.M109.075010

Christensen B, Zachariae ED, Scavenius C, Thybo M, Callesen MM, Kloverpris S, Oxvig C, Enghild JJ, Sorensen ES (2014) Identification of transglutaminase reactive residues in human osteopontin and their role in polymerization. PLoS One 9(11):e113650. doi:10.1371/journal.pone.0113650

Clapham DE (2007) Calcium signaling. Cell 131(6):1047–1058. doi:10.1016/j.cell.2007.11.028

Cornwall GA, Von Horsten HH, Whelly S (2011) Cystatin-related epididymal spermatogenic aggregates in the epididymis. J Androl 32(6):679–685. doi:10.2164/jandrol.111.012963

Denda S, Reichardt LF, Muller U (1998) Identification of osteopontin as a novel ligand for the integrin α8β1 and potential roles for this integrin-ligand interaction in kidney morphogenesis. Mol Biol Cell 9(6):1425–1435

Denhardt D, Noda M, O'Regan A, Pavlin D, Berman J (2001) Osteopontin as a means to cope with environmental insults: regulation of inflammation, tissue remodeling, and cell survival. J Clin Invest 107(9):1055–1061

Dierker T, Dreier R, Migone M, Hamer S, Grobe K (2009) Heparan sulfate and transglutaminase activity are required for the formation of covalently cross-linked hedgehog oligomers. J Biol Chem 284(47):32562–32571. doi:10.1074/jbc.M109.044867

el Alaoui S, Legastelois S, Roch AM, Chantepie J, Quash G (1991) Transglutaminase activity and N epsilon (gamma glutamyl) lysine isopeptide levels during cell growth: an enzymic and immunological study. Int J Cancer 48(2):221–226

Erickson HP (1993) Gene knockouts of c-src, transforming growth factor/ 1, and tenascin suggest superfluous, nonfunctional expression of proteins. J Cell Biol 120(5):1079–1081

Folk JE (1969) Mechanism of action of guinea pig liver transglutaminase. VI. Order of substrate addition. J Biol Chem 244(13):3707–3713

Folk JE, Cole PW (1965) Structural requirements of specific substrates for guinea pig liver transglutaminase. J Biol Chem 240(7):2951–2960

Green PM, Ludbrook SB, Miller DD, Horgan CM, Barry ST (2001) Structural elements of the osteopontin SVVYGLR motif important for the interaction with α4 integrins. FEBS Lett 503 (1):75–79

Grosso H, Mouradian MM (2012) Transglutaminase 2: biology, relevance to neurodegenerative diseases and therapeutic implications. Pharmacol Ther 133(3):392–410. doi:10.1016/j.pharmthera.2011.12.003

Hamamoto S, Yasui T, Okada A, Hirose M, Matsui Y, Kon S, Sakai F, Kojima Y, Hayashi Y, Tozawa K, Uede T, Kohri K (2011) Crucial role of the cryptic epitope SLAYGLR within osteopontin in renal crystal formation of mice. J Bone Miner Res 26(12):2967–2977. doi:10.1002/jbmr.495

Hartley DM, Zhao C, Speier AC, Woodard GA, Li S, Li Z, Walz T (2008) Transglutaminase induces protofibril-like amyloid β-protein assemblies that are protease-resistant and inhibit long-term potentiation. J Biol Chem 283(24):16790–16800. doi:M802215200 [pii] 10.1074/jbc.M802215200

Hasegawa M, Segawa T, Maeda M, Yoshida T, Sudo A (2011) Thrombin-cleaved osteopontin levels in synovial fluid correlate with disease severity of knee osteoarthritis. J Rheumatol 38 (1):129–134. doi:10.3899/jrheum.100637

Hebert SS, Daviau A, Grondin G, Latreille M, Aubin RA, Blouin R (2000) The mixed lineage kinase DLK is oligomerized by tissue transglutaminase during apoptosis. J Biol Chem 275 (42):32482–32490. doi:10.1074/jbc.M006528200

Higashikawa F, Eboshida A, Yokosaki Y (2007) Enhanced biological activity of polymeric osteopontin. FEBS Lett 581(14):2697–2701. doi:10.1016/j.febslet.2007.05.018

Hu DD, Lin EC, Kovach NL, Hoyer R, Smith JW (1995) A biochemical characterization of the binding of osteopontin to integrins αvβ1 and αvβ5. J Biol Chem 270:26232–26238

Huynh KW, Cohen MR, Chakrapani S, Holdaway HA, Stewart PL, Moiseenkova-Bell VY (2014) Structural insight into the assembly of TRPV channels. Structure 22(2):260–268. doi:10.1016/j.str.2013.11.008

Iismaa SE, Begg GE, Graham RM (2006) Cross-linking transglutaminases with G protein-coupled receptor signaling. Sci STKE 2006:pe34

Iwasaki W, Nagata K, Hatanaka H, Inui T, Kimura T, Muramatsu T, Yoshid K, Tasumi M, Inagak F (1997) Solution structure of midkine, a new heparin-binding growth factor. EMBO J 16 (23):6936–6946

Jelenska MM, Fesus L, Kopec M (1980) The comparative ability of plasma and tissue transglutaminases to use collagen as a substrate. Biochim Biophys Acta 616(2):167–178

Jones RA, Nicholas B, Mian S, Davies PJA, Griffin M (1997) Reduced expression of tissue transglutaminase in a human endothelial cell line leads to changes in cell spreading, cell adhesion and reduced polymerisation of fibronectin. J Cell Sci 110(19):2461–2472

Ju-Hong Jeon, Kyung-Ho Choi, Sung-Yup Cho, Chai-Wan Kim, Dong-Myung Shin, Joon-Cheol Kwon, Kye-Yong Song, Sang-Chul Park, In-Gyu Kim (2003) Transglutaminase 2 inhibits Rb binding of human papillomavirus E7 by incorporating polyamine. EMBO J 22(19):5273–5282

Kaartinen MT, Pirhonen A, Linnala-Kankkunen A, Maenpaa PH (1999) Cross-linking of osteopontin by tissue transglutaminase increases its collagen binding properties. J Biol Chem 274(3):1729–1735

Kaartinen MT, El-Maadawy S, Rasanen NH, McKee MD (2002) Tissue transglutaminase and its substrates in bone. J Bone Miner Res 17(12):2161–2173

Kaartinen MT, Murshed M, Karsenty G, McKee MD (2007) Osteopontin upregulation and polymerization by transglutaminase 2 in calcified arteries of matrix Gla protein-deficient mice. J Histochem Cytochem 55(4):375–386

Kayed R, Head E, Thompson JL, McIntire TM, Milton SC, Cotman CW, Glabe CG (2003) Common structure of soluble amyloid oligomers implies common mechanism of pathogenesis. Science 300(5618):486–489

Kayed R, Canto I, Breydo L, Rasool S, Lukacsovich T, Wu J, Albay R 3rd, Pensalfini A, Yeung S, Head E, Marsh JL, Glabe C (2010) Conformation dependent monoclonal antibodies distinguish different replicating strains or conformers of prefibrillar Aβ oligomers. Mol Neurodegener 5:57. doi:10.1186/1750-1326-5-57

Kedei N, Szabo T, Lile JD, Treanor JJ, Olah Z, Iadarola MJ, Blumberg PM (2001) Analysis of the native quaternary structure of vanilloid receptor 1. J Biol Chem 276(30):28613–28619. doi:10.1074/jbc.M103272200

Kiraly R, Demeny M, Fesus L (2011) Protein transamidation by transglutaminase 2 in cells: a disputed Ca^{2+}-dependent action of a multifunctional protein. FEBS J 278(24):4717–4739. doi:10.1111/j.1742-4658.2011.08345.x

Koh A, da Silva APB, Bansal AK, Bansal M, Sun C, Lee H, Glogauer M, Sodek J, Zohar R (2007) Role of osteopontin in neutrophil function. Immunology 122(4):466–475

Kohri K, Suzuki Y, Yoshida K, Yamamoto K, Amasaki N, Yamate T, Umekawa T, Iguchi M, Sinohara H, Kurita T (1992) Molecular cloning and sequencing of cDNA encoding urinary stone protein, which is identical to osteopontin. Biochem Biophys Res Commun 184 (2):859–864

Kojima S, Muramatsu H, Amanuma H, Muramatsu T (1995) Midkine enhances fibrinolytic activity of bovine endothelial cells. J Biol Chem 270(16):9590–9596

Kojima S, Inui T, Muramatsu H, Suzuki Y, Kadomatsu K, Yoshizawa M, Hirose S, Kimura T, Sakakibara S, Muramatsu T (1997) Dimerization of midkine by tissue transglutaminase and its functional implication. J Biol Chem 272(14):9410–9416

Kornguth SE, Waelsch H (1963) Protein modification catalysed by transglutaminase. Nature 198:188–189

Lorand L, Graham R (2003) Transglutaminases: crosslinking enzymes with pleiotropic functions. Nat Rev Mol Cell Biol 4(2):140–156

Lorand JB, Urayama T, Lorand L (1966) Transglutaminase as a blood clotting enzyme. Biochem Biophys Res Commun 23(6):828–834

Lorand L, Downey J, Gotoh T, Jacobsen A, Tokura S (1968) The transpeptidase system which crosslinks fibrin by gamma-glutamyle-episilon-lysine bonds. Biochem Biophys Res Commun 31(2):222–230

Mahoney S-A, Perry M, Seddon A, Bohlen P, Haynes L (1996) Transglutaminase forms midkine homodimers in cerebellar neurons and modulates the neurite-outgrowth response. Biochem Biophys Res Com 224(1):147–152

Mambole A, Bigot S, Baruch D, Lesavre P, Halbwachs-Mecarelli L (2010) Human neutrophil integrin α9β1: up-regulation by cell activation and synergy with β2 integrins during adhesion to endothelium under flow. J Leukoc Biol 88(2):321–327. doi:10.1189/jlb.1009704

Mamenko M, Zaika O, Boukelmoune N, O'Neil RG, Pochynyuk O (2015) Deciphering physiological role of the mechanosensitive TRPV4 channel in the distal nephron. Am J Physiol Renal Physiol 308(4):F275–F286. doi:10.1152/ ajprenal.00485.2014

Matacic S, Loewy AG (1966) Transglutaminase activity of the fibrin crosslinking enzyme. Biochem Biophys Res Commun 24(6):858–866

Miyauchi A, Alvarez J, Greenfield EM, Teti A, Grano M, Colucci S, Zanbinin-Zallone A, Ross FP, Teitelbaum SL, Cheresh D (1991) Recognition of osteopontin and related peptides by avb3 integrin stimulates immediate cell signals in osteoclasts. J Biol Chem 266:20369–20374

Mosher DF (1977) Labeling of a major fibroblast surface protein (fibronectin) catalyzed by blood coagulation factor XIIa. Biochim Biophys Acta 491(1):205–210

Mosher DF (1978) Cross-linking of plasma and cellular fibronectin by plasma transglutaminase. Ann N Y Acad Sci 312:38–42

Mosher DF, Schad PE (1979) Cross-linking of fibronectin to collagen by blood coagulation factor XIIIa. J Clin Invest 64(3):781–787. doi:10.1172/jci109524

Mosher DF, Schad PE, Kleinman HK (1979) Inhibition of blood coagulation factor XIIIa-mediated cross-linking between fibronectin and collagen by polyamines. J Supramol Struct 11(2):227–235. doi:10.1002/jss.400110212

Munger JS, Huang X, Kawakatsu H, Griffiths MJ, Dalton SL, Wu J, Pittet JF, Kaminski N, Garat C, Matthay MA, Rifkin DB, Sheppard D (1999) The integrin αvβ6 binds and activates latent TGF beta 1: a mechanism for regulating pulmonary inflammation and fibrosis. Cell 96 (3):319–328

Mycek MJ, Clarke DD, Neidle A, Waelsch H (1959) Amine incorporation into insulin as catalyzed by transglutaminase. Arch Biochem Biophys 84(2):528–540

Nasu K, Ishida T, Setoguchi M, Higuchi Y, Akizuki S, Yamamoto S (1995) Expression of wild-type and mutated rabbit osteopontin in Escherichia coli, and their effects on adhesion and migration of P388D1 cells. Biochem J 307(1):257–265

Nishimichi N, Higashikawa F, Kinoh HH, Tateishi Y, Matsuda H, Yokosaki Y (2009) Polymeric osteopontin employs integrin alpha9beta1 as a receptor and attracts neutrophils by presenting a de novo binding site. J Biol Chem 284(22):14769–14776. doi:10.1074/jbc.M901515200

Nishimichi N, Hayashita-Kinoh H, Chen C, Matsuda H, Sheppard D, Yokosaki Y (2011) Osteopontin undergoes polymerization in vivo and gains chemotactic activity for neutrophils mediated by integrin alpha9beta1. J Biol Chem 286(13):11170–11178. doi:10.1074/jbc.M110.189258

Nishimura H, Shibuya Y, Matsubara S, Tanase S, Kambara T, Yamamoto T (1996) Monocyte chemotactic factor in rheumatoid arthritis synovial tissue: probably a cross-linked derivative of S19 ribosomal protein. J Biol Chem 271(2):878–882

Nunes I, Gleizes P-E, Metz CN, Rifkin DB (1997) Latent transforming growthfactor-β binding protein domains involved in activation and transglutaminase dependent cross-linking of latent TGF-β. J Cell Biol 136(5):1151–1163

Oldberg A, Franzen A, Heinegard D (1986) Cloning and sequence analysis of rat bone sialoprotein (osteopontin) cDNA reveals an Arg-Gly-Asp cell-binding sequence. Proc Natl Acad Sci U S A 83(23):8819–8823

Oliverio S, Amendola A, Sano FD, Farrace MG, Fesus L, Nemes Z, Piredda L, Spinedi A, Piacentini M (1997) Tissue transglutaminase-dependent posttranslational modification of the retinoblastoma gene product in promonocytic cells undergoing apoptosis. Mol Cell Biol 17 (10):6040–6048

Palmer EL, Ruegg C, Ferrando R, Pytela R, Sheppard D (1993) Sequence and tissue distribution of the integrin α9 subunit, a novel partner of β1 that is widely distributed in epithelia and muscle. J Cell Biol 123(5):1289–1297

Patarca R, Freeman GJ, Singh RP, Wei FY, Durfee T, Blattner F, Regnier DC, Kozak CA, Mock BA, Morse HC 3rd et al (1989) Structural and functional studies of the early T lymphocyte activation 1 (Eta-1) gene. Definition of a novel T cell-dependent response associated with genetic resistance to bacterial infection. J Exp Med 170(1):145–161

Pinkas DM, Strop P, Brunger AT, Khosla C (2007) Transglutaminase 2 undergoes a large conformational change upon activation. PLoS Biol 5:e237. doi:10.1371/10.1371/journal.pbio.0050327.g001

Prince C, Dickie D, Krumdieck C (1991) Osteopontin, a substrate for transglutaminase and factor XIII activity. Biochem Biophys Res Commun 177(3):1205–1210

Robitaille K, Daviau A, Tucholski J, Johnson GV, Rancourt C, Blouin R (2004) Tissue transglutaminase triggers oligomerization and activation of dual leucine zipper-bearing kinase in calphostin C-treated cells to facilitate apoptosis. Cell Death Differ 11(5):542–549. doi:10.1038/sj.cdd.4401392

Ruse M, Lambert A, Robinson N, Ryan D, Shon K-J, Eckert RL (2001) S100A7, S100A10, and S100A11 are transglutaminase substrates. Biochemistry 40(10):3167–3173

Schack L, Stapulionis R, Christensen B, Kofod-Olsen E, Skov Sorensen UB, Vorup-Jensen T, Sorensen ES, Hollsberg P (2009) Osteopontin enhances phagocytosis through a novel osteopontin receptor, the αXβ2 integrin. J Immunol 182(11):6943–6950. doi:10.4049/jimmunol.0900065

Schmid AW, Condemi E, Tuchscherer G, Chiappe D, Mutter M, Vogel H, Moniatte M, Tsybin YO (2011) Tissue transglutaminase-mediated glutamine deamidation of beta-amyloid peptide increases peptide solubility, whereas enzymatic cross-linking and peptide fragmentation may serve as molecular triggers for rapid peptide aggregation. J Biol Chem 286(14):12172–12188. doi:10.1074/jbc.M110.176149

Senger D, Perruzzi C, Papadopoulous S, Van de Water L (1994) Adhesive properties of osteopontion: regulation by a naturally occurring thrombin-cleavage in close proximity to the GRDGS cell-binding domain. Mol Biol Cell 5(5):565–574

Shamas-Din A, Bindner S, Chi X, Leber B, Andrews DW, Fradin C (2015) Distinct lipid effects on tBid and Bim activation of membrane permeabilization by pro-apoptotic Bax. Biochem J 467:495–505. doi:10.1042/BJ20141291

Shang T, Yednock T, Issekutz AC (1999) α9β1 integrin is expressed on human neutrophils and contributes to neutrophil migration through human lung and synovial fibroblast barriers. J Leukoc Biol 66(5):809–816

Smith LL, Cheung HK, Ling LE, Chen J, Sheppard D, Pytela R, Giachelli CM (1996) Osteopontin N-terminal domain contains a cryptic adhesive sequence recognized by α9β1 integrin. J Biol Chem 271(45):28485–28491

Stamnaes J, Pinkas DM, Fleckenstein B, Khosla C, Sollid LM (2010) Redox regulation of transglutaminase 2 activity. J Biol Chem 285(33):25402–25409. doi:10.1074/jbc.M109.097162

Strnad P, Harada M, Siegel M, Terkeltaub RA, Graham RM, Khosla C, Omary MB (2007) Transglutaminase 2 regulates mallory body inclusion formation and injury-associated liver enlargement. Gastroenterology 132(4):1515–1526. doi:10.1053/j.gastro.2007.02.020

Sugimura Y, Hosono M, Wada F, Yoshimura T, Maki M, Hitomi K (2006) Screening for the preferred substrate sequence of transglutaminase using a phage-displayed peptide library: identification of peptide substrates for TGase 2 and factor XIIIA. J Biol Chem 281 (26):17699–17706. doi:10.1074/jbc.M513538200

Tatsukawa H, Fukaya Y, Frampton G, Martinez-Fuentes A, Suzuki K, Kuo TF, Nagatsuma K, Shimokado K, Okuno M, Wu J, Iismaa S, Matsuura T, Tsukamoto H, Zern MA, Graham RM, Kojima S (2009) Role of transglutaminase 2 in liver injury via cross-linking and silencing of transcription factor Sp1. Gastroenterology 136(5):1783–1795 e1710. doi:10.1053/j.gastro.2009.01.007

Thomas WG (2005) Double trouble for type 1 angiotensin receptors in atherosclerosis. N Engl J Med 352(5):506–508

Thomas V, Fournet G, Simonet F, Roch AM, Ceylan I, El Alaouia S, Quash G (2004) Definition of the fine specificity of the monoclonal antibody 81D4: its reactivity with lysine and polyamine isopeptide cross-links. J Immunol Methods 292(1–2):83–95. doi:10.1016/j.jim.2004.06.013

van den Brûle F, Liu F, Castronovo V (1998) Transglutaminase-mediated oligomerization of galectin-3 modulates human melanoma cell interactions with laminin. Cell Adhes Commun 5 (6):425–435

Verderio E, Gaudry C, Gross S, Smith C, Downes S, Griffin M (1999) Regulation of cell surface tissue transglutaminase: effects on matrix storage of latent transforming growth factor-binding protein-1. J Histochem Cytochem 47(11):1417–1432. doi:10.1177/002215549904701108

von Horsten HH, Johnson SS, SanFrancisco SK, Hastert MC, Whelly SM, Cornwall GA (2007) Oligomerization and transglutaminase cross-linking of the cystatin CRES in the mouse epididymal lumen: potential mechanism of extracellular quality control. J Biol Chem 282 (45):32912–32923. doi:10.1074/jbc.M703956200

Yamaguchi Y, Shao Z, Sharif S, Du XY, Myles T, Merchant M, Harsh G, Glantz M, Recht L, Morser J, Leung LL (2013) Thrombin-cleaved fragments of osteopontin are overexpressed in malignant glial tumors and provide a molecular niche with survival advantage. J Biol Chem 288(5):3097–3111. doi:10.1074/jbc.M112.362954

Yokasaki Y, Sheppard D (2000) Mapping of the cryptic integrin-binding site in osteopontin suggests a new mechanism by which thrombin can regulate inflammation and tissue repair. Trends Cardiovasc Med 10(4):155–159

Yokosaki Y, Monis H, Chen J, Sheppard D (1996) Differential effects of the integrins α9β1, αvβ3, and αvβ6 on cell proliferative responses to tenascin. Roles of the beta subunit extracellular and cytoplasmic domains. J Biol Chem 271(39):24144–24150
Yokosaki Y, Matsuura N, Sasaki T, Murakami I, Schneider H, Higashiyama S, Saitoh Y, Yamakido M, Taooka Y, Sheppard D (1999) The integrin α9β1 binds to a novel recognition sequence (SVVYGLR) in the thrombin-cleaved amino-terminal fragment of osteopontin. J Biol Chem 274(51):36328–36334
Yokosaki Y, Tanaka K, Higashikawa F, Yamashita K, Eboshida A (2005) Distinct structural requirements for binding of the integrins αvβ6, αvβ3, αvβ5, α5β1 and α9β1 to osteopontin. Matrix Biol 24(6):418–427

Chapter 7
Transglutaminase 2-Mediated Gene Regulation

Soo-Youl Kim

Abstract Two decades ago, transglutaminase 2 (TG2) was considered as an end of cellular signaling, such as in the case of factor XIIIa as the last step in the blood clotting cascade. Recently we have observed that TG2 cross-linking activity can change the gene expression through transcription factor activation via suppressor regulation. Many recent studies have elucidated a clearer picture of TG2 working on gene regulation under various cellular stresses. The stress-inducible nature of TG2 has helped to provide a deeper insight into the stress-related functions of TG2, providing contrast with its role under a resting/normal state. This chapter focuses on the molecular mechanisms related to TG2-mediated gene regulation under various stresses including inflammation, sepsis, and cancer, all of which are recognized as TG2-associated biological states in the field today. Under these conditions, it can be shown that TG2 directly commands gene regulation.

Keywords Transglutaminase 2 • Inflammation • Sepsis • Cancer • NF-κB • p53 • β-catenin

7.1 Introduction

7.1.1 What Is Transglutaminase 2 (TG2)?

Historically, the role of transglutaminase 2 (TG2; EC 2.3.2.13) has been superimposed with that of coagulating enzyme via cross-linking. About 70 years ago, Laki and Lorand reported that an unknown enzyme stabilizes blood fibrin clots (Laki and Lorand 1948), and referred to this unknown factor as fibrin-stabilizing factor, which was later renamed as blood clotting factor XIIIa, a member of the TG family. Around the same time, Folk and Cole identified an isozyme from guinea pig liver, called "TGase C" ("C" for cytosol and the -ase suffix to distinguish transglutaminase from triacylglycerol) or "TGase 2" ("2" from the second cloned TGase in the TGase family; the product of the gene named *TGM2*), and characterized the

S.-Y. Kim (✉)
Cancer Cell and Molecular Biology Branch, Division of Cancer Biology, Research Institute, National Cancer Center, Goyang, Gyonggi-Do 410-769, Republic of Korea
e-mail: kimsooyoul@gmail.com

K. Hitomi et al. (eds.), *Transglutaminases*, DOI 10.1007/978-4-431-55825-5_7

TGase enzymatic reaction as an acyltransfer of peptide-bound glutamine to polyamines or peptide-bound lysine (Folk 1980; Chung and Folk 1970; Folk and Chung 1973). [I prefer to use the term "TGase 2" in my paper to avoid confusion with an abbreviation of "TG" that commonly represents "triglyceride" in biochemistry. Here, however, I have been recommended by the Editor to adopt the term "TG2" for uniformity throughout this book.]

Since these early studies, many efforts focused on the physiological role(s) of TG2 have shown that TG2 is involved in many physiological processes including scar formation, apoptosis, inflammation, cancer, neuronal degenerative diseases, and fibrosis, to name a few [see reviews (Kim 2011; Iismaa et al. 2009; Piacentini et al. 2011)]. Although TG2 contributes to many physiological processes, the general physiological phenotype in three independent TG2 knockout mice has been reported to be normal (De Laurenzi and Melino 2001; Nanda et al. 2001; Kim et al. 2010b).

What is the primary function of TG? There is the most substantial evidence for the role of TG2 in wound healing. TG2 activity is seven-fold above that in the nucleus of liver cells in controls during liver regeneration (Haddox and Russell 1981). The increase in TG2 levels in the nucleus led the authors to propose that gene regulation occurs through aminylation during liver regeneration (Haddox and Russell 1981). It is still unclear how protein cross-linking contributes to tissue remodeling in the nucleus. TG2 may control gene expression through protein modification of proteins involved in gene regulation, by altering protein stability. TG2 is also induced by inflammatory stress such as those induced by anti-cancer drugs and cytokines, physical stress, including wounds and radiation, and biological stress, including viral and bacterial infections [for a review, see (Kim 2006)]. Typically, TG2 is not active until it is induced/activated by stress. Taken together, these data point to a role for TG2 in the host defense system, against infectious organisms or tissue damage; this role is consistent with that of blood clotting factor XIIIa. In this model for TG2 function, the enzyme is maintained in its inactive form until bleeding triggers FXIIIa-mediated fibrin clotting via thrombin activation. From a host defense perspective, FXIIIa is the armed soldier on the front lines and TG2 is the reservation army that is activated as a second-line of defense.

7.1.2 Mechanism of TG2 Induction

TG2 has been shown to be induced by retinoic acid (Moore et al. 1984). The TG2 promoter is activated by ligand activation of either retinoic acid receptor-retinoid X receptor (RAR/RXR) heterodimers or RXR homodimers (Nagy et al. 1996). TG2 induction is required for the clearance of apoptotic macrophages (Rebe et al. 2009).

TG2 expression is also induced by nuclear factor kappa-light-chain-enhancer of activated B cells (NF-κB) activation (Caccamo et al. 2005), as well as other signaling pathways, including the hypoxia-inducible factor-1 alpha (HIF-1α) (Jang et al. 2010) and microRNA mir1285 (Hidaka et al. 2012) pathways. TG2 expression is also induced by inflammatory stresses, including UV radiation, cytokines, and viral infection [for a review, see (Kim 2006)]. Most inflammatory signals are transduced through NF-κB activation. For example, glutamate exposure activates NF-κB, and in this context, one study showed that the TG2 promoter contains NF-κB-binding domains in astrocytes (Caccamo et al. 2005). Another study provides more evidence that TG2 inhibition reduces TG2 protein levels in microglia activated by bacterial lipopolysaccharide (LPS) treatment (Park et al. 2004). This study suggested that LPS-mediated NF-κB activation induces TG2 expression, and provided data supporting TG2-mediated NF-κB activation. TG2 over-expression resulted in extended activation of NF-κB in the presence of tumor necrosis factor alpha (TNF-α), for over 24 h, while in the absence of TG2 induction, NF-κB activity was restored to basal levels within 6 h of TNF-α treatment (Park et al. 2011). The duration of inflammation or the levels of the activated NF-κB may depend on the balance between TG2 and NF-κB inhibitor alpha (I-κBα), both of which are induced by NF-κB. TG2-mediated NF-κB activation is not noticeable under normal conditions. However, canonical NF-κB activation in a background of elevated TG2 levels exacerbates inflammation in a vicious feedback cycle, as suggested by the "Ouroboros" theory (Kim 2011).

Hypoxia induces TG2 expression through a HIF-1α – dependent pathway. A HIF response element (HRE) has been identified in the TG2 promoter, the induction of which also triggers TG2-mediated NF-κB activation (Jang et al. 2010). One other mechanism for TG2 induction is that HIF-1α is induced by von Hippel-Lindau tumor suppressor (pVHL) depletion through TG2-mediated polymerization (Kim et al. 2011). This observation suggests that TG2 over-expression may be responsible for the aberrant stabilization of HIF-1α under normoxic conditions in cancer cells.

There is some recent evidence for the role of miRNA in TG2 induction. Tumor-suppressive activities of 20 down-regulated micro RNAs (miRNAs) in renal cell carcinoma (RCC) specimens have been reported (Hidaka et al. 2012). Restoration of mature miRNAs in cancer cells showed that 14 of these miRNAs markedly inhibited cancer cell proliferation, suggesting that these miRNAs were candidate tumor-suppressive miRNAs in RCC. Among these, miR-1285 directly suppresses TG2 expression in RCC (Hidaka et al. 2012). Therefore, common loss of mir-1285 in RCC likely induces the high expression of TG2.

TG2 is also induced by interleukin-6 (IL-6) treatment in hepatoblastoma cells, and further studies revealed an IL-6-binding domain in the promoter region of TG2 (Ikura et al. 1994). Conversely, IL-6 production in human breast cancer cells is also dependent on TG2 expression (Oh et al. 2011a). TG2 is an important link in IL-6-mediated tumor aggressiveness, and it promotes distant metastasis in breast cancer (Oh et al. 2011a).

7.2 Inflammation

7.2.1 *TG2-Mediated NF-κB Activation*

NF-κB is a major inflammation inducer under stress. NF-κB expression is deregulated in various disease states, including chronic inflammation and cancer (Karin and Greten 2005). Interestingly, many inflammatory diseases are related to high expression of TG2, including celiac disease (Molberg et al. 1998), cystic fibrosis (Maiuri et al. 2008), allergic conjunctivitis (Sohn et al. 2003), asthma (Hallstrand et al. 2010; Hong et al. 2013), arthritis (Dzhambazov et al. 2009; Lauzier et al. 2012), and inclusion body myositis (Macaione et al. 2008; Choi et al. 2004). TG2−/− mice show an inflammatory early response but an arrested extended inflammatory response. NF-κB activity also sharply declines and arrests after challenge in TG2−/− mice [for a recent review, see (Kim 2011)]. These phenomena concur with a series of experiments in animal models. TG2−/− mice showed almost reversed inflammation in unilateral ureteral obstruction (Shweke et al. 2008; Kim et al. 2010b) and ovalbumin-induced allergic asthma (Hong et al. 2013). These results suggest that TG2 is the second driver gene in inflammation, and may be responsible for driving the resolution period in normal inflammatory response. Therefore, uncontrolled expression of TG2 may lead to chronic inflammation in various disease states. We have demonstrated extended NF-κB activity concomitant with TG2 over-expression (Park et al. 2011). TG2 is responsible for the extension of the primary signaling cascade for inflammation and survival, which may be the mechanism through which TG2 regulates the resolution period of inflammation.

7.2.2 *NF-κB Activation Mechanism: I-κBα Depletion*

We identified an I-κBα kinase (IKK) kinase-independent mechanism for NF-κB activation – TG2-mediated I-κBα depletion through proteasomal degradation (Lee et al. 2004; Park et al. 2006). TG2 catalyzes the formation of a strong covalent bond between I-κBα-bound glutamine and lysine residues, resulting in the formation of I-κBα polymers. TG2-mediated polymerization of I-κBα depletes the free I-κBα in the cytosol, resulting in activation of NF-κB (Lee et al. 2004). Subsequently, the polymerized I-κBα is degraded to macromolecules by calpain and completely degraded in the proteasome (Kim et al. 2010a). Biochemical studies of I-κBα revealed that a C-terminal glutamine cluster in I-κBα (a.a. 266–268) and lysine residues at positions 21, 22, and 177 play key roles in the formation of I-κBα polymers (Park et al. 2006). We have also demonstrated that mutation of these positions in I-κBα blocks TG2-mediated NF-κB activation almost completely (Park et al. 2006). Amino acid sequence homology analysis has highlighted some interesting differences between I-κBα and I-κBβ (Fig. 7.1). Although these proteins

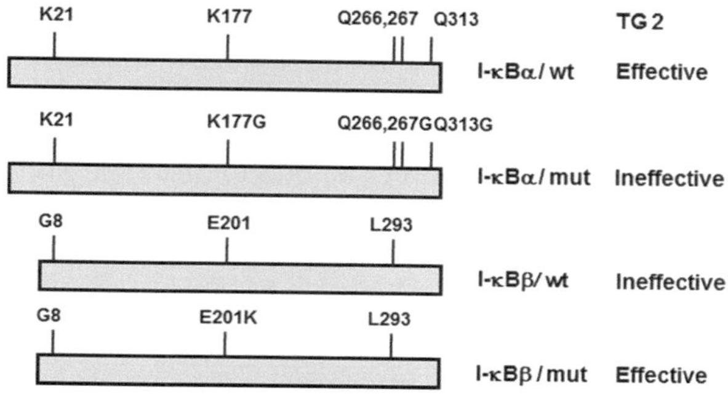

Fig. 7.1 TG2-mediated NF-κB activation through I-κBα polymerization. TG2 binds with I-κBα and forms homo-polymers by cross-linking glutamine and lysine residues (such as K21, K177, Q266, Q267, and Q313) in I-κBα (Park et al. 2006). Mutation of the targeted glutamine and lysine residues prevents TG2-mediated I-κBα polymerization, thereby reversing TG2-mediated NF-κB activation. We found that I-κBβ and I-κBα protein sequences are highly conserved and homologous, except for the TG2 target glutamine and lysine residues such as G8, E201, and L293 (Kim et al. 2008), which TG2 cannot cross-link, resulting in stable NF-κB inhibition upon TG2 overexpression. Mutation of E201 to lysine in I-κBβ renders it susceptible to TG2-mediated polymerization, which then promotes NF-κB activation

share all the important signaling sites, the latter does not contain the major sites targeted by TG2; I-κBβ is missing the glutamines at 266–268 and the lysines at 21 and 177 (Kim et al. 2008). We also found that I-κBβ is independent of TG2-mediated polymerization in vitro as well as suppressing TG2-mediated NF-κB activation (Kim et al. 2008). It is concurred that I-κBβ hardly degrades while I-κBα degrades immediately in response to most inflammatory stimuli. These data suggest that I-κB binding to TG2 is very efficient, and that NF-κB activation is dependent on TG2 activity.

7.2.3 NF-κB Activation Mechanism: PPAR-γ Sequestration

TG2 activation contributes to constant NF-κB activation through I-κBα depletion in inflammation. However, another pathway of TG2-mediated NF-κB regulation has been proposed in cystic fibrosis (Luciani et al. 2011). In this study, the authors reported that a mutated cystic fibrosis transmembrane conductance regulator (CFTR)-TG2-mediated aggresome initiates inflammation in cystic fibrosis, which accompanies the vicious cycle of an increase in NF-κB activity (Luciani et al. 2010). CFTR mutation leads to an increase of TG2 expression and activity that, in turn, down-regulates the anti-inflammatory effect of peroxisome proliferator-activated receptor-γ (PPAR-γ). TG2 induces dysfunction of PPAR-γ through cross-linking, possibly in conjunction with Beclin-1 and sequestration into aggresomes (Maiuri et al. 2008). Although there is direct cross-linking of TG2 to PPAR-γ, the detailed mechanism of PPAR-γ modification by TG2 remains to be elucidated. The F508del CFTR mutant increases the TG2 level, likely due to the associated increase of reactive oxygen species, but the mechanism of induction has not been clearly demonstrated yet. However, one study showed that TG2 inhibition reverses inflammation, together with recovery of PPAR-γ levels, as PPAR-γ is a known suppressor of NF-κB (Luciani et al. 2010). Therefore, TG2 inhibition in cystic fibrosis has a definite benefit, to control inflammation through reverse NF-κB activation.

7.3 Sepsis

7.3.1 TG2-Mediated Confusion in Sepsis

What is the consequence of bacterial infection in the absence of TG? This question has been answered in *Drosophila* using TG knock-down mutants (knock-down of the human FXIII homologs). *Drosophila* larvae with reduced TG levels show increased mortality after septic injury (Wang et al. 2010). The same larvae are also more susceptible to a natural infection involving entomopathogenic nematodes and their symbiotic bacteria, while neither phagocytosis nor the Toll pathway contribute to immunity. Akin to the effects of TG2 loss in flies, human TG2 also contributes to survival through apoptotic challenge. TG2 has been found to protect against the apoptotic process through depletion of caspase 3 (Yamaguchi and Wang 2006) and cathepsin D (Kim et al. 2013), by promoting protein polymerization, thus playing a key role in survival. Induction of TG2 through environmental stress thus appears a natural and essential consequence for ensuring cellular survival. Invading pathogens are recognized and agglutinated by lectins and immobilized by TG, which are finally killed by antimicrobial peptides from hemocytes in horseshoe crabs (Kawabata 2010). With regard to liver injury, TG2 exhibits a protective role in sepsis. Carbon tetrachloride (CCl_4) treatment of TG2−/− mice increases the incidence of post-injury death, relative to that in wild-type mice (Nardacci et al. 2003). The authors postulated that this enhanced mortality is due to the loss of the protective role TG2 plays in normal tissues, by supporting tissue stability and repair. In addition, lead nitrite ($PbNO_3$) treatment for 5 days in TG2−/− mice resulted in increased inflammation and apoptosis in the liver (Falasca et al. 2005), which was believed to be due to decreased apoptosis-dependent phagocytosis by macrophages (Falasca et al. 2005) or impaired maturation of autophagosomes (D'Eletto et al. 2009).

On the contrary, TG2−/− mice reversed LPS-induced septic shock as well as hypoxia-induced kidney damage (Yoo et al. 2013). TG2 plays a key role in enhancing NF-κB activation without activating the canonical inflammatory pathway (Kim 2011). Although NF-κB activation contributes to cellular survival, TG2−/− mice were resistant to LPS-induced septic shock. TG2−/− mice showed decreased hepatic apoptosis following treatment with an anti-Fas antibody or alcohol (Tatsukawa et al. 2009). Mallory body (MB) formation assays in TG2−/− mice revealed decreased MB formation and liver hypertrophy compared with wild-type mice (Tatsukawa et al. 2009). The authors proposed a model wherein TG2 induces hepatocyte apoptosis via Sp1 cross-linking and inactivation, with resultant inhibition of c-Met expression, which is required for hepatic cell viability (Tatsukawa et al. 2009). Sp1 depletion by TG2 might be one explanation for the increased survival of TG2−/− mice in MB formation assays (see Chap. 2). However, activation of immune cells such as Kupffer cells may be responsible for MB formation and liver hypertrophy induced in the MB assay or alcoholic liver hypertrophy, because both of these stressors significantly induce activation of immune cells. TG2 may contribute to activation of Kupffer cells, which may in turn lead to increased IL-6-mediated recruitment of immune cells to the damaged liver tissue (Tacke et al. 2009). Recently, another study reported that TG2 is responsible for IL-17 production by T cells following the induction of IL-6 in lung epithelial cells in a lung fibrosis model (Oh et al. 2011b). Therefore, the effect of TG2 activation in Kupffer cells warrants further investigation.

7.3.2 Differential Outcome in Various Septic Challenges

TG2 is one of the most important targets in inflammation and cancer. Many reports have shown that inflammation and malignant oncogenesis have been reversed by TG2 gene ablation or TG2 inhibition using synthetic inhibitors. However, there is a conflict in the outcome of TG2 contribution in sepsis. What factors drive differential outcomes of TG2-mediated sepsis?

Acute sepsis can be induced by cytokines such as TNF-α, and by biological products such as LPS (Fig. 7.2). All of these agents cause systemic inflammation, which is characterized by hemodynamic shock and liver toxicity. However, the outcomes of different septic shock models were contradictory in TG2−/− mice (Yoo et al. 2013). TNF-α-dependent septic shock resulted in increased liver damage in TG2−/− mice compared with wild-type mice (Yoo et al. 2013), and was accompanied by increased levels of caspase 3 and CTSD in the damaged liver. Conversely, LPS-induced septic shock resulted in ablation of inflammatory endotoxic shock in TG2−/− mice and decreased liver injury. We found that TG2 protected liver tissue from TNF-α-dependent septic shock by reducing the expression of caspase 3 and CTSD. However, TG2 differently participated in increasing the hemodynamic shock in LPS-induced septic shock through macrophage activation, rather than protecting direct liver damage.

TG2−/− mice respond differently to different types of septic shock. LPS/Gal-induced sepsis is more damaging to immune surveillance cells than to tissue cells, while TNF-α/ActD-induced sepsis affects liver tissue cells rather than immune cells. LPS/Gal-induced sepsis stimulates the production of multiple endotoxic cytokines through macrophage activation, leading to tissue damage, while TNF-α/ActD-induced sepsis triggers severe liver damage through activation of

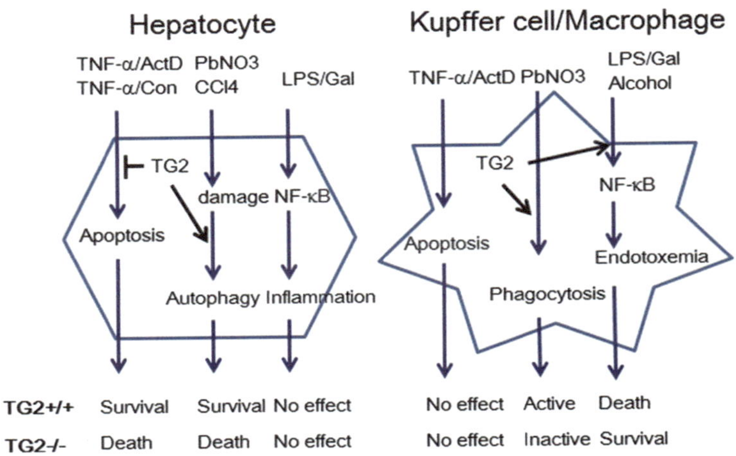

Fig. 7.2 Differential effects of TG2 in sepsis. TNF-α/ActD-induced septic shock results in severe liver damage in TG2−/− mice, compared with wild-type mice (Yoo et al. 2013), increasing caspase 3 and cathepsin D expression in the liver (Kim et al. 2013). In contrast, LPS/Gal-induced septic shock results in ablation of inflammatory endotoxic shock in TG2−/− mice (Yoo et al. 2013), because macrophage activation largely depends on TG2 expression (Schroff et al. 1981). Different septic shocks yield contradictory outcomes in TG2−/− mice; however, TG2 has a consistent function, depending on the cellular context, such as in hepatocytes or Kupffer cells. The most affected target of $PbNO_3$ (Falasca et al. 2005) or CCl_4 (Nardacci et al. 2003) shock appears to be hepatocytes, whereas LPS/Gal (Yoo et al. 2013) or alcohol (Tatsukawa et al. 2009) impact Kupffer cells or macrophages

apoptosis. Treatment with either LPS/Gal or TNF-α/ActD appears to induce TG2 expression through NF-κB activation (Kuncio et al. 1998). To mount an effective defense against agents such as LPS, TG2 must be induced in macrophages. To protect against chemically induced tissue damage, TG2 must be induced in the damaged tissue. Thus, TG2−/− mice are affected in different ways depending on the source of septic shock.

In summary, TG2 has a protective role against apoptotic stress in epithelial tissue, while it has a promoting role in inflammatory endotoxic shock. Therefore, it would be dangerous to apply TG2 inhibition against TNF-α – induced inflammation due to liver damage. However, it is beneficial to apply TG2 inhibition against LPS-induced inflammation because uncontrolled immune cell activation-mediated endotoxemia depends on TG2 activity.

7.4 Cancer

Cancer is very much a heterogenetic disease; therefore, it is not surprising that the roles of TG2 in cancer have been described in many different aspects of cancer progression, such as drug resistance, invasion, epithelial-to-mesenchymal transition

(Marktel et al. 2003), angiogenesis, and the cell cycle, among others [see reviews (Li et al. 2011; Brown 2013; Lentini et al. 2013; Bergamini et al. 2005) and Chap. 11]. In this section, we seek to clarify the role of TG2 in the specific context of cancer. A previous study proposed that the TG2 level is correlated with an increase of anti-cancer drug resistance, although the detailed mechanism for increased resistance was not clear (Han and Park 1999). The proposed roles of TG2 are important, because TG2 inhibition has a synergistic anti-cancer effect in conjunction with anti-cancer drugs (Budillon et al. 2013). In subsequent studies, another group applied a TG2 inhibitor, 3-halo-4,5-dihydroisoxazole, to glioblastoma for sensitization against the anti-cancer drug N,N′-bis(2-chloroethyl)-N-nitrosourea (Choi et al. 2005; Yuan et al. 2007). Another study reported that TG2 knock-down sensitizes BAX−/− colon cancer cells to thapsigargin treatment (Yamaguchi and Wang 2006). In this study, TG2 was shown to directly target caspase 3 for cross-linking, thus inhibiting caspase 3 activation. However, caspase 3 activation is downstream of gene regulation.

A recent study reported that over-expression of TG2 in neuroblastoma cells significantly protects against etoposide-induced cell death (Tucholski 2010). This report showed that only the cross-linking activity of TG2 is responsible for suppression of p53-mediated apoptosis. Although some roles of TG2 in cancer promotion have been suggested, specific signaling pathways regulated by TG2 or specific cross-linking targets of TG2 have not been clearly identified. TG2 potentiates epidermal growth factor receptor (EGFR) signaling and c-Src activation, but a specific target of TG2 in this model remains to be identified (Boroughs et al. 2014; Condello et al. 2013). TG2 was also shown to induce EMT in several cancer types, but, again, specific targets of TG2 in driving EMT are yet to be defined (Kumar et al. 2012; Oh et al. 2011a). Other studies have shown that TG2 induces Akt activation through as-yet unknown mechanisms. Along these lines, recently, we have found that TG2 knock-down induces complete reversal of RCC through revival of p53 level. The observation that restoration of p53 can be induced by ablation of TG2 is a significant breakthrough in further application of TG2 inhibition against cancer.

7.4.1 TG2-Mediated p53 Regulation

Decreased TG2 expression induces p53-mediated apoptosis in RCC (Ku et al. 2013, 2014). The down-regulation of TG2 stabilizes p53 expression, thereby inducing about a 10-fold increase in apoptosis in RCC cell lines (Ku et al. 2013). These data demonstrate, for the first time, that a single gene knock-down of TG2 is enough to induce apoptosis in RCC. Therefore, it is possible that a specific TG2 inhibitor may be useful for therapeutic approaches to RCC. TG2 inhibitors developed based on the quinoxaline derivatives of GK13 exhibited TG2 inhibitory activity, which was then optimized to GK921 [3-(phenylethynyl)-2-(2-(pyridin-2-yl)ethoxy)pyrido[3,2-b]pyrazine] (Lee et al. 2013). Treatment with the TG2 inhibitor GK921 alone was shown to reverse the levels of p53 in an RCC tumor growth study (Ku et al. 2014). This is an important finding because targeting TG2 may have benefits as a mono-therapeutic approach to RCC.

Loss of p53 activity in cancer is associated with more aggressive disease and poor prognosis. Many cancer types have p53 mutations that quench its tumor-suppressive role. In normal cells, the typical degradation pathway of p53 is mouse double minute 2 homolog (MDM2; HDM2 in human)-dependent ubiquitination and consequent proteasomal degradation. MDM2 is the principal E3 ligase for the tumor suppressor p53, and over-expression of MDM2 is related to tumorigenesis such as in acute leukemia and lymphoma (Xu-Monette et al. 2013). Several pharmacological approaches have been developed to disrupt MDM2-p53 binding, thus stabilizing p53 levels and increasing p53 activity (Crawford and Irvine 2013). However, fewer than 5 % of RCCs harbor p53 mutations, and there is no report for MDM2 over-expression in RCC, even though p53 levels are down-regulated (Cancer Genome Atlas Research 2013). TG2 expression depletes p53 by direct binding and cross-linking, which results in increased tumor cell survival (Ku et al. 2013). TG2 targets the DNA-binding domain (102–292 a.a.) of p53. Upon deletion of this region (102–292 a.a.), p53 completely loses its binding capacity to TG2 (Ku et al. 2013). Site-directed mutations of potential TG2 target lysine and glutamine residues in p53 (K164G and Q192G) increase p53 stability in RCC cell lines (Ku et al. 2013). TG2-p53 complexes are specifically transported to the autophagosome and are further processed in the lysosome (Ku et al. 2013). It is not clear whether p62 binds to the ubiquitinated-p53 or -TG2, although TG2-p53 complex clearly binds to p62. The detailed translocation mechanism of p53-TG2 complexes remains to be elucidated. TG2-mediated autophagic degradation of p53 is a novel mechanism, which may explain p53 regulation independent of MDM-2 expression in RCC and other cancers. TG2-mediated p53 depletion through autophagy provides a double advantage to RCC tumors, supplying biomass through p53 degradation as well as removing a critical tumor suppressor, thus promoting tumor proliferation.

7.4.2 TG2-Mediated HIF-1α Activation

HIF-1α is tightly regulated under normoxia. HIF-1α is hydroxylated at proline residues by HIF prolyl hydroxylase domain enzymes, which consequently leads to HIF1-α degradation by the E3 ligase VHL (Semenza 2003). However, the mechanisms that promote the increased HIF-1α levels under normoxic condition in cancer have not yet been characterized completely. HIF-1α protein appears to be stabilized, instead of an increase in gene expression (Kim et al. 2011). We observed that TG2 knock-down using siRNA reduces levels of insulin-like growth factor-1 receptor beta (IGF-1Rβ) and HIF-1α in breast cancer cells (Kim et al. 2011). Interestingly, a common suppressor of IGF-1Rβ and HIF-1α, pVHL, correlates inversely with TG2 expression levels (Kim et al. 2011). Moreover, one of the most down-regulated genes by pVHL is TG2 (Wykoff et al. 2000).

TG2 reduces pVHL protein, but does not affect pVHL at the transcription level (Kim et al. 2011). TG2 catalyzes the cross-linking of pVHL to polymers that are degraded in the proteasome. Increase of IGF-1Rβ through pVHL depletion by TG2 induces NF-κB activation (Kim et al. 2011). The most interesting aspect of this regulation is that HIF-1α levels are directly regulated by TG2 in breast cancer cells through TG2-mediated pVHL depletion (Kim et al. 2011). TG2 is also induced by HIF-1α through the HRE binding site in the TG2 promoter region (Jang et al. 2010). These data suggest a feedback model, in which TG2 over-expression in cancer cells stabilizes HIF-1α and induces TG2 expression, as suggested by the "Ouroboros" theory of NF-κB activation (Kim 2011). In another mode of regulation, it is known that TG2 is a pVHL target gene that is often targeted for suppression in cancers (Wykoff et al. 2000). Therefore, pVHL down-regulation by TG2 may synergistically induce TG2 expression in a feedback loop in cancer.

7.4.3 TG2-Mediated β-Catenin Activation

The Wnt/β-catenin signaling pathway is important in normal development. However, uncontrolled β-catenin activation due to suppressor mutations, including adenomatous polyposis coli (APC), contributes to the pathogenesis of several cancers [for a review, see (Mosimann et al. 2009)]. The canonical Wnt pathway is activated by the binding of a Wnt protein family ligand to a cell surface receptor complex with a Frizzled family protein and a co-receptor, typically, LRP5 or LRP6 (LRP5/6) (Tamai et al. 2000). Binding of a Wnt ligand leads to LRP5/6 oligomerization and phosphorylation of its cytoplasmic tail, which, in turn, results in recruitment of Disheveled, thereby disrupting the destruction complex containing axin, APC, and glycogen synthase kinase 3 that targets β-catenin for proteasomal degradation (Mosimann et al. 2009). By Disheveled recruitment, stabilized β-catenin can then translocate to the nucleus, where it forms a complex with transcription factors of the TCF/LEF family and activates expression of target genes (Mosimann et al. 2009).

TG2 can be secreted out to the extracellular matrix via macrovesicle-mediated transportation (Antonyak et al. 2011). The exact contribution of extracellular TG2 is not yet determined and it is not known whether TG2-containing vesicles can be taken up again by adjacent normal cells or remain localized in the matrix in a catalytically active form. Extracellular TG2 alone can activate β-catenin signaling by binding to the LRP5 receptor in primary mouse vascular smooth muscle cells (Faverman et al. 2008). Recently, TG2 has been shown to synergize with LRP6 in the activation of β-catenin-dependent gene expression in Cos-7 cells (Deasey et al. 2013). LRP6 is also a substrate for TG2-mediated cross-linking. Incubation of the extracellular domain of LRP6 with TG2 resulted in an extra higher molecular weight band (~200 kDa), the protein band of the highest molecular weight, likely an oligomer of LRP6 (Deasey et al. 2013). Further studies are needed to determine whether extracellular TG2 activates the ligand of the LRP5/6 receptors through cross-linking-mediated polymerization.

Another model for TG2-mediated β-catenin activation has been proposed in a recent report (Condello et al. 2013). TG2 is associated with fibronectin efficiently and forms a complex with β-integrin, which recruits c-Src, enabling phosphorylation of β-catenin at Tyr654 and its release from E-cadherin (Coluccia et al. 2006). The released β-catenin is available for translocation to the nucleus for activation of transcription. One report suggested that TG2 is often associated with c-Src in NIH3T3 fibroblasts (Boroughs et al. 2014). TG2 has also been shown to bind to

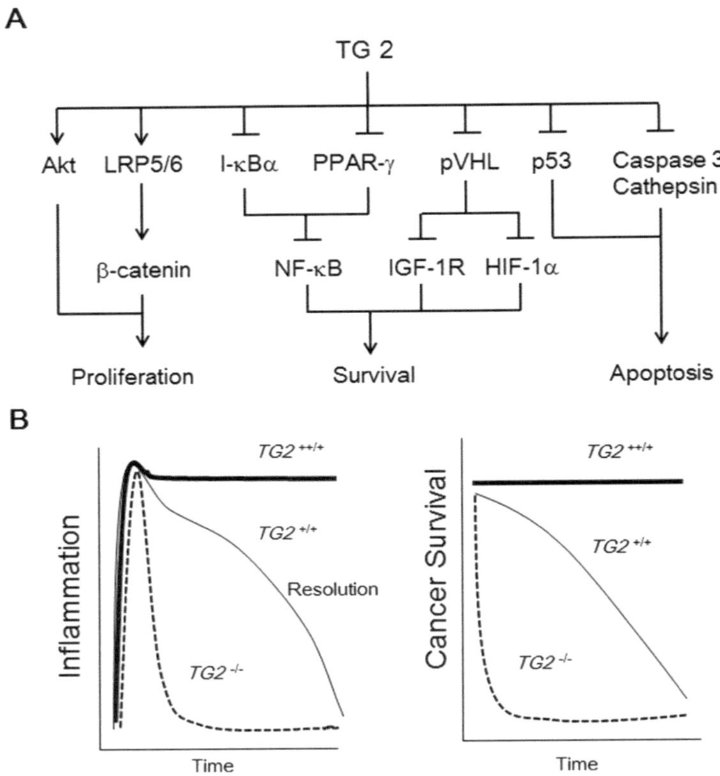

Fig. 7.3 TG2-mediated gene regulation. (**a**) TG2 directly regulates apoptosis signaling by targeting p53 and caspase 3. TG2 promotes survival through indirect regulation, by targeting suppressors, including I-κBα, PPAR-γ, and pVHL, which induce NF-κB, IGF-1R, and HIF-1α. TG2 is responsible for activation of β-catenin and Akt signaling, but the exact molecular mechanisms need to be elucidated. (**b**) Under normal physiological conditions, TG2 expression is regulated such that its levels promote survival in response to stress damage. However, uncontrolled over-expression of TG2 in pathological states contributes to aberrant activation of survival pathways and promotes pathogenesis of chronic inflammatory diseases and cancers

the c-Src-phosphoinositol 3-kinase (PI3K) complex, inducing PI3K phosphorylation by c-Src (Boroughs et al. 2014).

Although β-catenin activation can be explained by extracellular TG2-mediated signaling, intracellular TG2 may also participate in β-catenin activation through regulation of intracellular β-catenin regulators, because TG2 is most abundant in the cytosol (Chung et al. 1970).

7.5 Conclusion

Under normal conditions, the overall phenotype of TG2−/− mice is normal because primary defense systems such as innate immunity are still available and operational. The several lines of evidence discussed in this chapter support the conclusion that TG2 serves in the secondary response that follows the primary immune response, particularly because TG2 is induced by inflammatory signaling. TG2 is responsible for potentiating the defense system against threats from pathogens. In contrast, constant induction of TG2 by hypoxia or cytokines in cancer may contribute to the increase of pathogenesis through depletion of tumor suppressors such as p53 and pVHL, as well as by activating oncogenic genes, including HIF-1α and NF-κB (Fig. 7.3). In this report, I have also summarized the evidence that supports the application of TG2 inhibition or TG2 gene deletion in therapeutic strategies in cancer and inflammatory diseases. To this end, the specific mechanisms of TG2-mediated molecular changes in these disease models warrant further investigation. In the near future, a more detailed focus on TG2-mediated post-translational modification of key proteins will help elucidate the role of TG2 in the regulation of gene expression in cancer and inflammatory diseases.

Acknowledgments This work was supported by a research grant from the National Cancer Center of Korea (NCC1410280-2).

References

Antonyak MA, Li B, Boroughs LK, Johnson JL, Druso JE, Bryant KL, Holowka DA, Cerione RA (2011) Cancer cell-derived microvesicles induce transformation by transferring tissue transglutaminase and fibronectin to recipient cells. Proc Natl Acad Sci U S A 108(12):4852–4857. doi:10.1073/pnas.1017667108

Bergamini CM, Griffin M, Pansini FS (2005) Transglutaminase and vascular biology: physiopathologic implications and perspectives for therapeutic interventions. Curr Med Chem 12 (20):2357–2372

Boroughs LK, Antonyak MA, Cerione RA (2014) A novel mechanism by which tissue transglutaminase activates signaling events that promote cell survival. J Biol Chem 289 (14):10115–10125. doi:10.1074/jbc.M113.464693

Brown KD (2013) Transglutaminase 2 and NF-kappaB: an odd couple that shapes breast cancer phenotype. Breast Cancer Res Treat 137(2):329–336. doi:10.1007/s10549-012-2351-7

Budillon A, Carbone C, Di Gennaro E (2013) Tissue transglutaminase: a new target to reverse cancer drug resistance. Amino Acids 44(1):63–72. doi:10.1007/s00726-011-1167-9

Caccamo D, Campisi A, Curro M, Aguennouz M, Li Volti G, Avola R, Ientile R (2005) Nuclear factor-kappab activation is associated with glutamate-evoked tissue transglutaminase up-regulation in primary astrocyte cultures. J Neurosci Res 82(6):858–865. doi:10.1002/jnr.20683

Cancer Genome Atlas Research Network (2013) Comprehensive molecular characterization of clear cell renal cell carcinoma. Nature 499(7456):43–49. doi:10.1038/nature12222

Choi YC, Kim TS, Kim SY (2004) Increase in transglutaminase 2 in idiopathic inflammatory myopathies. Eur Neurol 51(1):10–14. doi:10.1159/000074911

Choi K, Siegel M, Piper JL, Yuan L, Cho E, Strnad P, Omary B, Rich KM, Khosla C (2005) Chemistry and biology of dihydroisoxazole derivatives: selective inhibitors of human transglutaminase 2. Chem Biol 12(4):469–475. doi:10.1016/j.chembiol.2005.02.007

Chung SI, Folk JE (1970) Mechanism of the inactivation of guinea pig liver transglutaminase by tetrathionate. J Biol Chem 245(4):681–689

Chung SI, Shrager RI, Folk JE (1970) Mechanism of action of guinea pig liver transglutaminase. VII Chemical and stereochemical aspects of substrate binding and catalysis. J Biol Chem 245(23):6424–6435

Coluccia AM, Benati D, Dekhil H, De Filippo A, Lan C, Gambacorti-Passerini C (2006) SKI-606 decreases growth and motility of colorectal cancer cells by preventing pp 60(c-Src)-dependent tyrosine phosphorylation of beta-catenin and its nuclear signaling. Cancer Res 66(4):2279–2286. doi:10.1158/0008-5472.CAN-05-2057

Condello S, Cao L, Matei D (2013) Tissue transglutaminase regulates beta-catenin signaling through a c-Src-dependent mechanism. FASEB J Off Publ Fed Am Soc Exp Biol 27(8):3100–3112. doi:10.1096/fj.12-222620

Crawford LJ, Irvine AE (2013) Targeting the ubiquitin proteasome system in haematological malignancies. Blood Rev 27(6):297–304. doi:10.1016/j.blre.2013.10.002

De Laurenzi V, Melino G (2001) Gene disruption of tissue transglutaminase. Mol Cell Biol 21(1):148–155. doi:10.1128/MCB.21.1.148-155.2001

Deasey S, Nurminsky D, Shanmugasundaram S, Lima F, Nurminskaya M (2013) Transglutaminase 2 as a novel activator of LRP6/beta-catenin signaling. Cell Signal 25(12):2646–2651. doi:10.1016/j.cellsig.2013.08.016

D'Eletto M, Farrace MG, Falasca L, Reali V, Oliverio S, Melino G, Griffin M, Fimia GM, Piacentini M (2009) Transglutaminase 2 is involved in autophagosome maturation. Autophagy 5(8):1145–1154

Dzhambazov B, Lindh I, Engstrom A, Holmdahl R (2009) Tissue transglutaminase enhances collagen type II-induced arthritis and modifies the immunodominant T-cell epitope CII260-270. Eur J Immunol 39(9):2412–2423. doi:10.1002/eji.200939438

Falasca L, Iadevaia V, Ciccosanti F, Melino G, Serafino A, Piacentini M (2005) Transglutaminase type II is a key element in the regulation of the anti-inflammatory response elicited by apoptotic cell engulfment. J Immunol 174(11):7330–7340

Faverman L, Mikhaylova L, Malmquist J, Nurminskaya M (2008) Extracellular transglutaminase 2 activates beta-catenin signaling in calcifying vascular smooth muscle cells. FEBS Lett 582(10):1552–1557. doi:10.1016/j.febslet.2008.03.053

Folk JE (1980) Transglutaminases. Annu Rev Biochem 49:517–531. doi:10.1146/annurev.bi.49.070180.002505

Folk JE, Chung SI (1973) Molecular and catalytic properties of transglutaminases. Adv Enzymol Relat Areas Mol Biol 38:109–191

Haddox MK, Russell DH (1981) Increased nuclear conjugated polyamines and transglutaminase during liver regeneration. Proc Natl Acad Sci U S A 78(3):1712–1716

Hallstrand TS, Wurfel MM, Lai Y, Ni Z, Gelb MH, Altemeier WA, Beyer RP, Aitken ML, Henderson WR (2010) Transglutaminase 2, a novel regulator of eicosanoid production in

asthma revealed by genome-wide expression profiling of distinct asthma phenotypes. PLoS One 5(1):e8583. doi:10.1371/journal.pone.0008583
Han JA, Park SC (1999) Reduction of transglutaminase 2 expression is associated with an induction of drug sensitivity in the PC-14 human lung cancer cell line. J Cancer Res Clin Oncol 125(2):89–95
Hidaka H, Seki N, Yoshino H, Yamasaki T, Yamada Y, Nohata N, Fuse M, Nakagawa M, Enokida H (2012) Tumor suppressive microRNA-1285 regulates novel molecular targets: aberrant expression and functional significance in renal cell carcinoma. Oncotarget 3(1):44–57
Hong GU, Park BS, Park JW, Kim SY, Ro JY (2013) IgE production in CD40/CD40L cross-talk of B and mast cells and mediator release via TGase 2 in mouse allergic asthma. Cell Signal 25 (6):1514–1525. doi:10.1016/j.cellsig.2013.03.010
Iismaa SE, Mearns BM, Lorand L, Graham RM (2009) Transglutaminases and disease: lessons from genetically engineered mouse models and inherited disorders. Physiol Rev 89 (3):991–1023. doi:10.1152/physrev.00044.2008
Ikura K, Shinagawa R, Suto N, Sasaki R (1994) Increase caused by interleukin-6 in promoter activity of guinea pig liver transglutaminase gene. Biosci Biotechnol Biochem 58 (8):1540–1541
Jang GY, Jeon JH, Cho SY, Shin DM, Kim CW, Jeong EM, Bae HC, Kim TW, Lee SH, Choi Y, Lee DS, Park SC, Kim IG (2010) Transglutaminase 2 suppresses apoptosis by modulating caspase 3 and NF-kappaB activity in hypoxic tumor cells. Oncogene 29(3):356–367. doi:10.1038/onc.2009.342
Karin M, Greten FR (2005) NF-kappaB: linking inflammation and immunity to cancer development and progression. Nat Rev Immunol 5(10):749–759. doi:10.1038/nri1703
Kawabata S (2010) Immunocompetent molecules and their response network in horseshoe crabs. Adv Exp Med Biol 708:122–136
Kim SY (2006) Transglutaminase 2 in inflammation. Front Biosci J Virtual Library 11:3026–3035
Kim SY (2011) Transglutaminase 2: a new paradigm for NF-kappaB involvement in disease. Adv Enzymol Relat Areas Mol Biol 78:161–195
Kim JM, Voll RE, Ko C, Kim DS, Park KS, Kim SY (2008) A new regulatory mechanism of NF-kappaB activation by I-kappaBbeta in cancer cells. J Mol Biol 384(4):756–765. doi:10.1016/j.jmb.2008.10.010
Kim DS, Han BG, Park KS, Lee BI, Kim SY, Bae CD (2010a) I-kappaBalpha depletion by transglutaminase 2 and mu-calpain occurs in parallel with the ubiquitin-proteasome pathway. Biochem Biophys Res Commun 399(2):300–306. doi:10.1016/j.bbrc.2010.07.078
Kim DS, Kim B, Tahk H, Kim DH, Ahn ER, Choi C, Jeon Y, Park SY, Lee H, Oh SH, Kim SY (2010b) Transglutaminase 2 gene ablation protects against renal ischemic injury by blocking constant NF-kappaB activation. Biochem Biophys Res Commun 403(3–4):479–484. doi:10.1016/j.bbrc.2010.11.063
Kim DS, Choi YB, Han BG, Park SY, Jeon Y, Kim DH, Ahn ER, Shin JE, Lee BI, Lee H, Hong KM, Kim SY (2011) Cancer cells promote survival through depletion of the von Hippel-Lindau tumor suppressor by protein crosslinking. Oncogene 30(48):4780–4790. doi:10.1038/onc.2011.183
Kim SJ, Kim KH, Ahn ER, Yoo BC, Kim SY (2013) Depletion of cathepsin D by transglutaminase 2 through protein cross-linking promotes cell survival. Amino Acids 44(1):73–80. doi:10.1007/s00726-011-1089-6
Ku BM, Kim DS, Kim KH, Yoo BC, Kim SH, Gong YD, Kim SY (2013) Transglutaminase 2 inhibition found to induce p53 mediated apoptosis in renal cell carcinoma. FASEB J Off Publ Fed Am Soc Exp Biol 27(9):3487–3495. doi:10.1096/fj.12-224220
Ku BM, Kim SJ, Kim N, Hong D, Choi YB, Lee SH, Gong YD, Kim SY (2014) Transglutaminase 2 inhibitor abrogates renal cell carcinoma in xenograft models. J Cancer Res Clin Oncol 140 (5):757–767. doi:10.1007/s00432-014-1623-5
Kumar A, Xu J, Sung B, Kumar S, Yu D, Aggarwal BB, Mehta K (2012) Evidence that GTP-binding domain but not catalytic domain of transglutaminase 2 is essential for epithelial-

to-mesenchymal transition in mammary epithelial cells. Breast Cancer Res BCR 14(1):R4. doi:10.1186/bcr3085

Kuncio GS, Tsyganskaya M, Zhu J, Liu SL, Nagy L, Thomazy V, Davies PJ, Zern MA (1998) TNF-alpha modulates expression of the tissue transglutaminase gene in liver cells. Am J Physiol 274(2 Pt 1):G240–245. http://www.ncbi.nlm.nih.gov/pubmed/9486175

Laki K, Lorand L (1948) On the solubility of fibrin clots. Science 108(2802):280. doi:10.1126/science.108.2802.280

Lauzier A, Charbonneau M, Paquette M, Harper K, Dubois CM (2012) Transglutaminase 2 cross-linking activity is linked to invadopodia formation and cartilage breakdown in arthritis. Arthritis Res Ther 14(4):R159. doi:10.1186/ar3899

Lee J, Kim YS, Choi DH, Bang MS, Han TR, Joh TH, Kim SY (2004) Transglutaminase 2 induces nuclear factor-kappaB activation via a novel pathway in BV-2 microglia. J Biol Chem 279 (51):53725–53735. doi:10.1074/jbc.M407627200

Lee SH, Kim N, Kim SJ, Song J, Gong YD, Kim SY (2013) Anti-cancer effect of a quinoxaline derivative GK13 as a transglutaminase 2 inhibitor. J Cancer Res Clin Oncol 139 (8):1279–1294. doi:10.1007/s00432-013-1433-1

Lentini A, Abbruzzese A, Provenzano B, Tabolacci C, Beninati S (2013) Transglutaminases: key regulators of cancer metastasis. Amino Acids 44(1):25–32. doi:10.1007/s00726-012-1229-7

Li B, Cerione RA, Antonyak M (2011) Tissue transglutaminase and its role in human cancer progression. Adv Enzymol Relat Areas Mol Biol 78:247–293

Luciani A, Villella VR, Esposito S, Brunetti-Pierri N, Medina D, Settembre C, Gavina M, Pulze L, Giardino I, Pettoello-Mantovani M, D'Apolito M, Guido S, Masliah E, Spencer B, Quaratino S, Raia V, Ballabio A, Maiuri L (2010) Defective CFTR induces aggresome formation and lung inflammation in cystic fibrosis through ROS-mediated autophagy inhibition. Nat Cell Biol 12 (9):863–875. doi:10.1038/ncb2090

Luciani A, Villella VR, Esposito S, Brunetti-Pierri N, Medina DL, Settembre C, Gavina M, Raia V, Ballabio A, Maiuri L (2011) Cystic fibrosis: a disorder with defective autophagy. Autophagy 7(1):104–106

Macaione V, Aguennouz M, Mazzeo A, De Pasquale MG, Russo M, Toscano A, De Luca G, Di Giorgio RM, Vita G, Rodolico C (2008) Expression of transglutaminase 2 does not differentiate focal myositis from generalized inflammatory myopathies. Acta Neurol Scand 117 (6):393–398. doi:10.1111/j.1600-0404.2007.00957.x

Maiuri L, Luciani A, Giardino I, Raia V, Villella VR, D'Apolito M, Pettoello-Mantovani M, Guido S, Ciacci C, Cimmino M, Cexus ON, Londei M, Quaratino S (2008) Tissue transglutaminase activation modulates inflammation in cystic fibrosis via PPARgamma down-regulation. J Immunol 180(11):7697–7705

Marktel S, Marin D, Foot N, Szydlo R, Bua M, Karadimitris A, De Melo VA, Kotzampaltiris P, Dazzi F, Rahemtulla A, Olavarria E, Apperley JF, Goldman JM (2003) Chronic myeloid leukemia in chronic phase responding to imatinib: the occurrence of additional cytogenetic abnormalities predicts disease progression. Haematologica 88(3):260–267

Molberg O, McAdam SN, Korner R, Quarsten H, Kristiansen C, Madsen L, Fugger L, Scott H, Noren O, Roepstorff P, Lundin KE, Sjostrom H, Sollid LM (1998) Tissue transglutaminase selectively modifies gliadin peptides that are recognized by gut-derived T cells in celiac disease. Nat Med 4(6):713–717

Moore WT Jr, Murtaugh MP, Davies PJ (1984) Retinoic acid-induced expression of tissue transglutaminase in mouse peritoneal macrophages. J Biol Chem 259(20):12794–12802

Mosimann C, Hausmann G, Basler K (2009) Beta-catenin hits chromatin: regulation of Wnt target gene activation. Nat Rev Mol Cell Biol 10(4):276–286. doi:10.1038/nrm2654

Nagy L, Saydak M, Shipley N, Lu S, Basilion JP, Yan ZH, Syka P, Chandraratna RA, Stein JP, Heyman RA, Davies PJ (1996) Identification and characterization of a versatile retinoid response element (retinoic acid receptor response element-retinoid X receptor response element) in the mouse tissue transglutaminase gene promoter. J Biol Chem 271(8):4355–4365

Nanda N, Iismaa SE, Owens WA, Husain A, Mackay F, Graham RM (2001) Targeted inactivation of Gh/tissue transglutaminase II. J Biol Chem 276(23):20673–20678. doi:10.1074/jbc.M010846200

Nardacci R, Lo Iacono O, Ciccosanti F, Falasca L, Addesso M, Amendola A, Antonucci G, Craxi A, Fimia GM, Iadevaia V, Melino G, Ruco L, Tocci G, Ippolito G, Piacentini M (2003) Transglutaminase type II plays a protective role in hepatic injury. Am J Pathol 162 (4):1293–1303. doi:10.1016/S0002-9440(10)63925-9

Oh K, Ko E, Kim HS, Park AK, Moon HG, Noh DY, Lee DS (2011a) Transglutaminase 2 facilitates the distant hematogenous metastasis of breast cancer by modulating interleukin-6 in cancer cells. Breast Cancer Res BCR 13(5):R96. doi:10.1186/bcr3034

Oh K, Park HB, Byoun OJ, Shin DM, Jeong EM, Kim YW, Kim YS, Melino G, Kim IG, Lee DS (2011b) Epithelial transglutaminase 2 is needed for T cell interleukin-17 production and subsequent pulmonary inflammation and fibrosis in bleomycin-treated mice. J Exp Med 208 (8):1707–1719. doi:10.1084/jem.20101457

Park KC, Chung KC, Kim YS, Lee J, Joh TH, Kim SY (2004) Transglutaminase 2 induces nitric oxide synthesis in BV-2 microglia. Biochem Biophys Res Commun 323(3):1055–1062. doi:10.1016/j.bbrc.2004.08.204

Park SS, Kim JM, Kim DS, Kim IH, Kim SY (2006) Transglutaminase 2 mediates polymer formation of I-kappaBalpha through C-terminal glutamine cluster. J Biol Chem 281 (46):34965–34972. doi:10.1074/jbc.M604150200

Park KS, Kim DS, Ko C, Lee SJ, Oh SH, Kim SY (2011) TNF-alpha mediated NF-kappaB activation is constantly extended by transglutaminase 2. Front Biosci (Elite Ed) 3:341–354

Piacentini M, D'Eletto M, Falasca L, Farrace MG, Rodolfo C (2011) Transglutaminase 2 at the crossroads between cell death and survival. Adv Enzymol Relat Areas Mol Biol 78:197–246

Rebe C, Raveneau M, Chevriaux A, Lakomy D, Sberna AL, Costa A, Bessede G, Athias A, Steinmetz E, Lobaccaro JM, Alves G, Menicacci A, Vachenc S, Solary E, Gambert P, Masson D (2009) Induction of transglutaminase 2 by a liver X receptor/retinoic acid receptor alpha pathway increases the clearance of apoptotic cells by human macrophages. Circ Res 105 (4):393–401. doi:10.1161/CIRCRESAHA.109.201855

Schroff G, Neumann C, Sorg C (1981) Transglutaminase as a marker for subsets of murine macrophages. Eur J Immunol 11(8):637–642. doi:10.1002/eji.1830110809

Semenza GL (2003) Targeting HIF-1 for cancer therapy. Nat Rev Cancer 3(10):721–732. doi:10.1038/nrc1187

Shweke N, Boulos N, Jouanneau C, Vandermeersch S, Melino G, Dussaule JC, Chatziantoniou C, Ronco P, Boffa JJ (2008) Tissue transglutaminase contributes to interstitial renal fibrosis by favoring accumulation of fibrillar collagen through TGF-beta activation and cell infiltration. Am J Pathol 173(3):631–642. doi:10.2353/ajpath.2008.080025

Sohn J, Kim TI, Yoon YH, Kim JY, Kim SY (2003) Novel transglutaminase inhibitors reverse the inflammation of allergic conjunctivitis. J Clin Invest 111(1):121–128. doi:10.1172/JCI15937

Tacke F, Luedde T, Trautwein C (2009) Inflammatory pathways in liver homeostasis and liver injury. Clin Rev Allergy Immunol 36(1):4–12. doi:10.1007/s12016-008-8091-0

Tamai K, Semenov M, Kato Y, Spokony R, Liu C, Katsuyama Y, Hess F, Saint-Jeannet JP, He X (2000) LDL-receptor-related proteins in Wnt signal transduction. Nature 407(6803):530–535. doi:10.1038/35035117

Tatsukawa H, Fukaya Y, Frampton G, Martinez-Fuentes A, Suzuki K, Kuo TF, Nagatsuma K, Shimokado K, Okuno M, Wu J, Iismaa S, Matsuura T, Tsukamoto H, Zern MA, Graham RM, Kojima S (2009) Role of transglutaminase 2 in liver injury via cross-linking and silencing of transcription factor Sp1. Gastroenterology 136(5):1783–1795 e1710. doi:10.1053/j.gastro.2009.01.007

Tucholski J (2010) TG2 protects neuroblastoma cells against DNA-damage-induced stress, suppresses p53 activation. Amino Acids 39(2):523–532. doi:10.1007/s00726-009-0468-8

Wang Z, Wilhelmsson C, Hyrsl P, Loof TG, Dobes P, Klupp M, Loseva O, Morgelin M, Ikle J, Cripps RM, Herwald H, Theopold U (2010) Pathogen entrapment by transglutaminase – a conserved early innate immune mechanism. PLoS Pathog 6(2):e1000763. doi:10.1371/journal.ppat.1000763

Wykoff CC, Pugh CW, Maxwell PH, Harris AL, Ratcliffe PJ (2000) Identification of novel hypoxia dependent and independent target genes of the von Hippel-Lindau (VHL) tumour suppressor by mRNA differential expression profiling. Oncogene 19(54):6297–6305. doi:10.1038/sj.onc.1204012

Xu-Monette ZY, Moller MB, Tzankov A, Montes-Moreno S, Hu W, Manyam GC, Kristensen L, Fan L, Visco C, Dybkaer K, Chiu A, Tam W, Zu Y, Bhagat G, Richards KL, Hsi ED, Choi WW, van Krieken JH, Huang Q, Huh J, Ai W, Ponzoni M, Ferreri AJ, Wu L, Zhao X, Bueso-Ramos CE, Wang SA, Go RS, Li Y, Winter JN, Piris MA, Medeiros LJ, Young KH (2013) MDM2 phenotypic and genotypic profiling, respective to TP53 genetic status, in diffuse large B-cell lymphoma patients treated with rituximab-CHOP immunochemotherapy: a report from the International DLBCL Rituximab-CHOP Consortium Program. Blood 122(15):2630–2640. doi:10.1182/blood-2012-12-473702

Yamaguchi H, Wang HG (2006) Tissue transglutaminase serves as an inhibitor of apoptosis by cross-linking caspase 3 in thapsigargin-treated cells. Mol Cell Biol 26(2):569–579. doi:10.1128/MCB.26.2.569-579.2006

Yoo H, Ahn ER, Kim SJ, Lee SH, Oh SH, Kim SY (2013) Divergent results induced by different types of septic shock in transglutaminase 2 knockout mice. Amino Acids 44(1):189–197. doi:10.1007/s00726-012-1412-x

Yuan L, Siegel M, Choi K, Khosla C, Miller CR, Jackson EN, Piwnica-Worms D, Rich KM (2007) Transglutaminase 2 inhibitor, KCC009, disrupts fibronectin assembly in the extracellular matrix and sensitizes orthotopic glioblastomas to chemotherapy. Oncogene 26(18):2563–2573. doi:10.1038/sj.onc.1210048

Chapter 8
The Role of Transglutaminase Type 2 in the Regulation of Autophagy

Manuela D'Eletto, Federica Rossin, Maria Grazia Farrace, and Mauro Piacentini

Abstract Eukaryotic cells are equipped with a very efficient quality control system to selectively eliminate misfolded and damaged proteins and organelles. Autophagy is the major intracellular degradation/recycling catabolic system for mutated/misfolded proteins and damaged organelles. It is a highly complex regulated process that plays a key role in cellular maintenance and development. Autophagy is recognized to play an important role in the pathogenesis of the major human diseases. Interestingly, recent studies have demonstrated that autophagy is not a simple metabolite recycling system, but also has the ability to degrade specific cellular targets, such as mitochondria, peroxisomes, cilia, and bacteria. In this chapter the involvement of TG2 in the autophagic pathway is discussed. Indeed, cells or mouse lacking the enzyme show impaired autophagy and accumulate ubiquitinated protein aggregates and damaged mitochondria. TG2 physically interacts with the autophagy cargo protein p62, and they are localized in cytosolic protein aggregates, which are then recruited into autophagosomes where TG2 is degraded. Interestingly, the enzyme's crosslinking activity is activated during autophagy and its inhibition leads to the accumulation of ubiquitinated proteins indicating that TG2 plays an important role in the assembly of protein aggregates as well as for the clearance of damaged organelles. Interestingly cells lacking the enzyme display impaired autophagy/mitophagy and as a consequence shift their metabolism to glycolysis.

Keywords Autophagy • Mitochondria • Mitophagy • Ubiquitinated proteins

M. D'Eletto • F. Rossin • M.G. Farrace
Department of Biology, University of Rome 'Tor Vergata', Rome 00133, Italy

M. Piacentini (✉)
Department of Biology, University of Rome 'Tor Vergata', Rome 00133, Italy

National Institute for Infectious Diseases I.R.C.C.S. 'Lazzaro Spallanzani', Rome 00149, Italy
e-mail: mauro.piacentini@uniroma2.it

K. Hitomi et al. (eds.), *Transglutaminases*, DOI 10.1007/978-4-431-55825-5_8

8.1 Introduction

The term "*autophagy*" coined by Christian de Duve in the early 1960s, who was awarded the Nobel Prize for the discovery of lysosomes, derives from the Greek expression meaning "eating its self". In biology autophagy identifies the cellular catabolic process that delivers intracellular components, including organelles, to lysosomes for degradation and recycling (Yang and Klionsky 2010). Christian de Duve also coined the word "autophagosome" to describe double-membrane vesicles containing cellular components at various degree of digestion localized in the cytoplasm (Ravikumar et al. 2010; Yang and Klionsky 2010). Thirty years afterward, by using genetic screenings in *Saccharomyces cerevisiae*, the molecular components of autophagy were described by Ohsumi (Ohsumi 2014). To date, more than 35 autophagy-related genes (ATG) have been identified and many of these have characterized mammalian orthologs thus indicating that the autophagic machinery is an evolutionarily highly conserved basic cellular process (Reggiori and Klionsky 2002).

It is now accepted that autophagy is acting in all eukaryotic cells, playing an essential role for cellular and energy homeostasis. Under physiological conditions, it functions in the clearance of misfolded/mutated proteins and damaged organelles (Rubinsztein 2006). Under stressful cellular conditions, such as nutrient and energy starvation, it recycles cytoplasmic components to supply cells with amino acids, sugars and lipids, thus contributing to keeping the metabolism compatible with cell survival (Virgin and Levine 2009). In keeping with this notion, it has been shown just after delivery in early neonatal stages that in fact, autophagy plays a critical role in counteracting nutrient deprivation taking place just after birth before the newborns start winning, allowing their survival (Kuma et al. 2004). In addition, autophagy has a protective role through its action in different aspects of immunity, including the elimination of invasive microbes and participation in antigen presentation (Virgin and Levine 2009; Levine and Deretic 2007). As a consequence of these important physiological activities, aberrant autophagy is implicated in the pathogenesis of most relevant diseases, including cancer, neurodegenerative and cardiovascular disorders, and infectious diseases (Mizushima et al. 2008).

Three types of autophagy mechanisms, differing in the way by which cytosolic components and organelles are delivered to lysosomes, have been described: macroautophagy, microautophagy and chaperone-mediated autophagy (CMA, only described in mammalian cells).

In microautophagy, cytosolic components are directly engulfed by the lysosome through invaginations of the lysosomal membrane (Li et al. 2012). In CMA, proteins exposing on the surface a specific pentapeptide motif are recognized by cytoplasmic chaperones, unfolded and then transported into the lysosome through a lysosomal channel formed by the multimerization of the lysosomal receptor, LAMP2A (Orenstein and Cuervo 2010). Macroautophagy (hereafter referred to as autophagy) is the most studied autophagic pathway. Unlike microautophagy and CMA, it goes through to the formation of an isolation membrane, called a phagophore, which is able to recognize and to engulf specific cargos present in the cytoplasm. The phagophore elongates and closes to form a double-membrane vesicle, the autophagosome. The autophagosome then fuses with lysosomes, forming a large and single-membrane-surrounded vesicle, called an autophago-lysosome, inside which the cargos are degraded by lysosomal hydrolases, providing the energy required to overcome stress conditions (Suzuki et al. 2007).

Although it has long been thought to be a nonselective degradative pathway, autophagy is now considered a selective process (Reggiori et al. 2012). In fact, it is now well established that the autophagosome can exclusively sequester and degrade specific substrates, such as damaged or redundant organelles and ubiquitinated protein aggregates. Different terms have been coined to describe the selectivity of each process according to the cargo, e.g. mitophagy is used for mitochondria degradation, pexophagy for peroxisomes, reticulophagy for surplus endoplasmic reticulum (ER), and ribophagy for ribosomes, while aggrephagy means the degradation by macroautophagy of aggregated or misfolded proteins (Kim et al. 2007; Lafontaine 2010; Sakai et al. 2006; Yorimitsu and Klionsky 2007). In addition, lipid and pathogen degradation have been described as specific forms of autophagy, called lipophagy and xenophagy, respectively (Levine 2005; Weidberg et al. 2009).

8.2 Macroautophagy Regulation in Mammals

In mammalian cells, autophagy occurs at basal low levels and is induced by a variety of physiological and stressful stimuli, like nutrient and growth factor depletion, hypoxia and drugs. The molecular components of autophagy were initially described in *Saccharomyces cerevisiae* by using genetic screenings. The mammalian orthologs for many of the 35 ATG identified in yeast have been found in the last 10 years (Reggiori and Klionsky 2002). These genes can be classified according to their functions during the key stages of the autophagy pathway, namely: initiation, elongation, maturation and fusion with the lysosomes.

8.2.1 Initiation

The best-characterized regulation of autophagy is achieved through the mTOR pathway, which negatively controls this process. mTOR is a serine/threonine protein kinase which regulates a large number of cellular responses such as cell growth, proliferation, protein synthesis and autophagy (Schmelzle and Hall 2000) (Fig. 8.1).

In mammalian cells, two distinct functional mTOR complexes exist, mTORC1 and mTORC2. mTORC1 protein complex (inhibited by rapamycin) includes mTOR, DEPTOR, mLST8, RAPTOR and PRAS40. When this complex is activated

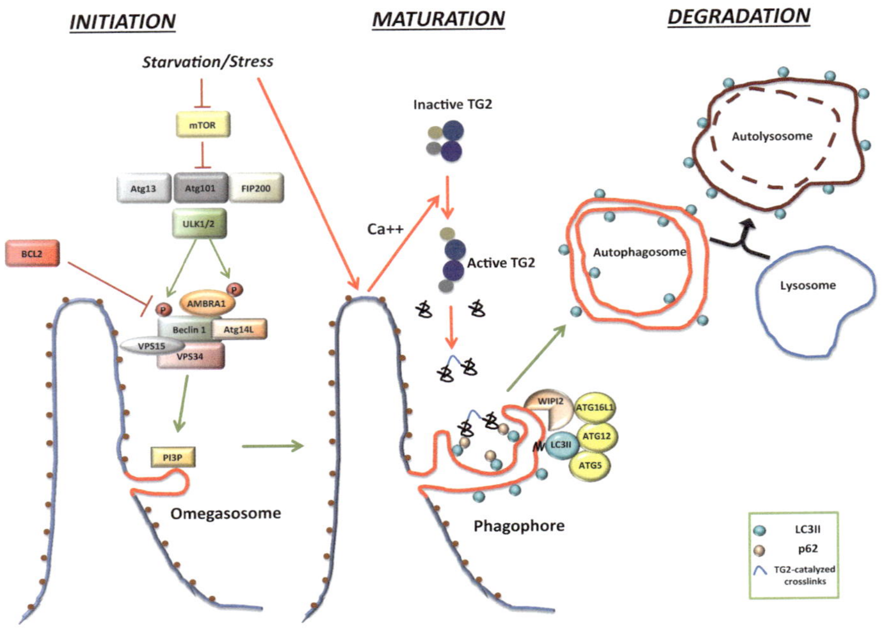

Fig. 8.1 TG2 involvement in the autophagic process. The main steps of autophagy (initiation, maturation and degradation) are regulated by different protein complexes. Following autophagy induction, calcium release leads to TG2 activation. During autophagosome maturation, TG2 transamidating activity is required to crosslink high molecular weight aggregates. In the final stage of autophagy protein aggregates are conveyed by cargo proteins to the LC3 protein family members for degradation

by amino acids, energy availability and growth factors, it positively regulates cell growth by inducing ribosomes biogenesis and protein synthesis and, at the same time it inhibits autophagy (He et al. 2009). By contrast mTORC2, known as rapamycin-insensitive complex, regulates autophagy via Akt-FoxO3 (Zhao et al. 2007).

The major signaling cascade that regulates mTORC1 is represented by the PI3K pathway. The binding of growth factors or insulin to their cell-surface receptors activates PI3 kinase (PI3K), which catalyzes on the plasma membrane the conversion of phosphatidylinositol-4,5-bisphosphate (PIP2) to phosphatidylinositol-3,4,5-trisphosphate (PIP3). PIP3 recruits to the plasma membrane pleckstrin homology (PH) domain proteins, such as the serine/threonine kinases phosphoinositide-dependent kinase 1 (PDK1) and Akt (Cantley 2002). Akt is activated by its phosphorylation on two different sites. Initially, mTORC1 phosphorylates Akt on Ser-473 facilitating the second phosphorylation by PDK1 on Thr-308 (Manning and Cantley 2003). Once phosphorylated Akt mediates an inhibitory phosphorylation of the tumor suppressor TSC (tuberous sclerosis complex), formed by TSC1/2 heterodimer (Cantley 2002; Sarbassov et al. 2005). TSC1/2 is a GTPase-activating protein (GAP) that inactivates the small G protein Ras homolog enriched in brain

(Rheb), which directly binds and activates the mTORC1 complex (Manning and Cantley 2003). Consequently Akt-dependent phosphorylation of TSC1/2 leads to the activation of Rheb, which in turn activates mTORC1 and inhibits autophagy (Huang and Manning 2008).

The protein kinase activated by AMP (AMPK) is a cellular energy sensor able to monitor changes in intracellular ATP concentration. Under low energy conditions (high AMP/ATP ratio), AMPK is active and phosphorylates TSC2 complex promoting its activity thus resulting in the inhibition of Rheb. These events lead to the inactivation of mTORC1 and promote autophagy (Alers et al. 2012). Moreover, AMPK inhibits RAPTOR, a member of the mTORC1 complex, causing its dissociation from mTORC1 (Gwinn et al. 2008). In addition, AMPK can also directly phosphorylate the autophagy kinase ULK1 on serine 317 and 777 leading to its dissociation from the mTORC1 complex and consequently enhancing autophagy (Kim et al. 2011).

The inhibition of mTORC1 is a key signal for the induction of autophagy and it is associated with reduced phosphorylation of the two downstream effectors, p70 S6K (ribosomal protein S6 kinase-1) and 4E-BP1 (translation initiation factor 4E-binding protein-1), whose phosphorylation status is therefore used as a marker to monitor TORC1 activity (Maiese et al. 2012).

The molecular link between mTORC1 signaling and the autophagic machinery is represented by the ULK1 complex, composed by ULK1/2 (Atg1 in yeast), mATG13 (Atg13 in yeast), FIP200 (Atg17 in yeast) and ATG101 (Atg101 in yeast) (Hosokawa et al. 2009) (Fig. 8.1). Under nutrient-rich conditions, mTORC1 associates with the ULK1 complex and represses autophagy by directly phosphorylating both ULK1 and mATG13, which results in ULK1 kinase activity inhibition (Alers et al. 2012; Jung et al. 2009). Instead, nutrient shortage or treatment with rapamycin leads to mTORC1 inactivation that causes its dissociation from the ULK1 complex and the subsequent loss of ULK1 phosphorylation on Ser757. These events result in ULK1 activation that mediates the phosphorylation of ATG13, FIP200 and ULK1 itself, which is then able to trigger the autophagic cascade (Jung et al. 2010).

The initiation stage of autophagosome biogenesis leads to the formation of a membranous cistern called a phagophore or isolation membrane (IM), whose nucleation site is still highly debated. Recent data suggest that there are several sources that may provide the components necessary for the formation of autophagosomes including ER, Golgi apparatus, mitochondria and plasma membrane (Hayashi et al. 2009; Takahashi et al. 2011; Hailey et al. 2010).

A key event in the formation of the phagophore is the production of phosphatidylinositol-3 phosphate (PtdIns3P) by the class III phosphatidylinositol-3-kinase (PI3K) complex. PtdIns3P is required for nucleation of particular structures, named "omegasomes", because of their Ω-like shape (Axe et al. 2008; Petiot et al. 2000). The omegasome acts as a platform for the recruitment of autophagic proteins and forms the curved phagophore by membrane invagination at the centre of the omegasome (Axe et al. 2008). When omegasomes reach their maximum size, the autophagosomal membranes close giving rise to a fully formed double-membrane autophagosome containing cytoplasmic material (Blommaart et al. 1997).

In mammalian cells, the core of the class III PI3K complex consists of the PI3K enzyme hVPS34 (vacuolar protein sorting 34) and BECLIN 1 (Bcl-2 interacting protein). This complex functions as a central platform regulated by different proteins, such as UVRAG and ATG14, at different stages of the autophagic signaling (Itakura et al. 2008; Zhong et al. 2009). For example, Atg14L is essential for autophagosome formation directing the PI3K complex to the endoplasmic reticulum from which most autophagosomes emerge (Zhong et al. 2009). On the other hand, the interaction of BECLIN 1 with UVRAG is highly debated since it was described to be involved both in autophagosome formation and maturation (Itakura et al. 2008; Liang et al. 2008; Matsunaga et al. 2009; Zhong et al. 2009). Moreover, a role of UVRAG-BECLIN 1-PI3K was also described in ligand/receptor degradation and cytokinesis (Thoresen et al. 2010). BCL2 was shown to inhibit autophagy by decreasing BECLIN 1 interaction with VPS34, and causing a reduction of BECLIN 1-associated VPS34 kinase activity (Pattingre et al. 2005).

Interestingly, additional BECLIN 1 interacting proteins, including AMBRA1, Bif-1 (endophilin B1), Rubicon (RUN domain and cysteine-rich domain containing BECLIN 1-interacting protein) and BCL2 have been described to stimulate or inhibit PI3K complex and, as consequence, autophagy (Fimia et al. 2007; Takahashi et al. 2007; Matsunaga et al. 2009; Zhong et al. 2009; Pattingre et al. 2005).

Bif-1 interacts with BECLIN 1 through UVRAG and functions as a positive regulator of the PI3K complex and autophagy induction in mammalian cells. Indeed, loss of Bif-1 significantly suppresses PI3K complex activation and the formation of autophagosomes (Takahashi et al. 2007).

Rubicon negatively regulates autophagy, likely through the inhibition of autophagosome maturation. In fact, it has been shown that overexpression of Rubicon remarkably reduces PI3K/Vps34 activity complex and impairs the acidification of LC3 (microtubule-associated protein 1 light chain 3)-associated vacuoles (Zhong et al. 2009). AMBRA1 (activating molecule in BECLIN 1-regulated autophagy) binds to BECLIN1 and stabilizes BECLIN1/Vps34 complex, thus potentiating its lipid kinase activity and promoting autophagosome formation (Fimia et al. 2007).

AMBRA1 is a WD40-containing protein conserved among vertebrates, crucial for central nervous system development. AMBRA1 is initially expressed in the neural plate during the first stages of neurulation, while it becomes ubiquitously expressed at the late stages of embryogenesis and in adult life (Fimia et al. 2007). AMBRA1 binds to BECLIN 1 and positively regulates the lipid kinase activity of BECLIN 1/VPS34 complex (Fimia et al. 2007; Gu et al. 2014). Under unstressed conditions, AMBRA1 activity is inhibited by multiple mechanisms including mTOR-mediate phosphorylation on Ser52 (Nazio et al. 2013) and its sequestration to microtubules through the interaction with the dynein complex (Di Bartolomeo et al. 2010), thus limiting autophagy initiation. The activation of the autophagic cascade allows the release of AMBRA1 from dynein complex through ULK1-

dependent phosphorylation. AMBRA1 is thus able to translocate to the ER together with BECLIN 1/VPS34 complex where it promotes autophagosomes nucleation (Di Bartolomeo et al. 2010). Moreover, AMBRA1 regulates autophagy by stimulating the kinase activity and the stability of ULK1 complex (Nazio et al. 2013; Nazio and Cecconi 2013). Interestingly, consequent to autophagy activation, AMBRA1 binds the E3-ubiquitin ligase TRAF6 and mediates K63-chain polyubiquitination of ULK1 leading to its self-association and increasing the kinase activity of the complex.

Another level of autophagy regulation by AMBRA1 has been recently identified. AMBRA1 appears to be also important in the temporal regulation of the autophagy response through dynamic interactions with different Cullin-RING E3 ubiquitin ligases (Antonioli et al. 2014). In non-autophagic conditions, AMBRA1 is ubiquitinated by Cullin 4-DDB1 (damage-specific DNA binding protein 1) E3-ligase complex, which leads to proteasomal degradation. Autophagic stimulation rapidly promotes AMBRA1 dissociation from Cullin4-DDB1 and consequent protein stabilization. Upon the release from Cullin4, AMBRA1 interacts and inhibits another Cullin complex (Cullin 5). The inhibition of Cullin 5 allows the stabilization of DEPTOR (DEP domain containing mTOR-interacting protein), an inhibitor of mTOR thus allowing the activation of the ULK complex. Prolonged stresses lead Cullin 4-DDB1 to bind back to AMBRA1, which causes its sharp degradation and the termination of autophagy response.

8.2.2 Elongation

Two evolutionary conserved ubiquitin-like (UBL) conjugation systems are necessary for the expansion of the phagophore to form autophagosome.

The first system involves the formation of Atg12-Atg5-Atg16/ATG16L1 complex (Fig. 8.1). Initially, the ubiquitin-like protein Atg12 is activated by Atg7 (E1 ubiquitin activating enzyme-like), and then transferred to Atg10 (E2 ubiquitin conjugating enzyme-like). Atg12 is finally covalently conjugated to an internal lysine of the substrate protein Atg5 (Mizushima et al. 1998). This process differs from ubiquitination in that the conjugation of Atg12 to Atg5 is irreversible and does not require an E3 ligase enzyme. Following Atg12–Atg5 conjugation, Atg16L1 non-covalently binds to Atg5, forming a complex containing Atg12–Atg5-Atg16L1 tetramers, which are linked via coiled-coil domain of Atg16L1 (Mizushima et al. 2003). This complex associates with the phagophore membrane, but dissociates following autophagosome closure; therefore components of this complex are used to monitor forming autophagosomes.

The second ubiquitin-like system, required in phagophore expansion, involves LC3, which is the human ortholog of yeast Atg8. LC3 is synthesized as a precursor form, which is cleaved at its carboxyl terminus by the cysteine protease Atg4, resulting in the cytosolic isoform LC3-I. Atg7 (E1 ubiquitin enzyme-like) activates LC3-I and transfers it to Atg3 (E2 ubiquitin enzyme-like). Finally, LC3 is covalently conjugated to the lipid phosphatidylethanolamine (PE) by the action of Atg12–Atg5 conjugate, which may function as E3 ligase-like (Hanada et al. 2007). Once lipidated, LC3-II localizes upon the initiation membrane and allows the tethering and hemifusion of membranes that are necessary for autophagosome expansion (Tanida et al. 2004). After autophagosome sealing, LC3 attached to the outer membrane is cleaved off by Atg4 and probably it can be reused. Instead, LC3 on the inner membrane is trapped inside sealed autophagosomes and degraded into lysosomes (Yoshimori and Noda 2008).

In nutrient-rich conditions the majority of LC3 is cytosolic but, upon autophagy induction, LC3 largely exists as the lipid-conjugated form (LC3-II) and localizes to autophagosomes (Kirisako et al. 1999) and for this reason it is widely used as a marker for microscopic analysis of autophagy (Mizushima and Yoshimori 2007). In addition, LC3II displays increased mobility in SDS-PAGE gel compared to its unconjugated form, and can be specifically detected by western blotting (Mizushima and Yoshimori 2007). It is well established that the amount of LC3II directly correlates with the number of autophagic vesicles and therefore it is widely used for the quantification of autophagic activity (Klionsky et al. 2012).

At least eight different Atg8 orthologs, belonging to two subfamilies (LC3 and GATE-16/GABARAP), are present in mammalian cells. Both subfamilies can localize with autophagosomes, but it has been proposed that they function at different steps in phagophore elongation and completion. In fact, recent work indicates that while LC3 is involved in elongation of the phagophore membrane, the GABARAP/GATE-16 subfamily acts at a later stage in autophagosome maturation, possibly in sealing the autophagosomes (Weidberg et al. 2010).

8.2.3 Autophagolysosome Formation

After the autophagasome formation, the outer membrane of the autophagosome fuses with the lysosome to form the autophagolysosome, also called a autolysosome (Fig. 8.1). During the maturation, the autophagosome contents and its inner membrane are degraded by lysosomal hydrolases released following the fusion with lysosomes.

This last step of autophagy depends on a series of molecules that regulate the maturation of autophagosomes, including their fusion with endosomal and/or lysosomal vesicles, the acidification of autophagic vescicles, and the release of degraded materials from the lysosomal compartment for their recycling. These events have a fundamental importance for the production of degrading material through the autophagic pathway. A blockage or an excessive autophagy flux could both have adverse consequences for cells homeostasis.

The fusion step is mediated by SNAREs, ESCRT, Rab7 and the class C Vps proteins (Atlashkin et al. 2003; Gutierrez et al. 2004; Lee et al. 2007; Rusten et al. 2007). Rab proteins are a family of monomeric GTPases involved in vesicular transport along the exo- and endocytic pathways in yeast, however some evidence shows also their involvement in the autophagic pathway (Gutierrez et al. 2004). In mammalian cells, Rab7 mediates the fusion between the autophagosome and the late endosome/lysosome. Thus, overexpression of a Rab7 dominant negative mutant leads to the accumulation of autophagosomes, with a concomitant decrease in long-lived protein degradation (Gutierrez et al. 2004).

Autophagosome maturation is also dependent on the interaction between class C Vps proteins and UVRAG. UVRAG, independently from its interaction with BECLIN 1, recruits the class C Vps proteins and via this interaction activates Rab7, thereby promoting fusion with lysosomes (Liang et al. 2008).

Another BECLIN 1 interacting protein, Rubicon, functions in the maturation of autophagosomes. Rubicon is thought to be a part of a distinct BECLIN 1 complex containing VPS34, Vps15 and UVRAG that suppresses autophagosome maturation (Matsunaga et al. 2009; Zhong et al. 2009).

The fusion event requires lysosomal membrane proteins such as LAMPs (Tanaka et al. 2000) and TECPR1 (TECtonin β-Propeller repeat containing 1) that bind the autophagosome-localized Atg12-Atg5 conjugate and recruit it to lysosomes (Chen and Zhong 2012).

In addition, a proper lysosomal activity is also essential to successfully mediate the fusion events. Inhibition of the lysosomal H^+-ATPase by chemicals (e. g. bafilomycin A1) blocks the lysosomal pumping of H^+ and consequently lysosomal enzymes which are active at low pH (Fass et al. 2006; Mousavi et al. 1998). It has been proposed that bafilomycin A1 may also block the late stage of autophagy by interfering with the fusion of autophagosomes with endosomes/lysosomes (Yamamoto et al. 1998). For this characteristic, bafilomycin A1 is a drug widely used to monitor the autophagic flux because it causes the accumulation of autophagic substrates into the lysosomal compartment (Klionsky et al. 2012).

Finally, microtubules are deeply involved in the maturation step. In fact, autophagosomes move bi-directionally along cytoskeletal structures to fuse with lysosomes and the centripetal movement of autophagic vesicles is dependent on the dynein motor complex. Accordingly, the destabilization of microtubules by drugs (e. g. vinblastine) blocks the maturation of autophagosomes (Kimura et al. 2008).

8.3 The Role of TG2 in Autophagy

The first indications that TG2 could be involved in autophagy was obtained by the use of the autofluorescent compound monodansylcadaverine (MDC) for in vivo labeling of autophagic vacuoles. When applied to various cell lines, spherical structures were detected which were predominantly located in the perinuclear region (Biederbick et al. 1995). The results reported in this study indicated that MDC selectively accumulates in autophagic vacuoles under in vivo conditions and is not present in the early and late endosome. Further indirect evidence of the relationships existing between the enzyme expression and autophagy was obtained in $TG2^{-/-}$/Huntington disease (HD) transgenic mice (Mastroberardino et al. 2002). In fact, $TG2^{-/-}$ HD transgenic mice showed that the HD onset is associated with a significant reduction in non-apoptotic cell death and with an increased number of nuclear protein inclusions clearly suggesting an impairment in the autophagic pathway (Piacentini et al. 2011).

The first direct involvement of TG2 in autophagy has been shown in highly metastatic pancreatic carcinoma cells where the knockdown of TG2 protein by siRNA in pancreatic cancer cells leads to the accumulation of LC3 II and autophagic vacuoles in the cytoplasm (Ozpolat et al. 2007). Two years later, we confirmed that the ablation of TG2 protein, both in vivo and in mouse embryonic fibroblasts, resulted in an evident accumulation of the cleaved isoform of LC3 (LC3 II) on pre-autophagic vesicles (Table 8.1). However, the formation of the acidic vesicular organelles in the same cells was very limited, indicating an impairment of the maturation of autophagosomes. This conclusion was also supported by the absence of acidification of the autophagolysosome observed in MEF cells lacking TG2. More recently, we have also demonstrated that TG2 regulation of autophagy occurs by its transamidating activity (Rossin et al. 2012). Interestingly, the ablation of TG2 leads to the accumulation of ubiquitinated protein aggregates, which are not cleared by autophagy. In keeping with these findings it has recently been shown that TG2 directly crosslinks the DNA binding domain of p53, leading to its degradation by autophagy in renal cell carcinoma cells (Ku et al. 2013). This is an interesting aspect of the TG2-mediated effects on autophagy and apoptosis; in fact, the nuclear pool of p53 can induce autophagy through the transcriptional activation of specific target genes while the cytoplasmic pool of p53 can inhibit autophagy by non-transcriptional mechanisms (Tasdemir et al. 2008; Maiuri et al. 2009).

TG2 is also required for the turnover and degradation of dysfunctional mitochondria. In fact, in the absence of TG2 and particularly under stressful conditions, we observed a marked accumulation of ubiquitinated proteins as well as damaged mitochondria, thus indicating a role for the enzyme in the selection of the autophagic substrates (D'Eletto et al. 2009, 2012).

It is well known that defects in the ubiquitin-proteasome machinery lead to the accumulation of misfolded proteins and protein aggregates removed by autophagy (Ding et al. 2003). In fact, once aggregated, misfolded proteins cannot be degraded efficiently by the proteasome as they cannot pass through the narrow structure, and it is macroautophagy that plays a critical role in the selective elimination of ubiquitinated protein aggregates in mammalian cells (Snyder et al. 2003). In this regard, it has been demonstrated that the TG2-dependent posttranslational modification of its substrate proteins is increased by autophagy induction, and further potentiated upon proteasome inhibition (D'Eletto el al. 2012). In connection with this, it has been shown that during the early stage of autophagy there is an accumulation of free calcium ions in the cytoplasm that may activate the TG2's transamidating activity (Høyer-Hansen et al. 2007).

The molecular mechanisms regulating the specific recognition and elimination of ubiquitinated protein aggregates by autophagy have recently begun to be

Table 8.1 TG2 protein effects on autophagy regulation

TG2 effects on autophagy regulation				
Experimental system	TG2	Stimulus	Autophagy	References
Knock out mouse heart and liver	Absent	Starvation	Inhibited	D'Eletto et al. *Autophagy* 2009; D'Eletto et al. *Cell Death Differ* 2012
MEFs wild type	Basal	Starvation	Activated	D'Eletto et al. *Autophagy* 2009; Rossin et al. *Amino Acids* 2012
MEFs knock out	Absent	Starvation	Inhibited	D'Eletto et al. *Autophagy* 2009; Rossin et al. *Amino Acids* 2012
Skin fibroblasts	Low	UVA	Inhibited	Lamore and Wondrack. *Photochem Photobiol Sci* 2012
Skin from epidermolysis bullosa	Low	Genetic disease	Inhibited	Küttner et al. *J Invest Dermatol* 2014
Renal cell carcinoma	High	Basal	Activated	Ku et al. *FASEB J* 2013
Cystic fibrosis	High	Genetic disease	Inhibited	De Stefano et al. *Autophagy* 2014
Pancreatic ductal carcinoma	High	Rottlerin	Activated	Akar et al. *Mol Cancer Res* 2007
Hela	High	Basal	Inhibited	Hamada et al. *Proc Natl Acad Sci U S A* 2014
Neuroblastoma SH-SY5Y	High	MPP(+)	Inhibited	Verhaar et al. *Neurochem Int* 2013

TG2 protein expression correlates with either autophagy activation or inhibition depending on different tissue localization

uncovered. This process requires autophagy cargo proteins, such as p62/SQSTM1 and NBR1, which are known to mediate the clearance of ubiquitinated proteins by autophagy. These cargo proteins contain UBA domains at their C-terminal regions, allowing them to interact with the aggregated protein complexes and organelles that have to be degraded (Fig. 8.2). Subsequently, by an LIR (LC3-interacting region) domain these cargo proteins interact with the ATG8-family members inside the nascent pre-autophagic vesicles (Weidberg et al. 2011; Lamark et al. 2009). Interestingly, TG2 is present in protein complexes containing either p62 or NBR1 and presents two putative LIR domains by which it can interact with LC3II family members (D'Eletto el al. 2012).

The accumulation of ubiquitinated proteins in cells lacking TG2 suggests that the enzyme is involved in the posttranslational modification of high molecular weight aggregates, which are then conveyed by cargo proteins to the autophagic machinery for degradation. Notably, the expression of the TG2 mutant in the transamidating activity compromises the clearance of ubiquitinated proteins, highlighting that the TG2's crosslinking activity has an important role in the formation/stabilization of substrates cleared by the autophagy pathway. These findings are in agreement with previous results showing that the transamidating

Detection of putative LIR motifs in the TG2

Carrier	LIR sequence	Binding partner
p62	GDDDWTHLSS	LC3
NBR1	SSEDYIIILP	LC3
NIX	LNSSWVELPM	GABARAP
CRT	LEDDWDFLPP	GABARAP
CHC	YTPDWIFLLR	GABARAP
ATG3	GVGDWEDLQD	ATG8
TG2	GQPFWLTLHF	LC3
TG2	GYEGWQALDP	LC3
Consensus	GxxDWxxLxP	

Fig. 8.2 Consensus sequences for LIR domains in the autophagy cargo proteins. The analysis of the TG2 sequence reveals two putative LIR motifs for LC3 binding

activity of the enzyme is required for the proper completion of the autophagic flux (D'Eletto et al. 2009; Rossin et al. 2012). In this regard, it is important to mention that TG2 has been involved in the pathogenesis of several degenerative diseases characterized by the accumulation of mutated/misfolded proteins such as Lewy bodies in PD (Parkinson's disease), neurofibrillary tangles in AD (Alzheimer's disease), and huntingtin aggregates in HD, as well as Mallory bodies in the liver (Malorni et al. 2008; Strnad et al. 2008) (see Chap. 13 "Transglutaminases and Neurological Diseases"). It has been suggested that TG2 acts as an important regulator of mitochondrial functions. Indeed, we observed that cells lacking TG2 display altered mitochondrial morphology and functionality as shown by the presence of more fragmented and depolarized mitochondria. We also demonstrated that these cells have high rates of aerobic glycolysis in comparison to the wild type and are more sensitive to the glycolytic inhibitor 2-DG, indicating an alteration of the mitochondrial metabolism when TG2 is absent (Rossin et al. 2015). This result correlates with a previous study demonstrating that deletion of TG2 in mice caused a significant deregulation of the respiratory complexes I and II and a reduction of ATP production (Battaglia et al. 2007). These findings seem to confirm a role for TG2 in the regulation of mitochondrial homeostasis and metabolism.

It is well known that autophagy has a crucial role in the turnover of cellular organelles; indeed it has been proposed that autophagy selectively degrades dysfunctional mitochondria, a process known as mitophagy. Cells lacking TG2 are not able to remove the dysfunctional mitochondria upon damage induction. In particular, we demonstrated that mitophagy, following CCCP treatment, is impaired in mice lacking TG2. In keeping with these results, cells knocked out for TG2 displayed higher levels of the fission protein DRP1 after mitochondrial damage (Rossin et al. 2015). By the way, fission seems to be a prerequisite for mitophagy to occur (Twig et al. 2008) and probably it is necessary to produce smaller mitochondrial fragments that can more easily be engulfed by autophagosomes.

Moreover, it is relevant to note that TG2 transamidating activity increased after mitophagy induction. The enzymatic activity seems to be specifically activated during the removal of dysfunctional mitochondria, as confirmed by the absence of transamidating activity in cells treated with the mitophagy inhibitor Mdivi-1. It is important to note that the accumulation of fragmented mitochondria, in the absence of TG2, leads to a caspase 3-dependent cell death while WT cells undergo apoptosis only when the degradation of mitochondria by autophagy is inhibited. These findings suggest an involvement of TG2 in the regulation of mitochondrial functionality as well as in the clearance of dysfunctional mitochondria (Rossin et al. 2015). In keeping with this assumption, it has recently been shown that PTEN-induced putative kinase 1 (PINK1) is a TG2 binding partner; PINK1 stabilizes the half-life of TG2 via inhibition of TG2 ubiquitination and subsequent proteasomal degradation. PINK1 is a serine/threonine protein kinase that protects cells from stress-induced mitochondrial dysfunction. It is well established that PINK1 gene mutations cause one form of autosomal recessive early-onset PD (Min et al. 2014). Interestingly, PINK1 controls Parkin-mediated mitophagy both in PD and HD pathogenesis (Narendra et al. 2012). These data suggest that the PINK1 positive regulation of TG2 activity might be an essential event in the mitophagy process and may also be closely associated with aggresome formation in neuronal cells.

8.3.1 TG2 and Autophagy Under Pathological Conditions

Defects in autophagy regulation have been shown to occur in a wide range of human diseases, including cancer as well as infectious, neurodegenerative, inflammatory, and metabolic diseases. These findings have led to the hypothesis that autophagy modulation may be very useful to prevent or treat many pathologies (Rubinsztein et al. 2012). In this regard, it is important to mention that TG2 has also been involved in the pathogenesis of several neurodegenerative diseases as well as inflammatory diseases and cancer (Malorni et al. 2008; Strnad et al. 2008).

8.3.1.1 Diseases Characterized by the Accumulation of Intracellular Protein Aggregates

Neurodegeneration is characterized by the accumulation of mutated/misfolded proteins such as Lewy bodies in PD, neurofibrillary tangles in AD, and huntingtin aggregates in HD. In all these nervous system age-related pathologies TG2 has been shown to be upregulated playing a putative pathogenic role by participating at the formation of the typical intracellular protein aggregates characterizing these diseases. However previous studies, carried out in a mouse model for HD, showed that in TG2$^{-/-/HD}$ transgenic mice the HD onset was associated with a large reduction in non-apoptotic cell death and with an increased number of nuclear protein inclusions, suggesting an impairment in their clearance by autophagy (Mastroberardino et al. 2002). These data are in some ways different from the findings obtained in an in vitro model of PD, using SH-SY5Y neurons treated with the PD-mimic MPP(+), which suggest that autophagy could be inhibited by TG2 activation via a mechanism leading to the crosslinking of BECLIN 1. In fact, under these in vitro conditions, accumulation of BECLIN 1 at the ER and the levels of the autophagy marker protein LC3II in MPP(+)-treated cells were significantly increased by the inhibition of TG2's transamidating activity (Verhaar et al. 2013). However, in this study the autophagic flux was not established, thus the accumulation of LC3II per se cannot be interpreted as a sign of increased autophagy (Table 8.1). A recent report suggests that TG2 negatively regulates the inositol 1,4,5-trisphosphate receptor (IP3R)-mediated calcium signaling and autophagy by the covalent posttranslational modification of the Gln2746 residue (Hamada et al. 2014). Modification of Gln2746 and IP3R1 function was observed in Huntington disease models, suggesting a pathological role of this modification in the neurodegenerative disease.

Mallory-Denk bodies (MDBs) are cytoplasmic inclusions found in several liver diseases and consist primarily of the cytoskeletal proteins keratins 8 and 18 (Hanada et al. 2012). Evidence indicated that the extent of stress-induced protein misfolding and TG2 activation promote MDB formation (Kwan et al. 2012). In fact, TG2(−/−) mice fed with 3,5-diethoxycarbonyl-1,4-dihydrocollidine (DDC) have dramatic decreases in MDB formation and liver hypertrophy compared with DDC-fed TG2 (+/+) mice. Inhibition of MDB formation in TG2(−/−) mice was associated with marked attenuation of ubiquitination and K8-containing protein crosslinking (Strnad et al. 2007). Interestingly, MDB formation is paralleled by the generation of high molecular weight ubiquitinated keratin-containing complexes and by accumulation of p62, indicating decreased activity of the protein degradation machinery (Hanada et al. 2012). Interestingly, activation of autophagy by rapamycin decreased the number of inclusions in bortezomib-treated K8 transgenic mice thus confirming the primary role of this catabolic process in the elimination of intracellular protein aggregates (Harada et al. 2008).

8.3.1.2 Cancer

TG2 expression is very high in renal cell carcinoma (RCC), the predominant form of kidney cancer. In RCC the inhibition of TG2 expression was found to stabilize p53 expression, thereby inducing a 3- to 10-fold increase in apoptosis. Interestingly, TG2 directly crosslinks the DNA binding domain of p53 leading to p53 depletion via autophagy (Ku et al. 2013). This finding implies that high TG2 expression might promote tumor cell survival through p53 depletion and autophagy. In disagreement with these finding in pancreatic cancer cells that are frequently insensitive to standard chemotherapeutic agents it was reported that protein kinase C delta (PKCdelta) regulates TG2 expression, which in turn inhibits autophagy (Akar et al. 2007). The PKC delta downregulation led to a decrease in the TG2 expression and induced growth inhibition without inducing apoptosis in pancreatic cancer cells (Akar et al. 2007).

8.4 Conclusions

The data reported in this review indicate that TG2 plays a key role in the regulation of autophagy by favoring the autophagosome maturation and in turn the autophagic flux (Fig. 8.1). In fact, in cells lacking TG2 the clearance of ubiquitylated protein aggregates as well as damaged mitochondria is impaired, thus suggesting an important contribution of the enzyme's transamidating activity in the late stages of autophagy, an effect that is particularly evident under cellular stressful conditions. It is becoming clear that the overexpression of TG2 leads to the inhibition of autophagy in the early stages by crosslinking the members of the Beclin1 complex and/or by favoring the degradation of cytoplasmic p53 (Luciani et al. 2010; Tucholski 2010), while the ablation of the enzyme impairs the final steps of the process by blocking the phagophore maturation (Fig. 8.3). From this evidence it is tempting to conclude that a physiological level of TG2 is crucial in order to assure functional autophagy (Fig. 8.3). These findings might also help to explain the complex role of TG2 in the pathogenesis of several degenerative diseases. For example, the complex role played by TG2 in oncogenesis can be at least partially explained by its involvement in autophagy regulation. In fact, autophagy induction is a key survival event for established tumors by providing nutrients under metabolic stress; by contrast autophagy can act as a tumor suppressor by enhancing the degradation of misfolded proteins and damaged organelles to maintain tissue homeostasis and genomic stability in normal cells or in the early stages of cancer development (Sun et al. 2013). Indeed, TG2 has been shown to be upregulated in cancer and might help tumor development by promoting the survival of tumors under nutrient stress via autophagy induction. In particular, TG2 can participate in oncogenesis via the p53-induced autophagy program. p53 has diverse roles in autophagy; nuclear p53 promotes autophagy through many of its target genes (DRAM, TIGAR, DAPK1, ULK1, PUMA, BAX, BAD, BNIP3) while cytoplasmic

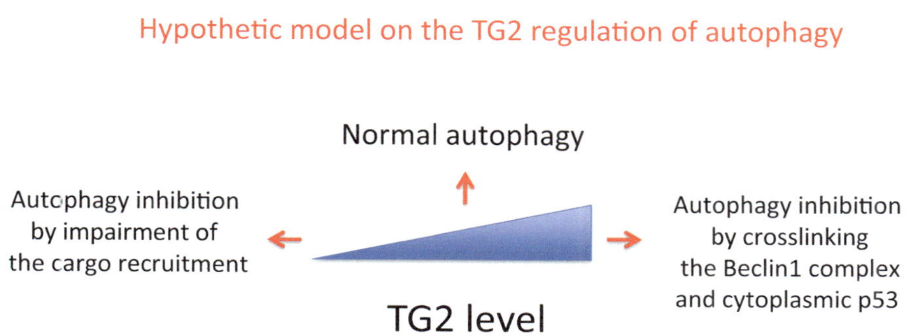

Fig. 8.3 Hypothetical scheme showing the relationships between TG2 protein expression and autophagy regulation. Under physiological condition TG2 is essential for autophagy induction, lower or higher TG2 protein levels in respect to physiological conditions both lead to autophagy inhibition

p53 inhibits autophagy (Liu et al. 2013). Considering that it has been shown that TG2 crosslinks the DNA binding domain of p53 leading to its depletion via autophagy, it is tempting to hypothesize that the enzyme might act as a rheostat by controlling the amount of p53 present in the cytoplasm and as a consequence allowing the normal autophagic flux. In cells lacking TG2 the persistent presence of p53 in the cytoplasm can lead to blocking the autophagic flux. In conclusion, it is now clear that TG2 plays a complex role in the regulation of autophagy acting both positively and negatively at various stage of the process. Its involvement in autophagy might help to decipher the complexity of its biological functions in both normal and pathological settings.

References

Akar U, Ozpolat B, Mehta K, Fok J, Kondo Y, Lopez-Berestein G (2007) Tissue transglutaminase inhibits autophagy in pancreatic cancer cells. Mol Cancer Res 5:241–249

Alers S, Löffler AS, Wesselborg S, Stork B (2012) Role of AMPK-mTOR-Ulk1/2 in the regulation of autophagy: cross talk, shortcuts, and feedbacks. Mol Cell Biol 32:2–11

Antonioli M, Albiero F, Nazio F, Vescovo T, Perdomo AB, Corazzari M, Marsella C, Piselli P, Gretzmeier C, Dengjel J, Cecconi F, Piacentini M, Fimia GM (2014) AMBRA1 interplay with Cullin E3 ubiquitin ligases regulates autophagy dynamics. Dev Cell 31:734–746

Atlashkin V, Kreykenbohm V, Eskelinen EL, Wenzel D, Fayyazi A, Fischer von Mollard G (2003) Deletion of the SNARE vti1b in mice results in the loss of a single SNARE partner, syntaxin 8. Mol Cell Biol 23:5198–5207

Axe EL, Walker SA, Manifava M, Chandra P, Roderick HL, Habermann A, Griffiths G, Ktistakis NT (2008) Autophagosome formation from membrane compartments enriched in phosphatidylinositol 3-phosphate and dynamically connected to the endoplasmic reticulum. J Cell Biol 182:685–701

Battaglia G, Farrace MG, Mastroberardino PG, Viti I, Fimia GM, Van Beeumen J, Devreese B, Melino G, Molinaro G, Busceti CL, Biagioni F, Nicoletti F, Piacentini M (2007) Transglutaminase 2 ablation leads to defective function of mitochondrial respiratory complex I affecting neuronal vulnerability in experimental models of extrapyramidal disorders. J Neurochem 100:36–49

Biederbick A, Kern HF, Elsässer HP (1995) Monodansylcadaverine (MDC) is a specific in vivo marker for autophagic vacuoles. Eur J Cell Biol 66:3–14

Blommaart EF, Luiken JJ, Meijer AJ (1997) Autophagic proteolysis: control and specificity. Histochem J 29:365–385

Cantley LC (2002) The phosphoinositide 3-kinase pathway. Science 296:1655–1657

Chen D, Zhong Q (2012) A tethering coherent protein in autophagosome maturation. Autophagy 8:985–986

D'Eletto M, Farrace MG, Falasca L, Reali V, Oliverio S, Melino G, Griffin M, Fimia GM, Piacentini M (2009) Transglutaminase 2 is involved in autophagosome maturation. Autophagy 5:1145–1154

D'Eletto M, Farrace MG, Rossin F, Strappazzon F, Giacomo GD, Cecconi F, Melino G, Sepe S, Moreno S, Fimia GM, Falasca L, Nardacci R, Piacentini M (2012) Type 2 transglutaminase is involved in the autophagy-dependent clearance of ubiquitinated proteins. Cell Death Differ 19:1228–1238

De Stefano D, Villella VR, Esposito S, Tosco A, Sepe A, De Gregorio F, Salvadori L, Grassia R, Leone CA, De Rosa G, Maiuri MC, Pettoello-Mantovani M, Guido S, Bossi A, Zolin A, Venerando A, Pinna LA, Mehta A, Bona G, Kroemer G, Maiuri L, Raia V (2014) Restoration of CFTR function in patients with cystic fibrosis carrying the F508del-CFTR mutation. Autophagy 10:2053–2074

Di Bartolomeo S, Corazzari M, Nazio F, Oliverio S, Lisi G, Antonioli M, Pagliarini V, Matteoni S, Fuoco C, Giunta L, D'Amelio M, Nardacci R, Romagnoli A, Piacentini M, Cecconi F, Fimia GM (2010) The dynamic interaction of AMBRA1 with the dynein motor complex regulates mammalian autophagy. J Cell Biol 191:155–168

Ding Q, Dimayuga E, Martin S, Bruce-Keller AJ, Nukala V, Cuervo AM, Keller JN (2003) Characterization of chronic low-level proteasome inhibition on neural homeostasis. J Neurochem 86:489–497

Fass E, Shvets E, Degani I, Hirschberg K, Elazar Z (2006) Microtubules support production of starvation-induced autophagosomes but not their targeting and fusion with lysosomes. J Biol Chem 281:36303–36316

Fimia GM, Stoykova A, Romagnoli A, Giunta L, Di Bartolomeo S, Nardacci R, Corazzari M, Fuoco C, Ucar A, Schwartz P, Gruss P, Piacentini M, Chowdhury K, Cecconi F (2007) Ambra1 regulates autophagy and development of the nervous system. Nature 447:1121–1125

Gu W, Wan D, Qian Q, Yi B, He Z, Gu Y, Wang L, He S (2014) Ambra1 is an essential regulator of autophagy and apoptosis in SW620 cells: pro-survival role of Ambra1. PLoS One 9:e90151

Gutierrez MG, Munafó DB, Berón W, Colombo MI (2004) Rab7 is required for the normal progression of the autophagic pathway in mammalian cells. J Cell Sci 117:2687–2697

Gwinn DM, Shackelford DB, Egan DF, Mihaylova MM, Mery A, Vasquez DS, Turk BE, Shaw RJ (2008) AMPK phosphorylation of raptor mediates a metabolic checkpoint. Mol Cell 30:214–226

Hailey DW, Rambold AS, Satpute-Krishnan P, Mitra K, Sougrat R, Kim PK, Lippincott-Schwartz J (2010) Mitochondria supply membranes for autophagosome biogenesis during starvation. Cell 141:656–667

Hamada K, Terauchi A, Nakamura K, Higo T, Nukina N, Matsumoto N, Hisatsune C, Nakamura T, Mikoshiba K (2014) Aberrant calcium signaling by transglutaminase-mediated posttranslational modification of inositol 1,4,5-trisphosphate receptors. Proc Natl Acad Sci 111:E3966–E3975

Hanada T, Noda NN, Satomi Y, Ichimura Y, Fujioka Y, Takao T, Inagaki F, Ohsumi Y (2007) The Atg12-Atg5 conjugate has a novel E3-like activity for protein lipidation in autophagy. J Biol Chem 282:37298–37302

Hanada S, Harada M, Abe M, Akiba J, Sakata M, Kwan R, Taniguchi E, Kawaguchi T, Koga H, Nagata E, Ueno T, Sata M (2012) Aging modulates susceptibility to mouse liver Mallory-Denk body formation. J Histochem Cytochem 60:475–483

Harada M, Hanada S, Toivola DM, Ghori N, Omary MB (2008) Autophagy activation by rapamycin eliminates mouse Mallory-Denk bodies and blocks their proteasome inhibitor-mediated formation. Hepatology 47:2026–2035

Hayashi-Nishino M, Fujita N, Noda T, Yamaguchi A, Yoshimori T, Yamamoto A (2009) A subdomain of the endoplasmic reticulum forms a cradle for autophagosome formation. Nat Cell Biol 11:1433–1437

He C, Bartholomew CR, Zhou W, Klionsky DJ (2009) Assaying autophagic activity in transgenic GFP-Lc3 and GFP-Gabarap zebrafish embryos. Autophagy 5:520–526

Hosokawa N, Hara T, Kaizuka T, Kishi C, Takamura A, Miura Y, Iemura S, Natsume T, Takehana K, Yamada N, Guan JL, Oshiro N, Mizushima N (2009) Nutrient-dependent mTORC1 association with the ULK1-Atg13-FIP200 complex required for autophagy. Mol Biol Cell 20:1981–1991

Høyer-Hansen M, Bastholm L, Szyniarowski P, Campanella M, Szabadkai G, Farkas T, Bianchi K, Fehrenbacher N, Elling F, Rizzuto R, Mathiasen IS, Jäättelä M (2007) Control of macroautophagy by calcium, calmodulin-dependent kinase kinase-beta, and Bcl-2. Mol Cell 25:193–205

Huang J, Manning BD (2008) The TSC1-TSC2 complex: a molecular switchboard controlling cell growth. Biochem J 412:179–190

Itakura E, Kishi C, Inoue K, Mizushima N (2008) Beclin 1 forms two distinct phosphatidylinositol 3-kinase complexes with mammalian Atg14 and UVRAG. Mol Biol Cell 19:5360–5372
Jung CH, Jun CB, Ro SH, Kim YM, Otto NM, Cao J, Kundu M, Kim DH (2009) ULK-Atg13-FIP200 complexes mediate mTOR signaling to the autophagy machinery. Mol Biol Cell 20:1992–2003
Jung CH, Ro SH, Cao J, Otto NM, Kim DH (2010) mTOR regulation of autophagy. FEBS Lett 584:1287–1295
Kim I, Rodriguez-Enriquez S, Lemasters JJ (2007) Selective degradation of mitochondria by mitophagy. Arch Biochem Biophys 462:245–253
Kim J, Kundu M, Viollet B, Guan KL (2011) AMPK and mTOR regulate autophagy through direct phosphorylation of Ulk1. Nat Cell Biol 13:132–141
Kimura S, Noda T, Yoshimori T (2008) Dynein-dependent movement of autophagosomes mediates efficient encounters with lysosomes. Cell Struct Funct 33:109–122
Kirisako T, Baba M, Ishihara N, Miyazawa K, Ohsumi M, Yoshimori T, Noda T, Ohsumi Y (1999) Formation process of autophagosome is traced with Apg8/Aut7p in yeast. J Cell Biol 147:435–446
Klionsky DJ, Abdalla FC, Abeliovich H, Abraham RT, Acevedo-Arozena A, Adeli K, Agholme L et al (2012) Guidelines for the use and interpretation of assays for monitoring autophagy. Autophagy 8:445–544
Ku BM, Kim DS, Kim KH, Yoo BC, Kim SH, Gong YD, Kim SY (2013) Transglutaminase 2 inhibition found to induce p53 mediated apoptosis in renal cell carcinoma. FASEB J 27:3487–3495
Kuma A, Hatano M, Matsui M, Yamamoto A, Nakaya H, Yoshimori T, Ohsumi Y, Tokuhisa T, Mizushima N (2004) The role of autophagy during the early neonatal starvation period. Nature 432:1032–1036
Küttner V, Mack C, Gretzmeier C, Bruckner-Tuderman L, Dengjel J (2014) Loss of collagen VII is associated with reduced transglutaminase 2 abundance and activity. J Invest Dermatol 134:2381–2389
Kwan R, Hanada S, Harada M, Strnad P, Li DH, Omary MB (2012) Keratin 8 phosphorylation regulates its transamidation and hepatocyte Mallory-Denk body formation. FASEB J 26:2318–2326
Lafontaine DL (2010) A 'garbage can' for ribosomes: how eukaryotes degrade their ribosomes. Trends Biochem Sci 35:267–277
Lamark T, Kirkin V, Dikic I, Johansen T (2009) NBR1 and p62 as cargo receptors for selective autophagy of ubiquitinated targets. Cell Cycle 8:1986–1990
Lamore SD, Wondrak GT (2012) Autophagic-lysosomal dysregulation downstream of cathepsin B inactivation in human skin fibroblasts exposed to UVA. Photochem Photobiol Sci 11:163–172
Lee JA, Beigneux A, Ahmad ST, Young SG, Gao FB (2007) ESCRT-III dysfunction causes autophagosome accumulation and neurodegeneration. Curr Biol 17:1561–1567
Levine B (2005) Eating oneself and uninvited guests: autophagy-related pathways in cellular defense. Cell 120:159–162
Levine B, Deretic V (2007) Unveiling the roles of autophagy in innate and adaptive immunity. Nat Rev Immunol 7:767–777
Li WW, Li J, Bao JK (2012) Microautophagy: lesser-known self-eating. Cell Mol Life Sci 69:1125–1136
Liang C, Lee JS, Inn KS, Gack MU, Li Q, Roberts EA, Vergne I, Deretic V, Feng P, Akazawa C, Jung JU (2008) Beclin1-binding UVRAG targets the class C Vps complex to coordinate autophagosome maturation and endocytic trafficking. Nat Cell Biol 10:776–787
Liu B, Bao JK, Yang JM, Cheng Y (2013) Targeting autophagic pathways for cancer drug discovery. Chin J Cancer 32:113–120
Luciani A, Villella VR, Esposito S, Brunetti-Pierri N, Medina D, Settembre C, Gavina M, Pulze L, Giardino I, Pettoello-Mantovani M, D'Apolito M, Guido S, Masliah E, Spencer B, Quaratino S, Raia V, Ballabio A, Maiuri L (2010) Defective CFTR induces aggresome formation and lung

inflammation in cystic fibrosis through ROS-mediated autophagy inhibition. Nat Cell Biol 12:863–875

Maiese K, Chong ZZ, Wang S, Shang YC (2012) Oxidant stress and signal transduction in the nervous system with the PI 3-K, Akt, and mTOR cascade. Int J Mol Sci 13:13830–13866

Maiuri MC, Tasdemir E, Criollo A, Morselli E, Vicencio JM, Carnuccio R, Kroemer G (2009) Control of autophagy by oncogenes and tumor suppressor genes. Cell Death Differ 16:87–93

Malorni W, Farrace MG, Rodolfo C, Piacentini M (2008) Type 2 transglutaminase in neurodegenerative diseases: the mitochondrial connection. Curr Pharm Des 14:278–288

Manning BD, Cantley LC (2003) United at last: the tuberous sclerosis complex gene products connect the phosphoinositide 3-kinase/Akt pathway to mammalian target of rapamycin (mTOR) signalling. Biochem Soc Trans 31:573–578

Mastroberardino PG, Iannicola C, Nardacci R, Bernassola F, De Laurenzi V, Melino G, Moreno S, Pavone F, Oliverio S, Fesus L, Piacentini M (2002) 'Tissue' transglutaminase ablation reduces neuronal death and prolongs survival in a mouse model of Huntington's disease. Cell Death Differ 9:873–880

Matsunaga K, Saitoh T, Tabata K, Omori H, Satoh T, Kurotori N, Maejima I, Shirahama-Noda K, Ichimura T, Isobe T, Akira S, Noda T, Yoshimori T (2009) Two BECLIN 1-binding proteins, Atg14L and Rubicon, reciprocally regulate autophagy at different stages. Nat Cell Biol 11:385–396

Min B, Kwon YC, Choe KM, Chung KC (2014) PINK1 phosphorylates transglutaminase 2 and blocks its proteasomal degradation. J Neurosci Res 93:722–735 doi: 10.1002/jnr.23535. [Epub ahead of print]

Mizushima N, Yoshimori T (2007) How to interpret LC3 immunoblotting. Autophagy 3:542–545

Mizushima N, Sugita H, Yoshimori T, Ohsumi Y (1998) A new protein conjugation system in human. The counterpart of the yeast Apg12p conjugation system essential for autophagy. J Biol Chem 273:33889–33892

Mizushima N, Kuma A, Kobayashi Y, Yamamoto A, Matsubae M, Takao T, Natsume T, Ohsumi Y, Yoshimori T (2003) Mouse Apg16L, a novel WD-repeat protein, targets to the autophagic isolation membrane with the Apg12-Apg5 conjugate. J Cell Sci 116:1679–1688

Mizushima N, Levine B, Cuervo AM, Klionsky DJ (2008) Autophagy fights disease through cellular self-digestion. Nature 451:1069–1075

Mousavi SA, Kjeken R, Berg TO, Seglen PO, Berg T, Brech A (1998) Effects of inhibitors of the vacuolar proton pump on hepatic heterophagy and autophagy. Biochim Biophys Acta 1510:243–257

Narendra D, Walker JE, Youle R (2012) Mitochondrial quality control mediated by PINK1 and Parkin: links to parkinsonism. Cold Spring Harb Perspect Biol 4

Nazio F, Cecconi F (2013) mTOR, AMBRA1, and autophagy: an intricate relationship. Cell Cycle 12:2524–2525

Nazio F, Strappazzon F, Antonioli M, Bielli P, Cianfanelli V, Bordi M, Gretzmeier C, Dengjel J, Piacentini M, Fimia GM, Cecconi F (2013) mTOR inhibits autophagy by controlling ULK1 ubiquitylation, self-association and function through AMBRA1 and TRAF6. Nat Cell Biol 15:406–416

Ohsumi Y (2014) Historical landmarks of autophagy research. Cell Res 24:9–23

Orenstein SJ, Cuervo AM (2010) Chaperone-mediated autophagy: molecular mechanisms and physiological relevance. Semin Cell Dev Biol 21:719–726

Ozpolat B, Akar U, Mehta K, Lopez-Berestein G (2007) PKC delta and tissue transglutaminase are novel inhibitors of autophagy in pancreatic cancer cells. Autophagy 3:480–483

Pattingre S, Tassa A, Qu X, Garuti R, Liang XH, Mizushima N, Packer M, Schneider MD, Levine B (2005) Bcl-2 antiapoptotic proteins inhibit Beclin 1-dependent autophagy. Cell 122:927–939

Petiot A, Ogier-Denis E, Blommaart EF, Meijer AJ, Codogno P (2000) Distinct classes of phosphatidylinositol 3′-kinases are involved in signaling pathways that control macroautophagy in HT-29 cells. J Biol Chem 275:12360

Piacentini M, D'Eletto M, Falasca L, Farrace MG, Rodolfo C (2011) Transglutaminase 2 at the crossroads between cell death and survival. Adv Enzymol Relat Areas Mol Biol 78:197–246
Ravikumar B, Moreau K, Jahreiss L, Puri C, Rubinsztein DC (2010) Plasma membrane contributes to the formation of pre-autophagosomal structures. Nat Cell Biol 12:747–757
Reggiori F, Klionsky DJ (2002) Autophagy in the eukaryotic cell. Eukaryot Cell 1:11–21
Reggiori F, Komatsu M, Finley K, Simonsen A (2012) Autophagy: more than a nonselective pathway. Int J Cell Biol 2012:219625
Rossin F, D'Eletto M, Macdonald D, Farrace MG, Piacentini M (2012) TG2 transamidating activity acts as a reostat controlling the interplay between apoptosis and autophagy. Amino Acids 42:1793–1802
Rossin F, D'Eletto M, Falasca L, Sepe S, Cocco S, Fimia GM, Campanella M, Mastroberardino PG, Farrace MG, Piacentini M (2015) Transglutaminase 2 ablation leads to mitophagy impairment associated with a metabolic shift towards aerobic glycolysis. Cell Death Differ 22:408–418
Rubinsztein DC (2006) The roles of intracellular protein-degradation pathways in neurodegeneration. Nature 443:780–786
Rubinsztein DC, Codogno P, Levine B (2012) Autophagy modulation as a potential therapeutic target for diverse diseases. Nat Rev Drug Discov 11:709–730
Rusten TE, Vaccari T, Lindmo K, Rodahl LM, Nezis IP, Sem-Jacobsen C, Wendler F, Vincent JP, Brech A, Bilder D, Stenmark H (2007) ESCRTs and Fab1 regulate distinct steps of autophagy. Curr Biol 17:1817–1825
Sakai Y, Oku M, van der Klei IJ, Kiel JA (2006) Pexophagy: autophagic degradation of peroxisomes. Biochim Biophys Acta 1763:1767–1775
Sarbassov DD, Guertin DA, Ali SM, Sabatini DM (2005) Phosphorylation and regulation of Akt/PKB by the rictor-mTOR complex. Science 307:1098–1101
Schmelzle T, Hall MN (2000) TOR, a central controller of cell growth. Cell 103:253–262
Snyder H, Mensah K, Theisler C, Lee J, Matouschek A, Wolozin B (2003) Aggregated and monomeric alpha-synuclein bind to the S6′ proteasomal protein and inhibit proteasomal function. J Biol Chem 278:11753–11759
Strnad P, Harada M, Siegel M, Terkeltaub RA, Graham RM, Khosla C, Omary MB (2007) Transglutaminase 2 regulates Mallory body inclusion formation and injury-associated liver enlargement. Gastroenterology 132:1515–1526
Strnad P, Zatloukal K, Stumptner C, Kulaksiz H, Denk H (2008) Mallory-Denk-bodies: lessons from keratin-containing hepatic inclusion bodies. Biochim Biophys Acta 1782:764–774
Sun K, Deng W, Zhang S, Cai N, Jiao S, Song J, Wei L (2013) Paradoxical roles of autophagy in different stages of tumorigenesis: protector for normal or cancer cells. Cell Biosci 3:35
Suzuki K, Kubota Y, Sekito T, Ohsumi Y (2007) Hierarchy of Atg proteins in pre-autophagosomal structure organization. Genes Cells 12:209–218
Takahashi Y, Coppola D, Matsushita N, Cualing HD, Sun M, Sato Y, Liang C, Jung JU, Cheng JQ, Mulé JJ, Pledger WJ, Wang HG (2007) Bif-1 interacts with BECLIN 1 through UVRAG and regulates autophagy and tumorigenesis. Nat Cell Biol 9:1142–1151
Takahashi Y, Meyerkord CL, Hori T, Runkle K, Fox TE, Kester M, Loughran TP, Wang HG (2011) Bif-1 regulates Atg9 trafficking by mediating the fission of Golgi membranes during autophagy. Autophagy 7:61–73
Tanaka Y, Guhde G, Suter A, Eskelinen EL, Hartmann D, Lüllmann-Rauch R, Janssen PM, Blanz J, von Figura K, Saftig P (2000) Accumulation of autophagic vacuoles and cardiomyopathy in LAMP-2-deficient mice. Nature 406:902–906
Tanida I, Ueno T, Kominami E (2004) LC3 conjugation system in mammalian autophagy. Int J Biochem Cell Biol 36:2503–2518
Tasdemir E, Maiuri MC, Galluzzi L, Vitale I, Djavaheri-Mergny M, D'Amelio M, Criollo A et al (2008) Regulation of autophagy by cytoplasmic p53. Nat Cell Biol 10:676–687

Thoresen SB, Pedersen NM, Liestøl K, Stenmark H (2010) A phosphatidylinositol 3-kinase class III sub-complex containing VPS15, VPS34, Beclin 1, UVRAG and BIF-1 regulates cytokinesis and degradative endocytic traffic. Exp Cell Res 316:3368–3378

Tucholski J (2010) TG2 protects neuroblastoma cells against DNA-damage-induced stress, suppresses p53 activation. Amino Acids 39:523–532

Twig G, Elorza A, Molina AJ, Mohamed H, Wikstrom JD, Walzer G, Stiles L, Haigh SE, Katz S, Las G, Alroy J, Wu M, Py BF, Yuan J, Deeney JT, Corkey BE, Shirihai OS (2008) Fission and selective fusion govern mitochondrial segregation and elimination by autophagy. EMBO J 27:433–446

Verhaar R, Drukarch B, Bol JG, Jongenelen CA, Wilhelmus MM (2013) Tissue transglutaminase cross-links Beclin 1 and regulates autophagy in MPP^+-treated human SH-SY5Y cells. Neurochem Int 62:486–491

Virgin HW, Levine B (2009) Autophagy genes in immunity. Nat Immunol 10:461–470

Weidberg H, Shvets E, Elazar Z (2009) Lipophagy: selective catabolism designed for lipids. Dev Cell 16:628–630

Weidberg H, Shvets E, Shpilka T, Shimron F, Shinder V, Elazar Z (2010) LC3 and GATE-16/GABARAP subfamilies are both essential yet act differently in autophagosome biogenesis. EMBO J 29:1792–1802

Weidberg H, Shpilka T, Shvets E, Abada A, Shimron F, Elazar Z (2011) LC3 and GATE-16 N termini mediate membrane fusion processes required for autophagosome biogenesis. Dev Cell 20:444–454

Yamamoto A, Tagawa Y, Yoshimori T, Moriyama Y, Masaki R, Tashiro Y (1998) Bafilomycin A1 prevents maturation of autophagic vacuoles by inhibiting fusion between autophagosomes and lysosomes in rat hepatoma cell line, H-4-II-E cells. Cell Struct Funct 23:33–42

Yang Z, Klionsky DJ (2010) Mammalian autophagy: core molecular machinery and signaling regulation. Curr Opin Cell Biol 22:124–131

Yorimitsu T, Klionsky DJ (2007) Endoplasmic reticulum stress: a new pathway to induce autophagy. Autophagy 3:160–162

Yoshimori T, Noda T (2008) Toward unraveling membrane biogenesis in mammalian autophagy. Curr Opin Cell Biol 20:401–407

Zhao J, Brault JJ, Schild A, Cao P, Sandri M, Schiaffino S, Lecker SH, Goldberg AL (2007) FoxO3 coordinately activates protein degradation by the autophagic/lysosomal and proteasomal pathways in atrophying muscle cells. Cell Metab 6:472–483

Zhong Y, Wang QJ, Li X, Yan Y, Backer JM, Chait BT, Heintz N, Yue Z (2009) Distinct regulation of autophagic activity by Atg14L and Rubicon associated with Beclin 1-phosphatidylinositol-3-kinase complex. Nat Cell Biol 11:468–476

Chapter 9
Transglutaminase 2 and Celiac Disease

Rasmus Iversen and Ludvig M. Sollid

Abstract The enzyme transglutaminase 2 (TG2) is deeply implicated in the development of celiac disease, an inflammatory disorder of the small intestine occurring in genetically susceptible individuals when they are exposed to dietary gluten proteins. TG2 is responsible for deamidating gluten peptides and thereby turns them into immunogenic T-cell antigens, leading to the activation of inflammatory gluten-reactive T cells. In addition, B-cell activation leads to the production of TG2-targeting autoantibodies which, together with antibodies against deamidated gluten, are highly specific to celiac disease. Recently, cloning of the antigen-specific receptor genes of the implicated T and B cells has provided new insight into the basis for the gluten-dependent immune response. Here, we review what is currently known about the involvement of TG2 in celiac disease, and we propose mechanisms that can explain the activation of T and B cells based on the special properties of gluten peptides which make them exceptional substrates for TG2.

Keywords Gluten • Deamidation • T cells • B cells • Antibodies • Antigen presentation

9.1 Introduction

Celiac disease is a common enteropathy affecting around 1 % of the population (Green and Cellier 2007). It is characterized by destruction of the intestinal villi and infiltration of immune cells in the gut epithelium and in the lamina propria (Sollid 2002). The condition is driven by ingestion of cereal gluten proteins. Hence, if gluten is excluded from the diet, the small intestinal lesion resolves and the structure of the villi returns to normal. Currently, a life-long gluten-free diet is the only available treatment for the disease.

There is a strong genetic component in celiac disease, and genome-wide association studies have identified several genetic variants that confer risk of disease

R. Iversen (✉) • L.M. Sollid
Centre for Immune Regulation and Department of Immunology, University of Oslo and Oslo University Hospital, N-0372 Oslo, Norway
e-mail: rasmus.iversen@rr-research.no; l.m.sollid@medisin.uio.no

K. Hitomi et al. (eds.), *Transglutaminases*, DOI 10.1007/978-4-431-55825-5_9

development (van Heel et al. 2007; Hunt et al. 2008; Dubois et al. 2010; Trynka et al. 2011). Many of the implicated genes are involved in regulation of the immune system, consistent with celiac disease being an immune-mediated disorder. By far the most important contribution to the genetic risk comes from the human leukocyte antigen (HLA) locus. Thus, celiac disease patients express certain variants of the MHC class II molecule HLA-DQ (HLA-DQ2.5, HLA-DQ8 or HLA-DQ2.2) (Sollid and Lie 2005; Mubarak et al. 2013). The strong association of MHC class II with celiac disease suggests that antigen presentation to CD4+ T cells is crucial for disease development. In line with this notion, CD4+ T cells reactive with gluten peptides presented in the context of the disease-associated HLA molecules are present in the gut mucosa of celiac disease patients (Lundin et al. 1993, 1994; Molberg et al. 1997; Bodd et al. 2012). It is believed that these cells through the release of pro-inflammatory cytokines are responsible for orchestrating the gluten-dependent killing of enterocytes, leading to villous atrophy (Abadie et al. 2011). In addition, the CD4+ T cells can provide activation signals to B cells causing the latter to differentiate into antibody-secreting plasma cells. Correspondingly, celiac disease is tightly associated with antibody production against gluten but also with antibodies targeting the self-antigen transglutaminase 2 (TG2) (Qiao et al. 2012). This ubiquitously expressed protein, which is found both intra- and extracellularly, has been implicated in a number of biological processes, including wound healing, cell migration, extracellular matrix remodeling and transmembrane signaling (Fesus and Piacentini 2002). Curiously, in celiac disease, TG2 is also responsible for creating T-cell epitopes, as gluten-reactive T cells recognize peptides in which certain glutamine residues have been converted to glutamic acid through TG2-catalyzed deamidation (Sollid et al. 2012). Thus, TG2 plays several roles in celiac disease, and the enzyme seems to be directly involved in the initiation of the disease-causing immune response through selective activation of T and B cells carrying highly specific surface receptors (Fig. 9.1).

9.2 TG2-Mediated Generation of Gluten T-Cell Epitopes

9.2.1 Gluten as a TG2 Substrate

Gluten constitutes a large and complex family of proteins. In wheat, gluten proteins are divided into gliadins and glutenins, which can each be further divided into subgroups. Many gluten proteins, however, have very similar amino acid sequences, and they are typically rich in proline and glutamine residues. The high proline content makes gluten proteins resistant to cleavage by gastrointestinal proteases and allows long peptides to reach the small intestine (Shan et al. 2002). Here, they cross the epithelial layer and are recognized by the immune system, leading to inflammation and destruction of the epithelium. Food proteins, although recognizable by the immune system, do not normally cause inflammation, since the

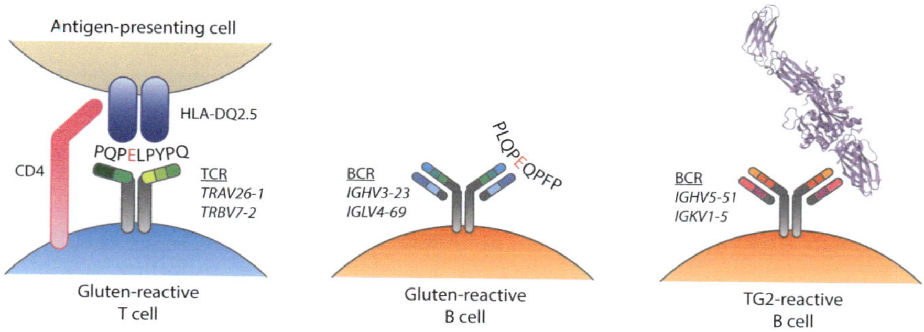

Fig. 9.1 Cells of the adaptive immune system involved in celiac disease. T cells and B cells recognize antigen via the T-cell receptor (TCR, comprising one α and one β chain) and the B-cell receptor (BCR, comprising two heavy and two light chains), respectively. The TCR on CD4+ T cells recognizes antigenic peptides bound in a complex with MHC class II molecules such as HLA-DQ2.5 on the surface of an antigen-presenting cell, whereas B cells recognize intact antigens. Both TCRs and BCRs are made up of constant regions (*gray*) and variable regions (*colored segments*). The variable regions, which determine the antigen reactivity of the cell, are generated through stochastic recombination of two or three individual gene segments from a large genetic repertoire. Typical combinations of variable region gene segments used by gluten-reactive T cells, gluten-reactive B cells and TG2-reactive B cells are indicated. The epitopes commonly targeted by these cells are shown as gluten peptide sequences with deamidated residues highlighted in *red* or on the crystal structure of TG2 in the open conformation (Pinkas et al. 2007) (PDB ID code 2Q3Z)

microenvironment of the gut suppresses this type of response, a phenomenon known as oral tolerance (Pabst and Mowat 2012). It is not known why oral tolerance to gluten is broken in celiac disease, but it has been suggested that altered cytokine expression in the gut mucosa of patients causes disruption of normal homeostasis leading to breaking of gluten tolerance (Monteleone et al. 2001; Mention et al. 2003; DePaolo et al. 2011). It is not clear what makes gluten special compared to other food proteins, though. A possible hint to answer this question might come from the finding that gluten-reactive CD4+ T cells in celiac disease recognize deamidated rather than native peptides (Molberg et al. 1998; van de Wal et al. 1998). Thus, the peptides that the immune system reacts to are most likely different from those initially taken up in the small intestine. As the deamidation process is mediated by TG2, the enzyme can be considered as the selective force responsible for picking out and generating immunogenic peptides. In this regard, it is noteworthy that many gluten peptides serve as excellent TG2 substrates (Bruce et al. 1985; Piper et al. 2002).

TG2 reacts with glutamine residue side chains in a Ca^{2+}-dependent manner, leading to formation of a covalent enzyme-substrate intermediate. This intermediate is resolved through a reaction either with water, leading to *de*amidation, or with a primary amine, leading to *trans*amidation. When the amino group is provided by the side chain of a lysine residue, the transamidation reaction leads to covalent cross-linking of two polypeptide chains through formation of an N^{ε}(γ-glutamyl) lysine isopeptide bond (Lorand and Graham 2003). The sequence context of a glutamine residue is decisive for its reactivity with TG2. In particular, if a proline residue is positioned close to and on the C-terminal side of a glutamine residue, it has been demonstrated that the spacing between the two residues determines whether the glutamine residue is targeted by TG2. Thus, glutamine residues residing in the sequence motif QXP (where X can be any amino acid) are preferred TG2 substrates, whereas the motifs QP and QXXP are not targeted (Vader et al. 2002; Fleckenstein et al. 2002). The large number of glutamine and proline residues in gluten proteins endows many gluten-derived peptides with exceptional TG2 substrate abilities. Interestingly, when offered a complex mixture of gluten peptides, TG2 selectively targeted peptides harboring known T-cell epitopes (Dørum et al. 2010), suggesting that the activity of TG2 profoundly shapes the anti-gluten T-cell response in celiac disease.

9.2.2 T-Cell Recognition of Deamidated Gluten

The conversion of particular gluten glutamine residues to glutamic acid through TG2-mediated deamidation increases the peptide binding affinity of the celiac disease-associated HLA molecules (Arentz-Hansen et al. 2000) and impedes the disassociation off-rate for the resulting complexes (Xia et al. 2005). The increased binding strength observed after deamidation reflects that the introduced negatively charged glutamate side chains fit into positively charged binding pockets in the peptide binding site of the HLA molecules (Kim et al. 2004; Henderson et al. 2007; Petersen et al. 2014). Thus, the preference for negative charges together with an ability to bind proline-rich gluten peptides (Bergseng et al. 2005) can explain the HLA association of celiac disease. As T cells recognize complexes of HLA and peptide, the increased peptide binding affinity and diminished off-rate observed upon deamidation potentially explain why TG2 activity seems to be the driver of the T-cell response. Correspondingly, gluten-reactive T cells in celiac disease target deamidated peptides that make stable interactions with the HLA molecules (Fallang et al. 2009; Bodd et al. 2012).

Gluten-reactive T cells, particularly those generated in HLA-DQ2.5 patients, are deamidation specific in the sense that they recognize the deamidated and not the native variant of peptides (Molberg et al. 2001). The preference is more pronounced than can be explained by the increased binding affinity to HLA molecules, indicating that sensing of the deamidated residues by the T-cell receptors (TCRs) is important for T-cell activation. The deamidation sensitivity, however, does not result from direct contact of the TCR with the glutamate residue, as shown by co-crystallization of the HLA molecule, peptide and TCR (Petersen et al. 2014; Broughton et al. 2012), suggesting that peptide deamidation has allosteric effects on the peptide-HLA structure. Notably, the specificity of T cells toward deamidated gluten peptides is reflected by a highly restricted usage of TCR gene segments, at least among T cells reactive with some epitopes (Qiao et al. 2011, 2014; Petersen et al. 2014; Broughton et al. 2012) (Fig. 9.1).

The role of the disease-associated HLA molecules in presentation of gluten peptides to T cells has been studied by comparing HLA-DQ2.5 and HLA-DQ2.2. Although the two molecules are highly homologous, HLA-DQ2.5 confers a much higher risk of celiac disease than HLA-DQ2.2. The reason for this difference appears to relate to distinct binding specificities of the two HLA molecules (Bergseng et al. 2015). Different repertoires of gluten peptides bind stably to HLA-DQ2.5 and HLA-DQ2.2 (Dorum et al. 2014), and, conceivably, the amount and number of distinct gluten peptides that can make a stable complex with HLA-DQ2.5 are higher than for HLA-DQ2.2, hence explaining the higher risk.

Antigen presentation and T-cell activation happen in secondary lymphoid tissues such as lymph nodes or the spleen. In celiac disease, gluten-reactive CD4+ T cells are most likely activated in the gut-draining mesenteric lymph nodes. After initial activation and proliferation, many of these T cells migrate to the small intestinal mucosa, where they start secreting pro-inflammatory cytokines like interferon (IFN)-γ and interleukin (IL)-21 (Nilsen et al. 1995; Bodd et al. 2010). Some gluten-reactive T cells, however, stay in the lymph node, where they can interact with B cells and provide activation signals, which drive B-cell proliferation and differentiation (Fig. 9.2). The outcome is high numbers of plasma cells, which can

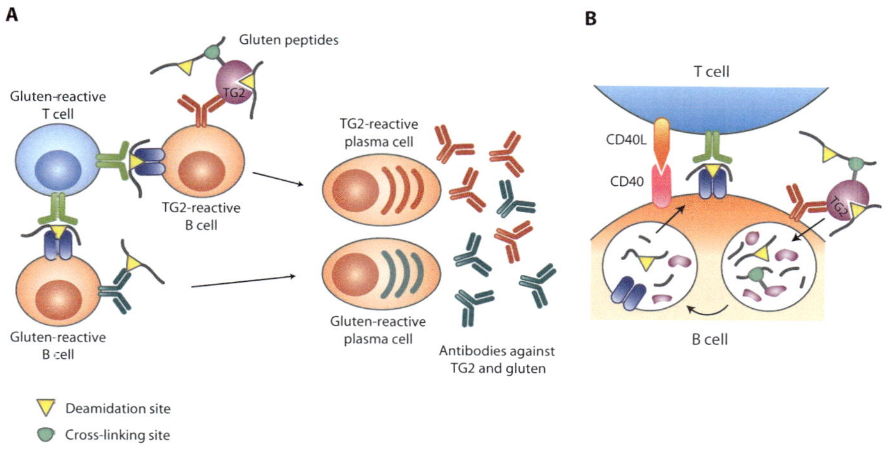

Fig. 9.2 Model showing the collaboration between a gluten-reactive T cell and gluten- and TG2-reactive B cells. (**a**) T cells reactive with deamidated gluten peptides can provide help to both gluten-reactive B cells and TG2-reactive B cells, leading to plasma cell differentiation and production of antibodies which are secreted variants of the gluten-reactive and TG2-reactive BCRs. TG2-reactive B cells can take up and present deamidated gluten peptides through formation of TG2-gluten peptide complexes in which the peptide is either cross-linked to a lysine residue on the surface of TG2 or bound in the active site of the enzyme. (**b**) After initial binding of TG2 and gluten peptide to the BCR, the complex is internalized and processed by proteases. Peptides that can form a stable complex with MHC class II molecules are subsequently presented on the surface of the B cell, allowing interaction with T cells that can provide the B cell with activation signals. An important signal is generated through the binding of CD40 on the B cell to CD40 ligand (CD40L) on the T cell

be found in the gut mucosa (Baklien et al. 1977). In celiac disease, plasma cells secreting antibodies to gluten (Steinsbø et al. 2014) or TG2 (Di Niro et al. 2012) can be visualized in patient gut biopsies. These antibodies are mainly of the IgA isotype and are destined for secretion into the intestinal lumen. In addition, both IgA and IgG antibodies specific for gluten or TG2 can be detected in patient serum samples.

9.3 Antibodies in Celiac Disease

9.3.1 Anti-Gluten Antibodies

Antibodies recognizing gluten proteins are present in the blood of celiac disease patients and have been employed in serological testing since the 1960s (Berger et al. 1964). These antibodies display relatively poor sensitivity and specificity for celiac disease, but, more recently, it was shown that the conversion of certain glutamine residues to glutamic acid increased the reactivity of gliadin peptides with serum antibodies, selectively for celiac disease patients (Osman et al. 2000; Aleanzi et al. 2001; Mothes 2007). The improved binding of celiac antibodies to deamidated gluten suggests that TG2 is not only involved in the generation of T-cell epitopes, but also plays an important role in B-cell epitope formation. Moreover, the deamidated gluten epitopes recognized by antibodies were overlapping with or situated in close proximity to known T-cell epitopes (Osman et al. 2000; Bateman et al. 2004). Thus, the same parts of gluten seem to be targeted by both T and B cells, and these are the parts that are efficiently deamidated by TG2.

Recently, our group successfully isolated single gluten-reactive plasma cells from small intestinal biopsies of celiac disease patients, allowing cloning of their antibody genes and recombinant expression of anti-gluten monoclonal antibodies (mAbs) (Steinsbø et al. 2014). Consistent with the results obtained with polyclonal serum antibodies, the gluten-specific mAbs recognized particular epitopes harboring TG2 deamidation sites. Some of the generated mAbs were deamidation dependent and did not show reactivity to native gluten peptide, strongly suggesting that gluten-reactive B cells in celiac disease are activated in response to peptides that have been deamidated by TG2. This notion is consistent with gluten-reactive T cells being specific to deamidated epitopes. B-cell activation normally requires recruitment of T-cell help through presentation of antigen in the form of peptide-MHC complexes, and as a B cell only presents antigens that bind to its B-cell receptor (BCR), a B cell can only interact with T cells that recognize the same antigen as itself (Fig. 9.2). Thus, although T- and B-cell epitopes are not necessarily identical, they must be physically linked in order for the cells to interact. The collaboration between T and B cells in celiac disease therefore explains why both cell types display specificity toward the same parts of gluten that have been deamidated by TG2. How gluten consumption induces an autoantibody response against TG2 is not as easily explained, but it has been suggested that TG2-reactive B cells can take up complexes of TG2 and gluten peptides, leading to antigen presentation to gluten-reactive T cells (Sollid et al. 1997) (Fig. 9.2).

9.3.2 Anti-TG2 Antibodies

Autoantibodies targeting connective tissue structures are a hallmark of celiac disease. Particularly, antibodies binding to the endomysium surrounding smooth muscle fibers are harnessed for testing of serum reactivity, usually by fluorescent staining of monkey esophagus sections (Chorzelski et al. 1984). In 1997, Dieterich and co-workers showed that the major antigen of the anti-endomysium antibodies is TG2 (Dieterich et al. 1997), allowing development of diagnostic tests measuring anti-TG2 serum antibodies by ELISA (Sulkanen et al. 1998; Dieterich et al. 1998). The available serological tests show very high sensitivity and specificity (Rostom et al. 2005; Leffler and Schuppan 2010), and, as a result, it is now possible to diagnose pediatric celiac disease based on the presence of high levels of serum antibodies, thereby excluding the need for taking a small intestinal biopsy (Husby et al. 2012). Recently, a similar approach has also been suggested for adult patients (Zanini et al. 2012; Hill et al. 2015).

In addition to the antibodies present in serum, deposits of IgA bound to extracellular TG2 can be detected beneath the epithelium in small intestinal biopsies from celiac disease patients (Korponay-Szabo et al. 2004). Such deposits were observed even in the small subset of patients who are negative for anti-TG2 serum antibodies (Salmi et al. 2006b). Moreover, intestinal deposits of IgA were found to predict development of celiac disease as they could be detected before histological changes in the gut occurred (Salmi et al. 2006a). Thus, the production of TG2-specific autoantibodies is an early event in celiac disease development, and the activation of TG2-reactive B cells appears to be tightly linked with the pathogenesis.

Although anti-TG2 antibodies are strongly associated with celiac disease, it is uncertain whether the antibodies have any pathogenic effect or can be considered as merely a side effect of the immune response. Since TG2 has multiple functions, activity-dependent as well as non-enzymatic, it is conceivable that TG2-targeting antibodies can have an impact on several biological processes and therefore be relevant to disease development. Despite inducing an antibody response to TG2, immunization of mice with recombinant human TG2 did not lead to changes in the gut, though (Freitag et al. 2004). In a similar type of experiment, a TG2-reactive mAb fragment was constitutively expressed in mice by adenoviral transduction, but, also in this case, no morphological changes were observed (Di Niro et al. 2008). These studies are consistent with the notion that the small intestinal lesion in celiac disease is mainly established through the action of T cells. Autoantibodies to TG2 could, however, have more subtle effects that contribute to the observed histological changes. Several in vitro studies examining the effect of celiac serum antibodies support this view. Hence, it has been reported that the antibodies in serum affect intestinal epithelial cells by inhibiting differentiation (Halttunen and Maki 1999), inducing proliferation (Barone et al. 2007) and increasing intestinal permeability (Zanoni et al. 2006). Similarly, celiac antibodies were shown to facilitate transport of gluten across the epithelium, thus allowing immunogenic peptides to reach the lamina propria (Matysiak-Budnik et al. 2008; Rauhavirta et al. 2011; Lebreton et al. 2012). Antibodies directed against TG2 have also been shown to inhibit angiogenesis in vitro (Myrsky et al. 2008) as well as in vivo when injected into mice (Kalliokoski et al. 2013). These findings are consistent with an altered vascular organization as has been observed in the small intestine of celiac disease patients (Myrsky et al. 2009b). Moreover, anti-TG2 antibodies were found to increase vascular permeability and could thereby enhance the infiltration of immune cells in the gut mucosa (Myrsky et al. 2009a).

It is not clear how binding of antibodies to TG2 would cause the observed biological effects. Different studies have reported that celiac antibodies inhibit (Esposito et al. 2002), enhance (Kiraly et al. 2006; Myrsky et al. 2009a) or do not affect (Dieterich et al. 2003) TG2 activity. The reason for this discrepancy is probably that different assays were used to measure TG2 activity, and the size and physical context of the substrate could play important roles. Thus, it is conceivable that antibody binding to TG2 could inhibit the access of large or solid phase-associated substrates to the active site, while allowing TG2 to act on smaller, soluble substrates. We recently generated a panel of TG2-specific mAbs by isolating single TG2-reactive plasma cells from intestinal biopsies of celiac disease patients followed by cloning of their antibody genes (Di Niro et al. 2012). Notably, when incubated with TG2 in the presence of Ca^{2+} before the addition of soluble peptide substrate, none of the mAbs were found to interfere with either deamidation or transamidation (Di Niro et al. 2012). However, by assessing the conformational state of TG2 using hydrogen/deuterium exchange monitored by mass spectrometry, binding of mAbs was found to cause structural changes in TG2 (Iversen et al. 2014). Thus, it is possible that TG2-reactive antibodies can exert a biological effect by inducing changes in the TG2 structure that alter some properties of the protein, such as its ability to bind interaction partners.

Epitope mapping studies revealed that TG2-specific mAbs target a few, partly overlapping epitopes that are clustered in the N-terminal half of the TG2 molecule (Iversen et al. 2013) (Fig. 9.3). Hence, 49 out of 57 mAbs could be placed into one of four distinct epitope groups (epitope 1–4) (Iversen et al. 2013). Further, the TG2 residues E8/K30 and R19 in the N-terminal domain of the protein were shown to be implicated in the binding of mAbs recognizing epitope 1 and 2, respectively (Iversen et al. 2014). Importantly, epitope 2 coincided with a previously reported epitope, which was demonstrated to be important for the binding of celiac serum antibodies (Simon-Vecsei et al. 2012). Our results indicate that the major epitopes targeted by celiac autoantibodies are located in the N-terminal domain of TG2, but because B-cell epitopes typically cover 650–900 Å^2 of protein surface area (Laver et al. 1990) and the length of the N-terminal domain is only around 35 Å, it is likely

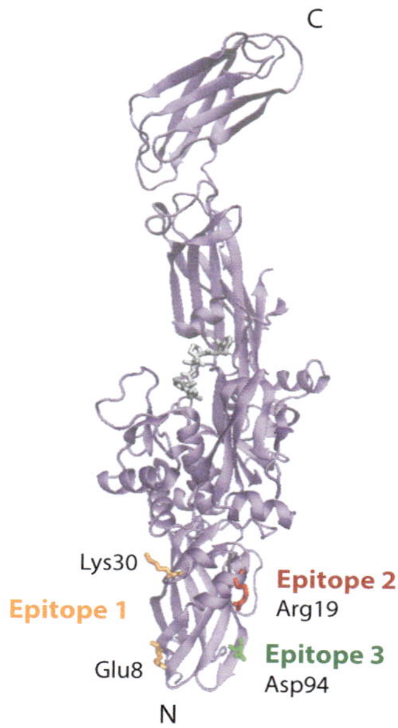

Fig. 9.3 Location of epitopes targeted by TG2-specific mAbs. mAbs generated from TG2-reactive plasma cells in gut biopsies of celiac disease patients (Di Niro et al. 2012) have been grouped according to epitope reactivity (Iversen et al. 2013), and amino acid residues that are part of the three main epitopes have been identified (Iversen et al. 2014). The epitopes are located close to each other in the N-terminal part of the TG2 molecule as indicated on the crystal structure of TG2 in the open conformation with a peptide inhibitor (*gray* stick representation) bound in the active site (Pinkas et al. 2007) (PDB ID code 2Q3Z)

that the epitopes stretch into the catalytic core domain of TG2. Correspondingly, mutations introduced into the core domain have also been shown to interfere with binding of serum antibodies (Kiraly et al. 2009; Simon-Vecsei et al. 2012). Moreover, testing of antibody reactivity to recombinant TG2 length variants has pointed to the N-terminal domain and the core domain as the main targets of celiac antibodies, although some patients also have antibodies reactive with the two C-terminal domains (Seissler et al. 2001; Nakachi et al. 2004). Analyzing the specificity of celiac anti-TG2 antibodies in more detail, however, is challenging as the antibodies recognize conformation-dependent epitopes that can be disrupted due to misfolding of truncated or mutated TG2 (Seissler et al. 2001; Di Niro et al. 2012; Simon-Vecsei et al. 2012).

9.3.3 Antibodies to Other Transglutaminases

Although antibodies targeting TG2 show the highest association with celiac disease and are the only autoantibodies harnessed for diagnostic purposes, antibodies recognizing other self-antigens have also been described (Alaedini and Green 2008). These include antibodies to TG3 in the skin (Sardy et al. 2002) and TG6 in the brain (Hadjivassiliou et al. 2008). Interestingly, anti-TG3 and anti-TG6 antibodies have been associated with the gluten-sensitive skin disease dermatitis herpetiformis and the neurological disorder gluten ataxia, respectively. Both conditions are dependent on a gluten-containing diet and can be considered extra-intestinal manifestations of celiac disease occurring in a subset of patients. Thus, the appearance of serum antibodies targeting transglutaminases other than TG2 correlates with spreading of the immune response against gluten from the gut to other parts of the body, reflecting the tissue expression profiles of the involved transglutaminases. Notably, the antibodies targeting TG3 and TG6 seem to be generated in an independent immune response involving B cells other than those targeting TG2, as TG2-reactive mAbs generated from gut plasma cells do not cross-react with the other transglutaminases (Iversen et al. 2013).

9.3.4 Comparison of the Antibody Responses to Gluten and TG2

Antibodies to deamidated gluten and antibodies to TG2 are both excellent markers for celiac disease, and both types of antibodies disappear from the circulation within months after patients commence a gluten-free diet (Sugai et al. 2006; Sulkanen et al. 1998). Moreover, the antibodies reappear with the same kinetics, if treated patients are challenged with gluten (Leffler et al. 2013), suggesting that B cells recognizing gluten and B cells recognizing TG2 respond in the same way to gluten exposure. Indeed, it has been suggested that the same gluten-reactive CD4+ T cells can provide activation help to both groups of B cells and thereby control both antibody responses (Sollid et al. 1997; Steinsbø et al. 2014) (Fig. 9.2).

The possibility of visualizing single gluten- (Steinsbø et al. 2014) or TG2-reactive (Di Niro et al. 2012) plasma cells in gut biopsies from untreated celiac disease patients has allowed enumeration of the cells secreting one type of antibody or the other. Remarkably, on average 10 % of all IgA-producing plasma cells in the small intestinal mucosa were found to be TG2 specific (Di Niro et al. 2012). In comparison, only around 1 % of the cells were specific to deamidated gliadin peptides (Steinsbø et al. 2014), suggesting that, at the B-cell level, the response against TG2 dominates over the response against gluten.

Cloning of the antibody genes from gluten- or TG2-reactive plasma cells has revealed that both populations use certain combinations of variable region gene segments to make up their antibody heavy and light chains (Di Niro et al. 2012;

Steinsbø et al. 2014; Marzari et al. 2001) (Fig. 9.1). Notably, for the TG2-specific plasma cells, the *IGHV* segment usage correlated with targeting of individual epitopes on TG2 (Iversen et al. 2013). Hence, half of the mAbs generated from TG2-specific plasma cells targeted a particular epitope (epitope 1), and most of these used the *IGHV5-51* gene segment, which was found to be overrepresented among both mAbs generated from single TG2-reactive plasma cells (Di Niro et al. 2012) and TG2-binding antibody fragments obtained from phage-display libraries (Marzari et al. 2001).

A peculiar aspect of both gluten-reactive and TG2-reactive plasma cells in the celiac lesion is the relatively few mutations present in their antibody genes (Di Niro et al. 2012; Steinsbø et al. 2014). The interactions between antigen-specific T and B cells in secondary lymphoid tissues normally result in the formation of structures known as germinal centers (GCs) in which B cells undergo rapid proliferation accompanied by the introduction of antibody mutations (Victora and Nussenzweig 2012). Some of these mutations result in increased affinity of the B cell for the driving antigen, leading to increased antigen presentation and reception of higher levels of survival signals compared with B cells of lower affinity. Thus, through repetitive rounds of divisions and selection, GC B cells gradually increase their affinity for antigen through the accumulation of antibody mutations. As a result, plasma cells derived from GCs usually secrete high-affinity antibodies carrying many mutations (Victora and Nussenzweig 2012). The apparent lack of mutations among gluten- and TG2-reactive plasma cells in celiac disease therefore suggests that these cells are generated in an atypical manner and that the B cells bypass or spend limited time in GCs.

9.4 Mechanisms for Activation of T and B Cells in Celiac Disease

9.4.1 Antigen Presentation to T Cells

As T cells in celiac disease recognize gluten peptides that have been deamidated by TG2, a key question in our understanding of T-cell activation is where this deamidation takes place. Since gluten peptides enter the body in the small intestine, and extracellular TG2 is found abundantly beneath the intestinal epithelium, an obvious possibility is that gluten peptides are deamidated after crossing the epithelium and facing TG2 in the extracellular matrix. It has been reported, however, that in celiac disease patients, TG2 is also expressed on the brush border facing the intestinal lumen (Molberg et al. 1998; Esposito et al. 2003), suggesting that gluten peptides could meet active TG2 already before crossing the epithelium and possibly be transported to the lamina propria in a complex with TG2 and secretory IgA through binding of IgA to the transferrin receptor (Matysiak-Budnik et al. 2008; Lebreton et al. 2012).

Compared to healthy control subjects, celiac disease patients express higher levels of TG2 in the gut mucosa (Molberg et al. 1998; Brusco et al. 1999; Esposito et al. 2003), and it was shown that this TG2 is catalytically active when incubating biopsy sections with cross-linking substrate in the presence of Ca^{2+} (Esposito et al. 2003; Bruce et al. 1985). In the small intestine of mice, however, catalytically active TG2 could not be detected in vivo under normal physiological conditions (Siegel et al. 2008), thus raising the question of whether the TG2 activity observed in tissue sections is an artifact arising during biopsy processing. Presumably, TG2 is inactivated through intramolecular disulfide bond formation in the oxidative extracellular environment (Stamnaes et al. 2010). Importantly, though, TG2 activity could be detected in vivo after induction of tissue injury or inflammation (Siegel et al. 2008; Dafik et al. 2012), possibly as a consequence of changes in the local redox environment (Jin et al. 2011). Hence, it is possible that TG2 in the small intestinal mucosa is active under the inflammatory conditions in celiac disease and engaged in deamidation of gluten peptides. However, as TG2 activity is directly involved in the initiation of the immune response, enzyme activity is also required before the celiac lesion is established. It is possible that a low level of TG2 activity present under normal physiological conditions is sufficient to modify gluten peptides. Alternatively, a type of intestinal inflammation not dependent on activation of gluten-specific T cells is needed to trigger TG2 activity. Components of wheat could potentially work as triggering agents, as it has been demonstrated that both gluten peptides (Maiuri et al. 2003) and wheat α-amylase/trypsin inhibitors (Junker et al. 2012) can activate cells of the innate immune system. Perhaps more likely, a gut infection, which would create an inflammatory environment, could be the cause of TG2 activation. In keeping with this idea, rotavirus infection has been associated with the development of celiac disease (Troncone and Auricchio 2007).

Foreign antigens that enter the body are typically picked up by dendritic cells, which subsequently migrate to draining lymph nodes where they present antigenic peptides to T cells. Thus, it is believed that gut-residing dendritic cells pick up gluten peptides that have been deamidated by TG2 and migrate to the mesenteric lymph nodes to activate gluten-reactive T cells through display of peptide-MHC complexes on their surface (Jabri and Sollid 2009) (Fig. 9.4). The phenotype of dendritic cells that pick up soluble antigen in the gut normally prevents activation of inflammatory antigen-specific T cells and thereby ensures that oral tolerance toward food proteins is maintained (Pabst and Mowat 2012). It has been suggested that increased levels of the cytokine IL-15 in the small intestinal mucosa of celiac disease patients in conjunction with the vitamin A metabolite retinoic acid change the phenotype of dendritic cells in the gut and cause them to induce an inflammatory rather than a tolerogenic anti-gluten T-cell response (DePaolo et al. 2011). It is possible that IFN-α, which is also expressed at increased levels in celiac disease, has a similar effect and contributes to the breaking of gluten tolerance (Monteleone et al. 2001). Thus, cytokine-mediated changes in the phenotype of the dendritic cells that take up gluten peptides and present them to T cells could explain the inflammatory T-cell response against gluten in celiac disease.

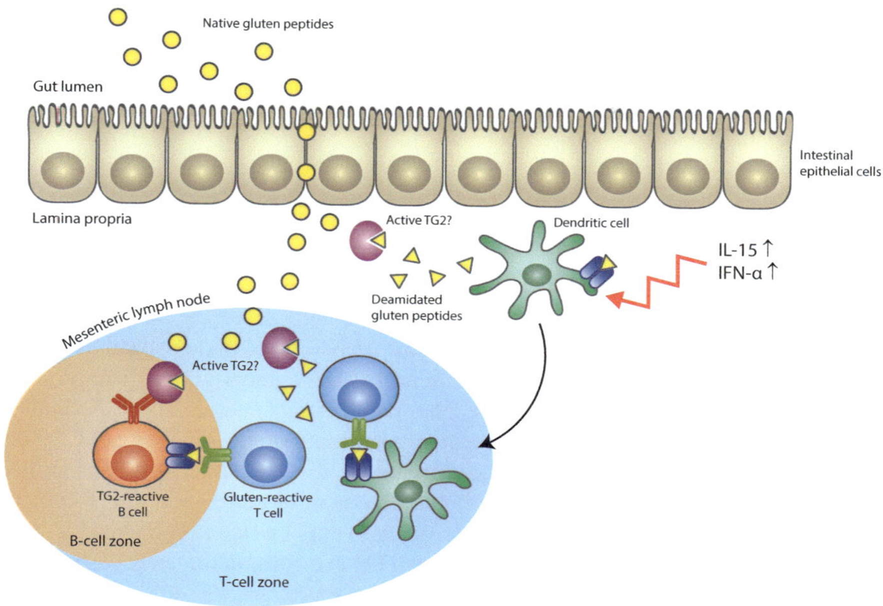

Fig. 9.4 Events leading to the activation of inflammatory T cells reactive with deamidated gluten peptides in celiac disease. TG2-mediated gluten deamidation is essential for T-cell activation, but we do not know under what conditions the deamidation reaction takes place. If TG2 in the intestinal mucosa is active, deamidation could happen locally after gluten peptides have crossed the epithelium. Alternatively, native gluten peptides flow via the lymphatic system to the mesenteric lymph nodes, where they are deamidated by TG2 and presented to T cells either by B cells or by dendritic cells present in the lymph node. Dendritic cells present in the gut mucosa can take up locally deamidated gluten peptides and migrate to the mesenteric lymph nodes to present them to T cells. It is possible that increased expression of the cytokines IL-15 and IFN-α in the intestinal mucosa of celiac disease patients reprograms these dendritic cells so that they induce an inflammatory rather than a tolerogenic T-cell response upon arrival in the lymph node

An alternative explanation for the breaking of gluten tolerance in celiac disease is that dendritic cells in the gut do not take up deamidated gluten, because the deamidation reaction happens elsewhere. If extracellular TG2 in the small intestine is largely inactive, the dendritic cells present there would only be able to take up and present peptides in their native state, thus enabling tolerance toward non-deamidated gluten. Since small antigens such as peptides can access lymph nodes directly via the lymph and do not need to be transported by cells (Gretz et al. 2000), gluten peptides could enter the mesenteric lymph nodes and be deamidated by TG2 present there, leading to uptake and presentation of deamidated gluten by inflammatory MHC class II-expressing cells in the lymph nodes (Fig. 9.4). Such a mechanism would clearly distinguish deamidated from non-deamidated gluten peptides and could explain the specificity of inflammatory T cells for deamidated epitopes. Importantly, we consider this mechanism likely to be operating in the early phase of celiac disease before intestinal damage is observed. Later on, when the inflammation is established, changes in the cytokine environment and local deamidation by activated TG2 in the gut mucosa could be important for further augmenting the immune response.

9.4.2 Gluten-Dependent B-Cell Activation

Activated CD4+ T cells can interact with so-called cognate B cells that recognize the same antigen, leading to generation of plasma cells secreting antigen-specific antibodies. It is therefore not surprising that the activation of gluten-reactive T cells in celiac disease is accompanied by serum antibodies against gluten. The production of antibodies targeting TG2, on the other hand, is unexpected. These antibodies are dependent on patients being on a gluten-containing diet (Sulkanen et al. 1998) and are only found in individuals who express the disease-associated HLA molecules (Bjorck et al. 2010), suggesting the involvement of gluten-reactive T cells in the activation of TG2-reactive B cells. In order to explain this non-cognate T- and B-cell interaction, it has been suggested that TG2 can form complexes with gluten peptides, either through cross-linking or by formation of an enzyme-substrate intermediate, thus allowing TG2-reactive B cells to take up and present gluten peptides after initial binding of TG2-gluten complexes to the BCR (Sollid et al. 1997) (Fig. 9.2). Indeed, it was shown in vitro that TG2 efficiently forms cross-links between glutamine residues in gluten and lysine residues on the surface of the enzyme itself (Fleckenstein et al. 2004). Moreover, cross-linked complexes of TG2 and gluten peptide could be taken up by retrovirally transduced B cells expressing HLA-DQ2.5 and a TG2-specific BCR followed by presentation of peptide to gluten-specific T cells (Di Niro et al. 2012).

Epitope mapping studies using TG2-specific mAbs indicated that the main epitopes targeted by anti-TG2 antibodies in celiac disease are clustered within a relatively small area on the protein surface (Iversen et al. 2013, 2014) (Fig. 9.3). A possible explanation for this finding is that B cells targeting other parts of TG2 are eliminated during B-cell development. Immature B cells reactive with membrane-bound self-antigens are usually efficiently removed from the repertoire, as antigen binding to such B cells causes BCR aggregation and induction of intracellular signaling, which leads either to apoptosis (Nemazee and Burki 1989; Hartley et al. 1991) or to a state of unresponsiveness known as anergy (Goodnow et al. 1988). B cells recognizing soluble self-antigens, on the other hand, can reach the mature state but will not normally become activated because self-reactive T cells that can provide B-cell help are absent (Taylor et al. 2012). In agreement with the mechanisms of self-tolerance induction in B cells, TG2-specific mAbs did not bind cell-surface TG2, suggesting that the epitopes targeted by celiac antibodies are hidden when TG2 is associated with cell membranes (Iversen et al. 2013). Thus, only the B cells that target these epitopes will complete their development and eventually be able to differentiate into plasma cells secreting anti-TG2 antibodies.

An alternative explanation for the limited number of epitopes targeted by TG2-specific antibodies is that B cells recognizing these particular epitopes have an advantage during B-cell activation. In this regard, the observation that TG2 can catalyze cross-linking of antibody molecules could be relevant (Di Niro et al. 2012). Thus, enzymatic cross-linking of BCR molecules on the surface of TG2-targting B cells could be important for B-cell activation, as aggregation of BCR molecules on mature B cells induces intracellular signaling, which, together with signals received from T cells, are necessary for B-cell activation. As a result, B cells recognizing TG2 epitopes that would allow the enzyme to carry out cross-linking of BCR molecules could be selectively activated. In line with this notion, we recently found that the location of epitope 1, the most frequently targeted epitope among TG2-specific mAbs, predicts that the active site of the enzyme will point toward the BCR upon binding (Iversen et al. 2014).

TG2-mediated antibody cross-linking was found to happen efficiently with antibody of the IgD isotype but not with IgA or IgG molecules (Di Niro et al. 2012). This finding has implications for B-cell activation, as IgD BCRs are expressed on all mature B cells before they are activated, whereas activated cells usually switch to a different BCR isotype (Pape et al. 2003). Hence, GC B cells do not express IgD (Bhan et al. 1981). Therefore, if TG2-mediated cross-linking of BCR molecules supports B-cell activation, IgD-expressing cells would have an advantage over cells that have switched to a different isotype, possibly explaining why TG2-specific B cells would avoid the GC and instead proceed straight to plasma cell differentiation. This mechanism could thus explain the surprisingly low number of antibody mutations observed in TG2-specific plasma cells (Di Niro et al. 2012). However, it does not explain why gluten-reactive plasma cells seem to have equally low mutation levels (Steinsbø et al. 2014). We can only speculate about the reasons for this peculiarity, but the lack of mutations in both plasma cell populations could be a consequence of their shared dependence on gluten-reactive T cells. Notably, skewing of the B-cell response toward production of anti-TG2 antibodies could imply that the location of the T cells providing B-cell help is largely dictated by the TG2-reactive B cells. As a result, also gluten-reactive B cells would be activated outside of GCs and consequently introduce few mutations in their antibody genes.

Activation of gluten-reactive and TG2-reactive B cells without GC formation is supported by the observation that serum antibody levels decline rapidly when patients commence a gluten-free diet, indicating that long-lived bone marrow-residing plasma cells are not generated. Such cells normally arise during a GC response and can continue to secrete antibodies into the circulation for years in the absence of the triggering antigen (Shlomchik and Weisel 2012). Thus, following vaccination, vaccine-specific IgG antibodies can be detected in the blood for decades (Crotty et al. 2003; Amanna et al. 2007). Notably, the levels of serum IgA and IgG against gluten or TG2 decrease with the same kinetics after exclusion of gluten from the diet (Basso et al. 2002; Bazzigaluppi et al. 2006; Sugai et al. 2006), indicating that there is no inherent difference between IgA- and

IgG-secreting plasma cells and that both cell types are generated by a similar mechanism.

9.4.3 Role of T- and B-Cell Interaction

The interactions between antigen-specific T and B cells not only result in B-cell activation and antibody production. Antigen presentation by the B cell also provides activation signals to the T cell, and the cellular contacts arising through binding of a common antigen can thus be seen as a collaboration resulting in mutual T-cell and B-cell activation (Lanzavecchia 1985). Initial activation of CD4+ T cells normally happens after TCR-mediated recognition of antigen presented in the context of MHC class II molecules on the surface of dendritic cells, whereas antigen presentation by B cells can serve to amplify the number of activated T cells later in the immune response (Rodriguez-Pinto 2005). However, in a mouse model of the autoimmune disease systemic lupus erythematosus, it was shown that B cells can play a primary role in the activation of T cells (Chan and Shlomchik 1998; Teichmann et al. 2013). Thus, it is possible that gluten- or TG2-reactive B cells, rather than dendritic cells, are responsible for the initial activation of gluten-reactive T cells in celiac disease (Fig. 9.4). Notably, gluten deamidation could be mediated by TG2 bound to the BCRs of TG2-reactive B cells. Such a mechanism would directly couple gluten deamidation to antigen uptake and presentation by TG2-reactive B cells followed by activation of gluten-reactive T cells. In addition, a primary role of TG2-reactive B cells in the presentation of antigen to gluten-reactive T cells would explain the tight association between celiac disease and autoantibodies against TG2 and is in agreement with the observation that the antibodies appear at an early stage in disease development.

9.5 Concluding Remarks

The activity of TG2 plays a central role in celiac disease pathogenesis, as the enzyme catalyzes the deamidation of gluten peptides and thereby transforms them into immunogenic T-cell antigens. It has therefore been suggested that TG2 inhibitors could be used in the treatment of celiac disease, offering an alternative to a gluten-free diet (Siegel and Khosla 2007). Still, we do not know where the initial enzymatic conversion of gluten peptides takes place. One possibility is the mesenteric lymph nodes, where TG2-reactive B cells could play an essential role in the presentation of deamidated gluten peptides to T cells. Another issue to be clarified is whether the activation of otherwise catalytically inactive TG2 is the triggering event that ultimately leads to celiac disease. If so, an infection or another type of inflammation is likely to be involved. It is also possible that a pool of TG2 can be kept in the active state through binding to the BCRs of TG2-reactive B cells, thus giving the B cells a unique role in the generation and presentation of deamidated peptides to T cells.

Our present knowledge only allows us to speculate about the detailed immunological mechanisms leading to celiac disease, but it is clear that TG2 activity is required, and the activation of TG2-reactive B cells recognizing certain epitopes seems to be directly linked with disease development. If we are to understand more about the in vivo activity of TG2 in celiac disease and the role of the putative collaboration between gluten-reactive T cells and TG2-reactive B cells, new animal models are needed. The cloning of TCR and BCR genes from the disease-relevant cells in patients allows the creation of transgenic mice having both gluten-reactive T cells and TG2-reactive B cells. Attempts to generate such mice are currently ongoing and will hopefully provide us with new information about the role of T and B cells in the initiation of celiac disease as well as the involvement of TG2.

References

Abadie V, Sollid LM, Barreiro LB et al (2011) Integration of genetic and immunological insights into a model of celiac disease pathogenesis. Annu Rev Immunol 29:493–525

Alaedini A, Green PH (2008) Autoantibodies in celiac disease. Autoimmunity 41(1):19–26

Aleanzi M, Demonte AM, Esper C et al (2001) Celiac disease: antibody recognition against native and selectively deamidated gliadin peptides. Clin Chem 47(11):2023–2028

Amanna IJ, Carlson NE, Slifka MK (2007) Duration of humoral immunity to common viral and vaccine antigens. N Engl J Med 357(19):1903–1915

Arentz-Hansen H, Korner R, Molberg O et al (2000) The intestinal T cell response to alpha-gliadin in adult celiac disease is focused on a single deamidated glutamine targeted by tissue transglutaminase. J Exp Med 191(4):603–612

Baklien K, Brandtzaeg P, Fausa O (1977) Immunoglobulins in jejunal mucosa and serum from patients with adult coeliac disease. Scand J Gastroenterol 12(2):149–159

Barone MV, Caputo I, Ribecco MT et al (2007) Humoral immune response to tissue transglutaminase is related to epithelial cell proliferation in celiac disease. Gastroenterology 132 (4):1245–1253

Basso D, Guariso G, Plebani M (2002) Serologic testing for celiac disease. Clin Chem 48 (11):2082–2083

Bateman EA, Ferry BL, Hall A et al (2004) IgA antibodies of coeliac disease patients recognise a dominant T cell epitope of A-gliadin. Gut 53(9):1274–1278

Bazzigaluppi E, Roggero P, Parma B et al (2006) Antibodies to recombinant human tissue-transglutaminase in coeliac disease: diagnostic effectiveness and decline pattern after gluten-free diet. Dig Liver Dis 38(2):98–102

Berger E, Buergin-Wolff A, Freudenberg E (1964) Diagnostische bewertung des nachweises von gliadin-antikörpern bei cöliakie. Klin Wochenschr 42:788–790

Bergseng E, Xia J, Kim CY et al (2005) Main chain hydrogen bond interactions in the binding of proline-rich gluten peptides to the celiac disease-associated HLA-DQ2 molecule. J Biol Chem 280(23):21791–21796

Bergseng E, Dorum S, Arntzen MO et al (2015) Different binding motifs of the celiac disease-associated HLA molecules DQ2.5, DQ2.2, and DQ7.5 revealed by relative quantitative proteomics of endogenous peptide repertoires. Immunogenetics 67(2):73–84

Bhan AK, Nadler LM, Stashenko P et al (1981) Stages of B cell differentiation in human lymphoid tissue. J Exp Med 154(3):737–749

Bjorck S, Brundin C, Lorinc E et al (2010) Screening detects a high proportion of celiac disease in young HLA-genotyped children. J Pediatr Gastroenterol Nutr 50(1):49–53
Bodd M, Raki M, Tollefsen S et al (2010) HLA-DQ2-restricted gluten-reactive T cells produce IL-21 but not IL-17 or IL-22. Mucosal Immunol 3(6):594–601
Bodd M, Kim CY, Lundin KE et al (2012) T-cell response to gluten in patients with HLA-DQ2.2 reveals requirement of peptide-MHC stability in celiac disease. Gastroenterology 142 (3):552–561
Broughton SE, Petersen J, Theodossis A et al (2012) Biased T cell receptor usage directed against human leukocyte antigen DQ8-restricted gliadin peptides is associated with celiac disease. Immunity 37(4):611–621
Bruce SE, Bjarnason I, Peters TJ (1985) Human jejunal transglutaminase: demonstration of activity, enzyme kinetics and substrate specificity with special relation to gliadin and coeliac disease. Clin Sci (Lond) 68(5):573–579
Brusco G, Muzi P, Ciccocioppo R et al (1999) Transglutaminase and coeliac disease: endomysial reactivity and small bowel expression. Clin Exp Immunol 118(3):371–375
Chan O, Shlomchik MJ (1998) A new role for B cells in systemic autoimmunity: B cells promote spontaneous T cell activation in MRL-lpr/lpr mice. J Immunol 160(1):51–59
Chorzelski TP, Beutner EH, Sulej J et al (1984) IgA anti-endomysium antibody. A new immunological marker of dermatitis herpetiformis and coeliac disease. Br J Dermatol 111(4):395–402
Crotty S, Felgner P, Davies H et al (2003) Cutting edge: long-term B cell memory in humans after smallpox vaccination. J Immunol 171(10):4969–4973
Dafik L, Albertelli M, Stamnaes J et al (2012) Activation and inhibition of transglutaminase 2 in mice. PLoS One 7(2):e30642
DePaolo RW, Abadie V, Tang F et al (2011) Co-adjuvant effects of retinoic acid and IL-15 induce inflammatory immunity to dietary antigens. Nature 471(7337):220–224
Di Niro R, Sblattero D, Florian F et al (2008) Anti-idiotypic response in mice expressing human autoantibodies. Mol Immunol 45(6):1782–1791
Di Niro R, Mesin L, Zheng NY et al (2012) High abundance of plasma cells secreting transglutaminase 2-specific IgA autoantibodies with limited somatic hypermutation in celiac disease intestinal lesions. Nat Med 18(3):441–445
Dieterich W, Ehnis T, Bauer M et al (1997) Identification of tissue transglutaminase as the autoantigen of celiac disease. Nat Med 3(7):797–801
Dieterich W, Laag E, Schopper H et al (1998) Autoantibodies to tissue transglutaminase as predictors of celiac disease. Gastroenterology 115(6):1317–1321
Dieterich W, Trapp D, Esslinger B et al (2003) Autoantibodies of patients with coeliac disease are insufficient to block tissue transglutaminase activity. Gut 52(11):1562–1566
Dorum S, Bodd M, Fallang LE et al (2014) HLA-DQ molecules as affinity matrix for identification of gluten T cell epitopes. J Immunol 193(9):4497–4506
Dørum S, Arntzen MO, Qiao SW et al (2010) The preferred substrates for transglutaminase 2 in a complex wheat gluten digest are peptide fragments harboring celiac disease T-cell epitopes. PLoS One 5(11):e14056
Dubois PC, Trynka G, Franke L et al (2010) Multiple common variants for celiac disease influencing immune gene expression. Nat Genet 42(4):295–302
Esposito C, Paparo F, Caputo I et al (2002) Anti-tissue transglutaminase antibodies from coeliac patients inhibit transglutaminase activity both in vitro and in situ. Gut 51(2):177–181
Esposito C, Paparo F, Caputo I et al (2003) Expression and enzymatic activity of small intestinal tissue transglutaminase in celiac disease. Am J Gastroenterol 98(8):1813–1820
Fallang LE, Bergseng E, Hotta K et al (2009) Differences in the risk of celiac disease associated with HLA-DQ2.5 or HLA-DQ2.2 are related to sustained gluten antigen presentation. Nat Immunol 10(10):1096–1101
Fesus L, Piacentini M (2002) Transglutaminase 2: an enigmatic enzyme with diverse functions. Trends Biochem Sci 27(10):534–539

Fleckenstein B, Molberg O, Qiao SW et al (2002) Gliadin T cell epitope selection by tissue transglutaminase in celiac disease. Role of enzyme specificity and pH influence on the transamidation versus deamidation process. J Biol Chem 277(37):34109–34116

Fleckenstein B, Qiao SW, Larsen MR et al (2004) Molecular characterization of covalent complexes between tissue transglutaminase and gliadin peptides. J Biol Chem 279 (17):17607–17616

Freitag T, Schulze-Koops H, Niedobitek G et al (2004) The role of the immune response against tissue transglutaminase in the pathogenesis of coeliac disease. Autoimmun Rev 3(2):13–20

Goodnow CC, Crosbie J, Adelstein S et al (1988) Altered immunoglobulin expression and functional silencing of self-reactive B lymphocytes in transgenic mice. Nature 334 (6184):676–682

Green PH, Cellier C (2007) Celiac disease. N Engl J Med 357(17):1731–1743

Gretz JE, Norbury CC, Anderson AO et al (2000) Lymph-borne chemokines and other low molecular weight molecules reach high endothelial venules via specialized conduits while a functional barrier limits access to the lymphocyte microenvironments in lymph node cortex. J Exp Med 192(10):1425–1440

Hadjivassiliou M, Aeschlimann P, Strigun A et al (2008) Autoantibodies in gluten ataxia recognize a novel neuronal transglutaminase. Ann Neurol 64(3):332–343

Halttunen T, Maki M (1999) Serum immunoglobulin A from patients with celiac disease inhibits human T84 intestinal crypt epithelial cell differentiation. Gastroenterology 116(3):566–572

Hartley SB, Crosbie J, Brink R et al (1991) Elimination from peripheral lymphoid tissues of self-reactive B lymphocytes recognizing membrane-bound antigens. Nature 353(6346):765–769

Henderson KN, Tye-Din JA, Reid HH et al (2007) A structural and immunological basis for the role of human leukocyte antigen DQ8 in celiac disease. Immunity 27(1):23–34

Hill P, Austin A, Forsyth J et al (2015) British Society of Gastroenterology guidelines on the diagnosis and management of coeliac disease. Gut 64(4):691–692

Hunt KA, Zhernakova A, Turner G et al (2008) Newly identified genetic risk variants for celiac disease related to the immune response. Nat Genet 40(4):395–402

Husby S, Koletzko S, Korponay-Szabo IR et al (2012) European Society for Pediatric Gastroenterology, Hepatology, and Nutrition guidelines for the diagnosis of coeliac disease. J Pediatr Gastroenterol Nutr 54(1):136–160

Iversen R, Di Niro R, Stamnaes J et al (2013) Transglutaminase 2-specific autoantibodies in celiac disease target clustered, N-terminal epitopes not displayed on the surface of cells. J Immunol 190(12):5981–5991

Iversen R, Mysling S, Hnida K et al (2014) Activity-regulating structural changes and autoantibody epitopes in transglutaminase 2 assessed by hydrogen/deuterium exchange. Proc Natl Acad Sci U S A 111(48):17146–17151

Jabri B, Sollid LM (2009) Tissue-mediated control of immunopathology in coeliac disease. Nat Rev Immunol 9(12):858–870

Jin X, Stamnaes J, Klock C et al (2011) Activation of extracellular transglutaminase 2 by thioredoxin. J Biol Chem 286(43):37866–37873

Junker Y, Zeissig S, Kim SJ et al (2012) Wheat amylase trypsin inhibitors drive intestinal inflammation via activation of toll-like receptor 4. J Exp Med 209(13):2395–2408

Kalliokoski S, Sulic AM, Korponay-Szabo IR et al (2013) Celiac disease-specific TG2-targeted autoantibodies inhibit angiogenesis ex vivo and in vivo in mice by interfering with endothelial cell dynamics. PLoS One 8(6):e65887

Kim CY, Quarsten H, Bergseng E et al (2004) Structural basis for HLA-DQ2-mediated presentation of gluten epitopes in celiac disease. Proc Natl Acad Sci U S A 101(12):4175–4179

Kiraly R, Vecsei Z, Demenyi T et al (2006) Coeliac autoantibodies can enhance transamidating and inhibit GTPase activity of tissue transglutaminase: dependence on reaction environment and enzyme fitness. J Autoimmun 26(4):278–287

Kiraly R, Csosz E, Kurtan T et al (2009) Functional significance of five noncanonical Ca^{2+} –binding sites of human transglutaminase 2 characterized by site-directed mutagenesis. FEBS J 276(23):7083–7096
Korponay-Szabo IR, Halttunen T, Szalai Z et al (2004) In vivo targeting of intestinal and extraintestinal transglutaminase 2 by coeliac autoantibodies. Gut 53(5):641–648
Lanzavecchia A (1985) Antigen-specific interaction between T and B cells. Nature 314 (6011):537–539
Laver WG, Air GM, Webster RG et al (1990) Epitopes on protein antigens: misconceptions and realities. Cell 61(4):553–556
Lebreton C, Menard S, Abed J et al (2012) Interactions among secretory immunoglobulin A, CD71, and transglutaminase-2 affect permeability of intestinal epithelial cells to gliadin peptides. Gastroenterology 143(3):698–707, e691–e694
Leffler DA, Schuppan D (2010) Update on serologic testing in celiac disease. Am J Gastroenterol 105(12):2520–2524
Leffler D, Schuppan D, Pallav K et al (2013) Kinetics of the histological, serological and symptomatic responses to gluten challenge in adults with coeliac disease. Gut 62(7):996–1004
Lorand L, Graham RM (2003) Transglutaminases: crosslinking enzymes with pleiotropic functions. Nat Rev Mol Cell Biol 4(2):140–156
Lundin KE, Scott H, Hansen T et al (1993) Gliadin-specific, HLA-DQ(alpha 1*0501, beta 1*0201) restricted T cells isolated from the small intestinal mucosa of celiac disease patients. J Exp Med 178(1):187–196
Lundin KE, Scott H, Fausa O et al (1994) T cells from the small intestinal mucosa of a DR4, DQ7/DR4, DQ8 celiac disease patient preferentially recognize gliadin when presented by DQ8. Hum Immunol 41(4):285–291
Maiuri L, Ciacci C, Ricciardelli I et al (2003) Association between innate response to gliadin and activation of pathogenic T cells in coeliac disease. Lancet 362(9377):30–37
Marzari R, Sblattero D, Florian F et al (2001) Molecular dissection of the tissue transglutaminase autoantibody response in celiac disease. J Immunol 166(6):4170–4176
Matysiak-Budnik T, Moura IC, Arcos-Fajardo M et al (2008) Secretory IgA mediates retrotranscytosis of intact gliadin peptides via the transferrin receptor in celiac disease. J Exp Med 205 (1):143–154
Mention JJ, Ben Ahmed M, Begue B et al (2003) Interleukin 15: a key to disrupted intraepithelial lymphocyte homeostasis and lymphomagenesis in celiac disease. Gastroenterology 125 (3):730–745
Molberg O, Kett K, Scott H et al (1997) Gliadin specific, HLA DQ2-restricted T cells are commonly found in small intestinal biopsies from coeliac disease patients, but not from controls. Scand J Immunol 46(3):103–109
Molberg O, McAdam SN, Korner R et al (1998) Tissue transglutaminase selectively modifies gliadin peptides that are recognized by gut-derived T cells in celiac disease. Nat Med 4 (6):713–717
Molberg O, McAdam S, Lundin KE et al (2001) T cells from celiac disease lesions recognize gliadin epitopes deamidated in situ by endogenous tissue transglutaminase. Eur J Immunol 31 (5):1317–1323
Monteleone G, Pender SL, Alstead E et al (2001) Role of interferon alpha in promoting T helper cell type 1 responses in the small intestine in coeliac disease. Gut 48(3):425–429
Mothes T (2007) Deamidated gliadin peptides as targets for celiac disease-specific antibodies. Adv Clin Chem 44:35–63
Mubarak A, Spierings E, Wolters V et al (2013) Human leukocyte antigen DQ2.2 and celiac disease. J Pediatr Gastroenterol Nutr 56(4):428–430
Myrsky E, Kaukinen K, Syrjanen M et al (2008) Coeliac disease-specific autoantibodies targeted against transglutaminase 2 disturb angiogenesis. Clin Exp Immunol 152(1):111–119

Myrsky E, Caja S, Simon-Vecsei Z et al (2009a) Celiac disease IgA modulates vascular permeability in vitro through the activity of transglutaminase 2 and RhoA. Cell Mol Life Sci 66 (20):3375–3385
Myrsky E, Syrjanen M, Korponay-Szabo IR et al (2009b) Altered small-bowel mucosal vascular network in untreated coeliac disease. Scand J Gastroenterol 44(2):162–167
Nakachi K, Powell M, Swift G et al (2004) Epitopes recognised by tissue transglutaminase antibodies in coeliac disease. J Autoimmun 22(1):53–63
Nemazee DA, Burki K (1989) Clonal deletion of B lymphocytes in a transgenic mouse bearing anti-MHC class I antibody genes. Nature 337(6207):562–566
Nilsen EM, Lundin KE, Krajci P et al (1995) Gluten specific, HLA-DQ restricted T cells from coeliac mucosa produce cytokines with Th1 or Th0 profile dominated by interferon gamma. Gut 37(6):766–776
Osman AA, Gunnel T, Dietl A et al (2000) B cell epitopes of gliadin. Clin Exp Immunol 121 (2):248–254
Pabst O, Mowat AM (2012) Oral tolerance to food protein. Mucosal Immunol 5(3):232–239
Pape KA, Kouskoff V, Nemazee D et al (2003) Visualization of the genesis and fate of isotype-switched B cells during a primary immune response. J Exp Med 197(12):1677–1687
Petersen J, Montserrat V, Mujico JR et al (2014) T-cell receptor recognition of HLA-DQ2-gliadin complexes associated with celiac disease. Nat Struct Mol Biol 21(5):480–488
Pinkas DM, Strop P, Brunger AT et al (2007) Transglutaminase 2 undergoes a large conformational change upon activation. PLoS Biol 5(12):e327
Piper JL, Gray GM, Khosla C (2002) High selectivity of human tissue transglutaminase for immunoactive gliadin peptides: implications for celiac sprue. Biochemistry 41(1):386–393
Qiao SW, Raki M, Gunnarsen KS et al (2011) Posttranslational modification of gluten shapes TCR usage in celiac disease. J Immunol 187(6):3064–3071
Qiao SW, Iversen R, Raki M et al (2012) The adaptive immune response in celiac disease. Semin Immunopathol 34(4):523–540
Qiao SW, Christophersen A, Lundin KE et al (2014) Biased usage and preferred pairing of alpha- and beta-chains of TCRs specific for an immunodominant gluten epitope in coeliac disease. Int Immunol 26(1):13–19
Rauhavirta T, Qiao SW, Jiang Z et al (2011) Epithelial transport and deamidation of gliadin peptides: a role for coeliac disease patient immunoglobulin A. Clin Exp Immunol 164 (1):127–136
Rodriguez-Pinto D (2005) B cells as antigen presenting cells. Cell Immunol 238(2):67–75
Rostom A, Dube C, Cranney A et al (2005) The diagnostic accuracy of serologic tests for celiac disease: a systematic review. Gastroenterology 128(4 Suppl 1):S38–S46
Salmi TT, Collin P, Jarvinen O et al (2006a) Immunoglobulin A autoantibodies against transglutaminase 2 in the small intestinal mucosa predict forthcoming coeliac disease. Aliment Pharmacol Ther 24(3):541–552
Salmi TT, Collin P, Korponay-Szabo IR et al (2006b) Endomysial antibody-negative coeliac disease: clinical characteristics and intestinal autoantibody deposits. Gut 55(12):1746–1753
Sardy M, Karpati S, Merkl B et al (2002) Epidermal transglutaminase (TGase 3) is the autoantigen of dermatitis herpetiformis. J Exp Med 195(6):747–757
Seissler J, Wohlrab U, Wuensche C et al (2001) Autoantibodies from patients with coeliac disease recognize distinct functional domains of the autoantigen tissue transglutaminase. Clin Exp Immunol 125(2):216–221
Shan L, Molberg O, Parrot I et al (2002) Structural basis for gluten intolerance in celiac sprue. Science 297(5590):2275–2279
Shlomchik MJ, Weisel F (2012) Germinal center selection and the development of memory B and plasma cells. Immunol Rev 247(1):52–63
Siegel M, Khosla C (2007) Transglutaminase 2 inhibitors and their therapeutic role in disease states. Pharmacol Ther 115(2):232–245

Siegel M, Strnad P, Watts RE et al (2008) Extracellular transglutaminase 2 is catalytically inactive, but is transiently activated upon tissue injury. PLoS One 3(3):e1861
Simon-Vecsei Z, Kiraly R, Bagossi P et al (2012) A single conformational transglutaminase 2 epitope contributed by three domains is critical for celiac antibody binding and effects. Proc Natl Acad Sci U S A 109(2):431–436
Sollid LM (2002) Coeliac disease: dissecting a complex inflammatory disorder. Nat Rev Immunol 2(9):647–655
Sollid LM, Lie BA (2005) Celiac disease genetics: current concepts and practical applications. Clin Gastroenterol Hepatol 3(9):843–851
Sollid LM, Molberg O, McAdam S et al (1997) Autoantibodies in coeliac disease: tissue transglutaminase--guilt by association? Gut 41(6):851–852
Sollid LM, Qiao SW, Anderson RP et al (2012) Nomenclature and listing of celiac disease relevant gluten T-cell epitopes restricted by HLA-DQ molecules. Immunogenetics 64(6):455–460
Stamnaes J, Pinkas DM, Fleckenstein B et al (2010) Redox regulation of transglutaminase 2 activity. J Biol Chem 285(33):25402–25409
Steinsbø Ø, Henry Dunand CJ, Huang M et al (2014) Restricted VH/VL usage and limited mutations in gluten-specific IgA of coeliac disease lesion plasma cells. Nat Commun 5:4041
Sugai E, Vazquez H, Nachman F et al (2006) Accuracy of testing for antibodies to synthetic gliadin-related peptides in celiac disease. Clin Gastroenterol Hepatol 4(9):1112–1117
Sulkanen S, Halttunen T, Laurila K et al (1998) Tissue transglutaminase autoantibody enzyme-linked immunosorbent assay in detecting celiac disease. Gastroenterology 115(6):1322–1328
Taylor JJ, Martinez RJ, Titcombe PJ et al (2012) Deletion and anergy of polyclonal B cells specific for ubiquitous membrane-bound self-antigen. J Exp Med 209(11):2065–2077
Teichmann LL, Schenten D, Medzhitov R et al (2013) Signals via the adaptor MyD88 in B cells and DCs make distinct and synergistic contributions to immune activation and tissue damage in lupus. Immunity 38(3):528–540
Troncone R, Auricchio S (2007) Rotavirus and celiac disease: clues to the pathogenesis and perspectives on prevention. J Pediatr Gastroenterol Nutr 44(5):527–528
Trynka G, Hunt KA, Bockett NA et al (2011) Dense genotyping identifies and localizes multiple common and rare variant association signals in celiac disease. Nat Genet 43(12):1193–1201
Vader LW, de Ru A, van der Wal Y et al (2002) Specificity of tissue transglutaminase explains cereal toxicity in celiac disease. J Exp Med 195(5):643–649
van de Wal Y, Kooy Y, van Veelen P et al (1998) Selective deamidation by tissue transglutaminase strongly enhances gliadin-specific T cell reactivity. J Immunol 161(4):1585–1588
van Heel DA, Franke L, Hunt KA et al (2007) A genome-wide association study for celiac disease identifies risk variants in the region harboring IL2 and IL21. Nat Genet 39(7):827–829
Victora GD, Nussenzweig MC (2012) Germinal centers. Annu Rev Immunol 30:429–457
Xia J, Sollid LM, Khosla C (2005) Equilibrium and kinetic analysis of the unusual binding behavior of a highly immunogenic gluten peptide to HLA-DQ2. Biochemistry 44 (11):4442–4449
Zanini B, Magni A, Caselani F et al (2012) High tissue-transglutaminase antibody level predicts small intestinal villous atrophy in adult patients at high risk of celiac disease. Dig Liver Dis 44 (4):280–285
Zanoni G, Navone R, Lunardi C et al (2006) In celiac disease, a subset of autoantibodies against transglutaminase binds toll-like receptor 4 and induces activation of monocytes. PLoS Med 3 (9):e358

Chapter 10
Transglutaminase II and Metastasis: How Hot Is the Link?

Kapil Mehta

Abstract The transient and dynamic nature of changes that cancer cells undergo during metastasis cannot be explained by simple progressive accumulation of irreversible genetic alterations in the primary tumor. The ability of cancer cells to switch on and off different cellular processes such as epithelial-to-mesenchymal transition (EMT) and the ability to de-/re-differentiate allows them to respond to different cellular environments and to disseminate from the primary tumor to distant organs. This and many other features of metastatic tumor cells such as dormancy, active DNA repair, and resistance to apoptosis are shared by normal stem cells. It has been theorized that a subpopulation of cancer cells can acquire stem cell-like traits and aid in the process of metastasis. Moreover, the EMT and stem cellness support the ability of cancer cells to migrate and survive conventional therapies that predominantly target actively proliferating cells. Accordingly, characterization of genes and signal transduction pathways that regulate the EMT and stemness within the tumor mass are likely to shed new light on the complex molecular events that promote metastasis, invasion and resistance to therapy and could offer novel therapeutic targets for developing tailor-made intervention strategies. In this chapter, I discuss the evidence that aberrant expression of a stress-response protein, transglutaminase II (TG2), reprograms the inflammatory signaling networks in cancer cells that play fundamental roles in promoting drug resistance and metastatic competence by inducing EMT and the stem cell phenotype.

Keywords Cancer • Drug resistance • Cancer stem cells • Inflammation • Metastasis • Hypoxia-inducing factor • Metabolism

10.1 Introduction

With an estimated 7.4 million new cases and 4.6 million deaths every year, cancer continues to be a major health problem and cause of mortality worldwide. Available anticancer therapies predominantly target rapidly dividing cells in the primary

K. Mehta (✉)
Department of Experimental Therapeutics, Unit 1950, The University of Texas MD Anderson Cancer Center, Houston, TX, USA
e-mail: kapilmeh@gmail.com

K. Hitomi et al. (eds.), *Transglutaminases*, DOI 10.1007/978-4-431-55825-5_10

tumor. Similarly, many recently launched targeted therapies are also based on identification of targets that are activated (due to genetic or epigenetic alterations) and drive proliferation of cells in the primary tumor. Although targeted therapies have been successful in reducing the toxicity, they have failed to yield appreciable survival benefits despite exuberant increases in treatment costs.

Due to advances in early diagnosis, surgery, and adjuvant treatments many early-stage or localized tumors are now considered curable. However, the same cannot be said of metastatic disease. In most cases, the prognosis of metastatic cancer remains grim. Metastatic tumors often exhibit resistance to most or all the available therapies. While it is often possible to moderately prolong the survival, eradication of metastatic disease is not possible, while side effects of treatment are significant. As a result of this, majority of these patients succumb to the disease. Although the clinical importance of metastasis has long been appreciated, understanding of molecular mechanisms that facilitate metastasis has lagged behind over the development of therapeutic modalities that target proliferating cells. This is partly due to the fact that metastasis is a highly complex and multistep process (Fig. 10.1) and each step requires coordinated action of many genes and pathways (Iiizumi et al. 2008).

Cancer can spread to areas near the primary site (regional metastasis), or to parts of the body that are farther away (distant metastasis). When a tumor grows to more than 1 mm^3 in size at the primary site, it induces growth of new blood vessels (neovascularization) in order to meet the increasing demand for oxygen and nutrients. Some tumor cells gain the growth advantage and acquire metastatic competence by accumulating additional genetic/epigenetic alterations. The first

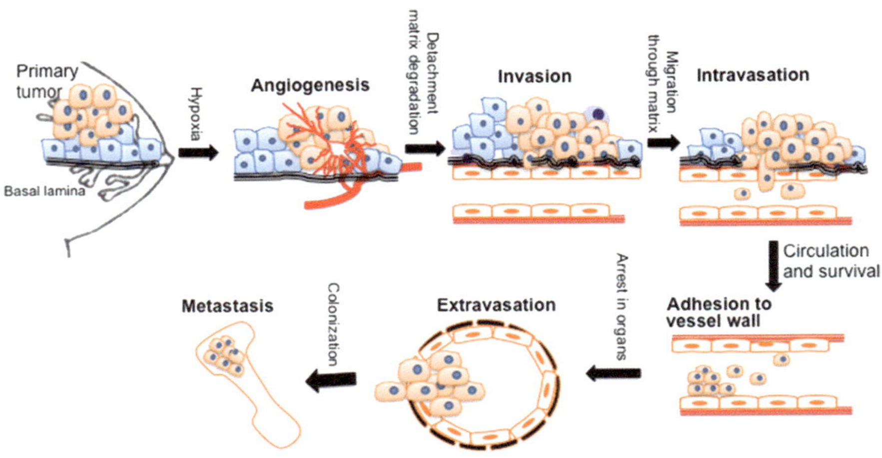

Fig. 10.1 Schematic representation of steps in the metastatic cascade. In order to metastasize, cancer cells must acquire the ability to break away from the primary tumor, break down the basement membrane, invade into the stroma (local invasion), enter into the blood circulation (intravasation) and manage to survive before they can arrest at a distant organ and grow into clinically detectable metastases

step in metastasis is the detachment of tumor cells from neighboring cells by deterring cell-cell and cell-extracellular matrix (ECM) adhesion, followed by degradation of the ECM (Slattum and Rosenblatt 2014). Once cancer cells break free from the main (primary) tumor they either enter the lymph system and end up in nearby lymph nodes (lymphatic spread) or enter the bloodstream (hematogenous spread) where they must survive the hostile environment including mechanical damage, lack of growth factors and the immune system. In the blood circulation, tumor cells often aggregate with platelets and fibrin and embolize in the capillaries or adhere to the endothelial cells like leukocytes at an inflammatory site. Surviving cells then extravasate and lodge at the secondary site to proliferate and colonize to make secondary/metastatic tumors (Fig. 10.1). Different processes during metastasis are controlled by a complex interplay between activated transcription factors and signaling pathways. In this chapter, I discuss evidence that aberrant expression of transglutaminase II (TG2) in epithelial cancer cells reprograms the inflammatory signaling networks that are implicated in conferring drug resistance and metastatic competence on cancer cells (Fig. 10.2).

10.2 Inflammation and Cancer

Although genetic alterations are essential for cancer development, they are not sufficient to cause disease progression. In addition to genetic alterations, early-stage tumors require some ancillary changes to become invasive (Radisky and Radisky 2007). It is now becoming clear that the inflammatory microenvironment favors the expansion of genomic aberrations and progression of cancer (Mantovani 2009). While acute or physiological inflammation is predominantly a self-limiting process with therapeutic significance, chronic or pathological inflammation frequently leads to the progression of various chronic diseases, including cancer (Grivennikov and Karin 2010). Many epidemiological, animal, and clinical studies have supported the concept that chronic inflammation promotes and exacerbates malignancy (Demaria et al. 2010). Several types of cancer arise in the setting of chronic inflammation, suggesting a strong link between inflammation and cancer and, for that reason, tumors are often referred to as wounds that never heal (Mantovani et al. 2008; Grivennikov et al. 2010). It is estimated that nearly 25 % of all cancers are etiologically linked to chronic inflammation and infection (Hussain and Harris 2007). Chronic inflammation can impact tumorigenesis during various phases such as cellular proliferation, transformation, apoptosis evasion, survival, invasion, angiogenesis and metastasis (Mantovani et al. 2008; Aggarwal et al. 2006). Immune cells that are continuously recruited to the tumor site produce a number of pro-inflammatory molecules within the tumor microenvironment that can provoke a complex signaling network and enable tumor cells to extravasate through the stroma, resulting in tumor progression (Colotta et al. 2009). My intent here is not to elaborate on details about cancer progression or inflammation, but rather to discuss the relationship between TG2 expression and tumor-promoting inflammation as

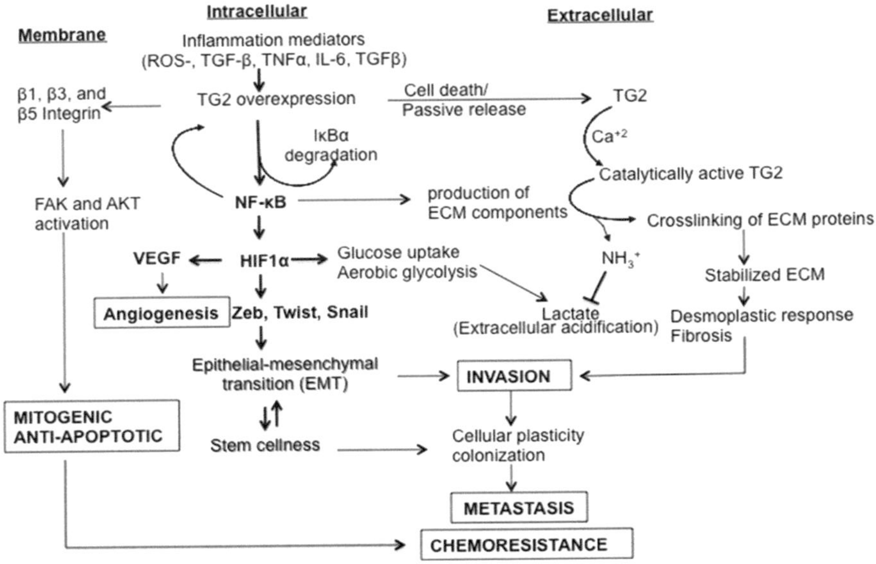

Fig. 10.2 TG2-regulated signaling promotes drug resistance and a metastatic phenotype. Smoldering inflammation due to recruitment of immune cells at the tumor site induces aberrant expression of TG2 in epithelial cancer cells. Owing to its scaffold function, TG2 binds to and degrades the inhibitory IκBα protein, resulting in the release and constitutive activation of NF-κB. Activated NF-κB, in complex with TG2, translocates to the nucleus where it binds to *HIF-1α* promoter and results in its transcriptional regulation and protein expression. Increased expression of HIF-1α, in turn, induces the expression of transcription repressors such as Snail, Zeb, and Twist. Collectively, TG2/NF-κB/HIF-1-induced alterations result in acquisition of an EMT phenotype and stem cell traits. Membrane-bound TG2, on the other hand, can interact with integrin and growth factor receptors (such as platelet-derived growth factor receptor and epithelial growth factor receptor) and induce mitogenic/cell survival signaling. Extracellular TG2 stabilizes the matrix by crosslinking its component proteins, adding to the shear force and promoting metastatic signaling. Ammonia, which is produced as a byproduct of TG2 catalytic activity, may have an important role in protecting tumor cells from acid toxicity caused by increased lactic production due to the Warburg effect (Katt et al. 2015). In a nutshell, aberrant expression of TG2 reprograms the inflammatory signaling networks which initiate a series of inside-out and outside-in signaling pathways to confer an aggressive phenotype in cancer cells

part of an effort to develop therapeutic approaches aimed at targeting TG2 to treat refractory cancers and to prevent metastatic progression of early-stage tumors.

10.2.1 TG2 and Cancer

TG2 is structurally and functionally a complex protein that comprises four structural domains (Fig. 10.3a). Each domain has a specific function, which permits TG2 to catalyze enzymatic as well as scaffold functions (Eckert et al. 2014). Its traditional role is that of a calcium-dependent enzymatic activity that crosslinks protein to form covalently linked protein complexes (Fig. 10.3b). However, it has also been shown to function as a G-protein intimately involved in cell signaling (Eckert et al. 2014), to modulate transcriptional complexes to drive gene expression (Tatsukawa et al. 2009; Kumar and Mehta 2012; Verma and Mehta 2007; Belkin 2011), to form complexes with other proteins to alter target protein function and to be secreted where it functions in the extracellular environment to regulate cell adhesion and motility (Eckert et al. 2014; Belkin 2011).

Three important physiological regulators of TG2 catalytic function include Ca^{2+}, GTP/GDP, and redox potential (Eckert et al. 2014; Klöck and Khosla 2012). Binding of Ca^{2+} to TG2 induces the catalytically active 'open' conformation, while binding to GTP or GDP promotes the catalytically inactive 'closed' conformation (Fig. 10.3c). Because most intracellular TG2 is GTP bound and intracellular calcium concentrations are low, TG2 stays in the catalytically inactive form under physiological conditions. In the 'closed' form, TG2 serves as a scaffold protein and acts as a signaling protein by binding and altering the function and stability of some key effector proteins. Increase in intracellular calcium due to cell damage or other stressors causes a major change in TG2 conformation to become catalytically active (extended), leading to TG2-catalyzed crosslinking of cellular proteins and apoptotic death. In contrast, in the extracellular environment, which has a considerably lower concentration of GTP/GDP and an abundance of free calcium, TG2 can be expected in a catalytically active (open) form. However, most TG2 even in extracellular environments is enzymatically inactive due to intramolecular disulfide bonding. Under oxidizing conditions, the disulfide bond between Cys230 and Cys370 facilitates the formation of the more stable Cys370-Cys371 disulfide bond that inactivates TG2 (see Chap. 14). The reactivation of extracellular TG2 can be achieved under favorable redox potential or alternatively by protein cofactor thioredoxin (Klöck and Khosla 2012). Once activated, TG2 can crosslink extracellular matrix (ECM) component proteins and stabilize the matrix for increased cell attachment, cell motility and outside-in signaling.

Recent studies have identified the important contribution of TG2 in cancer progression. TG2 expression is markedly increased in multiple tumors and tumor cell lines, especially in those selected for resistance to chemotherapy or isolated from metastatic sites (Mehta et al. 2010; Verma et al. 2006; Yuan et al. 2008; Satpathy et al. 2007; Park et al. 2002; Mehta et al. 2004; Budillon et al. 2011; Mangala et al. 2007; Oh et al. 2011). At a functional level, TG2 expression is associated with reduced expression of tumor suppressor genes, increased synthesis and deposition of fibronectin and collagen, stabilized ECM, reactivation of an embryonically regulated process, called epithelial-to-mesenchymal transition (EMT), enhanced drug resistance, and enhanced metastatic ability. Conversely, inhibition of TG2 by small-molecule inhibitors, antisense RNA or small inhibitory RNA (siRNA) results in reduced invasion and metastasis and increased sensitivity of cancer cells to chemotherapeutic drugs. These events are predominantly regulated via the scaffold function of TG2 whereby TG2 binds and modify the function and/or stability of key signaling proteins.

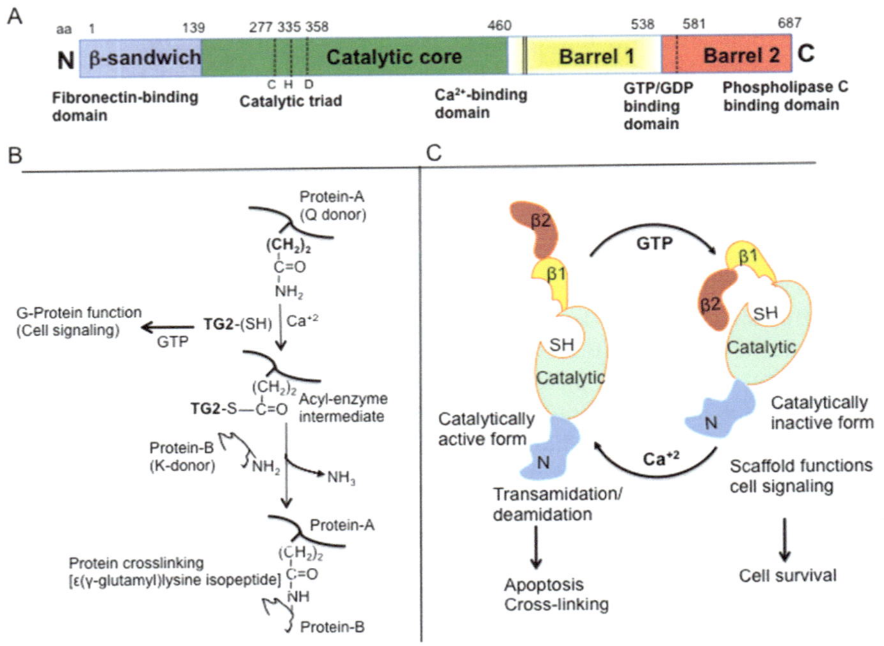

Fig. 10.3 Schematic representation of TG2 structure and functions. (**a**). TG2 comprises four major structural domains (depicted by *colored rectangles*). Each domain has a specific function (shown in *bold text* under the respective domain). (**b**). In the presence of calcium, TG2 catalyzes an irreversible crosslinking reaction between peptide-bound glutamine and peptide-bound lysine residues. In the GTP-bound form TG2 participates in multiple signaling pathways. (**c**). The TG2-catalyzed crosslinking reaction is regulated by Ca^{2+}, GTP, and the redox potential. Binding of Ca^{2+} (dissociation constant ~60 μM) is essential for TG2 to acquire a catalytically active or 'open' conformation. In contrast, binding of GTP/GDP (dissociation constant ~1.6 μM) renders TG2 in a catalytically inactive or 'closed' conformation. Under physiological conditions, high levels of GTP, a low redox potential, and low free Ca^{2+} (<0.5 μM) are likely to keep intracellular TG2 in a catalytically inactive state. Due to high calcium and low GTP levels in an extracellular environment, TG2 can be expected to be in a catalytically active state. However, even in extracellular environments a large fraction of TG2 remains in an inactive form due to disulfide bonding. Thioredoxin 1 has been suggested to be a physiological activator of oxidized TG2. In catalytically inactive state, TG2 acts as a scaffold protein and results in the activation of various transcription factors and signaling pathways. In a catalytically active state, it catalyzes highly stable protein crosslinking, resulting in apoptotic death if inside the cell or stabilizes the matrix if outside the cell

10.2.2 TG2-Regulated Inflammatory Signaling

Both intracellular and extracellular TG2 can regulate cell-signaling pathways that facilitate cell survival, cell motility, cell attachment, and invasive behavior (Fig. 10.2). Multiple reports have documented that aberrant expression of TG2 in epithelial cancer cells results in constitutive activation of focal adhesion kinase,

Akt, NF-κB, and HIF-1α signaling pathways (Eckert et al. 2014; Kumar and Mehta 2012; Verma and Mehta 2007; Belkin 2011), which play fundamental roles in cancer progression by inducing EMT and promoting drug resistance and metastasis (Lu et al. 2006; Monti and Gariboldi 2011; Kumar et al. 2011). For example, constitutively activated NF-κB is an important hallmark of advanced-stage cancer. It can regulate multiple downstream genes that are known to protect cells from cell death, promote invasiveness, or induce EMT and stem cell properties (Gupta et al. 2010). Moreover, activation of NF-κB is considered central to inflammation-induced tumor progression (Aggarwal et al. 2006). TG2 expression promotes constitutive activation of NF-κB via a canonical yet non-conventional pathway (see Chap. 5). TG2 binds to the inhibitory protein IκBα and results in its rapid degradation in a proteasomal-independent pathway (Kumar and Mehta 2012). This results in reduced levels of IκBα, permitting release and activation of NF-κB, its translocation to the nucleus and transactivation of multiple downstream target genes (Gupta et al. 2010). TG2-induced activation of NF-κB is independent of its catalytic functions as a catalytically inactive mutant (C277S) of TG2 was fully competent for activating NF-κB and downstream events (Grivennikov and Karin 2010; Gupta et al. 2010; Mantovani 2009). TG2 also binds to the p65 subunit of activated NF-κB and is recruited to the NF-κB binding site in the promoter sequence of *Snail* (Kim et al. 2010) and *HIF-1a* (Kumar and Mehta 2012), resulting in their transcriptional regulation. Downregulation of TG2 by a gene-specific siRNA attenuated NF-κB activation and inhibited HIF-1α expression (Kumar and Mehta 2012). Like NF-κB, HIF-1α expression is considered a negative prognostic factor because of its ability to promote chemoresistance, angiogenesis, invasiveness, metastasis, resistance to cell death, altered metabolism, and genomic instability (Wilson and Hay 2011). Moreover, TG2-regulated NF-κB and HIF-1α activation play an important role in initiation and induction of EMT due to their ability to regulate Snail, Twist, and Zeb1 transcription repressors (Kumar and Mehta 2012; Figure 2).

In addition to regulating intracellular signaling, TG2 in an extracellular environment can support outside-in signaling for increased cell growth, cell survival, and invasive functions. For example, it is well known that cell-surface TG2 can regulate integrin-mediated signaling through direct and stable interaction with β1, β3, and β5 integrin (Eckert et al. 2014; Belkin 2011; Nurminskaya and Belkin 2012). TG2 has a strong binding affinity for the gelatin-binding region of fibronectin and can interact strongly with fibronectin on one hand and integrin on the other to enhance integrin-fibronectin interaction. This interaction facilitates cell attachment to the matrix and activates integrin-mediated signaling. For example, integrin-TG2 interaction modulates the integrin-dependent activation of FAK, Src, p190RhoGAP, and increased levels of GTP-bound RhoA and its downstream target ROCK, leading to increased focal adhesion and actin stress fiber formation and enhanced actomyosin contractility (Nurminskaya and Belkin 2012). Based on this information, Dr. Matie's group ventured on screening a library of small-molecule inhibitors that could block fibronectin/integrin interaction (Yakubov et al. 2014). A compound identified in this screen (TG53) was able to inhibit adhesion of ovarian cancer cells to fibronectin and mitigated their migration and invasion. Contrary to the previous report, the authors observed that extracellular TG2 was able to activate NF-κB via a non-canonical pathway in ovarian tumors and the enzyme activity was required for this activation (Yakubov et al. 2013). TG2-induced activation of NF-κB was mediated through the hyaluronan receptor, CD44. The NF-κB activation by extracellular TG2 upregulated CD44 expression and induced EMT, contributing to increased invasiveness and peritoneal dissemination of ovarian cancer. These observations support the concept that extracellular TG2 can also play an important role in cancer progression and metastasis.

While intracellular TG2 can accelerate the synthesis and deposition of new ECM, extracellular TG2 can stabilize it by catalyzing the crosslinking of ECM component proteins (Fig. 10.2). It is likely that TG2-induced changes in the ECM contribute to a desmoplastic response in growing tumors. Indeed, the gene expression profile of TG2-overexpressing cells revealed increased expression of collagen, fibronectin, laminin, matrix metalloproteinases (MMP), growth factors such as TGFβ and platelet-derived growth factor and integrins (Kumar and Mehta, unpublished results) – the genes that represent key constituents of the desmoplastic response. TG2 can directly or indirectly regulate the expression and function of most of these effectors. ECM remodeling and stiffening are known to affect tumor behavior. Breast cancer progression, for example, requires collagen crosslinking, ECM stiffening, and increased focal adhesion formation (Levental et al. 2009). Lysyl oxidase-induced crosslinking of collagen, for example, results in ECM stiffening which is associated with increased focal adhesion formation, enhanced PI3 kinase (PI3K) activity, and increased tumor invasion. On the basis of these observations, it is tempting to speculate that TG2-catalyzed crosslinking and deposition of the ECM could play a role in promoting a malignant phenotype. Indeed, in a recent study, Lee et al. (2015) reported that extracellular TG2 catalyzes crosslinking of collagen in a pancreatic cancer milieu and promotes fibroblast proliferation and tumor growth. These results imply that inhibition of TG2 could be a promising target to block pancreatic cancer growth.

10.3 TG2-Regulates EMT, Stemness, and Glucose Metabolism

EMT is an evolutionarily conserved cellular process that plays crucial roles in the differentiation of tissue and organs during normal development of multicellular organisms (Thiery et al. 2009). It is characterized by loss of apical-basal cellular polarity, which results in the transition of polarized epithelial cells into a mesenchymal phenotype (Nieto and Cano 2012). In addition, EMT plays an important role during tissue repair but adversely can also promote organ fibrosis and cancer progression by conferring invasive and migratory properties on cancer cells (Thiery et al. 2009; Kalluri 2009). Recent evidence suggests that EMT can induce stem cell properties and prevent apoptosis (Dave et al. 2012).

During cancer progression, EMT is considered to be the first important step in metastatic dissemination of tumor cells. It allows tumor cells to detach from neighboring cells, invade tissues and survive in hostile environments. During EMT, the polarized epithelial cells undergo multiple morphological and biochemical changes that enable them to acquire a mesenchymal phenotype. Tumor cells with a mesenchymal phenotype exhibit enhanced migratory functions, invasiveness, and resistance to apoptosis, all of which are important hallmarks of metastatic tumors (Dave et al. 2012; Kalluri 2009).

Stable expression of TG2 in epithelial cancer cells is associated with morphological and molecular alterations that are characteristic of EMT (Kumar et al. 2010, 2012; Cao et al. 2012; Satpathy et al. 2007; Lin et al. 2011). Thus, overexpression of TG2 in mammary and ovarian cancer cells resulted in loss of epithelial markers and gain of mesenchymal markers. Moreover, TG2-induced EMT in cancer cells was accompanied by increased cell motility, invasiveness, and anchorage-independent growth (Fig. 10.4). TG2 overexpression in mammary epithelial cells resulted in loss of apico-basal polarity and disrupted their ability to form organized acinar structures when grown in 3D Matrigel cultures (Fig. 10.5). Unlike TG2-deficient cells, TG2-overexpressing cells grew into irregular spheroids with fragmented basement membrane and no defined lumen (Fig. 10.5). Conversely, suppression of TG2 by siRNA resulted in reversal of EMT (MET) as revealed by a gain in E-cadherin expression and loss of Snail1 and Zeb1 expression (Kumar et al. 2010).

It is now generally believed that small subset of cells within a tumor, termed cancer stem cells (CSCs) or tumor-initiating cells (TICc), are responsible for tumor sustenance and regrowth after chemotherapy (Lacerda et al. 2010). Because CSCs/TICs exhibit intrinsic resistance to therapy, their number would be expected to increase after chemotherapy. Indeed, the gene expression profile of residual tumors that survived after chemotherapy closely resembled the EMT gene signature (Creighton et al. 2009; Calcagno et al. 2010). Interestingly, TG2-induced EMT in breast cancer and ovarian cancer cells is accompanied by acquisition of stem cell characteristics (Kumar et al. 2011; Cao et al. 2012). TG2-expressing cells showed enhanced ability to form mammospheres, self-renewal ability, and cellular plasticity (Fig. 10.4) – important traits of mammary stem cells. Mammospheres derived from TG2-expressing cells differentiated into complex secondary structures when grown in Matrigel and treated with prolactin. Importantly, cells in these secondary structures differentiated into Muc1-positive (luminal marker) and/or integrin α6-positive (basal marker) cells (Kumar et al. 2012).

Increased glucose uptake and its metabolism via glycolysis (even in the presence of oxygen) is another important hallmark of metastatic cancers. This metabolic reprogramming is essential for successful growth and survival of tumor cells in distant tissues and as a source of substrates for biomass generation. TG2 expression reprograms glucose metabolism as a result of constitutive activation of NF-κB and HIF1α. Thus, TG2 overexpression resulted in increased glucose uptake, increased lactate production and decreased oxygen consumption rates by mitochondria

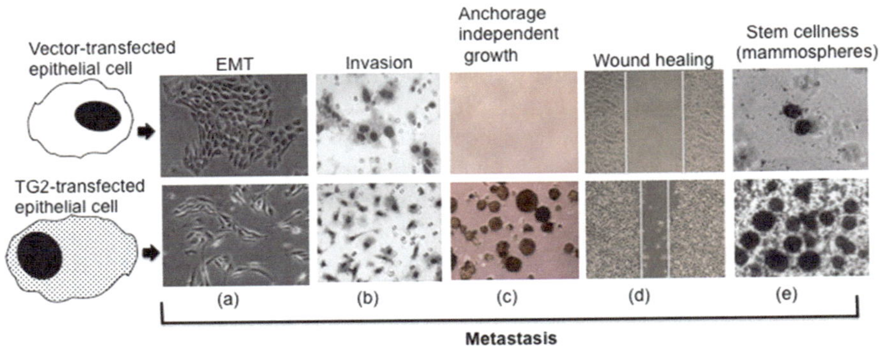

Fig. 10.4 TG2-induced changes in epithelial cells. Stable expression of TG2 in epithelial cancer cells is associated with their trans-differentiation into mesenchymal looking cells (EMT) (**a**). Such differentiation endows cancer cells with the ability to invade (**b**) and acquire the ability to grow in an anchorage-independent manner (**c**), increases cell motility as determined by a wound-healing scratch assay (**d**), and acquire a stem cell phenotype as established by enrichment of mammosphere-forming cells (**e**)

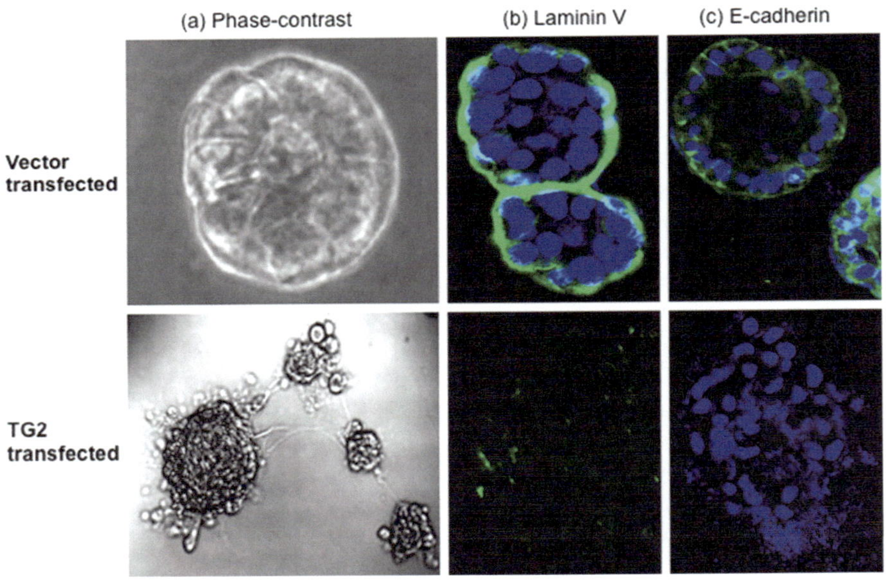

Fig. 10.5 TG2 induces tumorigenic phenotype. Morphology of acinar structures (**a**, *upper panel*) formed when TG2-deficient mammary epithelial cells (MCF-10A) were grown in Matrigel for 12 days. TG2 overexpression, however, resulted in loss of their ability to form such defined structures; instead cells grew as tumoroid structures with many cells in the periphery invading through the Matrigel (**a**, *lower panel*). Moreover, significant loss in basement membrane integrity (**b**) as well as in E-cadherin expression (**c**) was observed in TG2-overexpressing cells. *Blue* staining- DAPI

(Kumar et al. 2014). Experimental suppression of TG2 attenuated HIF-1α and reversed these downstream events in mammary epithelial cells. Moreover, downregulation of either TG2, p65/RelA or HIF-1α by gene-specific siRNAs enabled epithelial cells to restore normal glucose uptake, lactate production, mitochondrial respiration and glycolytic protein expression (Kumar et al. 2014). These results imply that aberrant expression of TG2 in epithelial cancer cells is an important modulator of glucose metabolism and it facilitates metabolic alterations owing to its ability to activate NF-κB and its downstream target HIF-1α. A TG2-induced shift in glucose metabolism helps cancer cells to survive under stressful conditions and makes them competent to survive and colonize in distant organs.

10.4 Conclusion and Clinical Perspective

Taken together, these observations suggest that aberrant expression of TG2 in cancer cells is a key event during cancer progression. It promotes drug resistance and metastatic competence in epithelial cancer cells owing to its ability to reprogram the inflammatory signaling networks as outlined in Fig. 10.2. These signaling pathways, in turn, play fundamental roles in inducing EMT, stemness, and altering glucose metabolism. Clinical evidence also supports this contention; elevated expression of TG2 in tumor samples is associated with poor survival rates in patients, resistance to therapy, and increased incidence of metastasis (Verma et al. 2006; Mehta et al. 2004; Hwang et al. 2008). Also, these patients show shorter relapse-free and metastasis-free survival after adjustment for known prognostic factors such as tumor size, lymph node metastasis, age, and hormone-receptor status (Ai et al. 2008). Therefore, inhibition of TG2 represents an attractive therapeutic option to reverse chemoresistance and intervention of metastatic progression. As a proof-of-concept, TG2-siRNA has been successfully used to reverse chemoresistance and to inhibit metastatic progression in preclinical models both in vitro and in vivo (Verma et al. 2008; Hwang et al. 2008). However, due to limited clinical evidence to support the effectiveness of siRNA as a therapeutic approach, its use in patients with cancer warrants further studies for optimization of siRNA delivery and safety in preclinical models.

Alternatively, the use of small-molecule inhibitors that can bind and inhibit TG2 may offer an alternative strategy to block TG2 signaling. Increased awareness about TG2's role in multiple pathological conditions (inflammation, tissue fibrosis, cancer, organ degenerative disorders, neurodegenerative disorders, celiac disease) has tickled interest by many groups to develop inhibitors against TG2 in a hope that some potent and selective compounds with therapeutic potential may soon be discovered (Keillor et al. 2015). A systematic study with larger cohort of patients to establish whether TG2 is a promising target for reversing chemoresistance and inhibiting metastatic progression is warranted. If TG2, as discussed in this chapter, turns out to be an important mediator of the metastatic cascade, it could not only offer a novel therapeutic target for treatment drug-resistant (refractory) and metastatic (recurrent) tumors (which together account for more than 90 % of cancer-related deaths) but also offer a promising diagnostic marker for early stratification of aggressive tumors.

Acknowledgments Supported in part by Bayer Healthcare Pharmaceutical Grants4Targets.

References

Aggarwal BB, Shishodia S, Sandur SK et al (2006) Inflammation and cancer: how hot is the link? Biochem Pharmacol 72:1605–1621
Ai L, Kim WJ, Demircan B et al (2008) The transglutaminase 2 gene (TGM2), a potential molecular marker for chemotherapeutic drug sensitivity, is epigenetically silenced in breast cancer. Carcinogenesis 29:510–518
Belkin AM (2011) Extracellular TG2- emerging functions and regulation. FEBS J 278:4704–4716
Budillon A, Carbone C, Di GE (2011) Tissue transglutaminase: a new target to reverse cancer drug resistance. Amino Acids 44:63–72
Calcagno AM, Salcido CD, Gillet JP et al (2010) Prolonged drug selection of breast cancer cells and enrichment of cancer stem cell characteristics. J Natl Cancer Inst 102:1637–1652
Cao L, Shao M, Schilder J et al (2012) Tissue transglutaminase links TGF-β, epithelial to mesenchymal transition and a stem cell phenotype in ovarian cancer. Oncogene 31:2521–2534
Colotta F, Allavena P, Sica A et al (2009) Cancer-related inflammation, the seventh hallmark of cancer: links to genetic instability. Carcinogenesis 30:1073–1081
Creighton CJ, Li X, Landis M et al (2009) Residual breast cancers after conventinal therapy display mesenchymal as well as tumro initiating features. Proc Natl Acad Sci U S A 106:13820–13825
Dave B, Mittal V, Tan NM, Chang JC (2012) Epithelial-mesenchymal transition, cancer stem cells and treatment reisistance. Breast Cancer Res 14:202
Demaria S, Pikarsky E, Karin M et al (2010) Cancer and inflammation: promise for biologic therapy. J Immunother 33:335–351
Eckert RL, Kaartinen MT, Nurminskaya M et al (2014) Transglutaminase regulation of cell function. Physiol Rev 94:383–417
Grivennikov SI, Karin M (2010) Inflammation and oncogenesis: a vicious connection. Curr Opin Genet Dev 20:65–71
Grivennikov SI, Greten FR, Karin M (2010) Immunity, inflammation, and cancer. Cell 140:883–899
Gupta SC, Kim JH, Prasad S, Aggarwal BB (2010) Regulation of survival, proliferation, invasion, angiogenesis, and metastasis of tumor cells through modulation of inflammatory pathways by nutraceuticals. Cancer Metastasis Rev 29:405–434
Hussain SP, Harris CC (2007) Inflammation and cancer: an ancient link with novel potentials. Int J Cancer 121:2373–2380
Hwang JY, Mangala LS, Fok JY et al (2008) Clinical and biological significance of tissue transglutaminase (TG2) in ovarian carcinoma. Cancer Res 68:5849–5858
Iizumi M, Liu W, Pai SK et al (2008) Drug development against metastasis-related genes and their pathways: a rationale for cancer therapy. Biochim Biophys Acta 1786:87–104
Kalluri R (2009) EMT: when epithelial cells decide to become mesenchymal-like cells. J Clin Invest 119:1417–1419

Katt WP, Antonyak MA, Cerione RA (2015) Simultaneously targeting tissue transglutaminase and kidney type glutaminase sensitizes cancer cells to acid toxicity and offers new opportunities for therapeutic intervention. Mol Pharm 12:46–55

Keillor JW, Apperley KYP, Akbar A (2015) Inhibitors of tissue transglutaminase. Trends Pharmacol Sci 36:32–40

Kim Y, Eom S, Kim K et al (2010) Transglutaminase II interacts with rac1, regulates production of reactive oxygen species, expression of snail, secretion of Th2 cytokines and mediates in vitro and in vivo allergic inflammation. Mol Immunol 47:1010–1022

Klöck C, Khosla C (2012) Regulation of the activities of the mammalian transglutaminase family of enzymes. Protein Sci 21:1781–1791

Kumar S, Mehta K (2012) Tissue transglutaminase constitutively activates HIF-1alpha promoter and nuclear factor-kappaB via a non-canonical pathway. PLoS One 7:e49321

Kumar A, Xu J, Brady S et al (2010) Tissue transglutaminase promotes drug resistance and invasion by inducing mesenchymal transition in mammary epithelial cells. PLoS One 5:e13390

Kumar A, Gao H, Xu J et al (2011) Evidence that aberrant expression of tissue transglutaminase promotes stem cell characteristics in mammary epithelial cells. PLoS One 6:e20701

Kumar A, Xu J, Sung B et al (2012) Evidence that GTP-binding domain but not catalytic domain of transglutaminase 2 is essential for epithelial-to-mesenchymal transition in mammary epithelial cells. Breast Cancer Res 14:R4

Kumar S, Donti TR, Agnihotri N, Mehta K (2014) Transglutaminase 2 reprogramming of glucose metabolism in mammary epithelial cells via activation of inflammatory signaling pathways. Int J Cancer 134:2798–2807

Lacerda L, Pusztai L, Woodward WA (2010) The role of tumor initiating cells in drug resistance of breast cancer: implications for future therapeutic approaches. Drug Resist Updat 13:99–108

Lee J, Condello S, Yakubov B et al (2015) Tissue transglutaminase mediated tumor-stroma interaction promotes pancreatic cancer progression. Clin Cancer Res 21:4482–4493

Levental KR, Yu H, Kass L et al (2009) Matrix crosslinking forces tumor progression by enhancing integrin signaling. Cell 139:891–906

Lin CY, Tsai PH, Kandaswami CC et al (2011) Role of tissue transglutaminase 2 in the acquisition of a mesenchymal-like phenotype in highly invasive A431 tumor cells. Mol Cancer 10:87

Lu H, Ouyang W, Huang C (2006) Inflammation, a key event in cancer development. Mol Cancer Res 4:221–233

Mangala LS, Fok JY, Zorrilla-Calancha IR et al (2007) Tissue transglutaminase expression promotes cell attachment, invasion and survival in breast cancer cells. Oncogene 26:2459–2470

Mantovani A (2009) Cancer: inflaming metastasis. Nature 457:36–37

Mantovani A, Allavena P, Sica A, Balkwill F (2008) Cancer-related inflammation. Nature 454:436–444

Mehta K, Fok J, Miller FR et al (2004) Prognostic significance of tissue transglutaminase in drug resistant and metastatic breast cancer. Clin Cancer Res 10:8068–8076

Mehta K, Kumar A, Kim HI (2010) Transglutaminase 2: a multi-tasking protein in the complex circuitry of inflammation and cancer. Biochem Pharmacol 80:1921–1929

Monti E, Gariboldi MB (2011) HIF-1 as a target for cancer chemotherapy, chemosensitization and chemoprevention. Curr Mol Pharmacol 4:62–77

Nieto MA, Cano A (2012) The epithelial-mesenchymal transition under control: global programs to regulate epithelial pasticity. Semin Cancer Biol 22:361–368

Nurminskaya MV, Belkin AM (2012) Cellular functions of tissue transglutaminase. Int Rev Cell Mol Biol 294:1–97

Oh K, Ko E, Kim HS et al (2011) Transglutaminase 2 facilitates the distant hematogenous metastasis of breast cancer by modulating interleukin-6 in cancer cells. Breast Cancer Res 13:R96

Park KS, Kim HK, Lee JH et al (2002) Transglutaminase 2 as a cisplatin resistance marker in non-small cell lung cancer. J Cancer Res Clin Oncol 136:493–502

Radisky E, Radisky DC (2007) Stromal induction of breast cancer: inflammation and invasion. Rev Endocr Metab Disord 8:279–287

Satpathy M, Cao L, Pincheira R et al (2007) Enhanced peritoneal ovarian tumor dissemination by tissue transglutaminase. Cancer Res 67:7194–7202

Slattum GM, Rosenblatt J (2014) Tumor cell invasion: an emerging role for basal epithelial cell extrusion. Nat Rev Cancer 14:495–501

Tatsukawa H, Fukaya Y, Frampton G et al (2009) Role of transglutaminase 2 in liver injury via cross-linking and silencing of transcription factor Sp1. Gastroenterology 136:1783–1795

Thiery JP, Acloque H, Huang RYJ, Nieto MA (2009) Epithelial-mesenchymal transitions in development and disease. Cell 139:871–890

Verma A, Mehta K (2007) Transglutaminase-mediated activation of nuclear transcription factor-kappaB in cancer cells: a new therapeutic opportunity. Curr Cancer Drug Targets 7:559–565

Verma A, Wang H, Manavathi B et al (2006) Increased expression of tissue transglutaminase in pancreatic ductal adenocarcinoma and its implications in drug resistance and metastasis. Cancer Res 66:10525–10533

Verma A, Guha S, Diagaradjane P et al (2008) Therapeutic significance of elevated tissue transglutaminase expression in pancreatic cancer. Clin Cancer Res 14:2476–2483

Wilson WR, Hay MP (2011) Targeting hypoxia in cancer therapy. Nat Rev Cancer 11:393–410

Yakubov B, Chelladurai B, Schmitt J et al (2013) Extracellular tissue transglutaminase activates non-canonical NF-κB signaling and promotes metastasis in ovarian cancer. Neoplasia 15:609–619

Yakubov B, Chen L, Belkin AM et al (2014) Small molecule inhibitors target the tissue transglutaminase and fibronectin interaction. PLoS One 9:e89285

Yuan L, Behdad A, Siegel M et al (2008) Tissue transgluaminase 2 expression in meningiomas. J Neurooncol 90:125–132

Chapter 11
Transglutaminases: Expression in Kidney and Relation to Kidney Fibrosis

Elisabetta A.M. Verderio, Giulia Furini, Izhar W. Burhan, and Timothy S. Johnson

Abstract Kidney fibrosis is regarded as a chronic wound response and tissue remodeling process consequent to a persistent tissue damage. The fibrosis is characterized by excessive extracellular matrix accumulation, fibroblast proliferation and chronic inflammation that leads to loss of tissue architecture and function. Transglutaminase-2 (TG2) is an essential component of wound repair and overexpression of TG2 and/or excessive crosslinking by TG2 have been notably linked to the pathogenesis of fibrosis in various organs. In this chapter, we discuss chronic kidney disease as one of the most prevalent chronic diseases, whose characteristic trait is fibrosis. We attempt to review recent thinking on the cellular and molecular causes of kidney fibrosis and review studies linking the role and enzymatic activity of TG2 with the pathogenesis of kidney fibrosis. Additionally we discuss how TG2 represents a new treatment target, which has catalyzed advances in the treatment of kidney disease.

Keywords CKD • Fibrosis • Transglutaminase-2 • TGF-beta • Syndecan-4 • Heparan sulfate proteoglycans • TG inhibitors

11.1 Chronic Kidney Disease as a New Health Burden

Chronic kidney disease (CKD) is a significant health issue worldwide affecting over 10 % of the world's population according to recent estimates, an epidemic of proportions similar to diabetes (Eckardt et al. 2013). The early signs of CKD have been long neglected by the scientific and health care community, who mainly focused on advanced severe CKD requiring dialysis and organ transplantation. It is well recognized now that even less severe CKD is associated with adverse

E.A.M. Verderio (✉) • G. Furini • I.W. Burhan
Biomedical Life and Health Science Research Centre, School of Science & Technology, Nottingham Trent University, Nottingham, UK
e-mail: elisabetta.verderio-edwards@ntu.ac.uk

T.S. Johnson (✉)
Academic Nephrology Unit, Sheffield Kidney Institute, University of Sheffield, Sheffield, UK
e-mail: t.johnson@sheffield.ac.uk

K. Hitomi et al. (eds.), *Transglutaminases*, DOI 10.1007/978-4-431-55825-5_11

outcomes. For instance, early stage CKD is an independent risk factor for cardiovascular conditions and therefore has an impact on non-renal tissues (Herzog et al. 2011; Gansevoort et al. 2013). An uniform definition of stages which characterize the progression of CKD has been agreed based on urine abnormalities (e.g. albuminuria) and function (e.g. decreased glomerular filtration rate, serum creatinine increases) (proposed by the National Kidney Foundation Kidney Disease Outcomes Quality Initiative K/DOQI in 2002). Kidney damage potentially impacts on a wide range of kidney functions affecting other organs including regulation of water, blood pressure, drug metabolism, potassium balance and calcium/phosphate metabolism, to name a few (Eckardt et al. 2013). As the disease progresses, glomerular filtration rate or GFR (ml/min/1.73 m^2) falls from GFR ≥ 90 (grade 1) to GFR 60–89 (grade 2), GFR 30–59 (grade 3), GFR 15–29 (grade 4), and GFR < 15 (grade 5). Typically, there are no symptoms at the early stages of the disease. The more exposed groups are the elderly, as we tend to lose kidney function at the rate of 1 % each year from about the age of 40, leading to an approximate loss of 30 % of kidney function at about the age of 70. Certain ethnic groups are more likely to be affected than others (e.g. Afro-Caribbean groups), and this is linked to higher risk of hypertension in these groups. In the early stages, CKD poses no threat but at the highest stages it requires ongoing dialysis treatment or surgery (kidney transplant).

Because of the involvement of the kidney in many extrarenal pathologies, kidney diseases are classically divided into primary diseases, that mainly manifest or origin in the kidney, and secondary diseases, for example systemic diseases which involve the kidney like obesity, metabolic syndromes including diabetes, hypertension, systemic infections. However, it is also true that many primary kidney diseases (like glomerulonephritis-types, glomerulosclerosis, ANCA-associated vasculitis, urinary tract infection, congenital diseases) either cause extrarenal pathologies or aggravate them. Therefore the attempt to classify kidney disease is challenging, but it is clear that the association with kidney diseases is both a poor prognostic and an aggravating factor for many extrarenal pathologies (Eckardt et al. 2013).

Another key point is the involvement of the kidney in age-related and lifestyle associated complex disorders such as obesity, diabetes, and hypertension. Obesity, defined as body mass index >30 kg/m^2, has plummeted in Western societies but also in emerging countries (Sharma 2014). Because of its association with an increase in diabetes and hypertension (both conditions linked to kidney disease), obesity may be regarded as a new risk factor for CKD. Early signs of CKD seems to develop before manifestation of hypertension and diabetes, especially when obesity is due to fat storage disorders (metabolic syndrome), suggesting that kidney damage appears early in individuals developing obesity and associated traits (Chen et al. 2004). This may be linked to oxidative stress, endothelial dysfunction and localized inflammation, which impact on glomerular capillary-wall function and may result in a decreased ability of the kidney to excrete salts, therefore causing hypertension. Indeed, in a 25-year follow-up study, the body mass index emerged as the second risk factor of end-stage renal disease, after proteinuria (Hsu et al. 2009).

A diminished ability of the kidney to excrete salts may also have a genetic component or be a consequence of acquired peritubular capillary loss/microvascular impairment, resulting in hypertension (defined as blood pressure above 140/90 mm Hg). Thus, arterial hypertension is closely related to kidney dysfunction. Concerning genetic components of CKD and their incidence, over 10 % of patients requiring dialysis are affected by some kind of monogenic disorders resulting in kidney failure, with polycystic kidney disease being the main genetic cause of kidney disease (Devuyst et al. 2012). Recent identification of risk variants, e.g. the MYH9/APOL1 region in the black population, have confirmed the evidence of an increase predisposition to kidney disease in certain ethnic groups (Kao et al. 2008; Kopp et al. 2011).

Diabetes and hypertension are the most frequent cause of end-stage renal disease in developed societies (Levey and Coresh 2012) and diabetic kidney disease is linked to increased mortality of diabetic patients (Afkarian et al. 2013). Regarded as a vascular complication of diabetes leading to a gradual increase in proteinuria and parallel decrease of kidney function, diabetic nephropathy is the main cause of end-stage renal disease in Western countries and in Japan, where it accounts for nearly half the newly induced hemodialysis (Maezawa et al. 2015). Glomerular injury/endothelium dysfunction have been recognized as the major components of diabetic nephropathy, although plenty of evidence has also shown changes in the tubulointerstitium (Nakagawa et al. 2011). The luminal surface of glomerular endothelial cells is covered by a network of membrane-bound proteoglycans such as syndecan, glypican, perlecan and versican, rich in heparan sulfate (HS) and chondroitin sulfate side-chains, which play an important role in the glomerular charge barrier (Dane et al. 2014); this network is reduced in diabetic animal models and type 1 and type 2 diabetic patients (Kuwabara et al. 2010; Nieuwdorp et al. 2006; Broekhuizen et al. 2010). Diabetes affects production of nitric oxide (NO), catalyzed by endothelial nitric oxide synthase (eNOS), which normally maintains the integrity of endothelial cells (Nakagawa et al. 2011). Hyperglycemia activates a wide number of additional pathways including the polyol pathway, involved in the metabolism of blood glucose, the renin-angiotensin-aldosterone system, reactive oxygen species (ROS) formation, transforming growth factor-beta (TGF-beta), angiopoietins, vascular endothelial growth factor (VEGF) and other pathways and systems. These not only affect endothelial cells, but also podocytes wrapping the glomerular capillaries and tubular epithelial cells (Nakagawa et al. 2011).

Lastly, it should not be forgotten that another cause of chronic nephropathy is kidney transplantation, the success of which is compromised by hypertension, proteinuria and progressive interstitial fibrosis leading to loss of renal function and graft loss. The interplay between "alloantigen-dependent" and "alloantigen-independent" factors can lead to irreversible tissue damage characterized by graft loss following graft fibrosis (Colvin 2003; Shrestha and Haylor 2014).

11.2 Kidney Scarring

Fibrosis or scarring is a key cause of the decline in renal function shown by all patients with CKD. A parallel may be drawn between the response to wound healing and the response to kidney injury. The tissue repair process involved in both situations is complex and dynamic, involving different cell types and factors and importantly, in the adult, may lead to the formation of fibrotic tissue in various amounts, depending on age and extent/recurrence of the damage (Hewitson 2009). Some wounds heal better and scar less than others, but there is not a complete understanding of how fibrosis develops. One key feature of scarring is the excessive accumulation of extracellular matrix (ECM) beyond the level occurring in the normal wound healing response. This "pathological matrix" consists of abundant fibrillary collagens (types I and II) and capillary basement membrane, comprising collagen IV and V and other ECM proteins like fibronectin, laminin and HS proteoglycans (Duffield 2014). Copious fibrillary matrix deposits in the tubular interstitial space between tubules and peritubular capillaries, impairing the waste and nutrients exchange function of tubules. As the disease progresses, further matrix expansion leads to loss of nephrons and capillaries, ultimately leading to kidney failure.

Scarring does seem to only partly depend on increased synthesis of ECM proteins. Other causes are the collapse of renal parenchyma, which is reminiscent of the wound contraction in skin, and leads to a greater density of glomeruli and deposited matrix, and a decreased-size fibrotic kidney (el-Khatib et al. 1987).

In addition to loss of volume and increased ECM density, a key pathophysiological role is played by matrix remodeling, a process largely modulated by matrix metalloproteinases (MMP) and members of the transglutaminase family (TG) (Lorand and Graham 2003). Indeed the role played by MMP is controversial being not only able to degrade but in some cases also to increase matrix deposition; hence recent thinking ascribes to MMP a dual role of both modulation of matrix turnover and synthesis. The TG involvement spans from matrix remodeling to increase in matrix deposition via TGF-beta activation and modulation of a plethora of cell-matrix interactions. The role played by TG2 in the pathogenesis of kidney scarring is the specific focus of this chapter and will be discussed in the next section.

At the cellular level activation of interstitial fibroblasts into myofibroblasts, capable of synthesizing alpha smooth muscle actin (alpha-SMA), is a recognized event in tubulointerstitial fibrosis, responsible for synthesis, deposition and remodeling of ECM proteins including fibronectin, collagens, elastin and laminin as well as proteoglycans, resulting in matrix contraction. Although the role of myofibroblasts in wound closure is well established, their role remains controversial in the kidney and may be driven as a response to tubule injury (Boor et al. 2010). To complicate matters, a low level of alpha-SMA positive cells is also detected in healthy adult kidneys of various species. Moreover ECM deposition may be contributed to by other mesenchymal cells such as pericytes and perivascular cells (Lin et al. 2008). A recent school of thought in particular maintains that most myofibroblasts appearing in kidney disease arise from resident fibroblasts, embedded in the stromal matrix, and pericites attached to capillaries and forming junctions with endothelial cells. These cell types have common ancestors being derived from the FOXD1-lineage mesenchymal progenitor cells during kidney development (reviewed by Duffield (2014)).

Other studies support the view that fibroblasts accumulate from various sources, resident fibroblasts, endothelium, epithelium and leukocytes. The hypothesis that they may derive from epithelial to mesenchymal transition (EMT) of tubular epithelial cells (Liu 2004) and/or endothelial cells (Zeisberg et al. 2008) has been put forward. Early work already suggested that tubular epithelial cells go through a typical transformation process upon damage, losing cell polarity, epithelial cell adhesion and cell to cell contacts, and undergo changes in cytoskeletal elements acquiring some of the features of fibroblasts (Strutz et al. 1995), expressing the typical marker fibroblast specific protein-1 (FSP-1). This transformation process is particularly relevant in tubulointerstitial fibrosis, which is characteristic of all progressive renal conditions. Among the mediators of EMT which circulate and originate from the tubular epithelial cells, TGF-beta1, platelet derived growth factor (PDGF) and angiopoietin-2 (ANG2) are key factors. MMP2 has also been recognized as having a driving role, by disintegrating the basement membrane of tubular epithelial cells (Cheng et al. 2006), and also as a stimulator of TGF-beta activity. An opposite role is played by hepatocyte growth factor (HGF), identified as being reno-protective and counteracting the activity of TGF-beta. EMT may be regarded as an adaptation to cell stress and indeed is induced by different signaling pathways, not only TGF-beta/Smad but also integrin/ILK and Wnt/beta catenin (Liu 2010). Typical EMT markers are new expressions of vimentin and/or nestin (Boor et al. 2010). However E-cadherin, which is typically downregulated in EMT, did not fall in the rat experimental model of fibrosis (UUO) (Docherty et al. 2009) suggesting that in vivo EMT does not always follow canonical pathways.

Central mediators of the overproliferation of fibroblasts and myofibroblasts seen in fibrosis are PDGF and TGF-beta. PDGF receptor alpha is found to be overexpressed in the renal interstitium and PDGF receptor beta in fibrotic renal tissue; furthermore de novo synthesis of corresponding PDGF ligands is observed (reviewed in Boor et al. (2010)). Integrin-mediated cell adhesion enhances mitogen activated protein kinase (MAPK) signaling induced by PDGF, via focal adhesion kinase (FAK) activation, and this leads to further proliferation/migration. PDGF is not only a potent mitogen factor for mesenchymal cells but also a stimulator of ECM synthesis. TGF-beta is the other main inducer of ECM protein expression in damaged tissue. This cytokine is activated via a complex step-wise process, starting from a well described latent TGF-beta complex, which accumulates in the matrix and consists of the TGF-beta dimer bound to latency associated peptide (LAP) (Annes et al. 2003).

Although myofibroblasts are the typical cells found in the diseased glomerulus and interstitium in CKD, other cell types are involved, as seen in skin wound healing. Leukocytes are abundant cells in CKD, predominantly macrophages and dendritic cells. Macrophages accumulate in the CKD interstitium alongside myofibroblasts (Duffield 2014) and a large number of T lymphocytes are recognized in immune-mediated renal diseases. Activated leukocytes and macrophages are producers of profibrotic cytokines including TGF-beta, PDGF and other growth factors which contribute to the formation and amplification of pathological matrix.

Another common tissue transformation event in patients with CKD leading to matrix deposition is vascular calcification. This phenomenon appears in 30–65 % of patients with stage 3–5 CKD and in the majority of end-stage CKD patients, but in patients with earlier stages of CKD, coronary artery calcification has been linked with higher levels of morbidity and mortality (Moe and Chen 2008). Current thinking ascribes the process to a response to hyperphosphatemia, uremia, inflammation and hyperglycemia, whereby vascular smooth muscle cells would transform to chondrocyte/osteoblast-like cells, laying down both a collagen and non-collagenous matrix, and ultimately forming "matrix vesicles" (from which calcification begins). This process has been recognized in chronic kidney disease-mineral bone disorder (CKD-MBD), a complex syndrome characterized by abnormal bone remodeling and vascular calcification (Moe et al. 2007).

11.3 Transglutaminases in the Pathology of CKD

11.3.1 Transglutaminases in Kidney

Several transglutaminases (TG) have been recognized with distinct genes, structures and physiological function (Lorand and Graham 2003). Mammalian TG are a family of proteins consisting of eight zymogens/enzymes, namely factor XIIIa and TG1-7, and a structural protein, protein 4.2, which lacks catalytic activity. In normal kidney, TG2 is the most abundantly expressed member of the transglutaminase family (Deasey et al. 2013). As discussed in subsequent sections, TG2 has a clear fibrogenic role, contributing to the stabilization and accumulation of the ECM in kidney fibrosis (Johnson et al. 1997, 1999, 2003), affecting TGF-beta activity (Kojima et al. 1993) and potentially being implicated in several other of the many cell-events leading to renal fibrosis. In the kidney TG2 is mainly recognized for catalyzing calcium-dependent transamidation of protein substrates, mainly ECM proteins (Johnson et al. 1997, 1999; Aeschlimann and Thomazy 2000; Lorand and Graham 2003), however TG2 is also known to mediate cell-matrix interactions and ArgGlyAsp (RGD)-independent cell adhesion separately from its catalytic activity and also to act as a structural protein in complex with fibronectin (Akimov et al. 2000; Verderio et al. 1998, 2003; Balklava et al. 2002; Telci et al. 2008). Despite this, the renal phenotype of TG2 knock-out (KO) mice is normal, the glomerular filtration rate is similar in wild type (WT) and TG2-KO mice and no histologically differences are observed (Shweke et al. 2008), suggesting that TG2 has mainly a pathological role in kidney.

Among the nine TG isoforms, TG1, TG2, TG5, and TG7 are expressed in most of epithelial tissues, including kidney (Lorand and Graham 2003; Iismaa et al. 2009). In renal epithelial cells TG1 was shown to be expressed at cadherin-based adherens junctions (Hiiragi et al. 1999) and it may have a role in the maintenance of structural integrity of simple epithelial cells. TG1, but not TG2, TG5 and TG7, is uniquely expressed in renal proximal tubular cells and is important in regulation of renal epithelial cell proliferation through the JAK2-Stat-3 signaling pathway (Zhang et al. 2009; Ponnusamy et al. 2009). The finding of TG1 upregulation in TG2-KO mice kidney (Deasey et al. 2013) compared to WT kidney could suggest a possible compensatory role in TG2-KO. Low levels of TG3, TG5, TG6 and FXIIIa have been shown to be expressed in kidney (Deasey et al. 2013), however literature searches show absence of studies on the expression of TG family members other than TG2 in kidney in pathophysiological conditions.

11.3.2 TG Expression and Activity in Experimental Kidney Fibrosis

The dependency of TG2-mediated protein transamidation on calcium and low GTP restricts the classic catalytic activity of TG2 mainly to the extracellular compartment, leading to formation of highly stable ε-(γ-glutamyl)lysine crosslinks (Smethurst and Griffin 1996; Aeschlimann and Thomazy 2000). In the presence of cell permeation to extracellular calcium or calcium release from intracellular stores, TG2 massively crosslinks intracellular substrates, leading to cell death (Verderio et al. 1998; Johnson et al. 1998; Nicholas et al. 2003). As a consequence, the extracellular environment is ideal for the activation of TG2, which has a well studied role in matrix remodeling/stabilization via covalent modification of extracellular proteins such as fibronectin, collagens, laminin and nidogen, and formation of ε-(γ-glutamyl)lysine crosslinks (Verderio et al. 1998; Jones et al. 1997; Bowness et al. 1987; Aeschlimann and Thomazy 2000). Pioneering work by Griffin et al. (1979) first suggested the involvement of TG2 in fibrotic disease in pulmonary fibrosis. Johnson and Griffin first suggested a role for TG2 in the rat subtotal nephrectomy model of kidney fibrosis, a model of chronic kidney scarring mimicking human disease (Johnson et al. 1997). In this disease model, progressive renal insufficiency (increases in proteinuria and serum creatinine and systolic blood pressure) was observed, as well as a progressive increase in tubular and glomerular scarring, with proteinuria strongly correlating with the scarring index. When the level of TG2 was assessed by immunohistochemistry and ε-(γ-glutamyl)lysine crosslinks along with measurement of TG2 activity and protein by western blotting in total kidney homogenates, a progressive increase was observed as the disease progressed. Notably, some ε-(γ-glutamyl)lysine crosslinks were detected in the tubular interstitial space, however most staining of TG2 and TG2-crosslinks was intracellular. This was likely due to the way TG2 immunohistochemistry was conducted as we now know that tissue section-embedding in paraffin and fixation are unsuitable for the detection of extracellular TG2, whose epitope is easily "masked" and lost by the fixative (Verderio et al. 1998). When ε-(γ-glutamyl)lysine crosslinks were determined by exhaustive proteolytic digestion and cation exchange chromatography (Griffin and Wilson 1984), a strong correlation was reported between the TG2 products and both tubular and glomerular scarring. Besides extracellular proteins, TG2 has been attributed a high number of intracellular substrates of the trasamidation reaction, including cytoskeletal and nuclear proteins (Lorand and Graham 2003). Primary amines such as putrescine, which may increase in kidney upon damage, can also be incorporated by TGs in γ-glutamyl groups of peptide bound glutamine residues (Campbell 1987). However, only a weak correlation was observed between scarring and intracellular TG2 in this first study (Johnson et al. 1997). Nevertheless, despite some methodology limits, this work suggested for the first time the possible involvement of TG2 in the pathogenesis of renal scarring and opened the way to mechanistic investigations. The hypothesis that qualitative changes in the ECM post-TG2 modification could slow down matrix degradation (e.g. by inhibiting the action of specific collagenases), thus moving the balance towards matrix deposition, was put forward. The analogy with the process of fibrin crosslinking by factor XIIIa, which renders fibrin more resistant to plasmin, was compelling. Furthermore, this initial work triggered further studies of TG2 and fibrosis in kidney and other organs.

Progress with the detection of extracellular TG2 was made when staining was undertaken prior to cell fixation in a culture of transfected mouse Swiss T3 fibroblasts inducible in TG2 expression (Verderio et al. 1998). Moreover, a convenient "in situ TG activity assay" was developed based on FITC-cadaverine incorporation into endogenous substrates in cultured cells (Verderio et al. 1998) which was adopted for immunohistochemical detection of TG2 antigen and TG activity in cryosections of kidney tissue (Johnson et al. 1999). TG2 labeling by anti-TG2 antibody prior to tissue fixation kept TG2 in its native state, thus allowing its recognition in the rat model of subtotal nephrectomy (SNx) (Johnson et al. 1999). Using these protocols, TG2 was recognized as abundantly located in the expanded ECM post-SNx and the activity of TG was visualized in the expanded peritubular ECM (Johnson et al. 1999). It became clear that TG2 accumulated outside the cells and was catalytically active in the matrix. TG2 mRNA hybridization identified the proximal tubules, especially cortical, as the main sites of TG2 expression in diseased kidney. Interestingly, a number of fibroblasts expressed TG2 in the expanded tubular interstitium post-SNx, and in blood vessels but independently of the disease. Evidence of a direct role played by TG2-mediated collagen crosslinking on ECM accumulation was provided by the observation that collagen fibrils, when formed in the presence of added purified TG2, were over 50 % more resistant to MMP degradation (using purified MMP-1) (Johnson et al. 1999). The crosslinking activity of TG2 emerged as a stabilizer of ECM, not only in kidney but in various pathophysiological contexts including lung development (Schittny et al. 1997) and liver fibrosis (Mirza et al. 1997; Chen et al. 2000).

Progressive accumulation of extracellular TG2-mediated isopeptide was also observed in an experimental model of diabetic nephropathy induced by administration of streptozoticin (STZ) to rats, mainly mimicking type 1 diabetes (Skill et al. 2001). Because of characteristic glomeruli changes in this model such as mesangium expansion, glomeruli were specifically isolated from hyperglycemic rat kidneys and scrutinized for accumulation of ε-(γ-glutamyl)lysine crosslinks/TG2 translocation in the glomerular basement membrane. Both damaged tubules and glomeruli presented increased TG2 crosslinks, however in the tubules this was not accompanied by de novo synthesis of TG2 protein, as observed in the rat SNx model (Johnson et al. 1997). Instead, "the same TG2" was found redistributed from the intracellular to the extracellular space in the hyperglycemic kidneys (Skill et al. 2001), suggesting its export in the diseased tissue.

In a subsequent study, when STZ was used to induce hyperglycemia in uninephrectomized (UNx) rats, both tubulointerstitial scarring and glomerulosclerosis established progressively up to 8 months post-STZ, and this was characterized by a several-fold increase in TG2 activity measured in situ at 8 months post-STZ in the UNx rats (Huang et al. 2009). A progressive accumulation of TG2 products was visualized by immunostaining with anti-isopeptide antibody, which was significantly higher than in normal and UNx-only rat kidneys (Huang et al. 2009).

In the rat model of focal and segmental glomerulosclerosis (FSGS), a primary kidney disease of glomerular origin, injection of puromycin aminonucleoside led to a significant increase of TG2 immunoreactivity and its product, predominantly in the glomeruli; this was accompanied by a reduction in MMP-9 protein levels (Liu et al. 2006).

A popular model of establishing renal interstitial fibrosis is unilateral ureteric obstruction (UUO), for the rapidity of its development in mice. In this model, TG2 expression which was faint in capillaries around normal tubules in control kidneys remarkably increased at 12 days post-UUO in C57BL/6 mice (Shweke et al. 2008). TG2 increased in the periglomerular and interstitial matrix, consistent with observations in other models, and paralleled the development of interstitial fibrosis, when assessed by total collagen measurement (hydroxyproline assay), macrophage infiltration and renal myofibroblast infiltration (Shweke et al. 2008). TG2 only partly co-localized with infiltrated macrophages in this model, when visualized by immunostaining, suggesting that TG2 is secreted by various cell types, not only macrophages. The increase in TG2 immunostaining was not species specific as it was also reported in the rat UUO model (Chen et al. 2005).

Although obstruction of the urinary tract is clinically relevant, the UUO induction of fibrosis is rapid and somehow less typical of human CKD. A recent study by Scarpellini et al. (2014) compared TG2 expression in two models of CKD in mice, the UUO and the aristolochic acid nephropathy model (AAN), which displays tubulointerstitial fibrosis more typical of human CKD (Huang et al. 2013). In both models, a progressive accumulation of interstitial TG2 was reported. Expression analysis by western blotting revealed that both the UUO and the AAN-induced kidney damage did not result in increased TG2 protein synthesis but led to a redistribution of TG2 from the cytosol to the cell membrane, leading to TG2 export in the extracellular space (Scarpellini et al. 2014) as previously observed in the STZ model (Skill et al. 2001).

As discussed in the previous section, vascular calcification is a common feature of advanced CKD and it is characteristic of CKD-MBD. In a recent study based on a novel rat model of progressive CKD-MBD (Cy/+ rat) displaying arterial calcification on a normal phosphorus diet (Moe et al. 2009), increased TG2 expression and activity was detected in Cy/+ rat-vascular smooth muscle cells and in pathological matrix vesicles ex vivo. Furthermore, the general TG inhibitor cystamine slowed down calcification in cells, matrix vesicles and aorta rings from the CKD rats (Chen et al. 2013).

From all these investigations in experimental models of fibrosis (summarized in Table 11.1), it has become clear that TG2 serves as a crosslinker predominantly in

Table 11.1 Studies of transglutaminase in CKD

Experimental models	Subtotal nephrectomy (SNx) in rat	Johnson et al. (1997, 1999, 2007) and Burhan et al. (2011)
	Unilateral ureteric obstruction (UUO) in rat	Chen et al. (2005)
	Unilateral ureteric obstruction (UUO) in mouse	Shweke et al. (2008) and Scarpellini et al. (2009)
	Aristolochic acid nephropathy (AAN) in mouse	Scarpellini et al. (2009)
	CKD-mineral bone disorder (CKD-MBD), Cy/+ rat	Chen et al. (2013)
	STZ induction of diabetic nephropathy in rat with or without UNx	Skill et al. (2001) and Huang et al. (2009, 2010)
	Rat Fisher-Lewis model of chronic allograft nephropathy	Shrestha et al. (2014)
	Puromycin aminonucleoside (PAN)-induced rat model of focal and segmental glomerulosclerosis	Liu et al. (2006)
Human biopsies	Human biopsies with different types of CKD	Johnson et al. (2003)
	Human biopsies at different stages of diabetic nephropathy	Johnson et al. (2003) and El Nahas et al. (2004)
	Human biopsies at different stages of allograft rejection	Johnson et al. (2004)
	Human biopsies at different stages of membranous nephropathy (MN)	Papasotiriou et al. (2012)
	Human biopsies of IgA nephropathy (IgAN)	Johnson et al. (2003), Ikee et al. (2007) and Takahashi et al. (2014)
Cell systems	Tubular epithelial cells	Skill et al. (2004), Fisher et al. (2009), Huang et al. (2010) and Chou et al. (2011)
	Vascular smooth muscle cells (VSMC) from Cy/+ rat	Chen et al. (2013)

the interstitial space, and its action site is both tubular and glomerular with variations depending on the type of kidney damage.

Although there are many expression studies of TG2, information on the role of the other TG family members is scarce. In a retrospective study of the rat subtotal nephrectomy model of progressive renal scarring, our group have shown that all the expressed TG members steadily increase post-SNx, with maximum expression by day 90 post-SNx. However TG2 expression exceeds that of all the other analyzed TG and it is the only member which correlates with fibrosis progression and loss of function (Burhan et al. 2011).

11.3.3 TG2 in Clinical Studies of CKD

The finding of the products of TG post-translational modification in human diseased kidneys established the importance of TG in humans. A strong correlation between interstitial fibrosis and in situ TG activity was reported in 33 biopsy samples from patients affected from mild to severe nephrotic syndrome (Johnson et al. 2003). A retrospective study of renal biopsies from 16 diabetic nephropathy patients with type 2 diabetes mellitus showed increased level of TG2 antigen and ε-(γ-glutamyl) lysine crosslinks in the peritubular and periglomerular areas, when these were visualized and quantified via immunofluorescence (El Nahas et al. 2004). Increased TG2 expression was reported predominantly in tubular epithelial cells but also in mesangial and interstitial cells, with TG2 emerging as a possible marker of disease progression. Intriguingly, in almost all the experimental and human disease models scrutinized, a number of tubular epithelial cells displayed high levels of TG2 crosslinks, which was accompanied by cell death. This process was first recognized in fibroblast cells induced to overexpress TG2 in the presence of a calcium ionophore, and it is recognized as a form of TG2-mediated cell death resembling neither apoptosis nor necrosis (Nicholas et al. 2003; Verderio et al. 1998).

Since then, more clinical studies of TG2 have been carried out in different centers in patients affected by various CKD types (listed in Table 11.1). In a clinical study of 22 patients affected by IgA nephropathy, the most common type of idiopathic glomerulonephritis, TG2 expression correlated with parameters of renal function when an immunohistochemistry approach was used. TG2 staining, predominant in interstitial fibrotic lesions and in proximity to vascular poles, correlated with renal pathology parameters, but it was not related to TGF-beta staining (Ikee et al. 2007).

A recent clinical study undertaken on renal biopsy specimens from 50 patients with biopsy-proven IgA nephropathy and 42 control patients (without predominant IgA deposition) measured TG2 activity with TG2-specific substrate probes. A conference paper reports that mesangial TG2 activity was measured in 56 % of IgA nephropathy (IgAN) patients and it was associated with a higher level of proteinuria and increased mesangial score, suggesting that mesangial TG2 activity may be related to mesangial proliferation in patients with IgAN (Takahashi et al. 2014).

Using a similar approach, biopsies from 32 patients affected by membranous nephropathy (MN), a primary cause of glomerular disease (Eckardt et al. 2013), displayed significantly higher TG2 immunostaining than those from healthy controls. However TG2 was not changed by immunosuppression therapy, suggesting that its expression was not downregulated in patients in remission at the end of the 5-year follow-up (Papasotiriou et al. 2012).

Curiously, a correlation between TG2 and tubulointerstitial fibrosis was also reported in a post-mortem study of 10 cats with primary renal azotemia compared with non-azotemic cats (Sanchez-Lara et al. 2014).

The combined findings clearly suggest that TG2 is exported and it is active as a matrix crosslinker in patients affected by a variety of CKD where it contributes to matrix stabilization. TG2 may provide a therapeutic target to decrease progression of renal fibrosis in humans and other animal species.

11.3.4 TG2 Expression and Activity in Kidney Allografts

With kidney transplantation being the ultimate treatment for end-stage CKD, chronic allograft nephropathy (CAN), a syndrome potentially associated with transplantation, and allograft loss are critical to the success of the intervention. The potential of TG2 as an early marker of chronic allograft dysfunction was first demonstrated in a retrospective study of adult cadaveric kidney recipients (Johnson et al. 2004). Biopsies from 23 patients showed absence of TG2 and TG2 crosslinks at implantation, however these were increased in approximately half of the biopsies as early as 3 months post-transplantation; moreover, all eight patients who developed established CAN displayed increased TG2 and TG2 crosslinks 36 months post-transplantation (Johnson et al. 2004). TG2 was recognized as most significantly linked to graft failure in early biopsies, more than collagen III and IV, alpha-SMA and general collagen staining (Masson's trichrome). Therefore, within the limits of the patient group number, this study highlighted an interesting potential application of TG2 as an early marker of allograft scarring.

In a larger clinical study examining 107 transplants biopsies taken sequentially at 6 weeks and 6 months, TG2 expression correlated well with that of TGF-beta and tissue inhibitor of matrix metalloproteinase-1 (TIMP-1) in subjects developing interstitial fibrosis and tubular atrophy. This finding suggests that matrix overproduction and limited degradation alter matrix equilibrium, leading to interstitial fibrosis and tubular atrophy in transplanted patients (Mengel et al. 2008).

Recently, consequences of CAN for TG2 and its product were corroborated in an experimental model of renal transplantation where a donor kidney was retrieved from a Lewis donor rat and transplanted into Fisher rats to generate an allogeneic renal transplant model (allograft), or Lewis rats to generate an isogenic transplant model (isograft) (Shrestha et al. 2014). The allograft model resulted in progressively increased parameters of renal insufficiency (hypertension, increased serum creatinine and decreased creatinine clearance) up to 52 weeks, or earlier if rats showed signs of distress. While no changes in TG2 were detected in the isograft model, a clear upregulation was found in the allograft model, accompanied by increased externalization in both glomeruli and tubulointerstitium. The distribution pattern of TG2 was similar to the distribution of myofibroblasts in peritubular and periglomerular spaces. Interestingly, ε-(γ-glutamyl)lysine crosslinks were found in the urine from the allografts compared to the isografts, and these significantly and progressively increased in the allogeneic rats; although it remains unclear what the protein substrates of the TG modification could be and what the impact of the post-translational modification on protein molecular mass versus excretion could be, this is a novel finding. Detection of TG2 hyperactivity in the urine could greatly facilitate exploitation of TG2 as an early marker of renal disease.

11.3.5 TG2 in the Urine of CKD Patients

Urine is a readily available body fluid accessible in large amounts. As current biomarkers are effective mostly for late stage CKD, there is a growing interest in urinary proteomics to predict and better classify CKD. Da Silva Lodge from the group of Tim Johnson at Sheffield Kidney Institute has recently presented data on urine TG2 expression in a large cohort of CKD patients (292) followed for at least 3 years (da Silva Lodge and Johnson 2013). When assessed by sandwich ELISA, urine TG2 was found to be over 40 times higher on average in patients than healthy individuals, with the highest expression found in the group of diabetic nephropathy patients (~7 μg/ml urine TG2) (da Silva Lodge and Johnson 2013). The ratio TG2/creatinine emerged as an even better predictor of progressive CKD than albumin/creatine in this conference report; moreover TG2 increases could be detected as early as from stage-2 CKD, suggesting that TG2 could be considered as a biomarker of progressive CKD. As mentioned in the previous section, TG2 isopeptides (ε-(γ-glutamyl)lysine crosslinks) could be detected in the urine of mice with allogeneic kidney transplants (Shrestha et al. 2014), suggesting that crosslinked substrates of TG2 may be regarded, under certain conditions, as a detectable marker for disease.

11.3.6 TG2 and TGF-Beta Pathway in CKD

It is well established that the fibrogenic role of TG2 is due to its direct role in the stabilization of the ECM via calcium-dependent transamidation (Table 11.2). The fibrogenic function of TG2 may also be due to TG2 contribution to the complex process of recruitment and activation of TGF-beta (Annes et al. 2003) since TG2 has been implicated in the binding of large latent TGF-beta1 to the ECM (Kojima

Table 11.2 Roles of transglutaminase in CKD

Formation of ε-(γ-glutamyl)lysine crosslinks (detected by exhaustive proteolytic digestion)	Johnson et al. (1997), Skill et al. (2001, 2004) and Shrestha et al. (2014)
Formation of ε-(γ-glutamyl)lysine crosslinks (detected by immunocytochemistry)	Johnson et al. (1997, 1999, 2003, 2004), Skill et al. (2001), El Nahas et al. (2004), Liu et al. (2006), Chen et al. (2013) and Shrestha et al. (2014)
In situ TG2 activity (TG2 substrate)	Takahashi et al. (2014)
ECM deposition	Skill et al. (2004), Chen et al. (2005, 2013), Johnson et al. (2007), Shweke et al. (2008), Fisher et al. (2009), Huang et al. (2009), and Scarpellini et al. (2014)
Matrix resistance to proteolytic turnover	Johnson et al. (1999) and Fisher et al. (2009)
Intracellular protein crosslinking	Johnson et al. (1997), El Nahas et al. (2004) and Johnson et al. (2004)
Activation of TGF-beta	Shweke et al. (2008), Huang et al. (2010), Scarpellini et al. (2014) and Furini et al. (2015)

et al. 1993; Nunes et al. 1997; Verderio et al. 1999). Evidence of TG2 involvement in TGF-beta activity in CKD was provided by Shweke et al. (2008) in the UUO murine model of interstitial fibrosis. UUO was performed in WT mice, as noted earlier, but also in a group of 10 age-matched TG2-KO mice (De Laurenzi and Melino 2001). The increase in renal fibrosis of the UUO kidneys was associated with an approximately 5-fold increase in active TGF-beta1. TG2-KO resulted in a milder fibrotic phenotype and lower collagen deposited; TG2-KO led to a considerably lower level of activation of TGF-beta 1 post-UUO. In a further study, TGF-beta activation was investigated in the rat subtotal nephrectomy model (SNx) (group of 5–6 rats) (Johnson et al. 2007). Here TGF-beta activity tended to increase post-SNx, but the change was not significant; furthermore, inhibition of TG2 activity did not decrease TGF-beta activation as expected. In another study, application of the same TG inhibitor (R281) to the streptozotocin model of diabetic nephropathy (UNx-STZ) lowered the level of active and total TGF-beta produced post-UNx-STZ (Huang et al. 2010).

Recently, our group has confirmed findings by Shweke et al. (2008) in the UUO model, by detecting a dramatic decrease in TGF-beta activation, measured by the mink lung cell bioassay, in kidney homogenates of TG2-KO UUO kidneys (fully backcrossed to C57BL/6) compared to WT (C57BL/6) kidneys (Furini et al. (2015) and manuscript in preparation). An increase in TGF-beta1 activity was also observed in kidney homogenates of mice with AAN-induced nephropathy compared to control kidneys (Scarpellini et al. 2014), suggesting that TGF-beta does increase in this model of CKD which gives fibrotic lesions similar to those in humans (Huang et al. 2013). Therefore, TG deletion may be renoprotective, by targeting not only direct deposition of the matrix but also TGF-beta1 activation. Interestingly, in the same study by Scarpellini et al. (2014), KO of the HS receptor syndecan-4 produced significant changes in the export and localization of TG2 in the tubulointerstitial space in both the UUO and the AAN model of CKD, and this was accompanied by decreased TGF-beta1 activity (Scarpellini et al. 2014). Although syndecan-4 may per se affect the activation of TGF-beta, e.g. through the distribution of its latent form (Ishiguro et al. 2001), there is no direct evidence for its role on TGF-beta1. Given the interplay between TG2 and syndecan-4 at the cell surface (Scarpellini et al. 2014) (discussed in the next section), lack of syndecan-4 may protect kidneys from fibrosis development via a mechanism involving control of TG2 externalization, which in turn impinges on matrix activation of TGF-beta1.

11.3.7 Heparan Sulfate Proteoglycans and TG2 Export in CKD

The HS chains of proteoglycans are synthesized in the Golgi apparatus as units of uronic acid (either iduronic or glucuronic acid) linked to N-acetylglucosoamine (GlcAC) and are found in the ECM, cell surface and basement membranes, acting as reservoirs of growth factors and cytokines (Bishop et al. 2007). HS chains have also high affinity binding for TG2 (Scarpellini et al. 2009; Lortat-Jacob et al. 2012) which is associated with the HS chains of syndecan-4 in a variety of cell types (Scarpellini et al. 2009; Lortat-Jacob et al. 2012; Wang and Griffin 2012; Nadella et al. 2015; Wang et al. 2011, 2012; Teesalu et al. 2012). HS proteoglycans have a physiopathological role in the kidney and have also been suggested to modulate TG2 trafficking in the kidney (Verderio et al. 2009).

Secreted HS proteoglycans such as perlecan, agrin and collagen VIII are typical glomerular basement membrane proteoglycans which play an important role in kidney filtration, maintaining the negative charge on the glomerular basement membrane and pore size (McCarthy and Wassenhove-McCarthy 2012). HS proteoglycans are also expressed in the tubule interstitium, and their amount is suggested to change in fibrotic lesions (Rienstra et al. 2010; Clayton et al. 2001). Cell surface HS proteoglycans such as the syndecan family of receptors mediate fibroblasts proliferation pathways, through aiding in fibroblast growth factor 2 (FGF2)-receptor interaction (Clayton et al. 2001). HS chains increase in the glomerulus and tubule interstitium of CKD patients and upregulation of syndecan-4 in both IgA nephropathy and diabetic nephropathy (Morita et al. 1994; Yung et al. 2001; Fan et al. 2003) are suggestive of its involvement in renal pathology. By assisting in recruitment of immune cells in the ECM, HS are also implicated in immune-mediated allograft rejection. HS chains typically increase during acute rejection in peritubular capillary vessels (Ali et al. 2003). Another critical factor is the degree and type of sulfation of HS chains. The 6-O-sulfation pattern for instance modulated by specific endosulfatases (SULF-1 and -2) can influence growth factor and cytokine binding sites (FGF, VEGF) affecting their distribution and function (Rosen and Lemjabbar-Alaoui 2010). The level of 6-O sulfation has been recently shown to increase in the UUO fibrotic kidney (Alhasan et al. 2014).

Data from our group and others has suggested that syndecan-4/HS contributes to TG2 cell surface trafficking and matrix distribution, since lack of syndecan-4/HS, or functional inhibition of HS, results in changes in cell surface TG2 antigen and TG2 transamidation in vitro, and a parallel accumulation of cytosolic TG2 (Scarpellini et al. 2009; Wang and Griffin 2012). Syndecan-4 also acts as a receptor for matrix TG2 once secreted and embedded in the fibronectin matrix (Telci et al. 2008), positively driving adhesion-mediated signaling and protecting cells from anoikis in situations of tissue damage (Verderio et al. 2003; Verderio and Scarpellini 2010). Membrane-proximal syndecan-4/HS may therefore either drive the unconventional secretion of TG2 by exerting a cell surface "molecular-trap" function – as proposed for FGF-2 (Zehe et al. 2006) – or affect TG2 localization/distribution in the matrix once exported given HS abundance in fibrotic kidney.

To investigate the role of syndecan-4 in vivo, we induced two models of kidney fibrosis, UUO (Vielhauer et al. 2001) and AAN (Huang et al. 2013), in syndecan-4-KO and WT mice (Ishiguro et al. 2000) and measured the development of tubulointerstitial fibrosis and TG2 externalization/extracellular activity. Deletion of syndecan-4 was protective in the development of fibrosis and also led to a lowering of extracellular TG2 protein and activity in the ECM, as binding of TG2 to the tubular interstitium depended on the HS chains (Scarpellini et al. 2014). These data have shown for the first time that syndecan-4 plays a critical role in the pathogenesis of kidney fibrosis by regulating TG2 trafficking and localization via HS chain-binding. Consistent with this finding, extracellular TG2 largely co-localizes with syndecan-4 in the tubular interstitium and basement membrane of fibrotic kidneys in the AAN (Scarpellini et al. 2014) and SNx model (Burhan et al. 2011). Moreover TG2 co-precipitates with syndecan-4 from SNx and UUO kidney homogenates (Burhan et al. 2011; Furini et al. 2015). In Fig. 11.1 we show dual immunofluorescent stainings of HS and extracellular TG2 in fibrotic kidney sections obtained as described before (Scarpellini et al. 2014). The staining reveals a large co-localization and proximity of signal of TG2 with the HS chains, which decorate peritubular and periglomerular regions and likely extend at variable length in the matrix (Fig. 11.1).

As lack of HS binding may result in less efficient TG2 export in the nephrotic kidney and as a consequence lower matrix stabilization and lower activation of TGF-beta1 (Scarpellini et al. 2014), the TG2-syndecan-4 interaction may provide a specific interventional target for limiting fibrosis progression.

11.3.8 TG2 as a Therapeutic Target

The first report of TG2 in fibrosis and thus as a potential target was in the lung in response to paraquat induced damage (Griffin et al. 1979). From those early observations TG2 has been investigated in numerous fibrotic diseases including

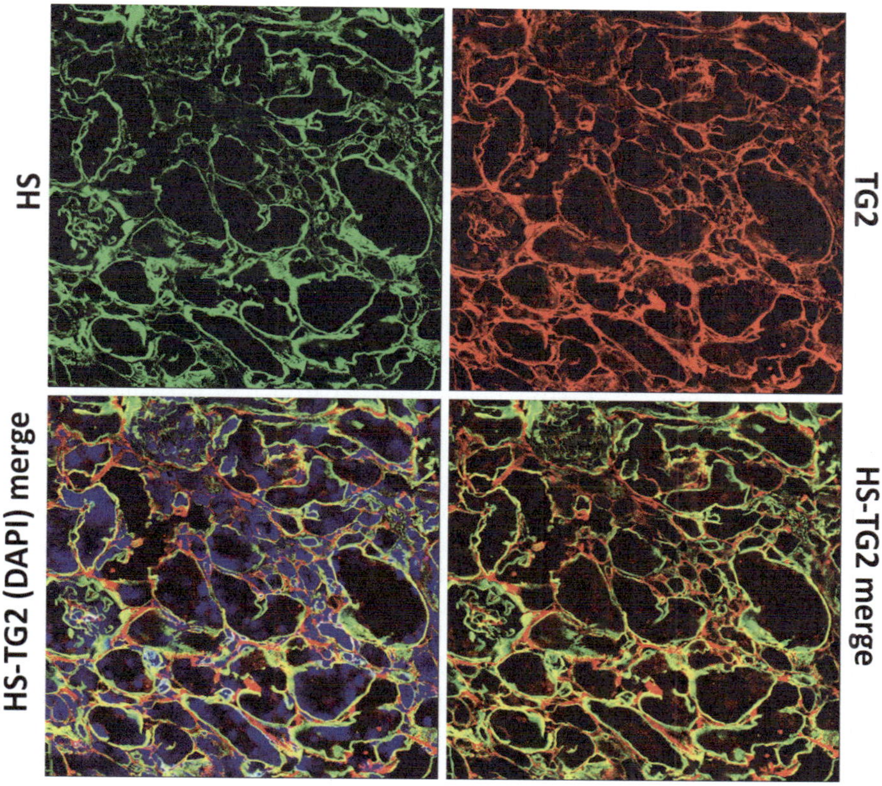

Fig. 11.1 TG2 largely co-localizes with HS chains in tubulointerstitial fibrotic lesions (AAN nephropathy mouse model of CKD). HS and TG2 immunostainings of cryostat sections were performed using mouse monoclonal anti-HS (IgM specific) antibody and rabbit polyclonal anti-TG2 antibody, respectively, followed by goat anti-mouse (IgM) FITC and donkey anti-rabbit IgG AlexaFluor 568, as described in Scarpellini et al. (2014). Representative pictures of HS (*green*) and TG2 (*red*) are shown separately and merged (with and without DAPI staining). Dual staining controls were carried out in TG2-KO kidney sections to rule out unspecific staining (not shown). $40 \times$ magnification

lung (Olsen et al. 2011), liver (Grenard et al. 2001), skin (Parsons et al. 2007) and heart (Small et al. 1999), however most of the compelling data for TG2 as a therapeutic target comes from studies in the kidney.

Early studies made use of the thiomidazole ring-based pan TG inhibitors that were developed by Merck in the late 1990s as factor XIIIa inhibitors (Freund et al. 1994). These compounds such as 1,3 dimethyl-2[(oxopropyl)thio] imidazolium chloride had little specificity for any TG family member over another, which is almost certainly why they had little clinical value as anti-thrombolytic agents, however what was very clear is that these were potent TG inhibitors and made great tool compounds. Application of these pan inhibitors in in vitro models of renal fibrosis such as a model of diabetic nephropathy for the first time showed that TG activity could be directly linked to a pro-ECM phenotype (Skill et al. 2004). In opossum kidney (OK) cortex proximal tubule epithelial cells, high glucose levels raised TG2 expression and activity, but also increased the amount of several ECM proteins. Application of the pan TG inhibitor 1,3 dimethyl-2[(oxopropyl)thio] imidazolium chloride normalized TG activity at 100 uM and had a marked reduction on the amount of mature ECM using both total and specific ECM measurements (Skill et al. 2004).

Based on these studies, the same thiomidazole compound and a CBZ-glutamine analogue inhibitor developed by the Griffin lab (N-benzyloxycarbonyl-L-phenylalanyl-6-dimethylsulfonium-5-oxo-L-norleucine or NTU281 (Griffin and Saint 2004)) were subsequently applied in in vivo models of kidney fibrosis (Johnson et al. 2007). The problem with these compounds was again specificity as both were pan inhibitors. Studies in human skin composites showed that such inhibitors could cause parakeratosis (most likely due to TG1 and 3 inhibition) (Harrison et al. 2007) whereas attempts to use these systemically led to extended bleed times in animals (TSJ personal communication). Thus to test the hypothesis that TG2 was a factor in tissue fibrosis a local delivery system was developed to allow direct intrarenal infusion of these compounds from a subcutaneous osmotic minipump (Huang et al. 2009).

In 2007, both compounds were infused into a 5/6th subtotal nephrectomy model of chronic kidney disease over a 3 month period. Both compounds despite differing models of action effectively lowered TG catalyzed crosslinking in the kidney, but perhaps most importantly they preserved kidney function (GFR) and prevented the increase in proteinuria in treated animals. The reason for this became clear when the kidneys were examined histologically as both had caused a dramatic amelioration of glomerulosclerosis and tubulointerstitial fibrosis development (Johnson et al. 2007).

A second study was performed with NTU281 in a much longer and more clinically relevant model in 2009 using a uninephrectomy/steptozotocin model of diabetic nephropathy (Huang et al. 2009). Again using direct intrarenal delivery, NTU281 was delivered for 8 months. Untreated animals developed extensive renal fibrosis and by 8 months were in end-stage renal failure. In contrast, animals receiving NTU281 had near normal serum creatinine, 50 % lower albuminuria and a dramatic reduction in both glomeruloscerosis and tubulointerstitial fibrosis (Huang et al. 2009).

These studies, matched with an emerging number of studies of TG2 in human renal disease, highlighted that TG2 could be a potential target in renal fibrosis, but it was also clear that specific TG2 inhibition would be needed due to the critical role of other TG family members. This raised the question as to whether specific TG2 inhibition would be enough to have these antifibrotic effects or whether compensation by other TG family members would overcome any benefit. This was largely unanswered by studies done using the TG2 KO mouse (Shweke et al. 2008). As noted earlier, using the UUO model, Schweke and colleagues (Shweke et al. 2008) showed that the TG2-KO had a remarkable effect on the development of kidney fibrosis. These data are even more impressive when one examines levels of other TGs in the TG2 KO mouse where TG1, 3, 6 and 7 increase in response to TG2 KO (unpublished data TSJ/EVE). What was also most interesting in this study was that the authors had assigned much of the benefit of TG2–KO to a reduction in active TGF-beta1 in the TG2 KO kidney. This was partly at odds with earlier studies which, using TG2 crosslinked collagen, suggested that TG2 stabilized the ECM to matrix metalloproteinase clearance (Johnson et al. 1999). Subsequently two studies looked more closely at this mechanism. The first by Fisher (Fisher et al. 2009) demonstrated convincingly, using a combination of TG2 overexpression and primary cells from the TG2-KO mouse, that TG2 was able to accelerate ECM deposition by 50 %, but that the matrix laid down in a TG2-rich environment was twice as resistant to proteolytic decay by MMP1 and 2 than that laid down with "normal" TG2 levels; this showed a direct TG2 action on matrix turnover. The second study by Huang (Huang et al. 2010) used a combination of tissue from a DN model treated with a TG2 inhibitor and co-culture systems with normal and TG2 overexpressing cells to readout the level of active TGF-beta1. This study showed convincingly that TG2 recruitment of large latent TGF-beta could be a rate limiting step. Thus, together these studies demonstrate that TG2 almost certainly has both a direct ECM processing and indirect TGF-beta1 activation role in wound repair/fibrosis in the kidney.

While studies in other organs did not address the detailed mechanism of action, parallel studies being run in the lung bleomycin model idiopathic pulmonary fibrosis (IPF) in the TG2 KO mouse showed a similar protection from fibrotic remodeling (Olsen et al. 2011) as did the siRNA knock-down of TG2 expression in the carbon tetrachloride model of liver fibrosis (Zhao et al. 2011). Thus there was strong evidence from multiple sources (Table 11.3) that TG2 inhibition alone was sufficient to slow down fibrosis development and that it would appear to be a conserved mechanism across multiple organs. The question then is how pharmacologically one can block TG2 specifically leaving the other TGs alone – a tough challenge given the 90 % homology in the catalytic core of TG family members and the highly conserved catalytic triad.

Table 11.3 Studies of transglutaminase inhibition in CKD progression

TG inhibitors	Skill et al. (2004), Chen et al. (2005), Johnson et al. (2007), Huang et al. (2009, 2010) and Chen et al. (2013)
TG2 knock-out	Shweke et al. (2008), Fisher et al. (2009) and Scarpellini et al. (2014)

A measure of how important TG2 is regarded as a therapeutic target may be judged by the huge number of patents published claiming TG2 inhibition. Table 11.4 summarizes the patent applications discussed below in this section. What is clear is that attempts to develop small molecular inhibitors have been hampered by extensive structural conservation in the active site of TG2 and other members of the transglutaminase enzyme family. While numerous warheads, typically irreversible, have been utilized to block TG activity, the recurrent problem seems to be how to dock the inhibitory model specifically on TG2 with minimal crossreactivity to other transglutaminases.

Several academics have registered numerous patents on TG2 small molecule inhibitors but systemic application remains sparse. The Khosla group at Stanford has generated several molecules which have led to the spin-out company Sitari Pharmaceuticals focusing on TG2 inhibition in celiac disease. Some of the molecules generated by this group such as KCC009 have been made widely available to the academic community and evaluated in vivo in mouse. The Keillor group at Ottawa have contributed to the field significantly and developed several cinnamoyl reversible TG2 inhibitors, which are competitive with the acyl donor substrate. Finally the Griffin Lab in Aston, UK, have been active in generating TG2 inhibitors and developed a range of interesting CBZ-glutamine analogues which did have some TG2 specificity, but not sufficient for clinical application. More recently they have filed two patents using both peptides and acylpiperazines as inhibitors of transglutaminase with claims of greater than 1000-fold specificity for TG2 (personal communication, MG). A detailed description of TG2 inhibition is provided in Chap. 16 by J. Keillor.

Zedira, a well know company in the TG field, have also developed a series of peptide based TG inhibitors that do show significant TG2 specificity but unknown PK stability for in vivo delivery. More recently Zedira have registered a patent claiming the use of pyridinone derivatives as TG2 inhibitors which offer an interesting development. In 2015 the biotech company announced a phase I clinical trial for the TG2 inhibitor ZED1227 targeting celiac disease. A long-standing relationship between Evotec and the Huntington's Trust has led to several patents being registered primarily for treatment of Huntington's chorea with TG2 inhibitors. Data presented at conferences suggested significant potency and specificity with a number of formulations listed in a recent patent showing promise. One of the world's largest pharmaceutical companies, Sanofi, has also recently registered two patents using sothiazolopyridine-2-carboxamides as TG2 inhibitors. While these claim TG2 as the primary target, again specificity data are not clear in the patent.

It is thus clear that small molecules targeting TG2 are of interest to academics and small/larger pharma alike. However, specificity remains an issue. A potential way of addressing this is with the development of biologics – antibodies able to target proteins highly specifically. The first attempt at this for TG2 was undertaken by Quark Biotechnology who in 2006 registered a patent for a TG2 inhibitory antibody generated from celiac patients by phage display, and presented data at the Transglutaminase and Crosslinking Enzymes conference in Lubeck, showing efficacy in the rat UUO model of kidney fibrosis. However it was difficult to predict from the patent the precise nature of this antibody and Quark let the patent elapse,

Table 11.4 Patents with claims towards inhibition of TG2

Company	Title	Type	Patent number	Pub date	Countries
Sanofi	Isothiazolopyridine-2-carboxamides and their use as pharmaceuticals	Small molecule	WO2013092574	27-Jun-2013	US, EP, +6 others
Sanofi	Isothiazolopyridine-2-carboxamides and their use as pharmaceuticals	Small molecule	TW201339166	20-Dec-2011	TW
Sankyo	New compounds F-11754	Small molecule	JPH9169693	30-Jun-1997	JP
CHDI, Evotec	TG2 inhibitors, pharmaceutical compositions and methods of use thereof	Small molecule	WO2014047288	27-Mar-2014	PCT, TW
CHDI, Evotec	TG2 inhibitors, pharmaceutical compositions and methods of use thereof	Small molecule	US2012302539	29-Nov-2012	US
CHDI, Evotec	TG2 inhibitors, pharmaceutical compositions and methods of use thereof	Small molecule	WO2011060321	19-May-2011	PCT, US
Stanford University	Modulation of tissue transglutaminase activation in disease	Small molecule	US20140322278	30-Oct-2014	PCT, US, EP, AU, CA, JP
Stanford University	Transglutaminase inhibitors and methods of use thereof	Small molecule	US2013102633	25-Apr-2013	US, EP + 9 others
Numerate, Stanford	3-Acylidene-2-oxoindole derivatives for inhibition of TG2	Small molecule	WO2012078519	14-Jun-2012	PCT
Quark Biotech	Recombinant antibodies against human TG2 and uses thereof	Antibodies	WO2006100679	28-Sep-2006	PCT
Zedira	Pyridinone derivatives as tTG inhibitors	Small molecule	WO2014012858	23-Jan-2014	EP
Zedira	Michael systems as TG inhibitors	Peptides and peptidomimetics	US2011229568	22-Nov-2011	US, EP, +6 others
Alvine Pharmaceuticals	Transglutaminase inhibitors and methods of use thereof	Small molecule	WO2007050795	03-May-2007	US, EP
Decode Genetics	Human transglutaminases	Nucleic acids and antibodies	WO02059265	01-Aug-2002	PCT

(continued)

Table 11.4 (continued)

Company	Title	Type	Patent number	Pub date	Countries
Toray Industries	Tissue TG inhibitor, chalcone derivative, and pharmaceutical application thereof	Small molecule	JP2013180955	12-Sep-2013	JP
National Cancer Center of Korea	Pharmaceutical composition for preventing or treating TG-related disease . . .	Small molecule	KR20120092538	21-Aug-2012	KR
National Cancer Center of Korea	Ethacrynic acid-containing composition for prevention or treatment of TG-related diseases . . .	Small molecule	US2011288176	24-Nov-2011	PCT, US, KR
National Cancer Center of Korea	Novel pyrazolodiazepine compounds as a transglutaminase inhibitor . . .	Small molecule	WO2010074480	01-Jul-2010	US, EP, +8 others
National Cancer Center of Korea	A pharmaceutical composition for preventing or treating TG-related disease . . .	Small molecule	KR2010056212	25-May-2010	KR
National Cancer Center of Korea	A pharmaceutical composition for preventing or treating TG-related disease . . .	Small molecule	KR2010056211	25-May-2010	KR
National Cancer Center of Korea	A composition for TG inhibitor comprising NDGA	Small molecule	KR20100011963	03-Feb-2010	KR
National Cancer Center of Korea	A composition for TG inhibitor comprising curcumin	Small molecule	KR20100011620	03-Feb-2010	KR
National Cancer Center of Korea	TG inhibitor comprising EGCG and a method for producing thereof	Small molecule	WO2008153319	18-Dec-2008	US, EP, CN, JP, KR
National Cancer Center of Korea	Glucosamine and derivatives useful as transglutaminase inhibitors	Small molecule	WO2007026996	08-Mar-2008	US, EP, AT, CN, JP
National Cancer Center of Korea	Peptides for inhibiting transglutaminase	Peptides	WO2007069817	21-Jun-2007	US, EP, AT, CN, JP, KR
National Cancer Center of Korea	Novel anti-inflammatory peptides	Peptides	US20020192780	19-Dec-2002	US, KR
Aston University	Novel polypeptides and uses thereof	Polypeptides	US20140199325	17-Jul-2014	PCT, US, EP, GB

Aston University	Acylpiperazines as inhibitors of transglutaminase and their use in medicine	Small molecule	WO2014057266	17-Apr-2014	PCT
Aston University	Novel compounds and methods for use in medicine	Small molecule	WO2012153125	15-Nov-2012	PCT, GB
Aston University		Polypeptides	WO2012146901	01-Nov-2012	PCT, US, EP, GB
Dongguk University	2-(Phenulethynol)thieno[3,4-B]pyrazine derivative and pharmaceutical composition comprising same for preventing or treating cancer	Small molecule	WO2014109530	17-Jul-2014	PCT, KR
Dongguk University	Method for inhibiting TG2 activity using 2-alkyloxyl-3-phenylethynyl-4A,5 dihydropyrido [2,3-B] pyrazine derivatives	Small molecule	WO2013016483	27-Jun-2013	US
Dongguk University	Composition for preventing, treating or alleviating atopic dermatitis comprising immunosuppressant and TG2 inhibitor	Drug combination	WO2013077617	20-May-2013	PCT, KR
Seoul National University Industry	Pharmaceutical compositions for preventing or treating fibrosis	Small molecule	WO2009038289	26-Mar-2009	PCT, KR
Seoul National University Industry	Pharmaceutical compositions for preventing or treating cataract	Small molecule	WO2009038288	26-Mar-2009	PCT, KR
European Institute for Cystic Fibrosis Research	Combined therapy for cystic fibrosis	Combo with TGi	US2013310329	21-Nov-2013	US, EP
INSERM	TG2 inhibitors for use in the prevention or treatment of rapidly progressive glomerulonephritis	Small molecule	WO2011151395	08-Dec-2011	US, EP
Japan Science and Technology Agency	Protein crosslinking inhibitor and use of same	Small molecule	US2012277423	01-Nov-2012	PCT, US, JP
MRC Technology	Anti-transglutaminase antibodies	Antibody	WO2013175229	28-Nov-2013	PCT
Nottingham Trent University	Novel compounds and methods of using the same	Small molecule	WO2004113363	29-Dec-2004	US, EPl, AT, CA, ES
Singapore Health Services	Transglutaminase-2 inhibitors and uses thereof	Small molecule	US2014050779	20-Feb-2014	PCT, US, SG

(continued)

Table 11.4 (continued)

Company	Title	Type	Patent number	Pub date	Countries
Université de Montréal	Cinnamoyl inhibitors of transglutaminase	Small molecule	WO2008147933	04-Dec-2008	PCT, US
University of Nigata	Remedy for CAG repeat expansion diseases	Small molecule	US6355690		
University of Nottingham	Inhibitors	Various	US2011237677	29-Sep-2011	PCT, US, GB
University of Queensland	Targets for growth factor signaling growth factor signaling, and methods of therapy	Various	WO2010105302	23-Sep-2010	US, EP, +7 others

the reason for which is unclear. Nevertheless it did raise the potential that using biologics to block TG2 activity was a viable option. In 2013, working with Medical Research Council Technology (MRCT), the authors' group filed a patent describing the identification of four regions in the TG2 catalytic core towards which an antibody response could be raised that was completely specific to TG2, offering enzyme inhibition at low nM IC_{50} values. Generated by classical immunization and hybridoma technology, the four sites were identified in human TG2 which seemed to have different inhibitor mechanisms, from blocking putative calcium binding sites, to binding to one of the amino acids in the catalytic triad, to steric hindrance of the active site. Application of some of these mouse anti-human antibodies in in vitro models of diabetic nephropathy showed strong efficacy in biological systems while early application in vivo in models of kidney fibrosis has been highly promising. Humanization of a number of these antibodies by MRCT served to increase both efficacy and affinity, generating inhibitory antibodies with characteristics suitable for clinical application. One would anticipate early human application with these antibodies which will finally allow evaluation of the true value of TG2 inhibition in renal fibrosis and other diseases.

While much work has gone into inhibiting the transamidation and deamidation activity of TG2, an alternative approach to controlling TG2 action would be to block its cellular trafficking. Many studies have shown that cellular stress can cause cytoplasmic TG2 to be rapidly exported into the extracellular environment leading to activation of TG2 and its subsequent pro-fibrotic action in the extracellular environment. This can be exemplified by studies in models of diabetic nephropathy (Skill et al. 2001), and recently in the UUO and AAN models (Scarpellini et al. 2014) whereby TG2 is seen to be elevated early in the tubular basement membrane with no overall increase in TG2 protein or mRNA. Work by Chou and colleagues investigated the mechanism of TG2 export in tubular epithelial cells and identified an amino acid sequence (96–108) that forms a loop in the beta-sandwich domain of TG2 (Chou et al. 2011). Mutations within this loop prevented TG2 from being exported while attachment of this sequence to other proteins was able to cause their cell export. This sequence is unique to TG2 which suggests that if compounds could be identified that would interfere with this export domain binding to its partner proteins then a very specific way of inhibiting TG2 export could be found. Yeast 2 hybrid studies identified over 30 proteins which were able to interact with this TG2 export motif (unpublished data, TSJ) however no agents have yet been developed that are able to show the practicality of interacting with this export domain in biological systems. That said, the biggest concern is that if one blocks TG2 export then high intracellular TG2 could affect cell viability, yet transfecting cells with large amounts of TG2 with mutations in its export domain failed to display any reduced viability (Chou et al. 2011).

In models of chronic kidney disease the HS proteoglycan receptor syndecan-4 also appears to be required for extracellular targeting, activity and extracellular trafficking of TG2 (Scarpellini et al. 2014), although there does not appear to be any direct link between the export domain and syndecan-4. Studies with the application of the HS blocking agent Surfen (Schuksz et al. 2008) clearly showed reduced extracellular TG2 (Scarpellini et al. 2009), suggesting that if specific interference with TG2-syndecan 4 interactions could be achieved, then it would also serve as a therapeutic intervention site. Several HS binding sites within TG2 have been proposed (Lortat-Jacob et al. 2012; Wang et al. 2012; Teesalu et al. 2012). Our group have identified two distant basic amino acid clusters, RRWK (262–265) and KQKRK (598–602) which would come together to form a conformationally dependent high affinity binding site (Lortat-Jacob et al. 2012). Targeting this TG2 conformational heparin binding site, e.g. with a monoclonal antibody, would not only prevent TG2 binding to HS, a critical step on its export, but would at the same time keep the matrix-unbound "soluble" TG2 in the closed inactive conformation, thus minimizing side effects of any potentially residual free TG2. We anticipate that disruption of TG2-HS binding would block association of secreted TG2 with the pericellular matrix environment and reduce profibrotic activity. Since this site forms when TG2 is in the closed conformation, it is a very attractive one; however, technically, generating an antibody to such a conformational site is a challenge.

11.4 Conclusion

Several independent studies have attributed to TG2 a pathological role in the development of kidney fibrosis, the ultimate cause of end-stage renal failure. TG2 has emerged as a marker of progression in some clinical studies and TG2 and its products can be identified in patients' biological fluids. It is clear that the action of TG2 is not only a direct one on extracellular matrix crosslinking and stabilization, but extracellular TG2 also seems to activate TGF-beta in various experimental models of CKD. A number of investigations have shown marked amelioration of kidney fibrosis in experimental models when TG2 or its activity are knocked-down, or its export reduced. As TG2 is externalized and active mainly in pathological kidney, it is an attractive pharmacological target, and there is an intense interest in the development of anti-scarring agents based on TG2 inhibition. Recent studies have recognized the HS chains of proteoglycans having the function of modulating the trafficking and extracellular role of TG2 in CKD models, but clues on the export mechanism of TG2 are needed to better exploit TG2 as an antifibrotic target in kidney. Advances in the "-omics" technologies and the ability to screen large volumes of patients' biological fluids directly could help establish TG2 as a disease marker and, at the same time, accelerate the discovery of partners of TG2 in fibrosis development.

Acknowledgments We are grateful to Kidney Research UK RP25/2012, the Wellcome Trust [087163], Nottingham Trent University and Sheffield Kidney Research Foundation for supporting our work.

References

Aeschlimann D, Thomazy V (2000) Protein crosslinking in assembly and remodelling of extracellular matrices: the role of transglutaminases. Connect Tissue Res 41(1):1–27

Afkarian M, Sachs MC, Kestenbaum B, Hirsch IB, Tuttle KR, Himmelfarb J, de Boer IH (2013) Kidney disease and increased mortality risk in type 2 diabetes. J Am Soc Nephrol 24 (2):302–308. doi:10.1681/ASN.2012070718 ASN.2012070718 [pii]

Akimov SS, Krylov D, Fleischman LF, Belkin AM (2000) Tissue transglutaminase is an integrin-binding adhesion coreceptor for fibronectin. J Cell Biol 148(4):825–838

Alhasan AA, Spielhofer J, Kusche-Gullberg M, Kirby JA, Ali S (2014) Role of 6-O-sulfated heparan sulfate in chronic renal fibrosis. J Biol Chem 289(29):20295–20306. doi:10.1074/jbc.M114.554691 M114.554691 [pii]

Ali S, Hardy LA, Kirby JA (2003) Transplant immunobiology: a crucial role for heparan sulfate glycosaminoglycans? Transplantation 75(11):1773–1782. doi:10.1097/01.TP.0000065805.97974.93

Annes JP, Munger JS, Rifkin DB (2003) Making sense of latent TGFbeta activation. J Cell Sci 116 (Pt 2):217–224

Balklava Z, Verderio E, Collighan R, Gross S, Adams J, Griffin M (2002) Analysis of tissue transglutaminase function in the migration of Swiss 3T3 fibroblasts: the active-state conformation of the enzyme does not affect cell motility but is important for its secretion. J Biol Chem 277(19):16567–16575. doi:10.1074/jbc.M109836200. doi:10.1074/jbc.M109836200 M109836200 [pii]

Bishop JR, Schuksz M, Esko JD (2007) Heparan sulphate proteoglycans fine-tune mammalian physiology. Nature 446(7139):1030–1037. doi:nature05817 [pii] 10.1038/nature05817

Boor P, Ostendorf T, Floege J (2010) Renal fibrosis: novel insights into mechanisms and therapeutic targets. Nat Rev Nephrol 6(11):643–656. doi:10.1038/nrneph.2010.120 nrneph.2010.120 [pii]

Bowness JM, Folk JE, Timpl R (1987) Identification of a substrate site for liver transglutaminase on the aminopropeptide of type III collagen. J Biol Chem 262(3):1022–1024

Broekhuizen LN, Lemkes BA, Mooij HL, Meuwese MC, Verberne H, Holleman F, Schlingemann RO, Nieuwdorp M, Stroes ES, Vink H (2010) Effect of sulodexide on endothelial glycocalyx and vascular permeability in patients with type 2 diabetes mellitus. Diabetologia 53 (12):2646–2655. doi:10.1007/s00125-010-1910-x

Burhan IW, Nutter FH, Johnson TS, Verderio E (2011) Association of Transglutaminase-2 and Syndecan-4 in the 5/6th subtotal nephrectomy (SNx) rat model of progressive renal scarring. In: Proceedings of the Renal Association (joint with British Renal Society) Congress-Birmingham, UK

Campbell RA (1987) Polyamines and uremia. Adv Exp Med Biol 223:47–54

Chen R, Gao B, Huang C, Olsen B, Rotundo RF, Blumenstock F, Saba TM (2000) Transglutaminase-mediated fibronectin multimerization in lung endothelial matrix in response to TNF-alpha. Am J Physiol Lung Cell Mol Physiol 279(1):L161–L174

Chen J, Muntner P, Hamm LL, Jones DW, Batuman V, Fonseca V, Whelton PK, He J (2004) The metabolic syndrome and chronic kidney disease in U.S. adults. Ann Intern Med 140 (3):167–174. doi:140/3/167 [pii]

Chen D, Huang HC, Yu L (2005) [Expression and implication of tissue transglutaminase and connective tissue growth factor at fibrotic tubulointerstitium in kidneys from UUO rats]. Beijing Da Xue Xue Bao 37(2):143–146

Chen NX, O'Neill K, Chen X, Kiattisunthorn K, Gattone VH, Moe SM (2013) Transglutaminase 2 accelerates vascular calcification in chronic kidney disease. Am J Nephrol 37(3):191–198. doi:10.1159/000347031 000347031 [pii]

Cheng S, Pollock AS, Mahimkar R, Olson JL, Lovett DH (2006) Matrix metalloproteinase 2 and basement membrane integrity: a unifying mechanism for progressive renal injury. FASEB J 20 (11):1898–1900. doi:fj.06-5898fje [pii] 10.1096/fj.06-5898fje

Chou CY, Streets AJ, Watson PF, Huang L, Verderio EA, Johnson TS (2011) A crucial sequence for transglutaminase type 2 extracellular trafficking in renal tubular epithelial cells lies in its N-terminal beta-sandwich domain. J Biol Chem 286(31):27825–27835. doi:10.1074/jbc.M111.226340 M111.226340 [pii]

Clayton A, Thomas J, Thomas GJ, Davies M, Steadman R (2001) Cell surface heparan sulfate proteoglycans control the response of renal interstitial fibroblasts to fibroblast growth factor-2. Kidney Int 59(6):2084–2094

Colvin RB (2003) Chronic allograft nephropathy. N Engl J Med 349(24):2288–2290. doi:10.1056/NEJMp038178 349/24/2288 [pii]

da Silva Lodge M ENM, Johnson TS (2013) Urinary transglutaminase 2 as a potential biomarker of chronic kidney disease detection and progression. Lancet 381(Supplement 1 (Spring Meeting for Clinician Scientists in Training, 27 February 2013)):S33 Poster Abstract

Dane M, van den Berg B, Rabelink T (2014) The endothelial glycocalyx: scratching the surface for cardiovascular disease in kidney failure. Atherosclerosis 235(1):56–57. doi:10.1016/j.atherosclerosis.2014.04.005 S0021-9150(14)00204-4 [pii]

De Laurenzi V, Melino G (2001) Gene disruption of tissue transglutaminase. Mol Cell Biol 21 (1):148–155. doi:10.1128/MCB.21.1.148-155.2001

Deasey S, Shanmugasundaram S, Nurminskaya M (2013) Tissue-specific responses to loss of transglutaminase 2. Amino Acids 44(1):179–187. doi:10.1007/s00726-011-1183-9

Devuyst O, Antignac C, Bindels RJ, Chauveau D, Emma F, Gansevoort R, Maxwell PH, Ong AC, Remuzzi G, Ronco P, Schaefer F (2012) The ERA-EDTA Working Group on inherited kidney disorders. Nephrol Dial Transplant 27(1):67–69. doi:10.1093/ndt/gfr764 gfr764 [pii]

Docherty NG, Calvo IF, Quinlan MR, Perez-Barriocanal F, McGuire BB, Fitzpatrick JM, Watson RW (2009) Increased E-cadherin expression in the ligated kidney following unilateral ureteric obstruction. Kidney Int 75(2):205–213. doi:10.1038/ki.2008.482 ki2008482 [pii]

Duffield JS (2014) Cellular and molecular mechanisms in kidney fibrosis. J Clin Invest 124 (6):2299–2306. doi:10.1172/JCI72267 72267 [pii]

Eckardt KU, Coresh J, Devuyst O, Johnson RJ, Kottgen A, Levey AS, Levin A (2013) Evolving importance of kidney disease: from subspecialty to global health burden. Lancet 382 (9887):158–169. doi:10.1016/S0140-6736(13)60439-0 S0140-6736(13)60439-0 [pii]

El Nahas AM, Abo-Zenah H, Skill NJ, Bex S, Wild G, Griffin M, Johnson TS (2004) Elevated epsilon-(gamma-glutamyl)lysine in human diabetic nephropathy results from increased expression and cellular release of tissue transglutaminase. Nephron Clin Pract 97(3):c108–c117. doi:10.1159/000078639 78639 [pii]

el-Khatib MT, Becker GJ, Kincaid-Smith PS (1987) Morphometric aspects of reflux nephropathy. Kidney Int 32(2):261–266

Fan Q, Shike T, Shigihara T, Tanimoto M, Gohda T, Makita Y, Wang LN, Horikoshi S, Tomino Y (2003) Gene expression profile in diabetic KK/Ta mice. Kidney Int 64(6):1978–1985

Fisher M, Jones RA, Huang L, Haylor JL, El Nahas M, Griffin M, Johnson TS (2009) Modulation of tissue transglutaminase in tubular epithelial cells alters extracellular matrix levels: a potential mechanism of tissue scarring. Matrix Biol 28(1):20–31. doi:10.1016/j.matbio.2008.10.003 S0945-053X(08)00732-4 [pii]

Freund KF, Doshi KP, Gaul SL, Claremon DA, Remy DC, Baldwin JJ, Pitzenberger SM, Stern AM (1994) Transglutaminase inhibition by 2-[(2-oxopropyl)thio]imidazolium derivatives: mechanism of factor XIIIa inactivation. Biochemistry 33(33):10109–10119

Furini G, Schroeder N, Huang L, Boocock D, Johnson TS, Verderio, EAM (2015) Quantitative proteomics by SWATH-MS reveals an endosomal transport hub of proteins which interact with TG2 in a model of experimental kidney fibrosis. Nephrol Dial Transplant 30(3 (Abstr)):FP013 (ERA-EDTA 052nd Congress, London, 28–31 May 2015)

Gansevoort RT, Correa-Rotter R, Hemmelgarn BR, Jafar TH, Heerspink HJ, Mann JF, Matsushita K, Wen CP (2013) Chronic kidney disease and cardiovascular risk: epidemiology, mechanisms, and prevention. Lancet 382(9889):339–352. doi:10.1016/S0140-6736(13)60595-4 S0140-6736(13)60595-4 [pii]

Grenard P, Bresson-Hadni S, El Alaoui S, Chevallier M, Vuitton DA, Ricard-Blum S (2001) Transglutaminase-mediated cross-linking is involved in the stabilization of extracellular matrix in human liver fibrosis. J Hepatol 35(3):367–375

Griffin M, Coutts IG, Saint R (2004) Novel compounds and methods of using the same. World Patent WO 2004/113363 A2

Griffin M, Wilson J (1984) Detection of epsilon(gamma-glutamyl) lysine. Mol Cell Biochem 58 (1–2):37–49

Griffin M, Smith LL, Wynne J (1979) Changes in transglutaminase activity in an experimental model of pulmonary fibrosis induced by paraquat. Br J Exp Pathol 60(6):653–661

Harrison CA, Layton CM, Hau Z, Bullock AJ, Johnson TS, MacNeil S (2007) Transglutaminase inhibitors induce hyperproliferation and parakeratosis in tissue-engineered skin. Br J Dermatol 156(2):247–257. doi:BJD7641 [pii] 10.1111/j.1365-2133.2006.07641.x

Herzog CA, Asinger RW, Berger AK, Charytan DM, Diez J, Hart RG, Eckardt KU, Kasiske BL, McCullough PA, Passman RS, DeLoach SS, Pun PH, Ritz E (2011) Cardiovascular disease in chronic kidney disease. A clinical update from Kidney Disease: Improving Global Outcomes (KDIGO). Kidney Int 80(6):572–586. doi:10.1038/ki.2011.223 ki2011223 [pii]

Hewitson TD (2009) Renal tubulointerstitial fibrosis: common but never simple. Am J Physiol Renal Physiol 296(6):F1239–F1244. doi:10.1152/ajprenal.90521.2008 90521.2008 [pii]

Hiiragi T, Sasaki H, Nagafuchi A, Sabe H, Shen SC, Matsuki M, Yamanishi K, Tsukita S (1999) Transglutaminase type 1 and its cross-linking activity are concentrated at adherens junctions in simple epithelial cells. J Biol Chem 274(48):34148–34154

Hsu CY, Iribarren C, McCulloch CE, Darbinian J, Go AS (2009) Risk factors for end-stage renal disease: 25-year follow-up. Arch Intern Med 169(4):342–350. doi:10.1001/archinternmed.2008.605 169/4/342 [pii]

Huang L, Haylor JL, Hau Z, Jones RA, Vickers ME, Wagner B, Griffin M, Saint RE, Coutts IG, El Nahas AM, Johnson TS (2009) Transglutaminase inhibition ameliorates experimental diabetic nephropathy. Kidney Int 76(4):383–394. doi:ki2009230 [pii] 10.1038/ki.2009.230

Huang L, Haylor JL, Fisher M, Hau Z, El Nahas AM, Griffin M, Johnson TS (2010) Do changes in transglutaminase activity alter latent transforming growth factor beta activation in experimental diabetic nephropathy? Nephrol Dial Transplant 25(12):3897–3910. doi:10.1093/ndt/gfq291. doi:10.1093/ndt/gfq291 gfq291 [pii]

Huang L, Scarpellini A, Funck M, Verderio EA, Johnson TS (2013) Development of a chronic kidney disease model in C57BL/6 mice with relevance to human pathology. Nephron Extra 3 (1):12–29. doi:10.1159/000346180 nne-0003-0012 [pii]

Iismaa SE, Mearns BM, Lorand L, Graham RM (2009) Transglutaminases and disease: lessons from genetically engineered mouse models and inherited disorders. Physiol Rev 89 (3):991–1023. doi:10.1152/physrev.00044.2008 89/3/991 [pii]

Ikee R, Kobayashi S, Hemmi N, Saigusa T, Namikoshi T, Yamada M, Imakiire T, Kikuchi Y, Suzuki S, Miura S (2007) Involvement of transglutaminase-2 in pathological changes in renal disease. Nephron Clin Pract 105(3):c139–c146. doi:10.1159/000098646. doi:000098646 [pii] 10.1159/000098646

Ishiguro K, Kadomatsu K, Kojima T, Muramatsu H, Tsuzuki S, Nakamura E, Kusugami K, Saito H, Muramatsu T (2000) Syndecan-4 deficiency impairs focal adhesion formation only under restricted conditions. J Biol Chem 275(8):5249–5252

Ishiguro K, Kadomatsu K, Kojima T, Muramatsu H, Iwase M, Yoshikai Y, Yanada M, Yamamoto K, Matsushita T, Nishimura M, Kusugami K, Saito H, Muramatsu T (2001) Syndecan-4 deficiency leads to high mortality of lipopolysaccharide-injected mice. J Biol Chem 276(50):47483–47488. doi:10.1074/jbc.M106268200 M106268200 [pii]

Johnson TS, Griffin M, Thomas GL, Skill J, Cox A, Yang B, Nicholas B, Birckbichler PJ, Muchaneta-Kubara C, Meguid El Nahas A (1997) The role of transglutaminase in the rat subtotal nephrectomy model of renal fibrosis. J Clin Invest 99(12):2950–2960

Johnson TS, Scholfield CI, Parry J, Griffin M (1998) Induction of tissue transglutaminase by dexamethasone: its correlation to receptor number and transglutaminase-mediated cell death in a series of malignant hamster fibrosarcomas. Biochem J 331(Pt 1):105–112

Johnson TS, Skill NJ, El Nahas AM, Oldroyd SD, Thomas GL, Douthwaite JA, Haylor JL, Griffin M (1999) Transglutaminase transcription and antigen translocation in experimental renal scarring. J Am Soc Nephrol 10(10):2146–2157

Johnson TS, El-Koraie AF, Skill NJ, Baddour NM, El Nahas AM, Njloma M, Adam AG, Griffin M (2003) Tissue transglutaminase and the progression of human renal scarring. J Am Soc Nephrol 14(8):2052–2062

Johnson TS, Abo-Zenah H, Skill JN, Bex S, Wild G, Brown CB, Griffin M, El Nahas AM (2004) Tissue transglutaminase: a mediator and predictor of chronic allograft nephropathy? Transplantation 77(11):1667–1675. doi:00007890-200406150-00009 [pii]

Johnson TS, Fisher M, Haylor JL, Hau Z, Skill NJ, Jones R, Saint R, Coutts I, Vickers ME, El Nahas AM, Griffin M (2007) Transglutaminase inhibition reduces fibrosis and preserves function in experimental chronic kidney disease. J Am Soc Nephrol 18(12):3078–3088. doi: ASN.2006070690 [pii] 10.1681/ASN.2006070690

Jones RA, Nicholas B, Mian S, Davies PJ, Griffin M (1997) Reduced expression of tissue transglutaminase in a human endothelial cell line leads to changes in cell spreading, cell adhesion and reduced polymerisation of fibronectin. J Cell Sci 110(Pt 19):2461–2472

K/DOQI (2002) K/DOQI clinical practice guidelines for chronic kidney disease: evaluation, classification, and stratification. Am J Kidney Dis 39(2 Suppl 1):S1–266. doi: S0272638602093563 [pii]

Kao WH, Klag MJ, Meoni LA, Reich D, Berthier-Schaad Y, Li M, Coresh J, Patterson N, Tandon A, Powe NR, Fink NE, Sadler JH, Weir MR, Abboud HE, Adler SG, Divers J, Iyengar SK, Freedman BI, Kimmel PL, Knowler WC, Kohn OF, Kramp K, Leehey DJ, Nicholas SB, Pahl MV, Schelling JR, Sedor JR, Thornley-Brown D, Winkler CA, Smith MW, Parekh RS (2008) MYH9 is associated with nondiabetic end-stage renal disease in African Americans. Nat Genet 40(10):1185–1192. doi:10.1038/ng.232 ng.232 [pii]

Kojima S, Nara K, Rifkin DB (1993) Requirement for transglutaminase in the activation of latent transforming growth factor-beta in bovine endothelial cells. J Cell Biol 121(2):439–448

Kopp JB, Nelson GW, Sampath K, Johnson RC, Genovese G, An P, Friedman D, Briggs W, Dart R, Korbet S, Mokrzycki MH, Kimmel PL, Limou S, Ahuja TS, Berns JS, Fryc J, Simon EE, Smith MC, Trachtman H, Michel DM, Schelling JR, Vlahov D, Pollak M, Winkler CA (2011) APOL1 genetic variants in focal segmental glomerulosclerosis and HIV-associated nephropathy. J Am Soc Nephrol 22(11):2129–2137. doi:10.1681/ASN.2011040388 ASN.2011040388 [pii]

Kuwabara A, Satoh M, Tomita N, Sasaki T, Kashihara N (2010) Deterioration of glomerular endothelial surface layer induced by oxidative stress is implicated in altered permeability of macromolecules in Zucker fatty rats. Diabetologia 53(9):2056–2065. doi:10.1007/s00125-010-1810-0

Levey AS, Coresh J (2012) Chronic kidney disease. Lancet 379(9811):165–180. doi:10.1016/S0140-6736(11)60178-5 S0140-6736(11)60178-5 [pii]

Lin SL, Kisseleva T, Brenner DA, Duffield JS (2008) Pericytes and perivascular fibroblasts are the primary source of collagen-producing cells in obstructive fibrosis of the kidney. Am J Pathol 173(6):1617–1627. doi:10.2353/ajpath.2008.080433 S0002-9440(10)61547-7 [pii]

Liu Y (2004) Epithelial to mesenchymal transition in renal fibrogenesis: pathologic significance, molecular mechanism, and therapeutic intervention. J Am Soc Nephrol 15(1):1–12

Liu Y (2010) New insights into epithelial-mesenchymal transition in kidney fibrosis. J Am Soc Nephrol 21(2):212–222. doi:10.1681/ASN.2008121226 ASN.2008121226 [pii]

Liu S, Li Y, Zhao H, Chen D, Huang Q, Wang S, Zou W, Zhang Y, Li X, Huang H (2006) Increase in extracellular cross-linking by tissue transglutaminase and reduction in expression of MMP-9 contribute differentially to focal segmental glomerulosclerosis in rats. Mol Cell Biochem 284 (1–2):9–17. doi:10.1007/s11010-005-9005-6

Lorand L, Graham RM (2003) Transglutaminases: crosslinking enzymes with pleiotropic functions. Nat Rev Mol Cell Biol 4(2):140–156. doi:10.1038/nrm1014 nrm1014 [pii]

Lortat-Jacob H, Burhan I, Scarpellini A, Thomas A, Imberty A, Vives RR, Johnson T, Gutierrez A, Verderio EA (2012) Transglutaminase-2 interaction with heparin: identification of a heparin binding site that regulates cell adhesion to fibronectin-transglutaminase-2 matrix. J Biol Chem. doi:M111.337089 [pii] 10.1074/jbc.M111.337089

Maezawa Y, Takemoto M, Yokote K (2015) Cell biology of diabetic nephropathy: roles of endothelial cells, tubulointerstitial cells and podocytes. J Diabetes Investig 6(1):3–15. doi:10.1111/jdi.12255

McCarthy KJ, Wassenhove-McCarthy DJ (2012) The glomerular basement membrane as a model system to study the bioactivity of heparan sulfate glycosaminoglycans. Microsc Microanal 18 (1):3–21. doi:10.1017/S1431927611012682 S1431927611012682 [pii]

Mengel M, Bock O, Priess M, Haller H, Kreipe H, Gwinner W (2008) Expression of pro- and antifibrotic genes in protocol biopsies from renal allografts with interstitial fibrosis and tubular atrophy. Clin Nephrol 69(6):408–416

Mirza A, Liu SL, Frizell E, Zhu J, Maddukuri S, Martinez J, Davies P, Schwarting R, Norton P, Zern MA (1997) A role for tissue transglutaminase in hepatic injury and fibrogenesis, and its regulation by NF-kappaB. Am J Physiol 272(2 Pt 1):G281–G288

Moe SM, Chen NX (2008) Mechanisms of vascular calcification in chronic kidney disease. J Am Soc Nephrol 19(2):213–216. doi:ASN.2007080854 [pii] 10.1681/ASN.2007080854

Moe SM, Drueke T, Lameire N, Eknoyan G (2007) Chronic kidney disease-mineral-bone disorder: a new paradigm. Adv Chronic Kidney Dis 14(1):3–12. doi:S1548-5595(06)00167-4 [pii] 10.1053/j.ackd.2006.10.005

Moe SM, Chen NX, Seifert MF, Sinders RM, Duan D, Chen X, Liang Y, Radcliff JS, White KE, Gattone VH 2nd (2009) A rat model of chronic kidney disease-mineral bone disorder. Kidney Int 75(2):176–184. doi:10.1038/ki.2008.456 ki2008456 [pii]

Morita H, Shinzato T, David G, Mizutani A, Habuchi H, Fujita Y, Ito M, Asai J, Maeda K, Kimata K (1994) Basic fibroblast growth factor-binding domain of heparan sulfate in the human glomerulosclerosis and renal tubulointerstitial fibrosis. Lab Invest 71(4):528–535

Nadella V, Wang Z, Johnson TS, Griffin M, Devitt A (2015) Transglutaminase 2 interacts with syndecan-4 and CD44 at the surface of human macrophages to promote removal of apoptotic cells. Biochim Biophys Acta 1853(1):201–212. doi:10.1016/j.bbamcr.2014.09.020 S0167-4889(14)00345-0 [pii]

Nakagawa T, Tanabe K, Croker BP, Johnson RJ, Grant MB, Kosugi T, Li Q (2011) Endothelial dysfunction as a potential contributor in diabetic nephropathy. Nat Rev Nephrol 7(1):36–44. doi:10.1038/nrneph.2010.152 nrneph.2010.152 [pii]

Nicholas B, Smethurst P, Verderio E, Jones R, Griffin M (2003) Cross-linking of cellular proteins by tissue transglutaminase during necrotic cell death: a mechanism for maintaining tissue integrity. Biochem J 371(Pt 2):413–422. doi:10.1042/BJ20021949 BJ20021949 [pii]

Nieuwdorp M, Mooij HL, Kroon J, Atasever B, Spaan JA, Ince C, Holleman F, Diamant M, Heine RJ, Hoekstra JB, Kastelein JJ, Stroes ES, Vink H (2006) Endothelial glycocalyx damage coincides with microalbuminuria in type 1 diabetes. Diabetes 55(4):1127–1132. doi:55/4/1127 [pii]

Nunes I, Gleizes PE, Metz CN, Rifkin DB (1997) Latent transforming growth factor-beta binding protein domains involved in activation and transglutaminase-dependent cross-linking of latent transforming growth factor-beta. J Cell Biol 136(5):1151–1163

Olsen KC, Sapinoro RE, Kottmann RM, Kulkarni AA, Iismaa SE, Johnson GV, Thatcher TH, Phipps RP, Sime PJ (2011) Transglutaminase 2 and its role in pulmonary fibrosis. Am J Respir Crit Care Med 184(6):699–707. doi:10.1164/rccm.201101-0013OC 201101-0013OC [pii]

Papasotiriou M, Kalliakmani P, Huang L, Gerolymos M, Goumenos DS, Johnson TS (2012) Does treatment with corticosteroids and cyclosporine reduce transglutaminase type 2 expression in the renal tissue of patients with membranous nephropathy? Nephron Clin Pract 121(1–2):c60–c67. doi:10.1159/000341116 000341116 [pii]

Parsons AC, Yosipovitch G, Sheehan DJ, Sangueza OP, Greenberg CS, Sane DC (2007) Transglutaminases: the missing link in nephrogenic systemic fibrosis. Am J Dermatopathol 29 (5):433–436. doi:10.1097/DAD.0b013e318156e43f 00000372-200710000-00001 [pii]

Ponnusamy M, Pang M, Annamaraju PK, Zhang Z, Gong R, Chin YE, Zhuang S (2009) Transglutaminase-1 protects renal epithelial cells from hydrogen peroxide-induced apoptosis through activation of STAT3 and AKT signaling pathways. Am J Physiol Renal Physiol 297 (5):F1361–F1370. doi:10.1152/ajprenal.00251.2009 00251.2009 [pii]

Rienstra H, Katta K, Celie JW, van Goor H, Navis G, van den Born J, Hillebrands JL (2010) Differential expression of proteoglycans in tissue remodeling and lymphangiogenesis after experimental renal transplantation in rats. PLoS One 5(2):e9095. doi:10.1371/journal.pone.0009095

Rosen SD, Lemjabbar-Alaoui H (2010) Sulf-2: an extracellular modulator of cell signaling and a cancer target candidate. Expert Opin Ther Targets 14(9):935–949. doi:10.1517/14728222.2010.504718

Sanchez-Lara AC, Elliott J, Syme HM, Brown CA, Haylor JL (2014) Feline chronic kidney disease is associated with upregulation of transglutaminase 2: a collagen cross-linking enzyme. Vet Pathol. doi:0300985814542811 [pii]. 10.1177/0300985814542811

Scarpellini A, Germack R, Lortat-Jacob H, Muramatsu T, Billett E, Johnson T, Verderio EA (2009) Heparan sulfate proteoglycans are receptors for the cell-surface trafficking and biological activity of transglutaminase-2. J Biol Chem 284(27):18411–18423. doi:10.1074/jbc.M109.012948 M109.012948 [pii]

Scarpellini A, Huang L, Burhan I, Schroeder N, Funck M, Johnson TS, Verderio EA (2014) Syndecan-4 knockout leads to reduced extracellular transglutaminase-2 and protects against tubulointerstitial fibrosis. J Am Soc Nephrol 25(5):1013–1027. doi:10.1681/ASN.2013050563 ASN.2013050563 [pii]

Schittny JC, Paulsson M, Vallan C, Burri PH, Kedei N, Aeschlimann D (1997) Protein cross-linking mediated by tissue transglutaminase correlates with the maturation of extracellular matrices during lung development. Am J Respir Cell Mol Biol 17(3):334–343. doi:10.1165/ajrcmb.17.3.2737

Schuksz M, Fuster MM, Brown JR, Crawford BE, Ditto DP, Lawrence R, Glass CA, Wang L, Tor Y, Esko JD (2008) Surfen, a small molecule antagonist of heparan sulfate. Proc Natl Acad Sci U S A 105(35):13075–13080. doi:10.1073/pnas.0805862105 0805862105 [pii]

Sharma K (2014) Obesity, oxidative stress, and fibrosis in chronic kidney disease. Kidney Int Suppl (2011) 4(1):113–117. doi:10.1038/kisup.2014.21

Shrestha B, Haylor J (2014) Experimental rat models of chronic allograft nephropathy: a review. Int J Nephrol Renov Dis 7:315–322. doi:10.2147/IJNRD.S65604 ijnrd-7-315 [pii]

Shrestha B, Butt I, Da Silva M, Sanchez-Lara A, Wagner B, Raftery A, Johnson T, Haylor J (2014) Upregulation of transglutaminase and epsilon (gamma-glutamyl)-lysine in the Fisher-Lewis rat model of chronic allograft nephropathy. Biomed Res Int 2014:651608. doi:10.1155/2014/651608

Shweke N, Boulos N, Jouanneau C, Vandermeersch S, Melino G, Dussaule JC, Chatziantoniou C, Ronco P, Boffa JJ (2008) Tissue transglutaminase contributes to interstitial renal fibrosis by favoring accumulation of fibrillar collagen through TGF-beta activation and cell infiltration. Am J Pathol 173(3):631–642. doi:S0002-9440(10)61637-9 [pii] 10.2353/ajpath.2008.080025

Skill NJ, Griffin M, El Nahas AM, Sanai T, Haylor JL, Fisher M, Jamie MF, Mould NN, Johnson TS (2001) Increases in renal epsilon-(gamma-glutamyl)-lysine crosslinks result from compartment-specific changes in tissue transglutaminase in early experimental diabetic nephropathy: pathologic implications. Lab Invest 81(5):705–716

Skill NJ, Johnson TS, Coutts IG, Saint RE, Fisher M, Huang L, El Nahas AM, Collighan RJ, Griffin M (2004) Inhibition of transglutaminase activity reduces extracellular matrix accumulation induced by high glucose levels in proximal tubular epithelial cells. J Biol Chem 279 (46):47754–47762. doi:10.1074/jbc.M402698200 M402698200 [pii]

Small K, Feng JF, Lorenz J, Donnelly ET, Yu A, Im MJ, Dorn GW 2nd, Liggett SB (1999) Cardiac specific overexpression of transglutaminase II (G(h)) results in a unique hypertrophy phenotype independent of phospholipase C activation. J Biol Chem 274(30):21291–21296

Smethurst PA, Griffin M (1996) Measurement of tissue transglutaminase activity in a permeabilized cell system: its regulation by Ca2+ and nucleotides. Biochem J 313 (Pt 3):803–808

Strutz F, Okada H, Lo CW, Danoff T, Carone RL, Tomaszewski JE, Neilson EG (1995) Identification and characterization of a fibroblast marker: FSP1. J Cell Biol 130(2):393–405

Takahashi KMT, Onouchi T, Akiyama S, Hall SD, Tatsukawa H, Kusaka M, Tsutsumi Y, Nagamatsu T, Maruyama S, Kitamura H, Novak J, Hitomi K, Yuzawa Y (2014) Mesangial tissue transglutaminase activity in patients with IgA nephropathy: implications for the pathogenesis. J Am Soc Nephrol 25:909

Teesalu K, Panarina M, Uibo O, Uibo R, Utt M (2012) Autoantibodies from patients with celiac disease inhibit transglutaminase 2 binding to heparin/heparan sulfate and interfere with intestinal epithelial cell adhesion. Amino Acids 42(2–3):1055–1064. doi:10.1007/s00726-011-1020-1

Telci D, Wang Z, Li X, Verderio EA, Humphries MJ, Baccarini M, Basaga H, Griffin M (2008) Fibronectin-tissue transglutaminase matrix rescues RGD-impaired cell adhesion through syndecan-4 and beta1 integrin co-signaling. J Biol Chem 283(30):20937–20947. doi:10.1074/jbc.M801763200. doi:M801763200 [pii] 10.1074/jbc.M801763200

Verderio E, Scarpellini A (2010) Significance of the syndecan-4-transglutaminase-2 interaction. ScientificWorldJournal 10:1073–1077. doi:10.1100/tsw.2010.102

Verderio E, Nicholas B, Gross S, Griffin M (1998) Regulated expression of tissue transglutaminase in Swiss 3T3 fibroblasts: effects on the processing of fibronectin, cell attachment, and cell death. Exp Cell Res 239(1):119–138. doi:S0014-4827(97)93874-X [pii] 10.1006/excr.1997.3874

Verderio E, Gaudry C, Gross S, Smith C, Downes S, Griffin M (1999) Regulation of cell surface tissue transglutaminase: effects on matrix storage of latent transforming growth factor-beta binding protein-1. J Histochem Cytochem 47(11):1417–1432

Verderio EA, Telci D, Okoye A, Melino G, Griffin M (2003) A novel RGD-independent cel adhesion pathway mediated by fibronectin-bound tissue transglutaminase rescues cells from anoikis. J Biol Chem 278(43):42604–42614. doi:10.1074/jbc.M303303200 M303303200 [pii]

Verderio EA, Scarpellini A, Johnson TS (2009) Novel interactions of TG2 with heparan sulfate proteoglycans: reflection on physiological implications. Amino Acids 36(4):671–677. doi:10.1007/s00726-008-0134-6

Vielhauer V, Anders HJ, Mack M, Cihak J, Strutz F, Stangassinger M, Luckow B, Grone HJ, Schlondorff D (2001) Obstructive nephropathy in the mouse: progressive fibrosis correlates with tubulointerstitial chemokine expression and accumulation of CC chemokine receptor 2- and 5-positive leukocytes. J Am Soc Nephrol 12(6):1173–1187

Wang Z, Griffin M (2012) TG2, a novel extracellular protein with multiple functions. Amino Acids 42(2–3):939–949. doi:10.1007/s00726-011-1008-x

Wang Z, Telci D, Griffin M (2011) Importance of syndecan-4 and syndecan -2 in osteoblast cell adhesion and survival mediated by a tissue transglutaminase-fibronectin complex. Exp Cell Res 317(3):367–381. doi:10.1016/j.yexcr.2010.10.015 S0014-4827(10)00481-7 [pii]

Wang Z, Collighan RJ, Pytel K, Rathbone DL, Li X, Griffin M (2012) Characterization of heparin-binding site of tissue transglutaminase: its importance in cell surface targeting, matrix deposition, and cell signaling. J Biol Chem 287(16):13063–13083. doi:10.1074/jbc.M111.294819 M111.294819 [pii]

Yung S, Woods A, Chan TM, Davies M, Williams JD, Couchman JR (2001) Syndecan-4 up-regulation in proliferative renal disease is related to microfilament organization. FASEB J 15(9):1631–1633

Zehe C, Engling A, Wegehingel S, Schafer T, Nickel W (2006) Cell-surface heparan sulfate proteoglycans are essential components of the unconventional export machinery of FGF-2.

Proc Natl Acad Sci U S A 103(42):15479–15484. doi:0605997103 [pii] 10.1073/pnas.0605997103

Zeisberg EM, Potenta SE, Sugimoto H, Zeisberg M, Kalluri R (2008) Fibroblasts in kidney fibrosis emerge via endothelial-to-mesenchymal transition. J Am Soc Nephrol 19(12):2282–2287. doi:10.1681/ASN.2008050513 ASN.2008050513 [pii]

Zhang Z, Xing J, Ma L, Gong R, Chin YE, Zhuang S (2009) Transglutaminase-1 regulates renal epithelial cell proliferation through activation of Stat-3. J Biol Chem 284(5):3345–3353. doi:10.1074/jbc.M808396200 M808396200 [pii]

Zhao G, Zhang ZQ, Zhang B, Luo M, Sun YW, Wu ZY (2011) Down-regulation of tTG expression by RNAi inhibits HSC proliferation and attenuates liver fibrosis. Int J Clin Exp Pathol 4(5):513–520

Chapter 12
Transglutaminases in Bone Formation and Bone Matrix Stabilization

Cui Cui and Mari T. Kaartinen

Abstract Bone is formed throughout a life span of an individual by osteoblasts in a process termed osteogenesis. Osteoblasts arise from mesenchymal stem cells and form type I collagen and plasma fibronectin rich extracellular matrix that mineralizes in the high phosphate concentration that is generated by alkaline phosphatase activity. Matrix quantity, quality and mineralization levels define the strength of bone. This chapter discusses the process of osteogenesis and the potential role of transglutaminases in this process. Osteoblasts express two transglutaminase enzymes, transglutaminase 2 and Factor XIII-A, which based on our current understanding appear to regulate osteoblast adhesion and their matrix deposition, respectively. Transglutaminase 2 acts on the cell surface and partners with FN (fibronectin) to promote integrin-mediated adhesion in a transamidase-independent manner. Factor XIII-A is mostly involved in promotion of extracellular matrix production and stabilization, the main substrate being plasma fibronectin. We will also briefly discuss the role of the enzymes in adipogenesis, the potential effect of serotonylation reaction to bone matrix quality and what transglutaminase mouse models have taught us on the enzymes in bone in vivo.

Keywords Bone • Bone formation • Osteogenesis • Mesenchymal stem cell • Osteoblast • Extracellular matrix stabilization • Transglutaminase 2 • Factor XIII-A • Plasma fibronectin

---

C. Cui
Division of Experimental Medicine, Department of Medicine, Faculty of Medicine, McGill University, Montreal, QC, Canada

School of Stomatology, Binzhou Medical University, 346 Guanhai Road, Laishan District, Yantai, Shandong 264003, China

M.T. Kaartinen, Ph.D. (✉)
Division of Experimental Medicine, Department of Medicine, Faculty of Medicine, McGill University, Montreal, QC, Canada

Division of Biomedical Sciences, Faculty of Dentistry, McGill University, 3640 University Street, Montreal, QC H3A 0C7, Canada
e-mail: mari.kaartinen@mcgill.ca

K. Hitomi et al. (eds.), *Transglutaminases*, DOI 10.1007/978-4-431-55825-5_12

12.1 Bone

Bone is a multifunctional mineralized connective tissue that protects the internal organs, provides locomotive function, acts as the storage site for calcium and phosphate, and serves as an endocrine organ and a maturation place for hematopoiesis (Karaplis 2002). Bone is formed by two distinct processes, endochondral ossification and intramembranous ossification. Endochondral ossification is a process involving cartilage formation which is followed by its degradation and replacement by bone (Ortega et al. 2004). In this process cartilage serves as a template for bone. Chondrocytes signal to osteoblasts and promote their differentiation (Wang et al. 2012a). Intramembranous bone is formed in the absence of cartilaginous anlagen and involves mesenchymal stem cell condensation and their direct differentiation into osteoblasts, which induces bone formation at the center of the cellular condensation (Thompson et al. 2002). Flat bones (the skull and scapulae) are created via intramembranous ossification, whereas long bones develop and grow via endochondral ossification (Karaplis 2002; Ortega et al. 2004; Thompson et al. 2002).

Bone is replenished postnatally through remodeling cycles that occur throughout the life span of vertebrates (Epstein et al. 1995). Bone remodeling involves two cellular phases: bone resorption, which is carried out by osteoclasts, and bone formation, which is orchestrated by osteoblasts (Seeman 2008). These events take place simultaneously throughout the skeleton. In the adult human skeleton, the resorption phase takes approximately 3 weeks and formation requires 3–4 months. Around 10 % of the skeleton is replaced annually (Beaupré et al. 1990). In order to maintain appropriate bone mass, bone remodeling must be balanced, i.e., resorbed bone should be replaced by an equal amount of new bone formation. Imbalance between these two phases leads to bone loss (osteopenia) and increased osteoporotic bone fracture risk (Epstein et al. 1995; Lerner 2006; Mundy 1993). Osteopenia/osteoporosis is often accompanied by increased adiposity in the bone marrow (Justesen et al. 2001). This review will summarize our current knowledge on TGs in bone formation and in bone matrix stabilization orchestrated by osteoblasts.

12.1.1 *Osteoblasts*

Osteoblasts arise from mesenchymal stem cells. These multipotent progenitors are capable of differentiating into many cell types such as osteoblasts, adipocytes, chondrocytes, myocytes, neurons, tendon tenocytes, and marrow stromal cells (Bianco et al. 2008; Pittenger et al. 1999) (Fig. 12.1). Osteoblasts are histologically recognized as cuboidal mononuclear cells residing on the bone surface, adhering to the extracellular matrix they have synthesized (osteoid). Bone formation involves osteoblast proliferation followed by type I collagen (COL I) matrix synthesis and

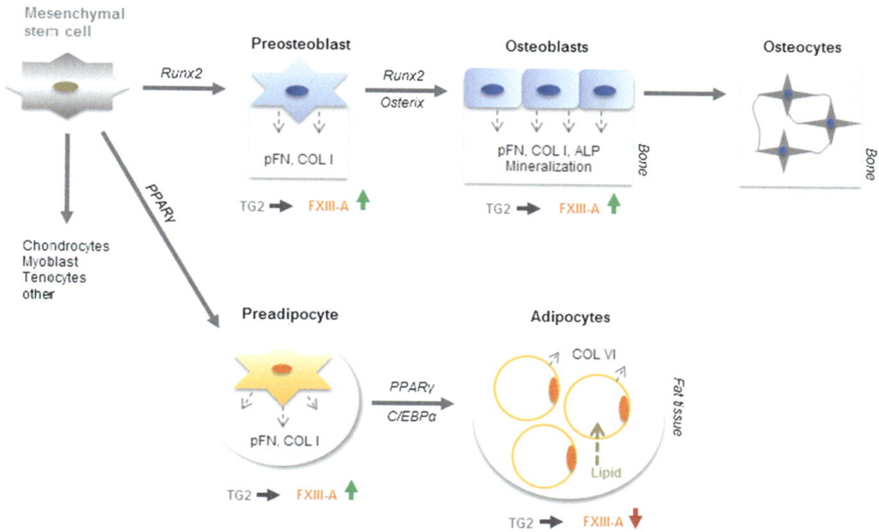

Fig. 12.1 Mesenchymal stem cell differentiation to osteoblasts and adipocytes and expression of TG2 and FXIII-A during the processes. Mesenchymal stem cells (MSCs) give rise to a number of cell types. Differentiation to preosteoblasts requires transcription factor Runx2 and maturation to osteoblasts is under the control of transcription factor Osterix. Preosteoblasts begin plasma FN (pFN) assembly and COL I production which both promote the maturation process. Mature osteoblasts begin expression of alkaline phosphatase (ALP) which initiates mineralization of the matrix. TG2 is expressed in both preosteoblasts and osteoblasts at steady levels and it likely mediates cellular adhesion to the surrounding matrix. FXIII-A is upregulated in preosteoblasts and osteoblasts and is responsible for the transamidase/crosslinking activity, based on an MC3T3-E1 osteoblast model where it promotes pFM assembly and COL I production in vitro. Inhibition of TG activity in osteoblast cultures results in decreased pFN matrix assembly, decreased COL I production, ALP activity and delayed mineralization making TG activity a positive regulator of osteogenesis in vitro. Adipocyte differentiation from preadipocytes involves coordinated action of the transcription factors PPARγ and C/EBPα to initiate lipid intake to cells. Preadipocytes proliferate in pFN and COL I matrix which is remodeled to allow major morphological and cytoskeletal changes that are also required for differentiation. The main matrix constituent of mature adipocytes is COL IV. TG activity, which arises from FXIII-A, is upregulated in preadipocytes and directed to assemble pFN for proliferation. FXIII-A is downregulated in mature adipocytes. Inhibition of TG activity results in decreased pFN matrix assembly and increased lipid accumulation in vitro, making TG activity a negative regulator of adipogenesis. This summary figure is based on references found in the text

maturation, and induction of alkaline phosphatase (ALP), which all lead to COL I matrix mineralization (Stein et al. 1989). With the basophilic cytoplasm and the outstanding Golgi apparatus, mature osteoblasts are capable of synthesizing high levels of protein. They produce large quantities of COL I as well as ALP and non-collagenous proteins including osteocalcin, osteopontin, bone sialoprotein and proteoglycans. These cell-secreted proteins, together with proteins from the extracellular fluid environment like fibronectin (FN), assemble into an organized matrix, which is suitable for allowing cell differentiation and hydroxyapatite deposition in the matrix template. A portion of osteoblasts differentiate into osteocytes and get buried into the bone matrix. Osteocytes serve a mechanosensory function and communicate with each other and with osteoblasts on the cell surface and regulate osteoclastogenesis (Bonewald 2006; Matsuo and Irie 2008).

Genetic studies with mice have advanced our understanding of the transcriptional control of osteoblast differentiation. Osteoblast differentiation is accomplished by the coordinated transcriptional activity of runt-related transcription factor 2/core-binding factor alpha 1 (Runx2/Cbfa1) (Fig. 12.1), which is a critical transcriptional regulator for the transition of mesenchymal stem cells to osteoprogenitor/preosteoblast cells. Runx2 null mice die at birth because they completely fail to form osteoblasts and thus bone (Karsenty 2001). Downstream of Runx2 is the zinc finger-containing transcription factor Osterix. Osterix is not expressed in Runx2 null mice. Osterix null mice show Runx2 expression, but do not show bone formation. This supports the concept that they regulate transition of osteoprogenitors to osteoblasts in a sequential manner (Nakashima et al. 2002) (Fig. 12.1). Osteoblasts share a common bipotential progenitor cell with adipocytes. In contrast to osteoblasts, adipocyte differentiation requires the expression of the key transcriptional regulators peroxisome proliferator-activated receptor gamma (PPARγ) (Barak et al. 1999) and CCAAT/enhancer binding protein alpha (C/EBPα) (Fig. 12.1) (Tanaka et al. 1997).

12.1.2 *Bone Matrix*

In the vertebrate skeleton, bone tissue created by osteoblasts comprises about 70 % inorganic mineral, 25 % organic matrix and 5 % water (Robey and Boskey 2008). The mineral crystals in bone are analogous to the natural geologic mineral hydroxyapatite ($Ca_{10}(PO_4)_6(OH)_2$) (Awonusi et al. 2007). The organic matrix comprises 90 % COL I and 10 % noncollagenous proteins. The mineral component contributes to the mechanical strength of the bone, which is required for structural support and protection of the organs, in addition to mechanical integrity for locomotion. It also serves as a reservoir involved in the metabolic pathways related to mineral homeostasis. The fibrillar collagen matrix acts as a scaffold to allow hydroxyapatite deposition and provides tensile strength to mineralized bone. The noncollagenous proteins have a number of functions in bone, which are not limited to the modulation of mineralization (Boskey 1989). Here we discuss these components in detail.

12.1.2.1 Inorganic Mineral

Bone mineralization is a carefully orchestrated process. Mineralization is initiated by calcium and inorganic phosphate accumulated in the matrix vesicles (MVs) to start the hydroxyapatite (HA) nucleation and deposition (Anderson 1969; 1995; Ali et al. 1970; Anderson et al. 1997, 2005; Register et al. 1986; Golub 2009). MVs bud from the osteoblast plasma membrane and, once in the extracellular matrix, they break and release the nanoscale HA crystals to the collagenous extracellular matrix microenvironment. This leads to HA growth and propagation in the extracellular matrix (Anderson 1995; Anderson et al. 2005). Experimental studies show that HA crystals deposit along the COL I fibrils (Glimcher 2006).

Other critical determinants of mineralization have been identified. ALP is the main enzyme, which raises extracellular phosphate ion concentration and allows HA growth (Millan 2013). ALP also promotes mineralization by eliminating the extracellular inorganic pyrophosphate (PPi) – which is a major mineralization inhibitor – by cleaving it into two phosphate molecules. PPi cleavage by ALP maintains the proper PPi/Pi ratio for HA crystal growth along the collagen fibrils (Johnson et al. 2000; Hessle et al. 2002; Murshed et al. 2005). ALP deficiency in mice results in the accumulation of extracellular PPi, which leads to the hypophosphatasia phenotype characterized as rickets and osteomalacia in humans (Whyte 2001; Narisawa et al. 1997; Fedde et al. 1999). Coexpression of ALP and a collagenous extracellular matrix scaffold is necessary for the progression of bone mineralization (Murshed et al. 2005). Also, osteoblasts do not produce ALP if COL I matrix is not present (Franceschi and Iyer 1992; Franceschi et al. 1994).

12.1.2.2 Organic Matrix

COL I is the main component of the bone organic matrix which acts as a scaffold for mineralization (Wassen et al. 2000). COL I is a triple-helical protein molecule approximately 300 nm in length. It contains two identical α1 chains and one α2 chain[$\alpha1(I)_2\alpha2(I)$], which are encoded by the *COL1A1* and *COL1A2* genes, respectively (Brodsky and Ramshaw 1997; Myllyharju and Kivirikko 2004). The α-chain amino acid sequence has a Gly-X-Y repeating module (X represents proline and Y represents hydroxyproline, commonly) which is critical for the triple-helix formation and stability. Collagen synthesis is a multistep process that involves intracellular as well as extracellular processing and assembly. COL I α-chain assembly into a triple-helical structure begins inside the cells, when the C-terminal ends of the three α-chains form intra- and intermolecular disulfide bonds, which initiate the formation of the helical procollagen. This procollagen is then transferred to the Golgi apparatus, packaged into the transport vesicles, and destined for the extracellular space (Myllyharju and Kivirikko 2004).

Once the procollagen is secreted to the extracellular space, membrane-bound collagen peptidases such as ADAMTS-2, 3, 14 and bone morphogenetic protein (BMP-1) cleave N- and C-terminal propeptides off the procollagen molecule. This creates tropocollagen molecules, which spontaneously arrange into a quarter-staggered array to form COL I fibrils. Lysyl oxidase (LOX) is required in the final step of COL I assembly. LOX mediates oxidation/deamination of lysines and hydroxylysines to form (hydroxy)allysines which then spontaneously form pyrrole and pyridinoline crosslinks between neighboring (hydroxy)allysines. The crosslinks occurring between tropocollagen molecules within the fibril create a highly stable COL I. These crosslinking events also condense and tighten the COL I fibrils and alter their rigidity, which is critical for the function and structural integrity of the tissues (Myllyharju and Kivirikko 2004; Kagan and Li 2003).

12.1.2.2.1 COL I and Bone Strength

For its locomotive function, bone must be able to resist mechanical forces. Bone strength depends on the quantity and quality of the bone. The shape and the geometry of the bone, the three-dimensional microarchitecture of the trabecular bone, and the bone turnover rate, as well as the properties of the mineral and the COL I matrix, all contribute to bone strength. The mineral provides the hardness (Currey 1988; Currey et al. 1996) and the COL I fibers provide bone with tensile strength. Studies on variations in bone strength show that the quality of the COL I matrix is related to the difference in bone strength (Boskey et al. 1999; Viguet-Carrin et al. 2006; Zioupos et al. 1999; Thompson et al. 2001). COL I contributes to 95 % of bone collagen content (Niyibizi and Eyre 1994). Genetic mutations of COL I like osteogenesis imperfecta lead to defective COL I fiber formation, which results in abnormal mineralization and fragile bones (Baum and Brodsky 1999). The organization and structure of COL I fibrils control the orientation and the size of the hydroxyapatite crystals. Woven bone is mechanically weaker than lamellar bone because of unorganized collagen fibrils, although the mineral content in the former is higher than in the latter (Weiner et al. 1999).

Studies on the molecular mechanisms suggest that COL I posttranslational modifications may be involved in regulating mechanical properties of COL I. Animals treated with vitamin B_6, copper-deficient diets or β-aminopropionitrile to inhibit LOX activity display a disorder defined as lathyrism, which shows reduced bone stiffness and strength (Oxlund et al. 1995; Massé et al. 1996; Opsahl et al. 1982). Therefore, the appropriate levels of LOX activity is important for bone strength and biomechanical properties.

12.1.2.2.2 Fibronectin

Fibronectin (FN) is a widely expressed extracellular matrix glycoprotein; it mediates a variety of cellular activities including cell adhesion, proliferation, migration and differentiation (Pankov and Yamada 2002; Wierzbicka-Patynowski and Schwarzbauer 2003; Mosher 1989; Magnusson and Mosher 1998; Hynes 1990). It plays a critical role in vertebrate development, as shown by the embryonic lethality in the global knockout of FN in mice (George et al. 1993). FN gene *(Fn)* encodes two forms of FN protein – plasma FN (pFN) and cellular FN (cFN) – which arise from alternative splicing of precursor mRNA. pFN is synthesized by hepatocytes in the liver and circulates in the blood at high concentrations (300 μg/ml in the human plasma; 600 μg/ml in the mouse plasma) (Hynes 1990). Cellular FN (cFN), produced by tissue-resident cells, contains extra type III domain A (EDA) and/or

B (EDB) which are not found in pFN. Accumulating evidence shows that different FN forms have different functions. cFN is required for embryogenesis, tissue remodeling and the late stage of wound healing while pFN is critical for thrombosis in the early stage of wound healing, extracellular matrix deposition, and immune responses (To and Midwood 2011).

FN is secreted as a soluble disulfide-bonded dimer containing three types of repeating modules (I, II and III). Sets of modules constitute different FN domains which bind to a number of biologically important cell surface and extracellular molecules/receptors, including integrin, collagen/gelatin, fibrin, and heparin. Several modules contain self-assembly sequences which are important for FN fibrillogenesis (Pankov and Yamada 2002). Soluble FN is in a globular, folded conformation, which is not capable of assembling into fibrils. FN fibrillogenesis involves cellular activity where FN binds to cell surface integrin receptor ($\alpha5\beta1$) via its RGD sequence which results in pulling of FN via cellular stretching and tension. The pulling opens the FN molecular structure which in turn exposes its self-assembly sites leading to fibril formation (Wierzbicka-Patynowski and Schwarzbauer 2003). The initial fibrils are deoxycholic acid-soluble (DOC-soluble) which are then converted to stable and detergent-insoluble (DOC-insoluble) FN fibrils. These two types of fibrils form the FN matrix network. The FN fibrillar network can act as a scaffold for further deposition of other matrix components, such as COL I and III, fibrinogen, fibrillin, fibulin, laminin and tenascin-C, as demonstrated in several cell types (Sottile and Hocking 2002; Sottile et al. 2007; Sabatier et al. 2009; Malara et al. 2011; Godyna et al. 1995; Chung et al. 1995).

12.1.2.2.3 FN Matrix Assembly and Bone Formation

Advances in osteoblast culture work and transgenic FN mouse models have shown a critical role for FN in osteoblast adhesion, proliferation, differentiation, and bone matrix quality (Sottile and Hocking 2002; Sottile et al. 2007; Moursi et al. 1996, 1997). FN fibrillogenesis is required for extracellular matrix deposition and stabilization, especially COL I in osteoblast cultures. Furthermore, the continuous supply of FN in the extracellular matrix is critical for matrix integrity (Sottile and Hocking 2002), i.e., FN matrix in several cell types is transient and highly dynamic and it is quickly endocytosed if pFN is not present in cell culture media (Sottile et al. 2007).

As outlined above, pFN is synthesized by hepatocytes in the liver, and it was recently shown to accumulate in the osteoblast extracellular matrix and facilitate COL I assembly for osteoblast mineralization. The effect of the FN network on bone quality may be due to its role as a scaffolding protein for COL I deposition. FN matrix was also shown to regulate bone morphogenic protein-1 (BMP-1) and LOX activity in vitro, both of which are critical for COL I deposition (Huang et al. 2009; Dallas et al. 2005). These all suggest pFN plays a vital role in bone formation.

Conditional knockout of the *Fn* gene in the liver in mice impaired bone matrix quality and decreased the bone mineral-to-matrix ratio (Bentmann et al. 2010).

Although osteoblasts can synthesize and secrete cFN, conditional *Fn* deletion in osteoblasts did not affect bone quality, nor altered FN protein levels in bone, demonstrating that the major portion of bone FN matrix is from circulating pFN. The *Fn* gene deletion in osteoblasts resulted in an increase in the number of differentiating osteoblasts, however, the authors showed that this did not result in more differentiated osteoblasts, because the new bone formation rate was not increased (Bentmann et al. 2010). Thus only pFN appears to have an impact on bone quality. As outlined above, it is well known that FN assembly is an integrin-mediated cell-dependent process. Five integrins are expressed on osteoblasts. They are α5β1, α8β1, αvβ1, αvβ3, and αvβ5; the latter one is capable of binding to RGD in the cell-binding module of FN (Moursi et al. 1997; Grzesik and Robey 1994; Cheng et al. 2000; Pistone et al. 1996; Saito et al. 1994). However, a transgenic mouse harboring an osteoblast specific mutation on the *Fn* gene that changes its cell-binding sequence RGD to RGE did not show any bone formation defects compared to control (Bentmann et al. 2010). Thus, FN-integrin binding via the RGD sequence in osteoblasts is not critical for its function.

Overall, FN is a major component in the noncollagenous bone matrix. Circulating pFN contributes to the majority of bone matrix and affects bone matrix quality. FN fibrillogenesis is required for osteoblast proliferation and differentiation, COL I deposition and bone mineralization in vitro.

12.2 The Expression of TGs in Bone

Only two TG family members, TG2 and FXIII-A, have been reported to exist in the cartilage and bone. TG2 was the first identified TG expressed in developing cartilage in rats (Aeschlimann et al. 1993), followed by FXIII-A detection in the avian embryonic growth plate (Aeschlimann et al. 1996). TG expression in chondrocytes is ordered spatially. The expression levels of TG2 and FXIII-A are downregulated in the proliferative and resting chondrocytes but are higher in the hypertrophic zone (Nurminskaya et al. 1998, 2002, 2003), where the chondrocytes are located at the border of the cartilaginous tissue and bony collar of endochondral bone formation.

TG2 and FXIII-A are also expressed both in mouse long bone in vivo and in osteoblasts in cultures in vitro (Kaartinen et al. 2002; Nakano et al. 2007; Al-Jallad et al. 2006). Both of them are found in the osteoblasts and in the osteoid layer, and colocalized on the osteoblast surface in the long bone (Al-Jallad et al. 2011). TG2 and FXIII-A are also found in the pericellular matrix of osteocytes (Kaartinen et al. 2002).

12.2.1 *The Expression of TGs by Osteoblasts In Vitro*

TG2 was the first TG enzyme described as being expressed in human osteoblast cultures and found both on the cell surface and in the extracellular matrix. It was reported that its crosslinking activity contributes to regulation of osteoblast adhesion and stabilization of osteoblast extracellular matrix (Heath et al. 2001). TG2 can form a complex with FN on the fibroblast and osteoblast surface and this TG2-FN complex can interact with heparan sulfate syndecan-4 and activate protein kinase Cα and affect subsequent β1 integrin signaling, which leads to the activation of cell survival focal adhesion kinase (FAK) and mitogen-activated protein kinases (MAPK). This maintains osteoblast viability in a manner that is not dependent on the RGD sequence of FN (Verderio et al. 2003; Telci et al. 2008). Similarly, TG2-FN matrix complex on the osteoblast surface can bind to syndecan-2 and syndecan-4 and activate the MAPK pathway through α5β1 integrin; this enhances osteoblast adhesion and survival without RGD mediation (Wang et al. 2011; Wang and Griffin 2012).

In addition to TG2, FXIII-A was described as being expressed both in mouse primary osteoblasts and in the MC3T3-E1 osteoblast culture model (Al-Jallad et al. 2006; Nurminskaya and Kaartinen 2006). These cells express Runx2 and upregulate Osterix once sufficient COL I matrix has been deposited (Nakashima et al. 2002; Franceschi and Iyer 1992). During the differentiation of MC3T3-E1 cells from preosteoblasts to mature, matrix depositing and mineralizing osteoblasts, TG2 is expressed at same level and it is found on the cell surface throughout the osteoblast differentiation process. FXIII-A expression on the other hand increases during differentiation and is regulated by ascorbic acid treatment, which activates COL I synthesis and secretion. In fact, the presence of the COL I matrix network appears to be required for FXIII-A production and externalization during osteoblast differentiation. Inhibition of COL I externalization or inhibition of its assembly in the extracellular space with dihydroxyproline results in decreased FXIII-A secretion and expression in osteoblasts (Piercy-Kotb et al. 2012). Also, the quality of the COL I network affects FXIII-A externalization as inhibition of LOX activity with β-aminopropionitrile reduces FXIII-A expression (Piercy-Kotb et al. 2012). The expression of FXIII-A in osteoblasts is modulated by the MAPK signaling pathway (Piercy-Kotb et al. 2012).

Upon its expression, FXIII-A is subsequently found on the osteoblast plasma membrane on the outer leaflet (Al-Jallad et al. 2011) and inside the cells in caveolae (Al-Jallad et al. 2011; Wang and Kaartinen 2014) and is secreted to the extracellular matrix where it is responsible for the crosslinking activity found in preosteoblast/osteoblast cultures. The caveolae-associated FXIII-A appears not to have transamidation activity and acts as a modulator of caveolae function that include endocytosis and c-Src activation (Wang and Kaartinen 2014). The functions that involve FXIII-A transamidation activity in preosteoblasts/osteoblast cultures is linked to promotion of pFN assembly and COL I matrix production (Al-Jallad et al. 2006, 2011; Piercy-Kotb et al. 2012). Inhibition of TG activity with cystamine or NC9 (which targets FXIII-A on the outer leaflet of the osteoblast membrane) leads to destabilization of microtubules and impaired secretory vesicle delivery to the plasma membrane (Al-Jallad et al. 2011; Wang et al. 2014). This decreases COL I secretion and deposition. NC9 inhibitor also causes a dramatic decrease in pFN assembly from cell culture serum into the osteoblast matrix (Cui et al. 2014). The absence of COL I and pFN matrix also results in decreased ALP, bone sialoprotein, and osteocalcin, and abrogates mineralization, which indicates a defect in preosteoblast-osteoblast differentiation in vitro (Al-Jallad et al. 2006; 2011).

12.2.2 TGs in Bone Matrix Stabilization – Substrates

FN is a well-established substrate and binding partner for TG2 and FXIII-A (Lorand and Graham 2003; Mosher 1975; Mosher and Schad 1979; Keski-Oja 1976). In vitro, FXIII-A can crosslink FN to fibrin (Mosher 1975), to itself (Keski-Oja 1976) and to COL I (Mosher and Schad 1979) and promote FN assembly in a test tube in vitro (Mosher 1975; Mosher and Schad 1979; Mosher et al. 1980). The TG-reactive glutamine residues of FN are Q3, 4, 6, 7, 9 which are located in the N-terminal 27 kDa fragment (Mosher et al. 1980). An additional TG-reactive Q246 residue is found in the N-terminal type I module of the FN molecule (Hoffmann et al. 2011). In MC3T3-E1 osteoblasts, the major TG and FXIII-A substrate has been identified as FN – more specifically, pFN (Al-Jallad et al. 2006; Cui et al. 2014). Cellular FN (EDA-FN) produced by osteoblasts appears not to be crosslinked in the cultures (Cui et al. 2014). Inhibition of TG activity blocks pFN matrix accumulation from cell culture serum and impairs both the quantity and quality of COL I matrix, and decreases LOX and ALP levels in osteoblast cultures, resulting in decreased mineralization of osteoblast cultures (Cui et al. 2014). Similar cellular effects on osteoblasts are seen when pFN is depleted from the cell culture serum (Cui et al. 2014). The TG responsible for the pFN assembly and crosslinking is FXIII-A (Cui et al. 2014). It is also possible that TG activity is required for crosslinking FN to COL I as this has been demonstrated to occur in vitro (Mosher and Schad 1979). FN-COL I crosslinks may occur between abundant K residues from COL I and Q residues from FN (Mosher et al. 1980). Furthermore, MDC labeling of heat-denatured COL I identified crosslinking residues in the triple-helical region (Stachel et al. 2010), that may indicate that TG-mediated crosslinking tightens COL I fibrils in a similar manner as LOX does. Our observation that COL I fibrils in osteoblasts cultures treated with TG inhibitor NC9 were thicker supports this concept. The "tightening" effect of TG-mediated crosslinking on COL I network was also demonstrated in in vitro-generated COL I scaffolds (Garcia et al. 2007).

In addition to COL I and pFN, bone matrix also contains many other noncollagenous proteins, such as small integrin binding ligand N-linked glycoproteins (SIBLING), including osteopontin, bone sialoprotein, osteonectin and fibrillin-1 (Kaartinen et al. 2002; Nurminskaya and Kaartinen 2006), which are all TG substrates in vitro. Their structures can also be modulated by TG crosslinking activity and this may contribute to their function in bone matrix assembly and bone cell function (Forsprecher et al. 2011).

12.3 TGs in Adipogenesis

Adipocytes arise from mesenchymal stem cells and share a common, bipotential progenitor cell with osteoblasts (Berendsen and Olsen 2014). Adipogenesis, i.e., preadipocyte differentiation to adipocytes, is a two-phase cellular process that involves a major morphological change from a cuboidal, matrix adherent preadipocyte to a rounded adipocyte that is capable of accumulating and storing lipids (Fig. 12.1) (Cristancho and Lazar 2011; Gregoire et al. 1998). Preadipocytes assemble pFN and COL I matrix whereas mature adipocytes downregulate these and begin generation of basement membrane type surroundings and COL VI rich matrix (Huang and Greenspan 2012). We have demonstrated recently that mouse white adipose tissue has TG activity and produces both TG2 and FXIII-A (Myneni et al. 2014). No other TG family members are expressed. Also a preadipocyte cell line, 3T3-L1, shows abundant TG activity at the early preadipocyte phase which is followed by gradual decrease in the activity as cells begin lipid storage (Myneni et al. 2014). Inhibition of TG activity with NC9 in 3T3-L1 cultures blocks pFN assembly, and enhances PPARγ and C/EBPα expression that leads to faster preadipocyte differentiation to lipid-accumulating adipocytes. The function of TG activity from FXIII-A is to promote pFN accumulation to matrix to assist preadipocyte proliferation and to regulate insulin sensitivity (Myneni et al. 2014).

12.4 TG-Mediated Protein Serotonylation and Effects on Matrix Stabilization in Vitro

Recent studies with genetic mouse models demonstrate that circulating peripheral serotonin is a negative regulator of bone mass, affecting both osteoblast and osteoclast function via its receptors in these cells (Yadav et al. 2008). In addition to receptor-mediated effects, serotonin, and other monoamines such as dopamine and histamine, can also act as a substrate for TGs (Walther et al. 2011). Monoamines can be crosslinked by TGs to glutamine residues of a TG substrate protein – the process is called monoaminylation (Walther et al. 2003). Monoaminylation of glutamine residues can regulate protein function in health and disease as demonstrated by serotonylation of small GTPases which is a vital signaling pathway that initiates platelet alpha-granule release (Walther et al. 2003) and by FN serotonylation that induces pulmonary artery smooth muscle cell proliferation and migration (Liu et al. 2011). Furthermore, the serotonylated FN level is higher in patients with pulmonary hypertension (Wei et al. 2012).

Serotonin can act as a competitive inhibitor to FXIII-A-mediated pFN fibrillogenesis and matrix assembly in vitro in MC3T3-E1 osteoblast cultures. Serotonin also decreases FXIII-A transamidation activity in osteoblast cultures and colocalized with the enzyme on the osteoblast surface. Serotonin-mediated pFN assembly defect impairs COL I deposition and decreases ALP activity and mineralization, and the effect can be rescued by adding extra, exogenous pFN into cultures. Matrix protein serotonylation presents a novel mechanism for how circulating peripheral serotonin may affect the quality and quantity of bone by interfering directly with TG activity and extracellular matrix protein networks.

12.5 Bone Phenotypes of TG2 and FXIII-A Mouse Models

The two enzymes expressed in bone and osteoblasts, TG2 and FXIII-A, are fairly different in their functionalities in vivo (Nakano et al. 2007; Al-Jallad et al. 2006, 2011; Piercy-Kotb et al. 2012). TG2 is the most ubiquitously expressed TG enzyme of the TG family, found in a large number of cells and all cellular compartments from the nucleus to the extracellular matrix (Eckert et al. 2014). One of the main functions of TG2 is to modulate cell adhesion in its role as a β1-integrin binding co-receptor to FN and as a binding partner of syndecan-2 and 4 (Telci et al. 2008; Wang et al. 2010, 2011, 2012b; Hang et al. 2005; Verderio et al. 2009; Lortat-Jacob et al. 2012). TG2 assists integrin clustering and inhibits c-Src signaling (Janiak et al. 2006). The *Tgm2*$^{-/-}$ mouse model was developed simultaneously by two groups in 2001 (De Laurenzi and Melino 2001; Nanda et al. 2001). The research groups showed that *Tgm2*$^{-/-}$ mice are viable and had no dramatic developmental phenotype. Few cellular phenotypes have been reported. For example, cultured *Tgm2*$^{-/-}$ thymocytes show significantly decreased viability with dexamethasone treatment-induced apoptosis, and primary fibroblasts in culture displayed decreased adherence. There has not been a report on any obvious skeletal phenotype and we reported that TG2 null bones have normal TG activity (Nurminskaya and Kaartinen 2006). Furthermore, *Tgm2*$^{-/-}$ mice show increased FXIII-A expression in bone, which suggests these two enzymes may share functions and act synergistically in maintaining bone homeostasis (Tarantino et al. 2009).

FXIII-A is best known as plasma TG and as a coagulation factor that stabilizes fibrin matrix as the last step of blood clot formation. This provides resistance to the thrombus against lytic enzymes, and gives stiffness to the clot (Iismaa et al. 2009). *F13a1*$^{-/-}$ mice have undetectable plasma fibrin γ-dimerization and show impaired clot stabilization, which leads to a blood coagulation defect. Although the *F13a1*$^{-/-}$ mice are fertile, the females tend to miscarry their pups and die during pregnancy. Mice often show spontaneous bleeding episodes, including hematothorax, hematoperitoneum and subcutaneous hemorrhage, contributing to the reduced survival of $F13a1^{-/-}$ mice (Lauer et al. 2002). No bone phenotype has been reported for $F13a1^{-/-}$ mice.

A mouse model lacking both TG2 and FXIII-A has been created by us. $Tgm2^{-/-}$; $F13a1^{-/-}$ mice show impaired bone biomechanical properties, severe bone loss (osteopenia), decreased pFN assembly into bone matrix and increased adipogenesis of bone marrow (unpublished results) (Mousa et al. 2013; 2014) reflecting the roles that have been suggested for the enzymes by in vitro studies.

Acknowledgments This work was supported by grants to MTK from the Canadian Institutes of Health Research (CIHR) and the CIHR Institute for Musculoskeletal Health. CC received studentship stipends from the China Scholarship Council (CSC) and the Faculty of Dentistry of McGill University. This manuscript is partly based on a PhD thesis submitted to McGill University in partial fulfillment of the requirements of the degree of Doctor of Philosophy in Dental Sciences.

References

Aeschlimann D, Wetterwald A, Fleisch H, Paulsson M (1993) Expression of tissue transglutaminase in skeletal tissues correlates with events of terminal differentiation of chondrocytes. J Cell Biol 120:1461–1470

Aeschlimann D, Mosher D, Paulsson M (1996) Tissue transglutaminase and factor XIII in cartilage and bone remodeling. Semin Thromb Hemost 22:437–443

Ali SY, Sajdera SW, Anderson HC (1970) Isolation and characterization of calcifying matrix vesicles from epiphyseal cartilage. Proc Natl Acad Sci U S A 67:1513–1520

Al-Jallad HF, Nakano Y, Chen JLY, McMillan E, Lefebvre C, Kaartinen MT (2006) Transglutaminase activity regulates osteoblast differentiation and matrix mineralization in MC3T3-E1 osteoblast cultures. Matrix Biol 25:135–148

Al-Jallad HF, Myneni VD, Piercy-Kotb SA, Chabot N, Mulani A, Keillor JW, Kaartinen MT (2011) Plasma membrane factor XIIIA transglutaminase activity regulates osteoblast matrix secretion and deposition by affecting microtubule dynamics. PLoS One 6:0015893

Anderson HC (1969) Vesicles associated with calcification in the matrix of epiphyseal cartilage. J Cell Biol 41:59–72

Anderson HC (1995) Molecular biology of matrix vesicles. Clin Orthop Relat Res 314:266–280

Anderson HC, Hsu HH, Morris DC, Fedde KN, Whyte MP (1997) Matrix vesicles in osteomalacic hypophosphatasia bone contain apatite-like mineral crystals. Am J Pathol 151:1555–1561

Anderson HC, Garimella R, Tague SE (2005) The role of matrix vesicles in growth plate development and biomineralization. Front Biosci 10:822–837

Awonusi A, Morris MD, Tecklenburg MM (2007) Carbonate assignment and calibration in the Raman spectrum of apatite. Calcif Tissue Int 81:46–52

Barak Y, Nelson MC, Ong ES, Jones YZ, Ruiz-Lozano P, Chien KR, Koder A, Evans RM (1999) PPARγ is required for placental, cardiac, and adipose tissue development. Mol Cell 4:585–595

Baum J, Brodsky B (1999) Folding of peptide models of collagen and misfolding in disease. Curr Opin Struct Biol 9:122–128

Beaupré G, Orr T, Carter D (1990) An approach for time-dependent bone modeling and remodeling—theoretical development. J Orthop Res 8:651–661

Bentmann A, Kawelke N, Moss D, Zentgraf H, Bala Y, Berger I, Gasser JA, Nakchbandi IA (2010) Circulating fibronectin affects bone matrix, whereas osteoblast fibronectin modulates osteoblast function. J Bone Miner Res 25:706–715

Berendsen A, Olsen B (2014) Osteoblast–adipocyte lineage plasticity in tissue development, maintenance and pathology. Cell Mol Life Sci 71:493–497

Bianco P, Robey PG, Simmons PJ (2008) Mesenchymal stem cells: revisiting history, concepts, and assays. Cell Stem Cell 2:313–319

Bonewald LF (2006) Mechanosensation and transduction in osteocytes. IBMS BoneKEy 3:7–15

Boskey AL (1989) Noncollagenous matrix proteins and their role in mineralization. Bone Miner 6:111–123

Boskey AL, Wright TM, Blank RD (1999) Collagen and bone strength. J Bone Miner Res 14:330–335

Brodsky B, Ramshaw JA (1997) The collagen triple-helix structure. Matrix Biol 15:545–554

Cheng SL, Lai CF, Fausto A, Chellaiah M, Feng X, McHugh KP, Teitelbaum SL, Civitelli R, Hruska KA, Ross FP, Avioli LV (2000) Regulation of alphaVbeta3 and alphaVbeta5 integrins by dexamethasone in normal human osteoblastic cells. J Cell Biochem 77:265–276

Chung CY, Zardi L, Erickson HP (1995) Binding of tenascin-C to soluble fibronectin and matrix fibrils. J Biol Chem 270:29012–29017

Cristancho AG, Lazar MA (2011) Forming functional fat: a growing understanding of adipocyte differentiation. Nat Rev Mol Cell Biol 12:722–734

Cui C, Wang S, Myneni VD, Hitomi K, Kaartinen MT (2014) Transglutaminase activity arising from Factor XIIIA is required for stabilization and conversion of plasma fibronectin into matrix in osteoblast cultures. Bone 59:127–138

Currey JD (1988) The effect of porosity and mineral content on the Young's modulus of elasticity of compact bone. J Biomech 21:131–139

Currey JD, Brear K, Zioupos P (1996) The effects of ageing and changes in mineral content in degrading the toughness of human femora. J Biomech 29:257–260

Dallas SL, Sivakumar P, Jones CJP, Chen Q, Peters DM, Mosher DF, Humphries MJ, Kielty CM (2005) Fibronectin regulates latent transforming growth factor-β (TGFβ) by controlling matrix assembly of latent TGFβ-binding protein-1. J Biol Chem 280:18871–18880

De Laurenzi V, Melino G (2001) Gene disruption of tissue transglutaminase. Mol Cell Biol 21:148–155

Eckert RL, Kaartinen MT, Nurminskaya M, Belkin AM, Colak G, Johnson GV, Mehta K (2014) Transglutaminase regulation of cell function. Physiol Rev 94:383–417

Epstein FH, Manolagas SC, Jilka RL (1995) Bone marrow, cytokines, and bone remodeling—emerging insights into the pathophysiology of osteoporosis. N Engl J Med 332:305–311

Fedde KN, Blair L, Silverstein J, Coburn SP, Ryan LM, Weinstein RS, Waymire K, Narisawa S, Millan JL, MacGregor GR, Whyte MP (1999) Alkaline phosphatase knock-out mice recapitulate the metabolic and skeletal defects of infantile hypophosphatasia. J Bone Miner Res 14:2015–2026

Forsprecher J, Wang Z, Goldberg HA, Kaartinen MT (2011) Transglutaminase-mediated oligomerization promotes osteoblast adhesive properties of osteopontin and bone sialoprotein. Cell Adh Migr 5:65–72

Franceschi RT, Iyer BS (1992) Relationship between collagen synthesis and expression of the osteoblast phenotype in MC3T3-E1 cells. J Bone Miner Res 7:235–246

Franceschi RT, Iyer BS, Cui Y (1994) Effects of ascorbic acid on collagen matrix formation and osteoblast differentiation in murine MC3T3-E1 cells. J Bone Miner Res 9:843–854

Garcia Y, Collighan R, Griffin M, Pandit A (2007) Assessment of cell viability in a three-dimensional enzymatically cross-linked collagen scaffold. J Mater Sci Mater Med 18:1991–2001

George EL, Georges-Labouesse EN, Patel-King RS, Rayburn H, Hynes RO (1993) Defects in mesoderm, neural tube and vascular development in mouse embryos lacking fibronectin. Development 119:1079–1091

Glimcher MJ (2006) Bone: nature of the calcium phosphate crystals and cellular, structural, and physical chemical mechanisms in their formation. Rev Mineral Geochem 64:223–282

Godyna S, Mann DM, Argraves WS (1995) A quantitative analysis of the incorporation of fibulin-1 into extracellular matrix indicates that fibronectin assembly is required. Matrix Biol 14:467–477

Golub EE (2009) Role of matrix vesicles in biomineralization. Biochim Biophys Acta 1790:1592–1598

Gregoire FM, Smas CM, Sul HS (1998) Understanding adipocyte differentiation. Physiol Rev 78:783–809

Grzesik WJ, Robey PG (1994) Bone matrix RGD glycoproteins: immunolocalization and interaction with human primary osteoblastic bone cells in vitro. J Bone Miner Res 9:487–496

Hang J, Zemskov EA, Lorand L, Belkin AM (2005) Identification of a novel recognition sequence for fibronectin within the NH2-terminal beta-sandwich domain of tissue transglutaminase. J Biol Chem 280:23675–23683

Heath DJ, Downes S, Verderio E, Griffin M (2001) Characterization of tissue transglutaminase in human osteoblast-like cells. J Bone Miner Res 16:1477–1485

Hessle L, Johnson KA, Anderson HC, Narisawa S, Sali A, Goding JW, Terkeltaub R, Millan JL (2002) Tissue-nonspecific alkaline phosphatase and plasma cell membrane glycoprotein-1 are central antagonistic regulators of bone mineralization. Proc Natl Acad Sci U S A 99:9445–9449

Hoffmann BR, Annis DS, Mosher DF (2011) Reactivity of the N-terminal region of fibronectin protein to transglutaminase 2 and factor XIIIA. J Biol Chem 286:32220–32230

Huang G, Greenspan DS (2012) ECM roles in the function of metabolic tissues. Trends Endocrinol Metab 23:16–22

Huang G, Zhang Y, Kim B, Ge G, Annis DS, Mosher DF, Greenspan DS (2009) Fibronectin binds and enhances the activity of bone morphogenetic protein 1. J Biol Chem 284:25879–25888

Hynes RO (1990) Fibronectins. Springer, New York

Iismaa SE, Mearns BM, Lorand L, Graham RM (2009) Transglutaminases and disease: lessons from genetically engineered mouse models and inherited disorders. Physiol Rev 89:991–1023

Janiak A, Zemskov EA, Belkin AM (2006) Cell surface transglutaminase promotes RhoA activation via integrin clustering and suppression of the Src-p190RhoGAP signaling pathway. Mol Biol Cell 17:1606–1619

Johnson KA, Hessle L, Vaingankar S, Wennberg C, Mauro S, Narisawa S, Goding JW, Sano K, Millan JL, Terkeltaub R (2000) Osteoblast tissue-nonspecific alkaline phosphatase antagonizes and regulates PC-1. Am J Physiol Regul Integr Comp Physiol 279:R1365–R1377

Justesen J, Stenderup K, Ebbesen EN, Mosekilde L, Steiniche T, Kassem M (2001) Adipocyte tissue volume in bone marrow is increased with aging and in patients with osteoporosis. Biogerontology 2:165–171

Kaartinen MT, El-Maadawy S, Räsänen NH, McKee MD (2002) Tissue transglutaminase and its substrates in bone. J Bone Miner Res 17:2161–2173

Kagan HM, Li W (2003) Lysyl oxidase: properties, specificity, and biological roles inside and outside of the cell. J Cell Biochem 88:660–672

Karaplis AC (2002) Embryonic development of bone and the molecular regulation of intramembranous and endochondral bone formation. Princ Bone Biol 1:33–58

Karsenty G (2001) Minireview: transcriptional control of osteoblast differentiation. Endocrinology 142:2731–2733

Keski-Oja J, Mosher DF, Vaheri A (1976) Cross-linking of a major fibroblast surface-associated glycoprotein (fibronectin) catalyzed by blood coagulation factor XIII. Cell 9:29–35

Lauer P, Metzner HJ, Zettlmeissl G, Li M, Smith AG, Lathe R, Dickneite G (2002) Targeted inactivation of the mouse locus encoding coagulation factor XIII-A: hemostatic abnormalities in mutant mice and characterization of the coagulation deficit. Thromb Haemost 88:967–974

Lerner UH (2006) Bone remodeling in post-menopausal osteoporosis. J Dent Res 85:584–595

Liu Y, Wei L, Laskin DL, Fanburg BL (2011) Role of protein transamidation in serotonin-induced proliferation and migration of pulmonary artery smooth muscle cells. Am J Respir Cell Mol Biol 44:548–555

Lorand L, Graham RM (2003) Transglutaminases: crosslinking enzymes with pleiotropic functions. Nat Rev Mol Cell Biol 4:140–156

Lortat-Jacob H, Burhan I, Scarpellini A, Thomas A, Imberty A, Vives RR, Johnson T, Gutierrez A, Verderio EA (2012) Transglutaminase-2 interaction with heparin: identification of a heparin binding site that regulates cell adhesion to fibronectin-transglutaminase-2 matrix. J Biol Chem 287:18005–18017

Magnusson MK, Mosher DF (1998) Fibronectin: structure, assembly, and cardiovascular implications. Arterioscler Thromb Vasc Biol 18:1363–1370

Malara A, Gruppi C, Rebuzzini P, Visai L, Perotti C, Moratti R, Balduini C, Tira ME, Balduini A (2011) Megakaryocyte-matrix interaction within bone marrow: new roles for fibronectin and factor XIII-A. Blood 117:2476–2483

Massé PG, Rimnac CM, Yamauchi M, Coburn SP, Rucker RB, Howell DS, Boskey AL (1996) Pyridoxine deficiency affects biomechanical properties of chick tibial bone. Bone 18:567–574

Matsuo K, Irie N (2008) Osteoclast–osteoblast communication. Arch Biochem Biophys 473:201–209

Millan JL (2013) The role of phosphatases in the initiation of skeletal mineralization. Calcif Tissue Int 93:299–306

Mosher DF (1975) Cross-linking of cold-insoluble globulin by fibrin-stabilizing factor. J Biol Chem 250:6614–6621

Mosher D (1989) Fibronectin: Access Online via Elsevier

Mosher DF, Schad PE (1979) Cross-linking of fibronectin to collagen by blood coagulation Factor XIIIa. J Clin Invest 64:781–787

Mosher DF, Schad PE, Vann JM (1980) Cross-linking of collagen and fibronectin by factor XIIIa. Localization of participating glutaminyl residues to a tryptic fragment of fibronectin. J Biol Chem 255:1181–1188

Moursi AM, Damsky CH, Lull J, Zimmerman D, Doty SB, Aota S, Globus RK (1996) Fibronectin regulates calvarial osteoblast differentiation. J Cell Sci 109:1369–1380

Moursi AM, Globus RK, Damsky CH (1997) Interactions between integrin receptors and fibronectin are required for calvarial osteoblast differentiation in vitro. J Cell Sci 110:2187–2196

Mousa A, Cui C, Song A, Myneni V, Li J, Melino G, Dickneite G, Murshed M, Kaartinen MT (2013) Severe osteopenia, increased bone marrow adipogenesis and fibronectin matrix changes in mice lacking both TG2 and FXIIIA transglutaminases. Presented at the European Calcified Tissue Society Conference ECTS 2013, Lisbon, Bone Abstracts Vol 1, OC4.3 2013

Mousa A, Cui C, Song A, Myneni V, Li J, Melino G, Dickneite G, Murshed M, Kaartinen MT (2014) Transglutaminase 2 and Factor XIIIA regulate bone remodeling and their deficiency in increased osteoclastogenesis, defective plasma FN stabilization to bone and increased bone marrow adiposity. Presented at 2014 Meeting of the American Society for Matrix Biology, Cleveland

Mundy GR (1993) Cytokines and growth factors in the regulation of bone remodeling. J Bone Miner Res 8:S505–S510

Murshed M, Harmey D, Millan JL, McKee MD, Karsenty G (2005) Unique coexpression in osteoblasts of broadly expressed genes accounts for the spatial restriction of ECM mineralization to bone. Genes Dev 19:1093–1104

Myllyharju J, Kivirikko KI (2004) Collagens, modifying enzymes and their mutations in humans, flies and worms. Trends Genet 20:33–43

Myneni VD, Hitomi K, Kaartinen MT (2014) Factor XIII-A transglutaminase acts as a switch between preadipocyte proliferation and differentiation. Blood 124:1344–1353

Nakano Y, Al-Jallad HF, Mousa A, Kaartinen MT (2007) Expression and localization of plasma transglutaminase factor XIIIA in bone. J Histochem Cytochem 55:675–685

Nakashima K, Zhou X, Kunkel G, Zhang Z, Deng JM, Behringer RR, de Crombrugghe B (2002) The novel zinc finger-containing transcription factor Osterix is required for osteoblast differentiation and bone formation. Cell 108:17–29

Nanda N, Iismaa SE, Owens WA, Husain A, Mackay F, Graham RM (2001) Targeted inactivation of Gh/tissue transglutaminase II. J Biol Chem 276:20673–20678

Narisawa S, Frohlander N, Millan JL (1997) Inactivation of two mouse alkaline phosphatase genes and establishment of a model of infantile hypophosphatasia. Dev Dyn 208:432–446

Niyibizi C, Eyre DR (1994) Structural characteristics of cross-linking sites in type V collagen of bone. Eur J Biochem 224:943–950

Nurminskaya M, Kaartinen MT (2006) Transglutaminases in mineralized tissues. Front Biosci 11:1591–1606

Nurminskaya M, Magee C, Nurminsky D, Linsenmayer TF (1998) Plasma transglutaminase in hypertrophic chondrocytes: expression and cell-specific intracellular activation produce cell death and externalization. J Cell Biol 142:1135–1144

Nurminskaya MV, Recheis B, Nimpf J, Magee C, Linsenmayer TF (2002) Transglutaminase factor XIIIA in the cartilage of developing avian long bones. Dev Dyn 223:24–32

Nurminskaya M, Magee C, Faverman L, Linsenmayer TF (2003) Chondrocyte-derived transglutaminase promotes maturation of preosteoblasts in periosteal bone. Dev Biol 263:139–152

Opsahl W, Zeronian H, Ellison M, Lewis D, Rucker RB, Riggins RS (1982) Role of copper in collagen cross-linking and its influence on selected mechanical properties of chick bone and tendon. J Nutr 112:708–716

Ortega N, Behonick DJ, Werb Z (2004) Matrix remodeling during endochondral ossification. Trends Cell Biol 14:86–93

Oxlund H, Barckman M, Ortoft G, Andreassen TT (1995) Reduced concentrations of collagen cross-links are associated with reduced strength of bone. Bone 17:365s–371s

Pankov R, Yamada KM (2002) Fibronectin at a glance. J Cell Sci 115:3861–3863

Piercy-Kotb SA, Mousa A, Al-Jallad HF, Myneni VD, Chicatun F, Nazhat SN, Kaartinen MT (2012) Factor XIIIA transglutaminase expression and secretion by osteoblasts is regulated by extracellular matrix collagen and the MAP kinase signaling pathway. J Cell Physiol 227:2936–2946

Pistone M, Sanguineti C, Federici A, Sanguineti F, Defilippi P, Santolini F, Querze G, Marchisio PC, Manduca P (1996) Integrin synthesis and utilization in cultured human osteoblasts. Cell Biol Int 20:471–479

Pittenger MF, Mackay AM, Beck SC, Jaiswal RK, Douglas R, Mosca JD, Moorman MA, Simonetti DW, Craig S, Marshak DR (1999) Multilineage potential of adult human mesenchymal stem cells. Science 284:143–147

Register TC, McLean FM, Low MG, Wuthier RE (1986) Roles of alkaline phosphatase and labile internal mineral in matrix vesicle-mediated calcification. Effect of selective release of membrane-bound alkaline phosphatase and treatment with isosmotic pH 6 buffer. J Biol Chem 261:9354–9360

Robey PG, Boskey AL (2008) The composition of bone. In: Rosen CJ (ed) Primer on the metabolic bone diseases and disorders of mineral metabolism, vol 7, 8th edn. American Society of Bone and Mineral Research, pp 32–38

Sabatier L, Chen D, Fagotto-Kaufmann C, Hubmacher D, McKee MD, Annis DS, Mosher DF, Reinhardt DP (2009) Fibrillin assembly requires fibronectin. Mol Biol Cell 20:846–858

Saito T, Albelda SM, Brighton CT (1994) Identification of integrin receptors on cultured human bone cells. J Orthop Res 12:384–394

Seeman E (2008) Modeling and remodeling: the cellular machinery responsible for the gain and loss of bone's material and structural strength. Princ Bone Biol 1:3–28

Sottile J, Hocking DC (2002) Fibronectin polymerization regulates the composition and stability of extracellular matrix fibrils and cell-matrix adhesions. Mol Biol Cell 13:3546–3559

Sottile J, Shi F, Rublyevska I, Chiang H-Y, Lust J, Chandler J (2007) Fibronectin-dependent collagen I deposition modulates the cell response to fibronectin. Am J Physiol Cell Physiol 293:C1934–C1946

Stachel I, Schwarzenbolz U, Henle T, Meyer M (2010) Cross-linking of type I collagen with microbial transglutaminase: identification of cross-linking sites. Biomacromolecules 11:698–705

Stein GS, Lian JB, Gerstenfeld LG, Shalhoub V, Aronow M, Owen T, Markose E (1989) The onset and progression of osteoblast differentiation is functionally related to cellular proliferation. Connect Tissue Res 20:3–13

Tanaka T, Yoshida N, Kishimoto T, Akira S (1997) Defective adipocyte differentiation in mice lacking the C/EBPbeta and/or C/EBPdelta gene. Embo J 16:7432–7443

Tarantino U, Oliva F, Taurisano G, Orlandi A, Pietroni V, Candi E, Melino G, Maffulli N (2009) FXIIIA and TGF-beta over-expression produces normal musculo-skeletal phenotype in TG2−/− mice. Amino Acids 36:679–684

Telci D, Wang Z, Li X, Verderio EA, Humphries MJ, Baccarini M, Basaga H, Griffin M (2008) Fibronectin-tissue transglutaminase matrix rescues RGD-impaired cell adhesion through syndecan-4 and β1 integrin co-signaling. J Biol Chem 283:20937–20947

Thompson JB, Kindt JH, Drake B, Hansma HG, Morse DE, Hansma PK (2001) Bone indentation recovery time correlates with bond reforming time. Nature 414:773–776

Thompson Z, Miclau T, Hu D, Helms JA (2002) A model for intramembranous ossification during fracture healing. J Orthop Res 20:1091–1098

To WS, Midwood KS (2011) Plasma and cellular fibronectin: distinct and independent functions during tissue repair. Fibrogenesis Tissue Repair 4:21

Verderio EAM, Telci D, Okoye A, Melino G, Griffin M (2003) A Novel RGD-independent Cell Adhesion Pathway Mediated by Fibronectin-bound Tissue Transglutaminase Rescues Cells from Anoikis. J Biol Chem 278:42604–42614

Verderio E, Scarpellini A, Johnson T (2009) Novel interactions of TG2 with heparan sulfate proteoglycans: reflection on physiological implications. Amino Acids 36:671–677

Viguet-Carrin S, Garnero P, Delmas PD (2006) The role of collagen in bone strength. Osteoporos Int 17:319–336

Walther DJ, Peter JU, Winter S, Holtje M, Paulmann N, Grohmann M, Vowinckel J, Alamo-Bethencourt V, Wilhelm CS, Ahnert-Hilger G, Bader M (2003) Serotonylation of small GTPases is a signal transduction pathway that triggers platelet alpha-granule release. Cell 115:851–862

Walther DJ, Stahlberg S, Vowinckel J (2011) Novel roles for biogenic monoamines: from monoamines in transglutaminase-mediated post-translational protein modification to monoaminylation deregulation diseases. FEBS J 278:4740–4755

Wang Z, Griffin M (2012) TG2, a novel extracellular protein with multiple functions. Amino Acids 42:939–949

Wang S, Kaartinen MT (2014) Cellular factor XIIIA transglutaminase localizes to caveolae in osteoblasts and regulates c-Src signaling in a transamidation-independent manner. Submitted to Amino Acids

Wang Z, Collighan RJ, Gross SR, Danen EHJ, Orend G, Telci D, Griffin M (2010) RGD-independent Cell Adhesion via a Tissue Transglutaminase-Fibronectin Matrix Promotes Fibronectin Fibril Deposition and Requires Syndecan-4/2 and α5β1 Integrin Co-signaling. J Biol Chem 285:40212–40229

Wang Z, Telci D, Griffin M (2011) Importance of syndecan-4 and syndecan-2 in osteoblast cell adhesion and survival mediated by a tissue transglutaminase – fibronectin complex. Exp Cell Res 317:367–381

Wang W, Lian N, Ma Y, Li L, Gallant RC, Elefteriou F, Yang X (2012a) Chondrocytic Atf4 regulates osteoblast differentiation and function via Ihh. Development 139:601–611

Wang Z, Collighan RJ, Pytel K, Rathbone DL, Li X, Griffin M (2012b) Characterization of heparin-binding site of tissue transglutaminase: its importance in cell surface targeting, matrix deposition, and cell signaling. J Biol Chem 287:13063–13083

Wang S, Cui C, Hitomi K, Kaartinen M (2014) Detyrosinated glu-tubulin is a substrate for cellular factor XIIIA transglutaminase in differentiating osteoblasts. Amino Acids 46(6):1513–1526

Wassen MHM, Lammens J, Tekoppele JM, Sakkers RJB, Liu Z, Verbout AJ, Bank RA (2000) Collagen structure regulates fibril mineralization in osteogenesis as revealed by cross-link patterns in calcifying callus. J Bone Miner Res 15:1776–1785

Wei L, Warburton RR, Preston IR, Roberts KE, Comhair SA, Erzurum SC, Hill NS, Fanburg BL (2012) Serotonylated fibronectin is elevated in pulmonary hypertension. Am J Physiol Lung Cell Mol Physiol 302:L1273–L1279

Weiner S, Traub W, Wagner HD (1999) Lamellar bone: structure-function relations. J Struct Biol 126:241–255

Whyte M (2001) Hypophosphatasia. In: Scriver C, Beaudet A, Sly W, Valle D, Childs B, Kinzler K (eds) The metabolic and molecular bases of inherited disease. McGraw-Hill, New York, pp 5313–5329

Wierzbicka-Patynowski I, Schwarzbauer JE (2003) The ins and outs of fibronectin matrix assembly. J Cell Sci 116:3269–3276

Yadav VK, Ryu J-H, Suda N, Tanaka KF, Gingrich JA, Schütz G, Glorieux FH, Chiang CY, Zajac JD, Insogna KL, Mann JJ, Hen R, Ducy P, Karsenty G (2008) Lrp5 controls bone formation by inhibiting serotonin synthesis in the duodenum. Cell 135:825–837

Zioupos P, Currey JD, Hamer AJ (1999) The role of collagen in the declining mechanical properties of aging human cortical bone. J Biomed Mater Res 45:108–116

Chapter 13
Transglutaminases and Neurological Diseases

Julianne Feola, Alina Monteagudo, Laura Yunes-Medina, and Gail V.W. Johnson

Abstract Transglutaminases, and predominantly transglutaminase 2 (TG2), have been implicated in the pathogenesis of a number of neurological conditions. In the context of older-age onset neurodegenerative diseases that are characterized by protein aggregates, it was initially postulated that transglutaminases contributed to the pathogenic process by crosslinking and facilitating aggregate formation. However, more recent studies have provided evidence that although TG2, and possibly other transglutaminases, likely contribute to the progression of certain neurodegenerative diseases, this probably does not involve crosslinking and the formation of insoluble protein aggregates. Indeed, there is growing evidence that TG2 could be a contributing factor in neurological diseases through mechanisms such as modulating transcriptional events, affecting cell migration and adhesion, or altering cellular differentiation processes. In this chapter, neurological diseases in which transglutaminases have been implicated as a contributing factor are discussed. The chapter will conclude with a discussion about cystamine, a non-specific in vitro inhibitor of transglutaminases, as a neuroprotective agent.

Keywords Transglutaminase 2 • Neurological disease • Cystamine • Transcription • Adhesion • Migration • Proliferation

---

These authors contribute equally to this chapter.

J. Feola • L. Yunes-Medina
Neuroscience Program, University of Rochester, Rochester, NY 14642, USA

A. Monteagudo
Department of Pharmacology and Physiology, University of Rochester, Rochester, NY 14642, USA

G.V.W. Johnson (✉)
Department of Anesthesiology, University of Rochester, Rochester, NY 14642, USA

Department of Pharmacology and Physiology, University of Rochester, Rochester, NY 14642, USA

Neuroscience Program, University of Rochester, Rochester, NY 14642, USA
e-mail: gail_johnsonvoll@urmc.rochester.edu

K. Hitomi et al. (eds.), *Transglutaminases*, DOI 10.1007/978-4-431-55825-5_13

13.1 Introduction

There has been a long-standing interest in the role of transglutaminases (TGs), and in particular transglutaminase 2 (TG2), in neurodegenerative diseases (De Vivo et al. 2009; Grosso and Mouradian 2012; Lesort et al. 2000; Green 1993; Selkoe et al. 1982). A neurodegenerative disease can be defined as a disease in which neurological function is compromised due to the loss of neurons/neuronal function. The particular patterns of degeneration and resulting symptoms are diverse among this class of diseases, which include Alzheimer's disease (AD), Huntington's disease (HD), amyotrophic lateral sclerosis (ALS) and Parkinson's disease (PD), among many others. In most cases, the mechanisms that underlie the onset and progression of these illnesses have not been fully elucidated.

A significant amount of research effort has been devoted to the study of neurodegenerative diseases, with the hopes of preventing or ameliorating the symptoms, if not the diseases themselves. Much of this work has focused on developing an understanding of the cellular mechanisms behind neurodegeneration. Intracellular aggregates of mutant proteins are a hallmark in a number of diseases. As TGs are widely known for their ability to crosslink proteins, early studies focused on examining their possible role in the formation of these aggregates (Green 1993; Selkoe et al. 1982). Although more recent studies suggest that the primary role of TGs in disease pathogenesis may not be the facilitation of protein crosslinking and aggregate formation, it is likely that TGs are still mediators of processes that contribute to neurological conditions. Indeed there is increasing evidence that TGs may be involved in the pathogenesis of neurological diseases by modulating a number of different cellular processes, including transcription, cell migration and adhesion, and cellular differentiation (De Vivo et al. 2009; Grosso and Mouradian 2012; Lesort et al. 2000; Gundemir et al. 2012; McConoughey et al. 2010; van Strien et al. 2011a),

Although TG2 is the predominant TG in the central nervous system (CNS), TG1 and TG3 have been identified in human brain (Kim et al. 1999). TG1 and TG3 are expressed in rather specific locations, with TG1 being most highly expressed in the corpus callosum and TG3 in the amygdala. Both TG1 and TG3 were found to a lesser extent in other regions of the cortex and cerebellum. On the other hand, TG2 was highly expressed in all regions examined (Kim et al. 1999). In addition to TG1-3, TG6 has also been identified in perivascular regions of the human cerebellum and medulla (Hadjivassiliou et al. 2008). A more detailed analysis in rat brain showed that TG6 is found in neurons throughout many brain regions involved in motor function, including the globus pallidus, substantia nigra and the cerebellum (Liu et al. 2013).

Given that TG2 is highly expressed in the CNS, its specific localization patterns have been investigated (Gundemir et al. 2012). TG2 is found in most cell types of the CNS, including neurons, astrocytes and endothelial cells of the brain vasculature (Kim et al. 1999; van Strien et al. 2011c; Wolf et al. 2013; Hilton et al. 1997; Monsonego et al. 1997). TG2 has also been found in microglia and oligodendrocyte precursor cells in mouse models that are representative of human diseases (Hwang et al. 2009; Kawabe et al. 2014; van Strien et al. 2011a). Its ubiquitous nature is also exemplified by its subcellular localization pattern; TG2 is found within numerous subcellular compartments, including the nucleus, cytosol, mitochondria, cell membrane and extracellular matrix (Grosso and Mouradian 2012; Gundemir et al. 2012; Lesort et al. 1998; Lorand and Graham 2003).

The widespread expression of TG2 throughout the brain permits it to take on multifaceted roles in both healthy and diseased states, as it can exhibit cell type and localization-specific functions. Indeed, TG2 has been shown to contribute to both cell death and survival, depending on its cellular context (Colak and Johnson 2012; Grosso and Mouradian 2012). This complexity has led to a broad interest in its role in neurological diseases. The aim of this chapter will be to describe the possible involvement of TGs, and in particular TG2, in the pathology and pathogenesis of certain neurological diseases.

13.2 Polyglutamine Diseases

Polyglutamine diseases are caused by an expansion of CAG repeats in the coding region of a particular gene, which leads to a polyglutamine (polyQ) expansion in the protein. This polyQ expansion results in aberrant function and, often, aggregation of the protein in particular intracellular compartments. These diseases are characterized not only by a polyQ expansion in a very specific protein, but also by neuronal degeneration in a very selective region of the brain. For example, in HD, the protein huntingtin (htt) has a polyQ repeat greater than 39 and there is significant neuronal atrophy in the striatal region of the brain and, as the disease progresses, in the cortical region as well (Margulis et al. 2013; Weber et al. 2014). In polyQ diseases in which different proteins contain the expansions, other areas, such as the cerebellum and spinal cord, are affected (Margulis et al. 2013).

Howard Green was the first to hypothesize that proteins with a polyQ expansion could be particularly suitable glutamyl donors for TG-catalyzed transamidation reactions (Green 1993). This idea was the basis for a number of initial in vitro studies investigating the propensity for TG2 to crosslink polyQ proteins. Peptides with polyQ stretches of various lengths were used as TG2 substrates in vitro (Cooper et al. 1997; Gentile et al. 1998; Kahlem et al. 1996). In addition, when COS-7 cells were transfected with GFP fusion proteins containing a polyQ chain of either 36 or 43 glutamines, the addition of TG2 increased aggregate formation (de Cristofaro et al. 1999). These initial studies supported the hypothesis that the aggregation of proteins with a polyQ expansion may be, at least in part, due to TG2 crosslinking. An important caveat, however, is that the highly truncated polyQ-containing constructs used in these experiments are likely not representative of the actual proteins found within diseased cells in the brain. Other factors such as the residues flanking the polyQ region, the tertiary and quaternary structure of the protein and the cellular context were not fully considered in these initial studies.

In contrast to the studies above, where TG2 was shown to decrease the solubility of the fusion proteins, another group demonstrated that the presence of TG2 actually *increased* the solubility of thioredoxin-polyQ fusion proteins with long polyQ chains. This effect was not observed in the presence of a transamidation-inactive TG2 construct. This suggests that TG2 might actually be preventing, rather than facilitating, the formation of insoluble aggregates, possibly by crosslinking these proteins into soluble oligomers and disrupting other intermolecular interactions (Lai et al. 2004). Though much emphasis has been placed on understanding the mechanisms behind protein aggregation in PolyQ expansion diseases, there has been an ongoing debate as to whether or not this aggregation is toxic or beneficial (Todd and Lim 2013), and the role of TG2 in this process is still unclear. Nevertheless, these studies set the stage for further experiments in specific disease models.

13.2.1 Huntington's Disease

HD, the most widely studied polyQ expansion disease, is an autosomal dominant neurodegenerative disease characterized by severe motor and cognitive deficits, as well as psychiatric difficulties (Bates et al. 1998). The disease is caused by a polyQ expansion in htt. PolyQ tracts greater than 39 alter the function of the protein and result in selective neuronal cell death. HD is defined by a pronounced loss of medium spiny neurons in the striatum and, at later stages, significant neuronal loss in the cortical region. The disease is further characterized by intracellular aggregates of mutant htt (Bates et al. 1998; Margulis et al. 2013).

Initial in vitro experiments demonstrating that polyQ peptides are substrates of TG2 (Cooper et al. 1997; Gentile et al. 1998; Kahlem et al. 1996) led to further in vitro studies in which TG2-mediated crosslinking of htt constructs was examined. High molecular weight bands of htt protein increased when constructs consisting of the N-terminus and polyQ expanded region of htt were incubated with purified guinea pig liver TG2. The same was true when extracts from cell lines derived from an HD patient were incubated with TG2. These studies indicated that TG2 facilitated the in vitro aggregation of htt (Karpuj et al. 1999; Kahlem et al. 1998).

Further evidence in support of a role for TG2 in HD pathology came from studies using tissue from HD cases. Brain extracts from HD and control patients were incubated with casein and radiolabeled putrescine (an amine acceptor and donor, respectively, for the TG-catalyzed transamidation reaction) in order to determine overall TG activity. There was a significant increase in TG activity in HD cortex and cerebellum relative to controls (Karpuj et al. 1999). Furthermore, TG2 protein and mRNA expression were increased in HD striatum and cortex, relative to control brain tissue (Lesort et al. 1999; Zainelli et al. 2005). In addition, γ-glutamylamines, markers of isopeptide linkages formed by TG-catalyzed transamidation reactions, were increased in the cerebrospinal fluid (CSF) of HD patients (Jeitner et al. 2001,

2008). These studies supported the hypothesis that TG-mediated transamidation reactions may be involved in HD pathology.

Studies using different cell models provided more insight into the possible role of TG2 in HD pathogenesis. In one study, human neuroblastoma (SH-SY5Y) cells stably overexpressing mutant truncated htt (N-terminal constructs containing 82 glutamine residues) formed intracellular aggregates, but the formation of these aggregates were not dependent on TG2. In fact, TG2 did not even colocalize or coprecipitate with the mutant htt construct (Chun et al. 2001). In contrast, in HEK293T cells with transient expression of both TG2 and an N-terminal mutant htt construct containing 148 glutamine repeats, TG2 and mutant htt appeared to colocalize. In addition, colocalization of the mutant htt construct with ε-(γ-glutamyl)lysine (GGEL) bonds, a product of TG2 transamidation reactions, was observed by immunocytochemistry. The GGEL antibody was also reported to immunoprecipitate with the mutant htt construct (Zainelli et al. 2003, 2005). It is important to note, however, that this antibody does not specifically recognize TG-catalyzed isopeptide linkages between or within full-length, intact proteins (Johnson and LeShoure 2004), and therefore caution must be exercised when interpreting these data. The evidence gathered from cellular models is therefore equivocal in terms of the role of TG2 in mediating intracellular aggregates by crosslinking the polyQ constructs. However, it needs to be emphasized that the cell studies have mostly been carried out with truncated versions of htt, or in some cases, constructs that only represent the polyQ domain, thus limiting data interpretation. In fact, the presence of full-length mutant htt in a striatal cell line actually reduced TG2 activity and prevented it from mediating proapoptotic responses to a cellular stressor. This suggests that mutant htt may be altering the cellular context such that conditions are not optimal for TG2 to exert its typical transamidation-dependent effects (Ruan et al. 2007).

Although TG2 may not be involved in facilitating aggregate formation in HD, other possible mechanisms by which TG2 could mediate HD pathology have been explored. In striatal cells, TG2 has been reported to play a transamidation activity-dependent role in the down-regulation of cytochrome c and PGC-1α transcript levels. In the presence of mutant but not wild-type htt (containing 111 or 7 glutamine residues, respectively), this effect is exacerbated. The authors thus propose that TG2 may be involved in HD disease progression through transcriptional dysregulation of key mitochondrial genes, though the exact mechanism is still unclear (McConoughey et al. 2010). In a separate study, the role of TG2 in aberrant actin-cofilin rod formation was examined. Rod formation is an important mechanism by which cells are able to conserve ATP during stress. This rod formation, however, is disrupted in striatal cell lines expressing mutant htt, which in turn could lead to widespread cellular dysfunction. Upon further examination, it was determined that TG2 colocalizes with these rods and that the overexpression of TG2 increases aberrant rod formation in the presence of wild-type htt (Munsie et al. 2011). The conclusions drawn from both of these studies, though still relatively preliminary, suggest two possible mechanisms by which TG2 could contribute to HD pathogenesis (see Fig. 13.1a).

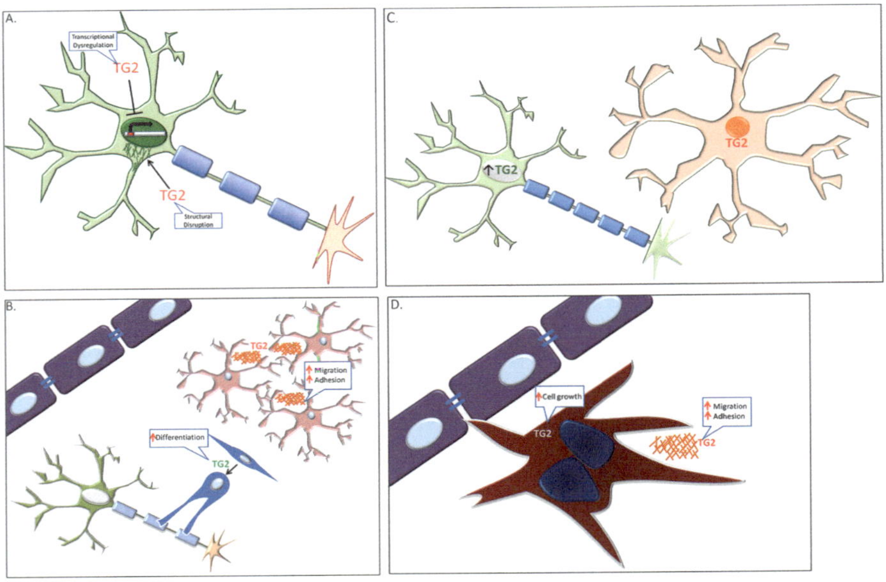

Fig. 13.1 Diagram of the current hypothetical framework of the role of TG2 in several neurological conditions. (**a**) *The role of TG2 in HD*. TG2 may be contributing to HD pathology by dysregulating transcription or facilitating aberrant actin-cofilin rod formation, thus leading to widespread cellular dysfunction (McConoughey et al. 2010; Munsie et al. 2011) (**b**) *The role of TG2 in MS*. TG2 can be beneficial through its role in facilitating the differentiation of oligodendrocyte precursors into mature oligodendrocytes, which is important for remyelination. On the other hand, it can also be detrimental by promoting formation of the astroglial scar through its role in adhesion and migration (Prins et al. 2013; van Strien et al. 2011a, b, c) (**c**) *The role of TG2 in ischemic stroke*. In neurons, the presence of TG2 is beneficial to their survival. Human TG2 increases in the nucleus in response to an ischemic injury and, in several cellular models, nuclear TG2 is protective (Filiano et al. 2010; Gundemir et al. 2013; Gundemir and Johnson 2009; Milakovic et al. 2004). On the other hand, when TG2 is expressed in astrocytes, it is detrimental to their survival, and to the survival of neurons (Colak and Johnson 2012) (**d**) *The role of TG2 in glial tumors*. TG2 enhances cellular proliferation and can also exhibit extracellular functions that increase tumor cell adhesion and migration (Yuan et al. 2011, 2007). *Red TG2* indicates a detrimental role; *green TG2* indicates a beneficial role

Some of the most intriguing evidence as to whether or not TG2 plays a role in HD pathology has come from in vivo studies using various mouse models of HD. In two studies, total body $TG2^{-/-}$ mouse lines were crossed with the R6/1 or R6/2 HD transgenic mouse lines, which express exon 1 of the human htt gene with 116 or 165–205 CAG repeats, respectively. In both cases, the deletion of TG2 led to improved lifespan and improved performance on the rotarod task, a widely used assessment of motor function. Interestingly, in both studies, TG2 deletion also led to an increase in neuronal htt aggregates (Bailey and Johnson 2005; Mastroberardino et al. 2002). This aligns with previous results that TG2 increased the solubility of polyQ proteins in vitro (Lai et al. 2004) and lends support to the hypothesis that aggregation in HD may actually be beneficial (Todd and Lim 2013), as an increase in aggregate load in these studies also coincided with improvements in disease progression.

Despite these initial in vivo experiments that identify TG2 as a potential contributor to HD pathology, several more recent studies have provided evidence to the contrary. In the first study, the R6/2 mouse line was crossed with a mouse line in which human TG2 was overexpressed in neurons only (Tucholski et al. 2006). This overexpression did not significantly affect motor performance, lifespan, striatal neuron atrophy or aggregate load. The authors thus concluded that either TG2 is not actually involved in any of these disease processes, or that any effects that TG2 may be exerting are not exacerbated by an excess of the protein (Kumar et al. 2012). The most significant piece of evidence that refutes the hypothesis that TG2 is involved in HD pathology is a recent study in which the R6/2 mouse line, this time with a CAG repeat length of ~240, and the zQ175 knock-in mouse line, which expresses human htt with a CAG repeat length of ~175, were crossed with the $TG2^{-/-}$ mouse model. In both mouse lines, the deletion of TG2 had no effect on HD pathology or functional outcomes. The differences in results could be due to several factors, including differences in housing conditions, more complex assessments of motor function and differences in CAG repeat length (Menalled et al. 2014). Overall, the most recent evidence indicates that TG2 may not be an important factor in HD disease pathology, despite initial evidence that TG2 can crosslink truncated versions of htt with pathological polyQ repeats in vitro and in cell culture systems. This may be due in part to the fact that relatively artificial htt constructs were used in many of the in vitro and in situ studies, thus limiting the translation of these findings to the in vivo situation. For a summary of in vivo studies in which the effect of TG2 manipulation on HD disease processes was examined, see Table 13.1.

13.2.2 *Other Polyglutamine Diseases*

The possible role of TGs in other polyQ expansion diseases has also been examined, although these studies are far less numerous than those that have focused on HD. One such disease, *dentorubral-pallidoluysian atrophy (DRPLA)*, is caused by a polyQ expansion in the atrophin-1 protein, which leads to motor and cognitive deficits (Margulis et al. 2013). Mutant truncated atrophin-1 formed intracellular aggregates when expressed in COS-7 cells. This aggregation was prevented by TG inhibitors, though the inhibitors used were non-specific and could have had off-target effects (Igarashi et al. 1998). In addition, ubiquitin-positive neuronal intranuclear inclusions were found in cerebellar tissue from DRPLA patients, and ~10 % of these inclusions were positive for TG activity. The incorporation of histidine-tagged GFP into the tissue was quantified and used as a measure for TG activity in this study, although this incorporation could have been non-specific and therefore may not accurately represent TG-mediated transamidation activity (Sato et al. 2002).

Table 13.1 Summary of in vivo experiments in HD models with genetic manipulation of TG2

Mouse Model		Effect of TG2 Manipulation on:				
HD mouse model	TG2 Manipulation	Motor function	Lifespan	Aggregate load	Neuronal atrophy	References
R6/1	Total knockout	Improved (rotarod task)	Improved	Increased	Reduced	Mastroberardino et al. (2002)
R6/2	Total knockout	Improved (rotarod task)	Improved	Increased	No effect	Bailey and Johnson (2005)
R6/2	Overexpression in neurons only	No effect	No effect	No effect	No effect	Kumar et al. (2012)
R6/2 and zQ175	Total knockout	No effect	No effect	No effect	No effect	Menalled et al. (2014)

Spinobulbar muscular atrophy (SBMA) is another polyQ expansion disease that leads to muscle weakness and atrophy. In this disease, the androgen receptor gene contains a CAG repeat expansion, which results in the degeneration of motor neurons in the spinal cord and brainstem (Margulis et al. 2013). It was shown that TG (isoform not specified) can crosslink N-terminal fragments of the androgen receptor in vitro, although its involvement in cellular or in vivo models is inconclusive (Mandrusiak et al. 2003).

Another subset of autosomal dominant polyQ expansion diseases known as *spinocerebellar ataxias (SCAs)* lead to the impairment of cerebellar functions, such as balance and coordination (Margulis et al. 2013). In spinocerebellar ataxia 1 (SCA1), the ataxin-1 protein contains a mutant polyQ expansion and has the tendency to aggregate in cerebellar nuclei. When lysates from HeLa cells expressing ataxin-1 with various polyQ lengths were exposed to TG2, protein crosslinking was evident, as high molecular weight bands appeared in a time-dependent manner. In addition, in a mouse model of SCA1, TG2 was elevated in the nucleus (D'Souza et al. 2006). This may suggest some involvement of TG2 in the aggregation of ataxin-1; however, much work remains to be done to confirm this role in cellular and in vivo systems.

Though most studies have focused on the role of TG2 in polyQ expansion diseases, a recent set of studies has come out about the possible role of TG6 in a specific form of spinocerebellar ataxia (SCA35). Two separate studies identified several missense mutations in the TGM6 gene in Chinese families with SCA35 (Li et al. 2013; Wang et al. 2010a). These mutations likely serve to destabilize TG6 and reduce its enzymatic activity (Guan et al. 2013a; Li et al. 2013). Two of these TG6 mutations, lysine (L) to tryptophan (W) at position 517 (L517W) and aspartic acid (D) to glycine (G) at position 327 (D327G), sensitized NIH3T3 cells to apoptosis (Guan et al. 2013a). In addition, TG6 was shown to contribute to the formation of ataxin-3 aggregates in HEK293 cells; however, the missense mutations did not affect its ability to do so (Guan et al. 2013b). This suggests that TG6 may contribute to the aggregation of mutant ataxin-3, but that the missense mutations likely have a separate effect that contributes to disease pathology.

Further work is necessary in order to determine whether or not TGs play a role in the processes contributing to polyQ expansion diseases.

13.3 Alzheimer's Disease

AD is the most common neurodegenerative disease and the most common cause of dementia in the elderly population. The pathology of this disease is characterized by the formation of neurofibrillary tangles (NFTs) composed of hyperphosphorylated tau protein and senile plaques composed of amyloid-β (Aβ). Since the pathological hallmarks are insoluble protein aggregates, it was initially thought that TGs were involved in their formation. TG2 levels and TG activity are increased in the cortex and hippocampus, but not in the cerebellum, of AD cases relative to aged matched controls (Johnson et al. 1997; Citron et al. 2001). These increases correlated with the pathology of AD since the hippocampus and cortex are affected, but not the cerebellum. However, other studies have found that TG1 and TG2 protein, mRNA and activity levels were increased in the cortex and in the cerebellum (Kim et al. 1999; Citron et al. 2001).

More than 30 years ago and prior to the discovery of the protein composition of the NFTs (Grundke-Iqbal et al. 1986a, b) (or the senile plaques), Selkoe and colleagues postulated that TGs may be responsible for the formation of the insoluble aggregates in the brains of AD patients (Selkoe et al. 1982). At the time of these studies, it was thought that neurofilaments were the primary constituents of the NFTs and thus the focus was on neurofilaments as the substrate. The formation of neurofilament polymers by TG2 was concentration dependent, with higher concentrations of TG2 resulting in more polymers (Miller and Anderton 1986). Furthermore, the formation of neurofilament polymers by TG2 was dependent on calcium and was inhibited by cystamine, a competitive substrate for the TG-catalyzed transamidation reaction (Jeitner et al. 2005). Additionally, once neurofilaments were crosslinked, they were no longer a substrate for TG2. However, upon closer examination, it was determined that the resulting filaments did not have the characteristic paired helical filament structure found in the aggregates from AD patients' brain, suggesting that TG2 was not responsible for NFT formation.

In 1986, it was discovered that tau is the primary protein of the NFTs (Grundke-Iqbal et al. 1986a, b). Subsequent to this discovery, the ability of TG2 to crosslink tau in vitro was investigated. Incubation of tau protein with guinea pig liver TG2 resulted in the rapid aggregation of tau (Dudek and Johnson 1993). The incubation with TG also rendered tau insoluble. Further studies revealed that all tau isoforms are substrates of TG2, although to varying degrees (Appelt and Balin 1997; Miller and Johnson 1995), and molecular experiments revealed that TGs modify tau protein at or near its microtubule binding domains (Miller and Johnson 1995; Murthy et al. 1998). These studies were able to clearly show that tau is an in vitro substrate for TGs.

Aβ, the peptide that forms the core of the senile plaques in AD brain, has also been studied as a substrate of TGs. In vitro experiments showed that synthetic Aβ, lengths 1–40, 1–28 and 1–42, were crosslinked by TG2 to form multimers (Dudek and Johnson 1994; Ikura et al. 1993; Rasmussen et al. 1994; Zhang et al. 1998; Hartley et al. 2008). The amyloid precursor protein (from which Aβ is formed) was also found to be a substrate of TG (isoform not specified) (Ho et al. 1994). However, all of these studies were done in vitro.

In order to better understand the role of TGs in AD, studies using AD brains were carried out. Paired helical filaments, which are formed by tau and are the primary structural components of NFTs, showed colocalization with TG2 in the hippocampus of AD cases (Appelt and Balin 1997). In another study, TG2 and tau colocalized in the hippocampus and frontal cortex of AD cases (Citron et al. 2001). Furthermore, TGs and Aβ colocalized in AD brains (Zhang et al. 1998). Since colocalization of proteins cannot be interpreted as a direct interaction, some researchers determined the presence of TG-mediated isopeptide bonds in AD brains. Various studies postulated that there was an increase in isopeptide bonds in AD cortex and spinal cord (Singer et al. 2002; Nemes et al. 2004; Wilhelmus et al. 2009). However, the antibodies used in these studies, 81D4 and 81D1C2, as noted above, are not specific for isopeptide bonds whenever they are within or between intact proteins (Johnson and LeShoure 2004), therefore immunocytochemical studies need to be interpreted with caution. However, the antibodies work well when the proteins are proteolyzed into short peptides and then used to isolate crosslinked fragments. Mass spectrometry can then be used to identify the proteins that have isopeptide bonds (Nemes et al. 2004). Interestingly, with this quantitative approach, neither Aβ nor tau was found to be crosslinked in AD brain. However, isopeptide bonds were found between ubiquitin and HSP27 in AD brain (Nemes et al. 2004). Finally, in the 3xTg-AD mouse model, TG2 did not colocalize with AD-like pathology in the cortex (Hohsfield et al. 2014).

After decades of studies, the data seem to indicate that TG2 may not be involved in the formation of senile plaques and NFTs, but there is still a possibility that it might contribute to the pathogenesis of AD. For example, the interaction of TG2 and the extracellular matrix (ECM) could be responsible for the accumulation of Aβ in blood vessel walls, which causes cerebral amyloid angiopathy (de Jager et al. 2013; Wilhelmus et al. 2012a). Finally, others have postulated that TGs can be used as a biological marker for AD since it was found to be elevated in the CSF of AD patients (Bonelli et al. 2002).

13.3.1 Other Tauopathies

Other diseases characterized by the aggregation of tau include progressive supranuclear palsy (PSP), Pick's disease and frontotemporal dementia with Parkinsonism linked to chromosome-17 (FTDP-17). The interaction between TGs and tau in these diseases has also been studied. In PSP brains, TG activity is increased in the globus pallidus and pons (Zemaitaitis et al. 2003). This increase in TG activity was accompanied by an increase in NFTs. Paired helical filaments in the pons also colocalized with isopeptide bonds, although the 81D4 antibody was used, which could result in an inaccurate detection of isopeptide bonds due to questions about specificity, as described above. mRNA levels of TG1 and TG2 were elevated in the globus pallidus and dentate of the cerebellum in PSP cases. However, only TG1 protein levels were increased in the globus pallidus; levels of TG2 were not increased in PSP in the brain regions examined as compared to controls (Zemaitaitis et al. 2000). Another study found colocalization of TG1 and TG-catalyzed crosslinked proteins in PSP, FTDP-17 and Pick's disease brain tissue. This study did not find TG2 to be colocalized with TG-catalyzed crosslinked proteins, which led to the conclusion that TG1, but not TG2, may be involved in the formation of NFTs in tauopathies (Wilhelmus et al. 2012b).

In a mouse model of FTDP-17, phosphorylated P301L tau colocalized with TGs in the cortex, spinal cord and hindbrain (Halverson et al. 2005). Phosphorylated tau and crosslinked proteins also coimmunoprecipitated in this mouse model. Furthermore, the activity of TGs was increased in the spinal cord. The immunohistochemical analysis and coimmunoprecipitation assay were performed, however, with the antibody 81D4. Further confirmatory studies are therefore needed to quantitatively determine the levels of isopeptide bonds and the proteins involved.

13.4 Parkinson's Disease

PD is a neurodegenerative condition that is defined by deficits in motor function due to the loss of dopaminergic neurons in the substantia nigra, a region of the midbrain responsible for controlling movement (Dawson and Dawson 2003). A pathological hallmark of the disease is the presence of cytoplasmic inclusions called Lewy bodies. Lewy bodies are mainly composed of α-synuclein, a presynaptic protein that forms amyloid-like fibers (Schmid et al. 2009). Although many pathological and physiological hallmarks of the disease are known, the available treatments attenuate the symptoms but do not affect disease progression. It has been hypothesized that inhibition of Lewy body formation could ameliorate disease progression. In addition, it has been suggested that within the Lewy bodies, α-synuclein is stabilized by crosslinking and that TG2 is the mediator of this stabilization process. The Mouradian group was the first to show, in vitro and in transfected COS-7 cells, that TG2 can crosslink α-synuclein (Junn et al. 2003). Additionally, the formation of the aggregates in COS-7 cells transfected with TG2 and α-synuclein was significantly increased when intracellular calcium was increased by treatment with the calcium ionophore A23187. In line with the findings from the Mouradian group, another group showed that TG2 has a high affinity for α-synuclein (Segers-Nolten et al. 2008), and the specific glutamine and lysine residues of α-synuclein that are crosslinked by TG2 ex vivo were identified (Schmid et al. 2009). Mutation of the glutamine residues in α-synuclein also abolished crosslinking by TG2. These initial studies suggested that TG2 could be a potential therapeutic target for the treatment of PD.

Recent studies have focused on determining if TG2 colocalizes with α-synuclein in culture and in vivo. In PD brains, TG2 is primarily found in the endoplasmic reticulum (ER) compartment of the cell and colocalizes with α-synuclein in melanized neurons (Wilhelmus et al. 2011). The same group also showed that inhibition of TG2 with an irreversible inhibitor, Z006, in α-synuclein transfected SH-SY5Y cells substantially reduced the formation of α-synuclein multimers (Verhaar et al. 2011). Furthermore, inhibition of TG2 attenuated α-synuclein aggregation induced by MPP treatment (a mitochondria complex I inhibitor used to model PD in mice). Moreover, TG2 and α-synuclein colocalize in neurons of double transgenic mice ($TG2^{Tg}/Syn^{Tg}$) (Grosso et al. 2014). The overexpression of TG2 increased the presence of higher order species of α-synuclein, which is consistent with an increase in the neuroinflammatory response and neurotoxicity. The $TG2^{Tg}/Syn^{Tg}$ mice also exhibited a behavioral phenotype that involved a decrease in their ability to form nests and balance (Grosso et al. 2014). Overall these findings suggest that TG2 may be involved in the formation of Lewy bodies in PD, although further studies are needed to confirm and extend these findings.

13.5 Amyotrophic Lateral Sclerosis

Amyotrophic lateral sclerosis (ALS) is a debilitating neurological disease that is caused by the progressive degeneration of motor neurons in the brain and spinal cord. In most cases, this deterioration first leads to a loss of motor function in the arms and legs, leaving patients unable to move. Respiratory failure is the most typical cause of death in ALS, as patients are eventually left unable to swallow and breathe. The prognosis for the disease is typically poor, with patients succumbing to the illness within several years of diagnosis. In some instances, respiratory failure occurs in the early stages of the disease, resulting in an even more dismal prognosis. Genetic mutations account for about 5–10 % of ALS cases; as such, the cause of most cases remains largely unknown (Kim et al. 2014; Ajroud-Driss and Siddique 2014).

The involvement of TGs in ALS pathology has not been extensively investigated, although some studies have been done. For example, in an early study, TG activity in postmortem spinal cord tissue from ALS patients and controls was examined. In this study, TG activity was reduced in the thoracic and lumbar spinal cord of ALS patients relative to that in controls (Fujita et al. 1995). In another study by the same group, TG activity in the CSF of later-stage ALS patients was lower than in controls, although TG activity in the CSF of earlier-stage ALS patients was higher. There was also a significant correlation between serum TG activity and the severity of the ALS case, with lower TG activity correlating with a more severe case of ALS (Fujita et al. 1998a). Taken together, these data suggest that in early stages of the disease, there is an increase in TG enzymatic activity, whereas in later stages this activity has been depleted, possibly due to the extensive motor neuron death that has occurred (Fujita et al. 1998a). These studies did not, however, demonstrate which TG isoform was contributing to these activity levels and whether or not this activity was contributing to disease progression.

TG activity has been measured in spinal cords from *mnd* mice, which are considered to be a model of ALS. *Mnd* mice harbor several genetic mutations that ultimately result in extensive degeneration of motor neurons in the anterior horn of the spinal cord, as well as in other regions of the CNS (Holmes and Haynes 1996). The incorporation of radiolabeled putrescine into exogenous or endogenous substrates was ~4 times higher in *mnd* mouse spinal cord than in control spinal cord at 8 months of age. At 6 months of age, there was still a significant elevation in TG activity in the *mnd* mice, but it was less robust than what was observed at 8 months. At 3 months of age, when signs of degeneration are not yet present, TG activity was not elevated (Holmes and Haynes 1996). In a similar study with the *mnd* mouse model, TG activity in the thoracic spinal cord exhibited a comparable pattern, in which activity continued to increase through 9 months of age relative to control spinal cord. In the lumbar spinal cord, however, TG activity was higher than in controls at early ages, peaking at 5 months of age and then beginning to decline. The authors suggest that this may be due to the fact that the lumbar spinal cord may be the first region to degenerate in this model. Thus, once degeneration has begun, TG enzymatic activity may begin to decline as well (Fujita et al. 1998b).

A more recent study focused on the mechanisms by which TG2 may be mediating pathology in a specific form of ALS. Approximately 2 % of ALS cases are caused by a mutation in the gene for superoxide dismutase 1 (SOD1), which can lead to the misfolding and oligomerization of the protein (Oono et al. 2014). In this study, HEK293A cells were transiently cotransfected with wild-type or mutant SOD1 constructs and TG2. TG2 coprecipitated with mutant but not wild-type forms of SOD1. In addition, in the presence of calcium, TG2 was able to induce the oligomerization of mutant SOD1 in vitro, which suggests that mutant SOD1 oligomers may at least in part be stabilized via TG2 crosslinking. They also showed that transgenic mice which express mutant human SOD1 (SOD1 with a glycine (G) to an alanine (A) mutation at position 93) had increased TG2 protein levels in the spinal cord relative to wild-type mice at the pre-symptomatic stage, but not at the early symptomatic stage (Oono et al. 2014). These findings are intriguing, but it is clear that further work is required in order to determine if TG2 is a key mediator of ALS pathology overall, or specifically in the forms of ALS that are caused by mutations in SOD1.

13.6 Multiple Sclerosis

Multiple sclerosis (MS) is an autoimmune demyelinating disease of the brain and spinal cord. Two main pathological hallmarks of the disease are demyelinated lesions and astroglial scars, both of which are a result of the infiltration of immune cells. The lack of myelinated neurons arises from the inability of oligodendrocytes to produce myelin sheaths. Astroglial scar formation occurs as a result of increased migration of astrocytes in response to inflammation. As astrocytes migrate to a given area, increased amount of matrix deposition impedes tissue repair and transport of nutrients (Espitia Pinzon et al. 2014; van Strien et al. 2011b). Current and past work from the van Dam group has shown how TG2 may play a role in both events.

13.6.1 Demyelinated Lesions

One of the main problems in MS is the inability of oligodendrocytes to form myelin sheaths. Myelin sheaths are critically important in the CNS, as they aid in the proper transmission of signals from the body of the neuron to the axonal terminal. This process is imperative for signal propagation by neurons. Mature oligodendrocytes are responsible for the production of myelin sheaths, and it has been hypothesized that the formation of demyelinated lesions may be due to a decrease in mature oligodendrocytes and an increase of oligodendrocyte precursor cells in MS patients (Goldschmidt et al. 2009). The van Dam group carried out studies to determine if TG2 played a role in the differentiation of oligodendrocyte precursor cells to mature oligodendrocytes (an Strien et al. 2011a). In their studies, the group used a standard cuprizone diet in order to kill mature oligodendrocytes. Cuprizone is a copper chelator that is thought to recruit a specific population of neutrophils to the CNS that target mature oligodendrocytes (Liu et al. 2010). Results showed that $TG2^{-/-}$ mice were not able to re-myelinate after cuprizone treatment compared to wild-type mice. Upon further examination, the $TG2^{-/-}$ mice showed decreases in MBP (mature oligodendrocyte marker), but did not have an increase in oligodendrocyte precursor cells. Additionally, TG2 expression was higher in oligodendrocyte precursor cells than mature oligodendrocytes, and inhibition of TG2 resulted in a decrease in oligodendrocyte precursor cell differentiation (van Strien et al. 2011a). In this context, TG2 appears to play a role in facilitating myelin formation.

13.6.2 *Astroglial Scar*

The formation of astroglial scars is the outcome of astrocyte migration to inflammation signals of the demyelinated lesions in the CNS. After migration, the changes in expression of ECM molecules create an environment that impedes tissue repair (van Strien et al. 2011c). The van Dam group hypothesized that TG2 is responsible for the crosslinking and dense ECM in the astroglial scar formation. The group first showed a decrease in migration and adhesion when TG2 was knocked down or inhibited, using U373 cells as the model system (van Strien et al. 2011c). The group then showed that there was an increase in TG2 expression in primary rat astrocytes upon pro-inflammatory cytokine treatment (van Strien et al. 2011b). Additionally, the interaction of TG2 with fibronectin facilitated adhesion and migration of astrocytes. Finally, the van Dam group was interested in translating these findings into an in vivo model and thus used the marmoset experimental MS model (Espitia Pinzon et al. 2014). In this study, they determined the differences in TG2, fibronectin and integrin expression in white versus gray matter. In white matter MS lesions, there is a disruption of the blood-brain barrier that leads to an infiltration of immune cells, which contributes to inflammation, demyelination and axonal damage. In gray matter lesions, there is a decrease in immune cell influx, but there is still damage (to a lesser extent than in the white matter region). Results showed that in white matter lesions, TG2 colocalized with β_1-integrin and was found near deposited fibronectin. This was an important finding, as they were able to show the interaction between TG2 and integrins in human brain (Prins et al. 2013). Finally, in gray matter lesions, TG2 and β_1-integrin were present but fibronectin was absent. More experiments are needed in order to assess if TG2 can promote proliferation and migration in the absence of fibronectin.

In conclusion, it seems that TG2 is a double-edged sword in MS. TG2 is needed in order to produce myelin sheaths but it also plays a key role in promoting astroglial scarring. Consequently, additional research is needed in order to dissect whether or not TG2 is a good therapeutic target for MS (see Fig. 13.1b).

13.7 Central Nervous System Injury

13.7.1 Stroke

Stroke affects approximately 15 million people worldwide every year (Organization 2002). Unfortunately, there are only limited treatments available to lessen the effects of a stroke and more effective therapeutic strategies are needed. Increased understanding of the molecular events that occur in the brain in response to ischemia will allow for the development of more effective treatments. One potential therapeutic target is TG2, which is emerging as a key modulator of ischemic cell death. In early studies using different animal models, it was demonstrated that the mRNA levels of TG2 are increased after ischemic injury. These increases in mRNA levels were also accompanied by an increase in protein levels (Ientile et al. 2004; Tolentino et al. 2004; Hwang et al. 2009; Basso et al. 2012; Fujita et al. 2006). Additional studies indicate that, in general, CNS injuries appear to result in increases in TG2. Although it is clear that TG2 increases subsequent to injury, what is not clear is whether this is detrimental or beneficial for neuronal survival. Early studies suggested that post-injury increases in TG2 facilitated cell death processes; however, more recent studies provide evidence that this may not always be the case.

In 2008 it was shown that in neurons, TG2 was protective against oxygen and glucose deprivation (OGD) and this protection was independent of the transamidating activity of TG2 (Filiano et al. 2008). These initial findings were then extended by several in vivo studies. Wild-type mice and mice in which human TG2 was overexpressed selectively in neurons were subjected to middle cerebral artery ligations (MCALs) followed by measurement of stroke volumes. These studies clearly demonstrated that increased neuronal expression of TG2 was protective, as infarct size was significantly reduced in the mice expressing human TG2 in neurons (Filiano et al. 2010). To extend these findings, a mouse model in which TG2 was knocked out in all tissues (TG2$^{-/-}$) (Nanda et al. 2001) was used. When these mice were subjected to MCALs, surprisingly their stroke volumes were significantly less than those of wild-type control mice (Colak and Johnson 2012). By culturing neurons and astrocytes from the wild-type and TG$^{2-/-}$ mice it was found that, as expected, neurons from TG2$^{-/-}$ mice showed increased cell death in response to OGD compared to wild-type neurons. However, astrocytes from $TG2^{-/-}$ mice were more resistant to ischemia-induced cell death. Additionally, astrocytes from $TG2^{-/-}$ mice protected neurons from OGD-induced cell death to a significantly greater extent than astrocytes from wild-type mice (Colak and Johnson 2012). These data would suggest that in the complete absence of TG2, the increased viability and protective effects in astrocytes is of greater influence in promoting survival of the neurons in ischemic conditions. The reasons why neuronal TG2 is pro-survival and astrocytic TG2 is detrimental to survival in response to ischemia are not yet known. Overall, this is an important finding because it clearly shows that the ability of TG2 to ameliorate or enhance cell death processes is dependent on cell type and context (see Fig. 13.1c).

In vitro studies have shown that exogenously expressed TG2 protected striatal cells, cortical cells and non-neuronal cells against OGD-induced cell death whether or not it was transamidating activity competent (Colak et al. 2011; Filiano et al. 2008; Gundemir and Johnson 2009). These data suggest that the neuroprotective role of TG2 after an ischemic injury is transamidation independent. Although the transamidating activity of TG2 may not play an important role in the ability of TG2 to protect against ischemic-induced cell death, cellular localization seems to be a factor. MCALs in mice overexpressing TG2 in neurons resulted in increases in nuclear levels of TG2 in the stroked area (Filiano et al. 2010). Additionally, neuronal nuclear TG2 levels were increased in the area of the lesion in postmortem brains of human stroke patients (Filiano et al. 2010). To examine the effects of subcellular localization on cell viability, neuronal and non-neuronal cells expressing nuclear localization sequence (NLS) tagged TG2, NLS-tagged transamidation-inactive TG2 or an NLS-tagged guanine nucleotide binding deficient form of TG2 were exposed to various cell stressors, such as OGD. In all cases, targeting TG2 to the nucleus was protective to a greater extent than untagged TG2 (Gundemir et al. 2013; Gundemir and Johnson 2009; Milakovic et al. 2004).

There is evidence that the increase in protein levels and nuclear presence of TG2 is pro-survival in neurons, but little is known about how it mediates this protection. Previous work in cortical neurons exposed to OGD showed a physical interaction between TG2 and HIF1β (Filiano et al. 2008; Gundemir et al. 2013). This interaction was accompanied by a decrease in HIF transcriptional activity in response to ischemic stress. However, subsequent studies revealed that the interaction between TG2 and HIF1β and its suppression of HIF activity is independent from TG2's pro-survival role (Gundemir et al. 2013). Further studies are needed to elucidate how TG2 confers neuroprotection after an ischemic event.

The vast majority of studies regarding the role of TGs in stroke have focused on TG2; however, the role of factor XIII in stroke has also been examined. Levels of factor XIII in the blood have been measured in many cohorts of stroke patients in order to determine if there is a correlation between levels and prognosis for the patient. One study found a decrease in the catalytic subunit A of factor XIII in stroke patients with higher rates of survival (Kohler et al. 2002). One well-studied mutation is the polymorphism in factor XIII, which results in a leucine (L) instead of a valine (V) at position 34 (V34L). This mutation can be either heterozygous or homozygous, and studies have focused on trying to link either one as protective in stroke patients. Although studies are still ongoing, one report suggested that patients with a homozygous mutation had a better outcome after ischemic or hemorrhagic strokes (Elbaz et al. 2000). Other studies have found that this mutation may actually predispose patients to suffer a stroke with poorer prognosis (Gemmati et al. 2001; Reiner et al. 2002; Catto et al. 1998). However, the majority of studies have found no correlation between this mutation, the predisposition, and prognosis after a stroke (Endler et al. 2003; Iniesta et al. 2001; Akar et al. 2007; Kain et al. 2005). The variability in the studies could be due to different number of patients per study, the grouping of many kinds of stroke instead of studying them individually, and failure to account for the fact that the environment of each population could also affect the outcome of a stroke. Whatever the reason for these disparate results, more controlled studies need to be carried out in order to understand the role or lack thereof of transglutaminase in ischemic injury.

13.7.2 Physical Injury of the Nervous System

There are also data to suggest that TGs may play a role in traumatic brain injury (TBI). When rats were injured using a cortical impact model, the levels of TG2 protein increased in the cortex, peaking at day 3 (Tolentino et al. 2002). In another study, a mild contusion to the rat spinal cord resulted in an increase in TG2 (Festoff et al. 2002). Physical injury to the peripheral nervous system by crushing the vagus nerve resulted in an increase in TG2 activity, peaking 30 min after injury near the nodose ganglion. However, 48 h after injury, TG2 activity was increased in the nodose ganglion itself, suggesting that TG2 activity increases as the injury progresses (Tetzlaff et al. 1988). Finally, in a study in which rats with injured optic nerves were injected with a nerve derived TG, a rescue of the visual evoked potential response to light previously lost due to injury was detected. Upon closer examination of the injury site, there was an increase in unmyelinated nerve fibers that resembled new axons in the TG-treated animals. The increase in these new nerve fibers peaked by week 6 after injury (Eitan et al. 1994). These findings show that the activity of TG2 is increased in injuries to the central and peripheral nervous system and it is possible that its administration in certain situations could reduce the injury.

13.8 Primary Brain Tumors

Among all the TGs, TG2 is the one that is most likely to play a critical role in primary brain tumor progression. It has been implicated as a contributing factor to increased adhesion, cell survival, and proliferation, primarily in meningioma and glioma type tumors.

13.8.1 Meningioma

Meningiomas arise from arachnoid cells located in the leptomeninges of the brain and account for approximately 20 % of CNS tumors (Huang et al. 2014). The role of TG2 has not been extensively characterized or studied in these tumors; however, there have been several interesting studies. Significant increases in TG2 activity in the most aggressive form of meningioma tumors compared to normal brain tissue have been reported (Yuan et al. 2008). Additionally extracellular TG2 was found to colocalize with fibronectin. This was expected, as previous work suggested that TG2 facilitates the interaction between fibronectin and β1-integrin (Wang et al. 2010b). The combination of radiation and TG2 inhibition with KCC009, an irreversible inhibitor of TG2, resulted in an increase in apoptosis that was dependent on the disruption of the TG2-fibronectin complex (Yuan et al. 2008). In a more recent study, a microarray analysis of meningiomas suggested that TG2 could be a possible molecular marker for high-grade tumors, as TG2 expression was consistently higher in the most aggressive grade (Huang et al. 2014). Moreover, knockdown of TG2 induced a significant increase in apoptosis in cultured meningiomas (Huang et al. 2014). Additional work is still needed in order to understand the molecular underpinnings of TG2 induction in high-grade meningioma and how this contributes to the aggressiveness of the tumor.

13.8.2 Glioma

Glial tumors are derived from a singular or a combination of several types of supportive brain cells; which includes astrocytes, oligodendrocytes, and ependymal cells. Interestingly, a study looking at all glial tumors showed that patients with a 2-fold down-regulation of TG2 had a better probability of survival than patients with a 2-fold up-regulation (Dyer et al. 2011). In another study, TG2 was found to be elevated in the most aggressive type of glioma, glioblastoma multiforme (GBM) (Zhang et al. 2013). GBM tumors are characterized as being highly diffusive, infiltrative, and having a high propensity for reoccurrence. Numerous studies have shown that TG2 is involved in the migration and adhesion of different types of cells (Nurminskaya and Belkin 2012; Gundemir et al. 2012). Given their infiltrative nature, this role of TG2 was investigated in GBMs. The TG2 inhibitor ERW1227B decreased GBM migration and increased cell death (Yuan et al. 2011). In a previous paper, this same group showed TG2 colocalization with fibronectin in frozen human GBM tumor tissue and cell lines (Yuan et al. 2007). Additionally, this interaction was disrupted when the TG2 inhibitor was used in the U87MG human glioblastoma cell line (Yuan et al. 2007). The inhibitor also disrupted the interaction between actin and integrin (Yuan et al. 2011). Finally, although inhibition of TG2 alone was not sufficient to increase survival in mice with intracranially implanted GBM cells, combining it with BCNU, an alkylating agent currently used in patients, increased survival to a greater extent than when only the alkylating agent was used (Yuan et al. 2011) (See Fig. 13.1d). Studies have indicated that TG2 may be able to function as a signal transducing G protein ($G\alpha_h$). Therefore, the ability of TG2 to regulate adenylate cyclase in GBMs was examined (Obara et al. 2013). Increased cyclic AMP production can contribute to cancer progression since it can cause increases in survival and proliferation. These studies indicated that TG2 regulates adenylate cyclase 8; however, the mechanism of this regulation was not clearly defined (Obara et al. 2013).

GBM tumors have been divided into several subtypes based on the molecular differences and cancer driver mutations. A common subtype is the classical subtype, which is characterized by having a mutation in and/or amplification of epidermal growth factor receptor (EGFR). The mutation results in the constitutive activation of the receptor. Recently a possible role of TG2 in EGFR internalization in GBM cells was examined (Zhang et al. 2013). One of the E3 ubiquitin ligases responsible for initiating EGFR internalization, cCbl, was implicated as a direct TG2 interactor. In the paper, the authors showed that, upon incubating with monodansyl cadaverine and stimulating the receptor with EGF, there was a decrease in EGFR total levels (Zhang et al. 2013). Moreover, there was a decrease in the TG2-cCbl interactions in the presence of monodansyl cadaverine. It was suggested that TG2 sequestered cCbl in the cytoplasm which impeded the ubiquitylation of EGFR. These findings are intriguing; however, additional studies are needed in order to elucidate the role of TG2 in other GBM models and subtypes.

It is now evident that chemotherapeutics do not work as effectively in certain cancer cell populations compared with others. GBM tumors are highly heterogeneous, and one of the identified cell populations within a GBM tumor that is resistant to chemotherapeutics is the glioma-initiating cell. Glioma-initiating cells are stem cell-like cells that are responsible for maintenance and re-population of cells within the tumor. A recent study shed light on the possible importance of TG2 in these cells. The authors found that glioma-initiating cells express significantly more TG2 than non-glioma-initiating cells and, when TG2 is knocked down or inhibited, there is an increase in apoptosis and a decrease in the ability to form colonies (Fu et al. 2013). Additionally, inhibition or knockdown of TG2 results in a decrease in inhibitor of DNA binding 1 protein (ID1), which is believed to be the mechanism by which TG2 increases proliferation in GBM cells. Finally, in vivo studies revealed that treatment with monodansyl cadaverine resulted in a decrease in tumor weight and an increase in survival (Fu et al. 2013). Unfortunately, monodansyl cadaverine is not a TG specific inhibitor. For example, it has been shown to inhibit EGF and fibrin-stabilizing factor (Haigler et al. 1980; Nilsson et al. 1971). Further experiments with specific TG inhibitors are needed in order to validate whether the effects of monodansyl cadaverine are due to TG2 inhibition.

13.9 Neuroprotection Mediated by Cystamine

The disulfide cystamine is a known TG inhibitor that acts as a competitive substrate for the TG-catalyzed transamidation reaction in vitro (Jeitner et al. 2005). Cystamine has therefore been considered as an attractive therapeutic option in the treatment of neurodegenerative diseases in which TGs may be involved. Indeed, cystamine is neuroprotective in several models of neurodegeneration (Hwang et al. 2009; Karpuj et al. 2002; Bailey and Johnson 2006). However, the exact mechanism by which cystamine is neuroprotective is a matter of debate. Cystamine can inhibit caspase-3 activity, in addition to acting as a non-specific TG inhibitor (Lesort et al. 2003). Cystamine can also act as an antioxidant by increasing the levels of glutathione (GSH) in cell culture and in rat tissue (Kovarova and Pulpanova 1979; Lesort et al. 2003; Fox et al. 2004). These additional roles for cystamine could be potential mechanisms for its neuroprotective effects in disease models. In addition, cystamine does not appear to be able to cross the blood-brain barrier. Instead, the disulfide bond is rapidly reduced, yielding cysteamine, which may be able to enter the brain (Pinto et al. 2005). Therefore, the neuroprotective effects may be due to cysteamine or its metabolite, L-cysteine (Bousquet et al. 2010; Gibrat and Cicchetti 2011; Fox et al. 2004; Pinto et al. 2005). An entirely separate peripheral antioxidant mechanism may also explain, at least partially, its protective effects. Some of the studies that have used cystamine in specific neurodegenerative disease models are discussed below.

13.9.1 *Huntington's Disease*

The possible efficacy of cystamine in the treatment of HD has been assessed in several mouse models (Dedeoglu et al. 2002; Karpuj et al. 2002; Van Raamsdonk et al. 2005; Wang et al. 2005). It was initially considered as a potential therapeutic due to its ability to inhibit TG2, which was thought to play a pathological role in HD progression (see previous section on Polyglutamine Diseases). In the YAC128 HD mouse, a model in which human mutant htt is expressed in the same tissues as endogenous mouse htt (Slow et al. 2003), cystamine treatment resulted in a decrease in striatal neuron loss but no functional improvements in motor skills (Van Raamsdonk et al. 2005). In a different HD mouse model, the R6/2 mouse (see Polyglutamine Diseases), cystamine slowed disease progression (Dedeoglu et al. 2002; Karpuj et al. 2002; Wang et al. 2005). There was a delay in weight loss as well as an improvement in motor function with cystamine treatment (Dedeoglu et al. 2002; Karpuj et al. 2002). In two studies, there was a decrease in intracellular aggregates and an improvement in the health of striatal neurons (Dedeoglu et al. 2002; Wang et al. 2005). In one study, however, there was no change in intracellular aggregates with cystamine treatment, despite improvements in motor function (Karpuj et al. 2002). This disparity could be due to differences in dosage and the time course for the administration of cystamine. Regardless, the presence of motor improvements with no concurrent effect on the formation of aggregates lends further support to the idea that aggregation may not actually be a significant contributing factor to HD pathogenesis (Todd and Lim 2013).

It is important to note that in previously mentioned studies (see Polyglutamine Diseases), a decrease in TG2 protein levels actually led to an increase in intracellular aggregates in HD mouse models (Bailey and Johnson 2005; Mastroberardino et al. 2002). If cystamine were functioning as a TG2 inhibitor in these models, it would be expected that the effect of TG2 inhibition on aggregate formation would be similar to that of TG2 depletion, but this is not the case. To address this discrepancy, Bailey and Johnson assessed the effects of cystamine treatment in R6/2 mice with or without TG2 deletion. Cystamine delayed the onset of motor dysfunction and improved the lifespan of R6/2 mice, whether or not TG2 was present (Bailey and Johnson 2006). The evidence therefore strongly suggests that cystamine exerts its neuroprotective effects independently of TG2 inhibition.

In several studies, cystamine treatment led to an increase in neuroprotective genes, such as Hsp40 in R6/2 mouse brain, and HSJ1b in striatal neurons (Borrell-Pages et al. 2006; Karpuj et al. 2002). The increase in HSJ1b led to an increase in neuronal BDNF secretion. Cystamine also led to an increase in brain BDNF levels in the Hdh^{109Q} knock-in HD mouse model (Borrell-Pages et al. 2006). In a separate study, L-cysteine, which functions as an antioxidant, was increased in R6/2 mouse brain following cystamine treatment (Fox et al. 2004). These changes are potential alternative explanations for the neuroprotective effects mediated by cystamine that are separate from its ability to inhibit TGs.

13.9.2 Parkinson's Disease

The first evidence of a putative role for TG2 in PD was published by the Mouradian group in 2003 (Junn et al. 2003). The group was interested in the transamidation activity of TG2 in α-synuclein oligomer formation, which is the primary component of Lewy bodies (see previous section on Parkinson's Disease). After in vitro characterization of the ability of TG2 to crosslink α-synuclein, cystamine was used in COS-7 cells transfected with TG2 and α-synuclein in order to see if inhibition of TG2 was sufficient to minimize α-synuclein crosslinking. Results in this cell model suggested that cystamine treatment decreased α-synuclein crosslinking.

13.9.3 Amyotrophic Lateral Sclerosis

There is a single study in which the effect of cystamine was investigated in a mouse model of ALS. In mice expressing mutant SOD1 (see previous section on

Amyotrophic Lateral Sclerosis), cystamine reduced disease progression, as evidenced by an attenuation of body weight loss, improvement in grip strength and an improved lifespan. The oligomerization of mutant SOD1 and microglial activation were also reduced (Oono et al. 2014). While treatment with cystamine is promising in this mouse model, there is no evidence to suggest that its effect is due to the inhibition of TGs.

13.9.4 Stroke

There have been various studies using cystamine treatment to attenuate injury in stroke models. The premise for these studies was based on the findings that TG2 activity and protein levels are increased after hypoxic stress and, initially, it was assumed that this contributed to neuronal death. Therefore, it was supposed that cystamine would be neuroprotective by inhibiting TG activity. Intracerebral hemorrhages (ICHs) in mice resulted in an increase in TG2 protein and mRNA levels on the ipsilateral side (Okauchi et al. 2009). Cystamine administered intraperitoneally after ICH resulted in decreased brain swelling and reduced cell death around the injury (Okauchi et al. 2009). The authors suggest that TG inhibition might be involved in the neuroprotective effects of cystamine after ICH, but they did not discard the possibility that other effects of cystamine are involved in this neuroprotection. However, they never specifically mention the possibility that cystamine might not be crossing the blood-brain barrier. In another study evaluating cystamine neuroprotection after stroke, gerbils were subjected to middle cerebral artery occlusions (MCAOs) followed by cystamine administration 30 min after the injury (Hwang et al. 2009). MCAOs increased TG2 levels, and cystamine treatment increased locomotion compared to vehicle-treated animals. However, the increase in locomotion due to cystamine treatment only occurred one day after MCAO and later decreased to control levels.

Mice subjected to a model of focal stroke and later treated with cystamine showed improved motor function, higher density of fiber tracks from motor cortex and an increase in neuronal survival around the injury site (Li et al. 2014). In order to understand the mechanism of action of cystamine, animals were treated with cystamine and an antagonist of the TrkB receptor, which is the receptor to which BDNF binds. Inhibition of the BDNF pathway resulted in the elimination of the neuroprotection conferred by cystamine. Upon closer examination, it was found that cystamine treatment did not result in an increase in BDNF protein levels, but rather an increase in the phosphorylation state of the TrkB receptor, which increased its activity. It is important to note that the activation or administration of BDNF correlates with better outcomes after a stroke (Zhang and Pardridge 2001; Chen et al. 2013; Almeida et al. 2005). Overall, cystamine treatment has a neuroprotective effect in stroke models, but consideration of the fact that it cannot be detected in the brain after peripheral administration suggests that a metabolite is likely the active agent, and that it is likely that the effects are not due to the inhibition of TGs.

13.10 Concluding Remarks

The potential contribution of TGs, and in particular TG2, to neurological conditions has been broadly investigated. TG2 is a multifunctional protein, and its ability to modulate numerous cellular functions that impact survival indicates that, in the various disease conditions in which it is involved, it likely plays different roles. The results from numerous studies have clearly indicated that we cannot define the effects of TGs in neurological disease processes solely in terms of its ability to crosslink proteins. Indeed, the formation and/or stabilization of protein aggregates are likely not the primary mechanisms by which TGs mediate disease progression. Nonetheless, it is evident that TG2 is a contributor to a number of different neurological diseases and further investigations are warranted to more fully define its role and to develop potential TG-targeted therapeutics to treat these diseases.

References

Ajroud-Driss S, Siddique T (2014) Sporadic and hereditary amyotrophic lateral sclerosis (ALS). Biochim Biophys Acta 1852:679–684

Akar N, Donmez B, Deda G (2007) FXIII gene Val34Leu polymorphism in Turkish children with cerebral infarct. J Child Neurol 22:222–224

Almeida RD, Manadas BJ, Melo CV, Gomes JR, Mendes CS, Graos MM, Carvalho RF, Carvalho AP, Duarte CB (2005) Neuroprotection by BDNF against glutamate-induced apoptotic cell death is mediated by ERK and PI3-kinase pathways. Cell Death Differ 12:1329–1343

Appelt DM, Balin BJ (1997) The association of tissue transglutaminase with human recombinant tau results in the formation of insoluble filamentous structures. Brain Res 745:21–31

Bailey CD, Johnson GV (2005) Tissue transglutaminase contributes to disease progression in the R6/2 Huntington's disease mouse model via aggregate-independent mechanisms. J Neurochem 92:83–92

Bailey CD, Johnson GV (2006) The protective effects of cystamine in the R6/2 Huntington's disease mouse involve mechanisms other than the inhibition of tissue transglutaminase. Neurobiol Aging 27:871–879

Basso M, Berlin J, Xia L, Sleiman SF, Ko B, Haskew-Layton R, Kim E, Antonyak MA, Cerione RA, Iismaa SE, Willis D, Cho S, Ratan RR (2012) Transglutaminase inhibition protects against oxidative stress-induced neuronal death downstream of pathological ERK activation. J Neurosci 32:6561–6569

Bates GP, Mangiarini L, Wanker EE, Davies SW (1998) Polyglutamine expansion and Huntington's disease. Biochem Soc Trans 26:471–475

Bonelli RM, Aschoff A, Niederwieser G, Heuberger C, Jirikowski G (2002) Cerebrospinal fluid tissue transglutaminase as a biochemical marker for Alzheimer's disease. Neurobiol Dis 11:106–110

Borrell-Pages M, Canals JM, Cordelieres FP, Parker JA, Pineda JR, Grange G, Bryson EA, Guillermier M, Hirsch E, Hantraye P, Cheetham ME, Neri C, Alberch J, Brouillet E,

Saudou F, Humbert S (2006) Cystamine and cysteamine increase brain levels of BDNF in Huntington disease via HSJ1b and transglutaminase. J Clin Invest 116:1410–1424

Bousquet M, Gibrat C, Ouellet M, Rouillard C, Calon F, Cicchetti F (2010) Cystamine metabolism and brain transport properties: clinical implications for neurodegenerative diseases. J Neurochem 114:1651–1658

Catto AJ, Kohler HP, Bannan S, Stickland M, Carter A, Grant PJ (1998) Factor XIII Val 34 Leu: a novel association with primary intracerebral hemorrhage. Stroke 29:813–816

Chen A, Xiong LJ, Tong Y, Mao M (2013) The neuroprotective roles of BDNF in hypoxic ischemic brain injury. Biomed Rep 1:167–176

Chun W, Lesort M, Tucholski J, Ross CA, Johnson GV (2001) Tissue transglutaminase does not contribute to the formation of mutant huntingtin aggregates. J Cell Biol 153:25–34

Citron BA, Santacruz KS, Davies PJ, Festoff BW (2001) Intron-exon swapping of transglutaminase mRNA and neuronal Tau aggregation in Alzheimer's disease. J Biol Chem 276:3295–3301

Colak G, Johnson GV (2012) Complete transglutaminase 2 ablation results in reduced stroke volumes and astrocytes that exhibit increased survival in response to ischemia. Neurobiol Dis 45:1042–1050

Colak G, Keillor JW, Johnson GV (2011) Cytosolic guanine nucleotide binding deficient form of transglutaminase 2 (R580a) potentiates cell death in oxygen glucose deprivation. PLoS One 6: e16665

Cooper AJ, Sheu KF, Burke JR, Onodera O, Strittmatter WJ, Roses AD, Blass JP (1997) Polyglutamine domains are substrates of tissue transglutaminase: does transglutaminase play a role in expanded CAG/poly-Q neurodegenerative diseases? J Neurochem 69:431–434

Dawson TM, Dawson VL (2003) Molecular pathways of neurodegeneration in Parkinson's disease. Science 302:819–822

de Cristofaro T, Affaitati A, Cariello L, Avvedimento EV, Varrone S (1999) The length of polyglutamine tract, its level of expression, the rate of degradation, and the transglutaminase activity influence the formation of intracellular aggregates. Biochem Biophys Res Commun 260:150–158

de Jager M, van der Wildt B, Schul E, Bol JG, van Duinen SG, Drukarch B, Wilhelmus MM (2013) Tissue transglutaminase colocalizes with extracellular matrix proteins in cerebral amyloid angiopathy. Neurobiol Aging 34:1159–1169

de Vivo G, di Lorenzo R, Ricotta M, Gentile V (2009) Role of the transglutaminase enzymes in the nervous system and their possible involvement in neurodegenerative diseases. Curr Med Chem 16:4767–4773

Dedeoglu A, Kubilus JK, Jeitner TM, Matson SA, Bogdanov M, Kowall NW, Matson WR, Cooper AJ, Ratan RR, Beal MF, Hersch SM, Ferrante RJ (2002) Therapeutic effects of cystamine in a murine model of Huntington's disease. J Neurosci 22:8942–8950

D'Souza DR, Wei J, Shao Q, Hebert MD, Subramony SH, Vig PJ (2006) Tissue transglutaminase crosslinks ataxin-1: possible role in SCA1 pathogenesis. Neurosci Lett 409:5–9

Dudek SM, Johnson GV (1993) Transglutaminase catalyzes the formation of sodium dodecyl sulfate-insoluble, Alz-50-reactive polymers of tau. J Neurochem 61:1159–1162

Dudek SM, Johnson GV (1994) Transglutaminase facilitates the formation of polymers of the beta-amyloid peptide. Brain Res 651:129–133

Dyer LM, Schooler KP, Ai L, Klop C, Qiu J, Robertson KD, Brown KD (2011) The transglutaminase 2 gene is aberrantly hypermethylated in glioma. J Neurooncol 101:429–440

Eitan S, Solomon A, Lavie V, Yoles E, Hirschberg DL, Belkin M, Schwartz M (1994) Recovery of visual response of injured adult rat optic nerves treated with transglutaminase. Science 264:1764–1768

Elbaz A, Poirier O, Canaple S, Chedru F, Cambien F, Amarenco P (2000) The association between the Val34Leu polymorphism in the factor XIII gene and brain infarction. Blood 95:586–591

Endler G, Funk M, Haering D, Lalouschek W, Lang W, Mirafzal M, Wagner O, Mannhalter C (2003) Is the factor XIII 34Val/Leu polymorphism a protective factor for cerebrovascular disease? Br J Haematol 120:310–314
Espitia Pinzon N, Stroo E, T Hart BA, Bol JG, Drukarch B, Bauer J, Van Dam AM (2014) Tissue transglutaminase in marmoset experimental multiple sclerosis: discrepancy between white and grey matter. PLoS One 9:e100574
Festoff BW, Santacruz K, Arnold PM, Sebastian CT, Davies PJ, Citron BA (2002) Injury-induced "switch" from GTP-regulated to novel GTP-independent isoform of tissue transglutaminase in the rat spinal cord. J Neurochem 81:708–718
Filiano AJ, Bailey CD, Tucholski J, Gundemir S, Johnson GV (2008) Transglutaminase 2 protects against ischemic insult, interacts with HIF1beta, and attenuates HIF1 signaling. FASEB J 22:2662–2675
Filiano AJ, Tucholski J, Dolan PJ, Colak G, Johnson GV (2010) Transglutaminase 2 protects against ischemic stroke. Neurobiol Dis 39:334–343
Fox JH, Barber DS, Singh B, Zucker B, Swindell MK, Norflus F, Buzescu R, Chopra R, Ferrante RJ, Kazantsev A, Hersch SM (2004) Cystamine increases L-cysteine levels in Huntington's disease transgenic mouse brain and in a PC12 model of polyglutamine aggregation. J Neurochem 91:413–422
Fu J, Yang QY, Sai K, Chen FR, Pang JC, Ng HK, Kwan AL, Chen ZP (2013) TGM2 inhibition attenuates ID1 expression in CD44-high glioma-initiating cells. Neuro Oncol 15:1353–1365
Fujita K, Ando M, Yamauchi M, Nagata Y, Honda M (1995) Alteration of transglutaminase activity in rat and human spinal cord after neuronal degeneration. Neurochem Res 20:1195–1201
Fujita K, Honda M, Hayashi R, Ogawa K, Ando M, Yamauchi M, Nagata Y (1998a) Transglutaminase activity in serum and cerebrospinal fluid in sporadic amyotrophic lateral sclerosis: a possible use as an indicator of extent of the motor neuron loss. J Neurol Sci 158:53–57
Fujita K, Shibayama K, Yamauchi M, Kato T, Ando M, Takahashi H, Iritani K, Yoshimoto N, Nagata Y (1998b) Alteration of enzymatic activities implicating neuronal degeneration in the spinal cord of the motor neuron degeneration mouse during postnatal development. Neurochem Res 23:557–562
Fujita K, Kato T, Shibayama K, Imada H, Yamauchi M, Yoshimoto N, Miyachi E, Nagata Y (2006) Protective effect against 17beta-estradiol on neuronal apoptosis in hippocampus tissue following transient ischemia/recirculation in mongolian gerbils via down-regulation of tissue transglutaminase activity. Neurochem Res 31:1059–1068
Gemmati D, Serino ML, Ongaro A, Tognazzo S, Moratelli S, Resca R, Moretti M, Scapoli GL (2001) A common mutation in the gene for coagulation factor XIII-A (VAL34Leu): a risk factor for primary intracerebral hemorrhage is protective against atherothrombotic diseases. Am J Hematol 67:183–188
Gentile V, Sepe C, Calvani M, Melone MA, Cotrufo R, Cooper AJ, Blass JP, Peluso G (1998) Tissue transglutaminase-catalyzed formation of high-molecular-weight aggregates in vitro is favored with long polyglutamine domains: a possible mechanism contributing to CAG-triplet diseases. Arch Biochem Biophys 352:314–321
Gibrat C, Cicchetti F (2011) Potential of cystamine and cysteamine in the treatment of neurodegenerative diseases. Prog Neuropsychopharmacol Biol Psychiatry 35:380–389
Goldschmidt T, Antel J, Konig FB, Bruck W, Kuhlmann T (2009) Remyelination capacity of the MS brain decreases with disease chronicity. Neurology 72:1914–1921
Green H (1993) Human genetic diseases due to codon reiteration: relationship to an evolutionary mechanism. Cell 74:955–956
Grosso H, Mouradian MM (2012) Transglutaminase 2: biology, relevance to neurodegenerative diseases and therapeutic implications. Pharmacol Ther 133:392–410
Grosso H, Woo JM, Lee KW, Im JY, Masliah E, Junn E, Mouradian MM (2014) Transglutaminase 2 exacerbates alpha-synuclein toxicity in mice and yeast. FASEB J 28:4280–4291

Grundke-Iqbal I, Iqbal K, Quinlan M, Tung YC, Zaidi MS, Wisniewski HM (1986a) Microtubule-associated protein tau. A component of Alzheimer paired helical filaments. J Biol Chem 261:6084–6089
Grundke-Iqbal I, Iqbal K, Tung YC, Quinlan M, Wisniewski HM, Binder LI (1986b) Abnormal phosphorylation of the microtubule-associated protein tau (tau) in Alzheimer cytoskeletal pathology. Proc Natl Acad Sci U S A 83:4913–4917
Guan WJ, Wang JL, Liu YT, Ma YT, Zhou Y, Jiang H, Shen L, Guo JF, Xia K, Li JD, Tang BS (2013a) Spinocerebellar ataxia type 35 (SCA35)-associated transglutaminase 6 mutants sensitize cells to apoptosis. Biochem Biophys Res Commun 430:780–786
Guan WJ, Xia KD, Ma YT, Liu YT, Shi YT, Jiang H, Shen L, Xia K, Li JD, Tang BS, Wang JL (2013b) Transglutaminase 6 interacts with polyQ proteins and promotes the formation of polyQ aggregates. Biochem Biophys Res Commun 437:94–100
Gundemir S, Johnson GV (2009) Intracellular localization and conformational state of transglutaminase 2: implications for cell death. PLoS One 4:e6123
Gundemir S, Colak G, Tucholski J, Johnson GV (2012) Transglutaminase 2: a molecular Swiss army knife. Biochim Biophys Acta 1823:406–419
Gundemir S, Colak G, Feola J, Blouin R, Johnson GV (2013) Transglutaminase 2 facilitates or ameliorates HIF signaling and ischemic cell death depending on its conformation and localization. Biochim Biophys Acta 1833:1–10
Hadjivassiliou M, Aeschlimann P, Strigun A, Sanders DS, Woodroofe N, Aeschlimann D (2008) Autoantibodies in gluten ataxia recognize a novel neuronal transglutaminase. Ann Neurol 64:332–343
Haigler HT, Maxfield FR, Willingham MC, Pastan I (1980) Dansylcadaverine inhibits internalization of 125I-epidermal growth factor in BALB 3T3 cells. J Biol Chem 255:1239–1241
Halverson RA, Lewis J, Frausto S, Hutton M, Muma NA (2005) Tau protein is cross-linked by transglutaminase in P301L tau transgenic mice. J Neurosci 25:1226–1233
Hartley DM, Zhao C, Speier AC, Woodard GA, Li S, Li Z, Walz T (2008) Transglutaminase induces protofibril-like amyloid beta-protein assemblies that are protease-resistant and inhibit long-term potentiation. J Biol Chem 283:16790–16800
Hilton DA, Love S, Barber R (1997) Increased endothelial expression of transglutaminase in glioblastomas. Neuropathol Appl Neurobiol 23:507–511
Ho GJ, Gregory EJ, Smirnova IV, Zoubine MN, Festoff BW (1994) Cross-linking of beta-amyloid protein precursor catalyzed by tissue transglutaminase. FEBS Lett 349:151–154
Hohsfield LA, Daschil N, Oradd G, Stromberg I, Humpel C (2014) Vascular pathology of 20-month-old hypercholesterolemia mice in comparison to triple-transgenic and APPSwDI Alzheimer's disease mouse models. Mol Cell Neurosci 63:83–95
Holmes FE, Haynes LW (1996) Superactivation of transglutaminase type 2 without change in enzyme level occurs during progressive neurodegeneration in the mnd mouse mutant. Neurosci Lett 213:185–188
Huang YC, Wei KC, Chang CN, Chen PY, Hsu PW, Chen CP, Lu CS, Wang HL, Gutmann DH, Yeh TH (2014) Transglutaminase 2 expression is increased as a function of malignancy grade and negatively regulates cell growth in meningioma. PLoS One 9:e108228
Hwang IK, Yoo KY, Yi SS, Kim IY, Hwang HS, Lee KY, Choi SM, Lee IS, Yoon YS, Kim SY, Won MH, Seong JK (2009) Expression of tissue-type transglutaminase (tTG) and the effect of tTG inhibitor on the hippocampal CA1 region after transient ischemia in gerbils. Brain Res 1263:134–142
Ientile R, Caccamo D, Marciano MC, Curro M, Mannucci C, Campisi A, Calapai G (2004) Transglutaminase activity and transglutaminase mRNA transcripts in gerbil brain ischemia. Neurosci Lett 363:173–177
Igarashi S, Koide R, Shimohata T, Yamada M, Hayashi Y, Takano H, Date H, Oyake M, Sato T, Sato A, Egawa S, Ikeuchi T, Tanaka H, Nakano R, Tanaka K, Hozumi I, Inuzuka T, Takahashi H, Tsuji S (1998) Suppression of aggregate formation and apoptosis by

transglutaminase inhibitors in cells expressing truncated DRPLA protein with an expanded polyglutamine stretch. Nat Genet 18:111–117
Ikura K, Takahata K, Sasaki R (1993) Cross-linking of a synthetic partial-length (1–28) peptide of the Alzheimer beta/A4 amyloid protein by transglutaminase. FEBS Lett 326:109–111
Iniesta JA, Corral J, Gonzalez-Conejero R, Diaz Ortuno A, Martinez Navarro ML, Vicente V (2001) Role of factor XIII Val 34 Leu polymorphism in patients with migraine. Cephalalgia 21:837–841
Jeitner TM, Bogdanov MB, Matson WR, Daikhin Y, Yudkoff M, Folk JE, Steinman L, Browne SE, Beal MF, Blass JP, Cooper AJ (2001) N(epsilon)-(gamma-L-glutamyl)-L-lysine (GGEL) is increased in cerebrospinal fluid of patients with Huntington's disease. J Neurochem 79:1109–1112
Jeitner TM, Delikatny EJ, Ahlqvist J, Capper H, Cooper AJ (2005) Mechanism for the inhibition of transglutaminase 2 by cystamine. Biochem Pharmacol 69:961–970
Jeitner TM, Matson WR, Folk JE, Blass JP, Cooper AJ (2008) Increased levels of gamma-glutamylamines in Huntington disease CSF. J Neurochem 106:37–44
Johnson GV, Leshoure R Jr (2004) Immunoblot analysis reveals that isopeptide antibodies do not specifically recognize the epsilon-(gamma-glutamyl)lysine bonds formed by transglutaminase activity. J Neurosci Methods 134:151–158
Johnson GV, Cox TM, Lockhart JP, Zinnerman MD, Miller ML, Powers RE (1997) Transglutaminase activity is increased in Alzheimer's disease brain. Brain Res 751:323–329
Junn E, Ronchetti RD, Quezado MM, Kim SY, Mouradian MM (2003) Tissue transglutaminase-induced aggregation of alpha-synuclein: Implications for Lewy body formation in Parkinson's disease and dementia with Lewy bodies. Proc Natl Acad Sci U S A 100:2047–2052
Kahlem P, Terre C, Green H, Djian P (1996) Peptides containing glutamine repeats as substrates for transglutaminase-catalyzed cross-linking: relevance to diseases of the nervous system. Proc Natl Acad Sci U S A 93:14580–14585
Kahlem P, Green H, Djian P (1998) Transglutaminase action imitates Huntington's disease: selective polymerization of Huntingtin containing expanded polyglutamine. Mol Cell 1:595–601
Kain K, Bamford J, Bavington J, Young J, Catto AJ (2005) Factor XIII--circulating levels and Val34Leu polymorphism in relatives of South Asian patients with ischemic stroke. J Thromb Haemost 3:171–173
Karpuj MV, Garren H, Slunt H, Price DL, Gusella J, Becher MW, Steinman L (1999) Transglutaminase aggregates huntingtin into nonamyloidogenic polymers, and its enzymatic activity increases in Huntington's disease brain nuclei. Proc Natl Acad Sci U S A 96:7388–7393
Karpuj MV, Becher MW, Springer JE, Chabas D, Youssef S, Pedotti R, Mitchell D, Steinman L (2002) Prolonged survival and decreased abnormal movements in transgenic model of Huntington disease, with administration of the transglutaminase inhibitor cystamine. Nat Med 8:143–149
Kawabe K, Takano K, Moriyama M, Nakamura Y (2014) Lipopolysaccharide-Stimulated Transglutaminase 2 Expression Enhances Endocytosis Activity in the Mouse Microglial Cell Line BV-2. Neuroimmunomodulation 22:243–249
Kim SY, Grant P, Lee JH, Pant HC, Steinert PM (1999) Differential expression of multiple transglutaminases in human brain. Increased expression and cross-linking by transglutaminases 1 and 2 in Alzheimer's disease. J Biol Chem 274:30715–30721
Kim C, Lee HC, Sung JJ (2014) Amyotrophic lateral sclerosis – cell based therapy and novel therapeutic development. Exp Neurobiol 23:207–214
Kohler HP, Ariens RA, Catto AJ, Carter AM, Miller GJ, Cooper JA, Mansfield MW, Standeven KF, Grant PJ (2002) Factor XIII A-subunit concentration predicts outcome in stroke subjects and vascular outcome in healthy, middle-aged men. Br J Haematol 118:825–832
Kovarova H, Pulpanova J (1979) Effect of cystamine on rat tissue GSH level and glutathione reductase activity. Strahlentherapie 155:875–878

Kumar A, Kneynsberg A, Tucholski J, Perry G, van Groen T, Detloff PJ, Lesort M (2012) Tissue transglutaminase overexpression does not modify the disease phenotype of the R6/2 mouse model of Huntington's disease. Exp Neurol 237:78–89

Lai TS, Tucker T, Burke JR, Strittmatter WJ, Greenberg CS (2004) Effect of tissue transglutaminase on the solubility of proteins containing expanded polyglutamine repeats. J Neurochem 88:1253–1260

Lesort M, Attanavanich K, Zhang J, Johnson GV (1998) Distinct nuclear localization and activity of tissue transglutaminase. J Biol Chem 273:11991–11994

Lesort M, Chun W, Johnson GV, Ferrante RJ (1999) Tissue transglutaminase is increased in Huntington's disease brain. J Neurochem 73:2018–2027

Lesort M, Tucholski J, Miller ML, Johnson GV (2000) Tissue transglutaminase: a possible role in neurodegenerative diseases. Prog Neurobiol 61:439–463

Lesort M, Lee M, Tucholski J, Johnson GV (2003) Cystamine inhibits caspase activity. Implications for the treatment of polyglutamine disorders. J Biol Chem 278:3825–3830

Li M, Pang SY, Song Y, Kung MH, Ho SL, Sham PC (2013) Whole exome sequencing identifies a novel mutation in the transglutaminase 6 gene for spinocerebellar ataxia in a Chinese family. Clin Genet 83:269–273

Li P-C, Jiao Y, Ding J, Chen Y-C, Cui Y, Qian C, Yang X-Y, Ju S-H, Yao H-H, Teng G-J (2014) Cystamine improves functional recovery via axon remodeling and neuroprotection after stroke in mice. CNS Neurosci Ther 21:231–240

Liu L, Belkadi A, Darnall L, Hu T, Drescher C, Cotleur AC, Padovani-Claudio D, He T, Choi K, Lane TE, Miller RH, Ransohoff RM (2010) CXCR2-positive neutrophils are essential for cuprizone-induced demyelination: relevance to multiple sclerosis. Nat Neurosci 13:319–326

Liu YT, Tang BS, Lan W, Song NN, Huang Y, Zhang L, Guan WJ, Shi YT, Shen L, Jiang H, Guo JF, Xia K, Ding YQ, Wang JL (2013) Distribution of transglutaminase 6 in the central nervous system of adult mice. Anat Rec (Hoboken) 296:1576–1587

Lorand L, Graham RM (2003) Transglutaminases: crosslinking enzymes with pleiotropic functions. Nat Rev Mol Cell Biol 4:140–156

Mandrusiak LM, Beitel LK, Wang X, Scanlon TC, Chevalier-Larsen E, Merry DE, Trifiro MA (2003) Transglutaminase potentiates ligand-dependent proteasome dysfunction induced by polyglutamine-expanded androgen receptor. Hum Mol Genet 12:1497–1506

Margulis BA, Vigont V, Lazarev VF, Kaznacheyeva EV, Guzhova IV (2013) Pharmacological protein targets in polyglutamine diseases: mutant polypeptides and their interactors. FEBS Lett 587:1997–2007

Mastroberardino PG, Iannicola C, Nardacci R, Bernassola F, de Laurenzi V, Melino G, Moreno S, Pavone F, Oliverio S, Fesus L, Piacentini M (2002) 'Tissue' transglutaminase ablation reduces neuronal death and prolongs survival in a mouse model of Huntington's disease. Cell Death Differ 9:873–880

Mcconoughey SJ, Basso M, Niatsetskaya ZV, Sleiman SF, Smirnova NA, Langley BC, Mahishi L, Cooper AJ, Antonyak MA, Cerione RA, Li B, Starkov A, Chaturvedi RK, Beal MF, Coppola G, Geschwind DH, Ryu H, Xia L, Iismaa SE, Pallos J, Pasternack R, Hils M, Fan J, Raymond LA, Marsh JL, Thompson LM, Ratan RR (2010) Inhibition of transglutaminase 2 mitigates transcriptional dysregulation in models of Huntington disease. EMBO Mol Med 2:349–370

Menalled LB, Kudwa AE, Oakeshott S, Farrar A, Paterson N, Filippov I, Miller S, Kwan M, Olsen M, Beltran J, Torello J, Fitzpatrick J, Mushlin R, Cox K, Mcconnell K, Mazzella M, He D, Osborne GF, Al-Nackkash R, Bates GP, Tuunanen P, Lehtimaki K, Brunner D, Ghavami A, Ramboz S, Park L, Macdonald D, Munoz-Sanjuan I, Howland D (2014) Genetic deletion of transglutaminase 2 does not rescue the phenotypic deficits observed in R6/2 and zQ175 mouse models of Huntington's disease. PLoS One 9:e99520

Milakovic T, Tucholski J, Mccoy E, Johnson GV (2004) Intracellular localization and activity state of tissue transglutaminase differentially impacts cell death. J Biol Chem 279:8715–8722

Miller CC, Anderton BH (1986) Transglutaminase and the neuronal cytoskeleton in Alzheimer's disease. J Neurochem 46:1912–1922
Miller ML, Johnson GV (1995) Transglutaminase cross-linking of the tau protein. J Neurochem 65:1760–1770
Monsonego A, Shani Y, Friedmann I, Paas Y, Eizenberg O, Schwartz M (1997) Expression of GTP-dependent and GTP-independent tissue-type transglutaminase in cytokine-treated rat brain astrocytes. J Biol Chem 272:3724–3732
Munsie L, Caron N, Atwal RS, Marsden I, Wild EJ, Bamburg JR, Tabrizi SJ, Truant R (2011) Mutant huntingtin causes defective actin remodeling during stress: defining a new role for transglutaminase 2 in neurodegenerative disease. Hum Mol Genet 20:1937–1951
Murthy SN, Wilson JH, Lukas TJ, Kuret J, Lorand L (1998) Cross-linking sites of the human tau protein, probed by reactions with human transglutaminase. J Neurochem 71:2607–2614
Nanda N, Iismaa SE, Owens WA, Husain A, Mackay F, Graham RM (2001) Targeted inactivation of Gh/tissue transglutaminase II. J Biol Chem 276:20673–20678
Nemes Z, Devreese B, Steinert PM, van Beeumen J, Fesus L (2004) Cross-linking of ubiquitin, HSP27, parkin, and alpha-synuclein by gamma-glutamyl-epsilon-lysine bonds in Alzheimer's neurofibrillary tangles. FASEB J 18:1135–1137
Nilsson JL, Stenberg P, Ljunggren C, Eriksson O, Lunden R (1971) Fibrin-stabilizing factor inhibitors. 3. Sulphonamides related to monodansylcadaverine. Acta Pharm Suec 8:497–504
Nurminskaya MV, Belkin AM (2012) Cellular functions of tissue transglutaminase. Int Rev Cell Mol Biol 294:1–97
Obara Y, Yanagihata Y, Abe T, Dafik L, Ishii K, Nakahata N (2013) Galpha(h)/transglutaminase-2 activity is required for maximal activation of adenylylcyclase 8 in human and rat glioma cells. Cell Signal 25:589–597
Okauchi M, Xi G, Keep RF, Hua Y (2009) Tissue-type transglutaminase and the effects of cystamine on intracerebral hemorrhage-induced brain edema and neurological deficits. Brain Res 1249:229–236
Oono M, Okado-Matsumoto A, Shodai A, Ido A, Ohta Y, Abe K, Ayaki T, Ito H, Takahashi R, Taniguchi N, Urushitani M (2014) Transglutaminase 2 accelerates neuroinflammation in amyotrophic lateral sclerosis through interaction with misfolded superoxide dismutase 1. J Neurochem 128:403–418
Organization WH (2002) The world health report. WHO, Geneva
Pinto JT, van Raamsdonk JM, Leavitt BR, Hayden MR, Jeitner TM, Thaler HT, Krasnikov BF, Cooper AJ (2005) Treatment of YAC128 mice and their wild-type littermates with cystamine does not lead to its accumulation in plasma or brain: implications for the treatment of Huntington disease. J Neurochem 94:1087–1101
Prins M, Eriksson C, Wierinckx A, Bol JG, Binnekade R, Tilders FJ, van Dam AM (2013) Interleukin-1beta and interleukin-1 receptor antagonist appear in grey matter additionally to white matter lesions during experimental multiple sclerosis. PLoS One 8:e83835
Rasmussen LK, Sorensen ES, Petersen TE, Gliemann J, Jensen PH (1994) Identification of glutamine and lysine residues in Alzheimer amyloid beta A4 peptide responsible for transglutaminase-catalysed homopolymerization and cross-linking to alpha 2M receptor. FEBS Lett 338:161–166
Reiner AP, Frank MB, Schwartz SM, Linenberger ML, Longstreth WT, Teramura G, Rosendaal FR, Psaty BM, Siscovick DS (2002) Coagulation factor XIII polymorphisms and the risk of myocardial infarction and ischaemic stroke in young women. Br J Haematol 116:376–382
Ruan Q, Quintanilla RA, Johnson GV (2007) Type 2 transglutaminase differentially modulates striatal cell death in the presence of wild type or mutant huntingtin. J Neurochem 102:25–36
Sato K, Murakami T, Hamakawa Y, Kamada H, Nagano I, Shoji M, Takata H, Nobukuni K, Ihara Y, Namba R, Hayabara T, Hirose S, Abe K (2002) Selective colocalization of transglutaminase-like activity in ubiquitinated intranuclear inclusions of hereditary dentatorubral-pallidoluysian atrophy. Brain Res 952:327–330

Schmid AW, Chiappe D, Pignat V, Grimminger V, Hang I, Moniatte M, Lashuel HA (2009) Dissecting the mechanisms of tissue transglutaminase-induced cross-linking of alpha-synuclein: implications for the pathogenesis of Parkinson disease. J Biol Chem 284:13128–13142

Segers-Nolten IM, Wilhelmus MM, Veldhuis G, van Rooijen BD, Drukarch B, Subramaniam V (2008) Tissue transglutaminase modulates alpha-synuclein oligomerization. Protein Sci 17:1395–1402

Selkoe DJ, Abraham C, Ihara Y (1982) Brain transglutaminase: in vitro crosslinking of human neurofilament proteins into insoluble polymers. Proc Natl Acad Sci U S A 79:6070–6074

Singer SM, Zainelli GM, Norlund MA, Lee JM, Muma NA (2002) Transglutaminase bonds in neurofibrillary tangles and paired helical filament tau early in Alzheimer's disease. Neurochem Int 40:17–30

Slow EJ, van Raamsdonk J, Rogers D, Coleman SH, Graham RK, Deng Y, Oh R, Bissada N, Hossain SM, Yang YZ, Li XJ, Simpson EM, Gutekunst CA, Leavitt BR, Hayden MR (2003) Selective striatal neuronal loss in a YAC128 mouse model of Huntington disease. Hum Mol Genet 12:1555–1567

Tetzlaff W, Gilad VH, Leonard C, Bisby MA, Gilad GM (1988) Retrograde changes in transglutaminase activity after peripheral nerve injuries. Brain Res 445:142–146

Todd TW, Lim J (2013) Aggregation formation in the polyglutamine diseases: protection at a cost? Mol Cells 36:185–194

Tolentino PJ, Deford SM, Notterpek L, Glenn CC, Pike BR, Wang KK, Hayes RL (2002) Up-regulation of tissue-type transglutaminase after traumatic brain injury. J Neurochem 80:579–588

Tolentino PJ, Waghray A, Wang KK, Hayes RL (2004) Increased expression of tissue-type transglutaminase following middle cerebral artery occlusion in rats. J Neurochem 89:1301–1307

Tucholski J, Roth KA, Johnson GV (2006) Tissue transglutaminase overexpression in the brain potentiates calcium-induced hippocampal damage. J Neurochem 97:582–594

van Raamsdonk JM, Pearson J, Bailey CD, Rogers DA, Johnson GV, Hayden MR, Leavitt BR (2005) Cystamine treatment is neuroprotective in the YAC128 mouse model of Huntington disease. J Neurochem 95:210–220

van Strien ME, Baron W, Bakker EN, Bauer J, Bol JG, Breve JJ, Binnekade R, van der Laarse WJ, Drukarch B, van Dam AM (2011a) Tissue transglutaminase activity is involved in the differentiation of oligodendrocyte precursor cells into myelin-forming oligodendrocytes during CNS remyelination. Glia 59:1622–1634

van Strien ME, Breve JJ, Fratantoni S, Schreurs MW, Bol JG, Jongenelen CA, Drukarch B, van Dam AM (2011b) Astrocyte-derived tissue transglutaminase interacts with fibronectin: a role in astrocyte adhesion and migration? PLoS One 6:e25037

van Strien ME, Drukarch B, Bol JG, van der Valk P, van Horssen J, Gerritsen WH, Breve JJ, van Dam AM (2011c) Appearance of tissue transglutaminase in astrocytes in multiple sclerosis lesions: a role in cell adhesion and migration? Brain Pathol 21:44–54

Verhaar R, Jongenelen CA, Gerard M, Baekelandt V, van Dam AM, Wilhelmus MM, Drukarch B (2011) Blockade of enzyme activity inhibits tissue transglutaminase-mediated transamidation of alpha-synuclein in a cellular model of Parkinson's disease. Neurochem Int 58:785–793

Wang X, Sarkar A, Cicchetti F, Yu M, Zhu A, Jokivarsi K, Saint-Pierre M, Brownell AL (2005) Cerebral PET imaging and histological evidence of transglutaminase inhibitor cystamine induced neuroprotection in transgenic R6/2 mouse model of Huntington's disease. J Neurol Sci 231:57–66

Wang JL, Yang X, Xia K, Hu ZM, Weng L, Jin X, Jiang H, Zhang P, Shen L, Guo JF, Li N, Li YR, Lei LF, Zhou J, Du J, Zhou YF, Pan Q, Wang J, Wang J, Li RQ, Tang BS (2010a) TGM6 identified as a novel causative gene of spinocerebellar ataxias using exome sequencing. Brain 133:3510–3518

Wang Z, Collighan RJ, Gross SR, Danen EH, Orend G, Telci D, Griffin M (2010b) RGD-independent cell adhesion via a tissue transglutaminase-fibronectin matrix promotes fibronectin fibril deposition and requires syndecan-4/2 alpha5beta1 integrin co-signaling. J Biol Chem 285:40212–40229

Weber JJ, Sowa AS, Binder T, Hubener J (2014) From Pathways to Targets: Understanding the Mechanisms behind Polyglutamine Disease. Biomed Res Int 2014:701758

Wilhelmus MM, Grunberg SC, Bol JG, van Dam AM, Hoozemans JJ, Rozemuller AJ, Drukarch B (2009) Transglutaminases and transglutaminase-catalyzed cross-links colocalize with the pathological lesions in Alzheimer's disease brain. Brain Pathol 19:612–622

Wilhelmus MM, Verhaar R, Andringa G, Bol JG, Cras P, Shan L, Hoozemans JJ, Drukarch B (2011) Presence of tissue transglutaminase in granular endoplasmic reticulum is characteristic of melanized neurons in Parkinson's disease brain. Brain Pathol 21:130–139

Wilhelmus MM, de Jager M, Drukarch B (2012a) Tissue transglutaminase: a novel therapeutic target in cerebral amyloid angiopathy. Neurodegener Dis 10:317–319

Wilhelmus MM, de Jager M, Rozemuller AJ, Breve J, Bol JG, Eckert RL, Drukarch B (2012b) Transglutaminase 1 and its regulator tazarotene-induced gene 3 localize to neuronal tau inclusions in tauopathies. J Pathol 226:132–142

Wolf J, Jager C, Lachmann I, Schonknecht P, Morawski M, Arendt T, Mothes T (2013) Tissue transglutaminase is not a biochemical marker for Alzheimer's disease. Neurobiol Aging 34:2495–2498

Yuan L, Siegel M, Choi K, Khosla C, Miller CR, Jackson EN, Piwnica-Worms D, Rich KM (2007) Transglutaminase 2 inhibitor, KCC009, disrupts fibronectin assembly in the extracellular matrix and sensitizes orthotopic glioblastomas to chemotherapy. Oncogene 26:2563–2573

Yuan L, Behdad A, Siegel M, Khosla C, Higashikubo R, Rich KM (2008) Tissue transgluaminase 2 expression in meningiomas. J Neurooncol 90:125–132

Yuan L, Holmes TC, Watts RE, Khosla C, Broekelmann TJ, Mecham R, Zheng H, Izaguirre EW, Rich KM (2011) Novel chemo-sensitizing agent, ERW1227B, impairs cellular motility and enhances cell death in glioblastomas. J Neurooncol 103:207–219

Zainelli GM, Ross CA, Troncoso JC, Muma NA (2003) Transglutaminase cross-links in intranuclear inclusions in Huntington disease. J Neuropathol Exp Neurol 62:14–24

Zainelli GM, Dudek NL, Ross CA, Kim SY, Muma NA (2005) Mutant huntingtin protein: a substrate for transglutaminase 1, 2, and 3. J Neuropathol Exp Neurol 64:58–65

Zemaitaitis MO, Lee JM, Troncoso JC, Muma NA (2000) Transglutaminase-induced cross-linking of tau proteins in progressive supranuclear palsy. J Neuropathol Exp Neurol 59:983–989

Zemaitaitis MO, Kim SY, Halverson RA, Troncoso JC, Lee JM, Muma NA (2003) Transglutaminase activity, protein, and mRNA expression are increased in progressive supranuclear palsy. J Neuropathol Exp Neurol 62:173–184

Zhang Y, Pardridge WM (2001) Neuroprotection in transient focal brain ischemia after delayed intravenous administration of brain-derived neurotrophic factor conjugated to a blood-brain barrier drug targeting system. Stroke 32:1378–1384

Zhang W, Johnson BR, Suri DE, Martinez J, Bjornsson TD (1998) Immunohistochemical demonstration of tissue transglutaminase in amyloid plaques. Acta Neuropathol 96:395–400

Zhang J, Antonyak MA, Singh G, Cerione RA (2013) A mechanism for the upregulation of EGF receptor levels in glioblastomas. Cell Rep 3:2008–2020

Chapter 14
Regulation of Transglutaminase 2 by Oxidative Stress

Eui Man Jeong and In-Gyu Kim

Abstract Transglutaminase 2 (TG2) is an enzyme that transamidates glutamine residue in its substrates, resulting in the accumulation of crosslinked, monoaminated, polyaminated, or deamidated protein products. Recent evidence has suggested that aberrant regulation of TG2 by oxidative stress causes various disorders including cataract formation, inflammation, fibrosis, celiac disease, neurodegenerative diseases, and diabetes mellitus. In this chapter, we describe how TG2 responds to oxidative stress and participates in disease pathogeneses in the following three ways. First, the TG2 expression level is increased by oxidative stress-mediated activation of transcription factors, NF-κB and hypoxia-inducible factor 1. Second, oxidative stress induces post-translational modifications (PTMs) of TG2 by forming intramolecular disulfides, SUMOylation, or ubiquitination, while nitrosative stress promotes TG2 *S*-nitrosylation. This regulates transamidation activity, protein stability, or subcellular localization of TG2. Finally, oxidative stress-responsive cellular mediators like intracellular calcium ions, transforming growth factor β, and newly identified endogenous inhibitor proteins can modulate TG2 enzymatic activity. Elucidating the oxidative stress-responsive regulatory mechanisms of TG2 will increase our understanding of its precise pathophysiological role in related disorders and also help in establishing therapeutic strategies.

Keywords Transglutaminase 2 • Oxidative stress • ROS • NF-κB • Hypoxia-inducible factor 1 • Vicinal disulfide • Ubiquitination • SUMOylation • Transforming growth factor β • Endogenous inhibitor protein

E.M. Jeong • I.-G. Kim (✉)
Department of Biochemistry and Molecular Biology, Seoul National University College of Medicine, Seoul, South Korea

Institute of Human-Environment Interface Biology, Seoul National University College of Medicine, Seoul, South Korea
e-mail: igkim@plaza.snu.ac.kr

K. Hitomi et al. (eds.), *Transglutaminases*, DOI 10.1007/978-4-431-55825-5_14

14.1 Oxidative Stress and Human Diseases

Living cells in an aerobic environment are inevitably exposed to reactive oxygen species (ROS). These are generated endogenously during various oxidative metabolic processes or in response to exogenous environmental stimuli (Fig. 14.1). In the process of mitochondrial oxidative phosphorylation, oxygen (O_2) is converted into water (H_2O) by transfer of four electrons, leading to the production of ATP. However, when an oxygen molecule receives only one electron via various endogenous and exogenous factors, it is reduced to a superoxide anion ($\cdot O_2^-$), which is immediately converted into relatively stable hydrogen peroxide (H_2O_2) by superoxide dismutase (SOD) or spontaneous dismutation using another single electron. Hydrogen peroxide can be reduced further by one electron transfer, generating a highly reactive hydroxyl radical ($\cdot OH$) and a non-reactive hydroxyl ion (OH^-) via Fenton's reaction or Harber-Weiss reaction. Thus, superoxide anion, hydrogen peroxide, and hydroxyl radical are the representative ROS generated in live cells. Since ROS are very reactive, there are a number of enzymatic and non-enzymatic antioxidant systems such as SOD, catalase, thioredoxin, and glutathione to eliminate ROS. When the rate of ROS generation exceeds cellular antioxidant capacity, oxidative stress occurs (Fig. 14.1). Under these conditions, ROS cause oxidative damage to biomolecules and/or induce oxidative modification of signaling proteins, resulting in the activation of redox signaling pathways (D'Autreaux and Toledano 2007). Therefore, oxidative stress elicits a wide range of cellular responses from apoptosis to cell proliferation.

Epidemiological, pathophysiological, and mechanistic data have strongly suggested that oxidative stress is a major cause of or contributes to the development of numerous diseases including cancer, inflammatory and degenerative diseases (Holmstrom and Finkel 2014; Hybertson et al. 2011; Mayne 2003). Most of the ROS intervention therapies have failed to show preventive or therapeutic effects on

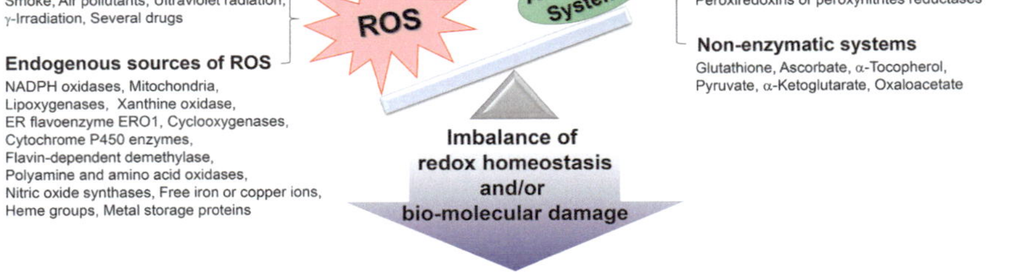

Fig. 14.1 Imbalance between ROS generation and antioxidant defense systems causes oxidative stress. Disruption of cellular redox balance leads to oxidative stress. The lists comprising ROS sources and antioxidant systems have been mainly compiled from a review article (Nathan and Cunningham-Bussel 2013). *ROS* reactive oxygen species, *ER* endoplasmic reticulum

the development of disease phenotypes in human and animal models, largely due to an inadequate understanding of ROS biochemistry (Steinhubl 2008). These results also imply that using ROS scavengers is not sufficient and targeting ROS effectors or mediators to interrupt ROS-mediated signaling is necessary for efficient ROS intervention therapies. Thus, efforts to elucidate precise mechanisms of ROS-related diseases are required to develop effective therapeutic medicines.

14.2 Role of Transglutaminase 2 (TG2) in Oxidative Stress-Related Diseases

TG2 is a member of the enzyme family that modify glutamine residues in proteins with primary amines or water, forming crosslinked, monoaminated, polyaminated or deamidated proteins (Iismaa et al. 2009). TG2 is activated in cells under various oxidative stress conditions. Activation of TG2 induces the polyamination or crosslinking of a number of substrates including caspase 3, IκBα, PPARγ, crystallins, collagen, vimentin, β-amyloid, α-synuclein, and Beclin 1, which results in the modulation of biological function(s) of these substrate proteins. Thus, aberrant activation of TG2 by ROS leads to changes in cell survival processes, inflammatory responses, extracellular matrix formation, the architecture of the cytoskeleton, and metabolic processes. These changes related to oxidative stress are involved in the pathogenesis of various diseases (Table 14.1 and Fig. 14.2).

Table 14.1 Causal role of TG2 in human diseases through oxidative stress-induced substrate modifications

Diseases	ROS generating conditions	TG2 substrates	References
Cancer	Hypoxia, genotoxic stress	Caspase 3, IκBα	Cao et al. (2008) and Jang et al. (2010)
Inflammatory and fibrotic diseases	Bleomycin, $CFTR^{\Delta F508}$	Fibronectin, collagen, IκBα, PPARγ, Beclin 1	Luciani et al. (2010a), Luciani et al. (2009), Maiuri et al. (2008) and Oh et al. (2011)
Celiac disease	Lysosomal accumulation of gliadin p31–43 peptide	PPARγ	Luciani et al. (2010b)
Cataracts	UV irradiation, H_2O_2, TGFβ, selenite	β-Crystallins, αB-Crystallin, vimentin	Lee et al. (2012) and Shin et al. (2008)
Neurodegenerative diseases	AT2 receptor oligomers, glutamate, stroke, mitochondrial dysfunction	Huntingtin, β-amyloid, α-Synuclein, prohibitin, ATP synthase β-chain	AbdAlla et al. (2009), Basso et al. (2012), Battaglia et al. (2007), Lai et al. (2004), Segers-Nolten et al. (2008) and Tucholski et al. (2006)
Diabetes mellitus	Hyperglycemia	Unknown	Bhatt et al. (2013)

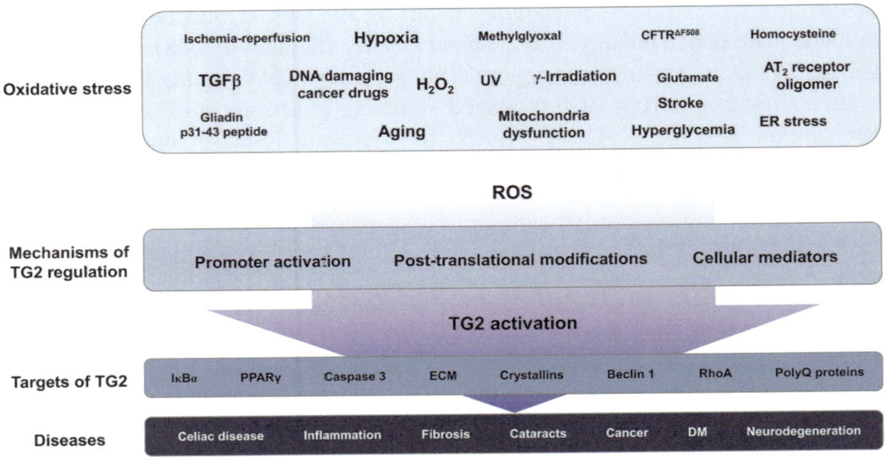

Fig. 14.2 Role of TG2 in the development of oxidative stress-associated diseases. TG2 regulation by oxidative stress is involved in the pathogeneses of various diseases as described in Table 14.1. *ROS* reactive oxygen species, *ER* endoplasmic reticulum, *AT₂ receptor* angiotensin II receptor type 2, *ECM* extracellular matrix, *DM* diabetes mellitus

Moreover, a causal role of TG2 has been verified in various animal models using TG2-deficient mice.

These findings indicate that TG2 is a downstream effector of the redox signaling pathway, even though the exact molecular mechanism by which TG2 is activated has not been thoroughly elucidated. In the following sections, we evaluate the body of literature describing the regulation of TG2 by oxidative stimuli, especially focusing on TG2 regulation by transcriptional controls (Sect. 14.3), post-translational modifications of TG2 (Sect. 14.4), and regulation by cellular mediators (Sect. 14.5). A full understanding of ROS-mediated TG2 activation could help to design targeted therapeutics to prevent ROS-associated diseases.

14.3 ROS-Sensitive Transcription Factors Involved in the Regulation of TG2 Promoter Activity

14.3.1 Nuclear Factor κ-Light-Chain-Enhancer of Activated B Cells (NF-κB)

NF-κB is a transcription factor that plays a pivotal role in inflammation, immunity, and cancer biology (Perkins 2007). NF-κB is activated via signaling pathways under diverse oxidative stress conditions. The mechanisms of oxidative stress-dependent NF-κB activation are depicted in Fig. 14.3. All pathways are dependent on IκBα phosphorylation by three kinds of kinases: tyrosine kinase, casein kinase II

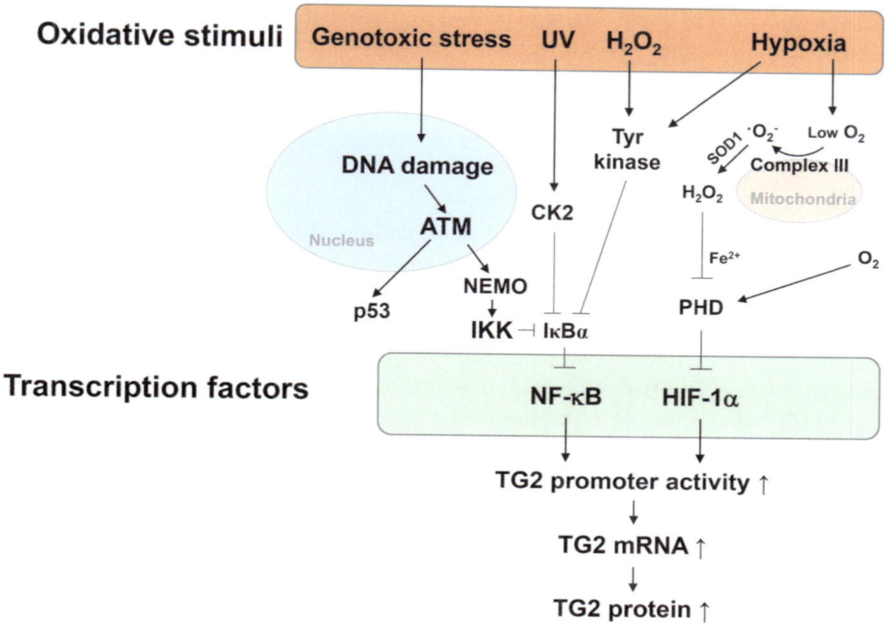

Fig. 14.3 Transcriptional regulation of TG2 via oxidative stress signaling pathways. (1) NF-κB; NF-κB is activated via three different signaling pathways in response to oxidative stress (Pinkas et al. 2007). H_2O_2 or hypoxia induces phosphorylation at Tyr^{42} of IκBα. UV activates CK2, which phosphorylates at the carboxyl terminal PEST domain of IκBα. Finally, genotoxic stress activates ATM, a primary DNA damage sensor kinase that phosphorylates NEMO, leading to IKK activation and subsequent IκBα phosphorylation at Ser^{32} and Ser^{36}. Phosphorylated IκBα proteins are degraded via the ubiquitin proteasomal pathway, favoring NF-κB activation. TG2 transcription has been reported to be activated by all oxidative stimuli described above (Jeong et al. 2009). However, until now, only that genotoxic stress/ATM/IKK/NF-κB signal pathway was proven to be involved in TG2 expression (Ai et al. 2012). (2) HIF-1; TG2 promoter is occupied by HIF-1 complex that consists of HIF-1α and HIF-1β subunits, resulting in its transcription activation in response to hypoxia (Jang et al. 2010). HIF-1β is constitutively expressed, whereas HIF-1α is hydroxylated by PHD and ubiquitinated by pVHL under normoxic conditions, resulting in its destruction. PHD is not active in the absence of O_2 and is inactivated by ROS generated from mitochondrial complex III under low O_2 conditions, favoring HIF-1 complex formation and initiation of target gene transcription (Sabharwal and Schumacker 2014). However, it has not been reported yet whether ROS are involved in HIF-1α-dependent TG2 expression under hypoxic conditions. *CK2* casein kinase 2, *ATM* ataxia telangiectasia mutated, *NEMO* NF-κB essential modulator, *IKK* IκB kinase, *pVHL* Von Hippel–Lindau tumor suppressor, *PEST* a peptide sequence rich in proline (*P*), glutamic acid (*E*), serine (*S*), and threonine (*T*), *HIF-1* hypoxia-inducible factor-1, *SOD1* superoxide dismutase 1, *PHD* prolyl hydroxylase

(CK2), and IκB kinase (IKK), each responding to different oxidative stimuli (Perkins 2007). Genotoxic stress induced by ionizing radiation (IR) or some cancer drugs activates ataxia telangiectasia mutated (ATM) kinase that phosphorylates NF-κB essential modulator (NEMO) and forms an active IκB kinase (IKK) complex to phosphorylate IκBα (Wu et al. 2006). UV induces CK2-dependent phosphorylation of IκBα in its peptide sequence rich in proline, glutamic acid, serine, and threonine (PEST) domain (Perkins and Gilmore 2006). H_2O_2 and hypoxia promote phosphorylation of IκBα at tyrosine 42 residue, leading to its degradation or separation from NF-κB (Perkins and Gilmore 2006). In all the above cases, phosphorylated IκBα is eventually destroyed, thus resulting in NF-κB activation.

The TG2 mRNA level increases in response to cytokines like TNFα and IL-6 via the NF-κB-dependent signaling pathway (Kuncio et al. 1998; Suto et al. 1993). The NF-κB signaling pathway seems to be also involved in TG2 expression induced by various oxidative stimuli such as H_2O_2, UV, and doxorubicin (Jeong et al. 2009; Mehta 1994). Actually, TG2 transcription increases via the ATM-dependent NF-κB signaling pathway under genotoxic stress conditions (Ai et al. 2012) (Fig. 14.3). A member of the NF-κB family, p65, binds to two functional sites located at -2121 and -1361 relative to the translation start site of TG2 promoter in human breast cancer cells (Ai et al. 2012), indicating that TG2 is a bona fide target gene of NF-κB.

NF-κB-dependent expression of TG2 may be crucial to the pathogenesis of oxidative stress-related diseases such as chemo- and radio-resistant cancer (Cao et al. 2008; Herman et al. 2006), pathogenic inflammatory responses, and fibrosis (Falasca et al. 2008; Oh et al. 2011). Interestingly, TG2 is known to activate NF-κB signaling through its crosslinking activity for IκBα or PPARγ (Lee et al. 2004; Maiuri et al. 2008). This implies that there might exist a positive feedback loop generated by NF-κB and TG2, causing constitutive NF-κB activation. On the other hand, a TG2-ablated mouse shows no obvious phenotype (De Laurenzi and Melino 2001), indicating that TG2 may be a responder to oxidative stress in a variety of pathological contexts. Thus, TG2 can be considered as an effective druggable target for NF-κB-related diseases (Kim 2006; Verma and Mehta 2007).

14.3.2 Hypoxia-Inducible Factor-1 (HIF-1)

HIF-1 is an oxygen-sensitive heterodimeric transcription factor comprising HIF-1α and HIF-1β. Although HIF-1β is constitutively expressed, HIF-1α is only stabilized at the protein level under hypoxic conditions. Subsequently, HIF-1α binds to the promoter region of target genes and activates their transcription. In the presence of oxygen, HIF prolyl hydroxylases (PHD) catalyzes the hydroxylation of HIF-1α at proline residues. The hydroxylated HIF-1α is targeted by a component of the ubiquitin E3 ligase complex, von-Hippel–Lindau-tumor-suppressor (pVHL) and subsequently destroyed by the proteasome (Schofield and Ratcliffe 2004). Under hypoxic conditions, PHD is inactive and HIF-1α destruction pathway is suppressed, leading to the formation of transcriptionally active HIF-1 complex.

ROS are crucially involved in the regulation of HIF-1α protein stability under hypoxic conditions (Fig. 14.3). Hyperoxia is generally regarded as an inducer of ROS, which are generated by incomplete reduction of oxygen during respiration in the mitochondrial matrix. Paradoxically, a low oxygen concentration also produces superoxide anions from the outer ubiquinone-binding site of the mitochondrial complex III into the intermembrane space. This is probably due to the extended presence of the semiquinone radical in complex III (Sabharwal and Schumacker 2014) (Fig. 14.3). Hypoxia-induced ROS inactivate PHD and subsequently increase the HIF-1α expression level, allowing HIF-1 complex to activate the target gene promoter.

TG2 expression is induced by hypoxia in a HIF-1α-dependent manner in various cancer cells (Fig. 14.3) (Jang et al. 2010). TG2 promoter contains three potential hypoxia response elements (HRE), which are conserved sequences found in the promoter regions of hypoxia-inducible genes. Among them, in response to hypoxia, the HIF-1 complex binds to only one functional HRE located at -371/-368 relative to the translation start site of TG2 promoter in U373MG human glioblastoma cells (Jang et al. 2010). Thus, TG2 is an actual target gene of HIF-1.

Hypoxia-responsive TG2 might play a crucial role in tumor cell survival, invasion, and chemo- and radio-resistance (Jang et al. 2010). In addition, according to recent reports, HIF-1α-dependent expression of TG2 is associated with hypoxia-induced smooth muscle cell proliferation in the pulmonary artery in a pulmonary hypertension model (Penumatsa et al. 2014). Since hypoxia plays a central role in numerous pathophysiologies, it is suspected that HIF-1α-dependent expression of TG2 might also be involved in other cellular responses to hypoxia such as metabolic adaptation and angiogenesis (Basso and Ratan 2013; Wang et al. 2013).

14.4 Post-Translational Modifications (PTMs) of TG2 in Response to Oxidative Stress

14.4.1 Vicinal Disulfides

Studies have reported that TG2 is regulated by several PTMs including vicinal disulfide formation, SUMOylation, and ubiquitination through the ROS-mediated signaling pathway (Fig. 14.4). Of the PTMs, oxidation of cysteine residues in TG2 is considered to be a crucial regulatory mechanism for its activation because human TG2 has 20 cysteines in the amino acid sequence and even the active site has cysteine (Cys^{277}). The sulfhydryl group (-SH) of cysteine is highly reactive to ROS or RNS, forming oxidative modifications such as disulfide bridges, *S*-glutathionylation, *S*-nitrosylation, *S*-sulfenylation, *S*-sulfinylation, *S*-sulfonylation, *S*-cysteinylation, and *S*-sulfhydration (Grek et al. 2013). Thus, it often acts as a sensor in redox-sensitive proteins (Green and Paget 2004).

Previously, Fork and colleagues showed that purified guinea pig liver transglutaminase, TG2, was inactivated by treatment with various thiol oxidants or alkylating agents (Chung and Folk 1970; Connellan and Folk 1969). On the other hand, oxidized TG2 was reactivated in vitro by treatment of thiol-based reducing

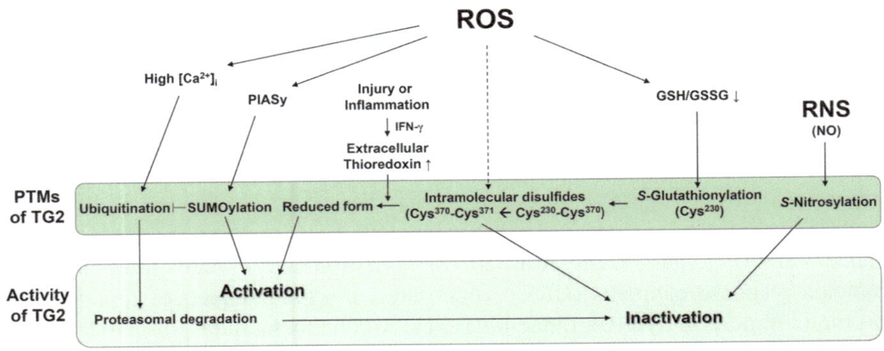

Fig. 14.4 ROS induce post-translational modifications of TG2. (1) Intramolecular disulfides; oxidative stress induces the formation of vicinal disulfide bond (Cys^{230}-Cys^{370}) in TG2, which facilitates another vicinal disulfide (Cys^{370}-Cys^{371}) formation (Bogeski et al. 2011). TG2 with vicinal disulfides is inactive, but can be reduced and activated by thioredoxin, which may be secreted from IFN-γ-stimulated monocytes or possibly the celiac intestinal mucosa (Jin et al. 2011). *S*-Glutathionylation at Cys^{230} of TG2 is suspected to be a transient modification to promote vicinal disulfide formation (Bogeski et al. 2011). (2) *S*-Nitrosylation; under nitrosative stress, TG2 can be modified by *S*-nitrosylation at up to 15 cysteine residues, resulting in its inactivation in vitro and in vivo (Lai et al. 2001; Santhanam et al. 2010). (3) Ubiquitination; a high level of ROS induces intracellular calcium overload that promotes TG2 degradation via the ubiquitin proteasome pathway (Jeong et al. 2009). (4) SUMOylation; CFTR$^{\Delta F508}$-induced ROS induce TG2 interaction with PIASy, a SUMO E3 ligase. Consequently, TG2 is modified by SUMO-1, which increases TG2 protein level via inhibition of ubiquitin proteasomal degradation, favoring TG2 activation (Luciani et al. 2009). *ROS* reactive oxygen species, *RNS* reactive nitrogen species, $[Ca^{2+}]_i$ intracellular calcium concentration, *GSH* reduced glutathione, *GSSG* oxidized glutathione, *IFN-γ* interferon gamma, *PTM* post-translational modification, *PIASy* protein inhibitor of activated STAT y, *SUMO-1* small ubiquitin-like modifier-1

reagents such as dithiothreitol (DTT), whereas alkylated TG2 was not. The active site Cys^{277} was not involved in the oxidation process. They suggested that TG2 might be inactivated through the formation of an intramolecular disulfide bridge between certain regulatory cysteine residues.

TG2 conformation changes substantially upon activation. TG2 exhibits an open form during calcium-mediated activation, however, it presents a closed inactive form that is induced by binding to the nucleotide (see Chap. 1). Recently, an intramolecular disulfide bond between Cys^{370} and Cys^{371} was found by solving crystal structures of the open form of TG2 in complex with a reactive gluten peptide mimetic inhibitor (Pinkas et al. 2007). Meanwhile, the x-ray crystal structures of GTP or ATP-bound TG2, which have closed forms, revealed another vicinal disulfide between Cys^{230} and Cys^{370} (Han et al. 2010; Jang et al. 2014). However, the vicinal disulfides did not exist in the crystal structures of the GDP-bound form (Liu et al. 2002). It seems that Cys^{370}-Cys^{371} stabilizes the open structure of TG2 and Cys^{230}-Cys^{370} stabilizes the nucleoside triphosphate binding closed conformation of TG2. Notably, residues of Cys^{230}, Cys^{370}, and Cys^{371} are proximate to each other in the crystal structures of both the open and closed forms of TG2

(Jang et al. 2014; Pinkas et al. 2007). This suggests that Cys^{370} can readily change neighboring cysteine residues in the process of TG2 oxidation.

Sollid and colleagues eventually proved that Cys^{230}, Cys^{370}, and Cys^{371} are bona fide redox-sensitive cysteine residues and disulfide bond formation between these residues renders TG2 inactive (Stamnaes et al. 2010). They showed that oxidative stress induces Cys^{230}-Cys^{370} disulfide bond formation, which facilitates the Cys^{370}-Cys^{371} disulfide bond, stabilizing the open structure (Fig. 14.4). Given that *S*-glutathionylation at Cys^{230} is observed early on in the oxidation process, vicinal disulfide bond formation might be initiated by Cys^{230} oxidation (Fig. 14.4), suggesting that Cys^{230} works as a redox sensor. Interestingly, Cys^{230} is not conserved in other TG isotypes. Thus, TG2 might be uniquely regulated by redox in this manner. Khosla and his colleagues calculated the redox potential of these redox-sensitive triad cysteines in human TG2 to be approximately −190 mV (Jin et al. 2011). This makes disulfide bond formation between TG2 cysteine residues a biologically relevant regulatory mechanism. Under normal conditions, the redox potential of the extracellular space through GSH/GSSG and cysteine/cystine redox couples is approximately −140 mV and −80 mV, respectively (Banerjee 2012). This implies that most of the secreted TG2 is probably oxidized and inactivated. However, in inflamed regions, TG2 can be reduced and activated by an increase of extracellular thioredoxin, which is secreted from monocytes triggered by interferon-γ (Jin et al. 2011). Because thioredoxin has lower redox potential (−230 mV) than the TG2 reactive cysteine residues (Watson et al. 2003), it has been proposed that this is the key activation mechanism of extracellular TG2 under pathological conditions, especially in celiac disease.

Apparently, the redox regulation should work in the endoplasmic reticulum (ER) as well. ER contains 100–500 μM of free Ca^{2+} (Monteith et al. 2007) which is enough to activate TG2 (see Sect. 14.5.1). However, the ER redox potential is between approximately −190 and −130 mV (Sarkar et al. 2013) and it is possible that unwanted TG2 activation is prevented by the oxidation of the reactive cysteines in the ER. On the other hand, it has been well proven that intracellular TG2 is aberrantly activated under oxidative environments, as they are generally favorable for vicinal disulfide formation. Thus, it is not clear whether the vicinal disulfide bonds control TG2 in the cytosol and other cell organelles.

14.4.2 S-Nitrosylation

It is well known that the thiol group of cysteine residues in protein can be reversibly oxidized by reactive nitrogen species (RNS) as well as ROS. TG2 can be modified with *S*-nitrosylation by RNS in vitro and in vivo, inhibiting its crosslinking activity (Fig. 14.4) (Lai et al. 2001; Santhanam et al. 2010). NO is also known to inhibit non-classical secretion of TG2 into the extracellular space in human aortic endothelial cells (Santhanam et al. 2010), Swiss 3T3 fibroblasts (Telci et al. 2009), and IMR90 fibroblasts (Jandu et al. 2013), which suggests that *S*-nitrosylation may be one of the mechanisms involved in inhibiting extracellular TG2 (Santhanam et al. 2011). In a model of age-associated vascular stiffness, it has been demonstrated that the level of *S*-nitrosylated TG2 is reduced in the aorta of aged mice. This is due to a decrease of NO production, leading to vascular stiffness by an increase of TG2-mediated extracellular crosslinking (Santhanam et al. 2010). In addition to *S*-nitrosylation of cysteine residues, NO is also known to increase nitrosylation of tyrosine residues in TG2 (Telci et al. 2009). Given that TG2 has a tyrosine residue that is critical (Tyr^{274} in human TG2) for activation (Begg et al. 2006a), TG2 can be inhibited by tyrosine nitrosylation (Telci et al. 2009) as well as *S*-nitrosylation under nitrosative stressed conditions. However, the biological role(s) of tyrosine nitrosylation in TG2 is/are not well known. Moreover, whether *S*-nitrosylation is directly involved in TG2 externalization remains unknown (Santhanam et al. 2011). Hence, the role of nitrosylation in TG2 regulation needs to be clarified.

14.4.3 Ubiquitination

In response to low or moderate oxidative stress, the TG2 expression level increases via its promoter activation (Sect. 14.3) or SUMOylation (Sect. 14.4.4), which favors its intracellular enzymatic activation. TG2 activity may inhibit apoptosis through crosslinking of caspase-3 and IκBα (Jang et al. 2010); however, under high or incessant oxidative conditions, TG2 can promote cell death, probably due to accumulation of crosslinked cellular proteins (Fesus and Tarcsa 1989). Thus, TG2 activity should be negatively regulated to avoid molecular aggregation under extremely stressed situations. One powerful mechanism to achieve this is ubiquitin-dependent proteasomal degradation. High oxidative stress triggers intracellular calcium overload that renders TG2 inactive via the polyubiquitination and proteasomal degradation pathway (Jeong et al. 2009) (Fig. 14.4). However, the ubiquitination sites in TG2 and the associated E3 ligase have not yet been identified.

14.4.4 SUMOylation

The cystic fibrosis transmembrane conductance regulator (CFTR) is a chloride ion channel formed at the apical membrane of epithelial cells in many organs including the lung, kidney, intestine, and pancreas. Mutations in the CFTR gene are the major cause of cystic fibrosis. $CFTR^{\Delta F508}$, a phenylalanine-deleted CFTR mutant at position 508, is the most common mutation to cause cystic fibrosis (Cutting 2015). $CFTR^{\Delta F508}$ is retained in the ER due to its misfolded structure, which in turn leads to ER stress and an increase in the production of ROS. Mairui and colleagues showed that $CFTR^{\Delta F508}$-mediated ROS induce the interaction of TG2 with one of the protein inhibitors of activated STAT (PIAS) proteins, PIASy. It is a small ubiquitin-like modifier (SUMO) E3 ligase and it drives TG2 to be modified by SUMO-1 (Fig. 14.4) (Luciani et al. 2009). SUMOylation in TG2 hinders its ubiquitination and proteasomal degradation like many other SUMOylated proteins such as nuclear factor (NF)-κB essential modulator (NEMO), IκBα, and proliferating cell nuclear antigen (PCNA) (Geiss-Friedlander and Melchior 2007). Therefore, TG2 expression can be sustained at high levels in epithelial cells of cystic fibrosis patients harboring the $CFTR^{\Delta F508}$ mutation.

SUMOylated TG2 plays a pathogenic role in cystic fibrosis by modulating IκBα, PPARγ, and Beclin 1 pathways. TG2 crosslinks PPARγ and probably IκBα to prevent their modification by SUMO-1. This results in their degradation via the ubiquitin proteasome system, as well as in aberrant activation of the NF-κB signal and an unrestrained inflammatory response (Luciani et al. 2009; Maiuri et al. 2008). Additionally, TG2 also catalyzes crosslinking of Beclin 1, a mammalian homolog of the yeast autophagy-related gene (Atg) 6, which plays a central role in the initial stages of autophagy (Marino et al. 2014). TG2-mediated crosslinking of Beclin 1 promotes the sequestration of Beclin 1-VPS34 complex into HDAC6-positive aggresomes, hindering the autophagy process (Luciani et al. 2010a). Overall, TG2 SUMOylation exacerbates cystic fibrosis by promoting inflammation and inhibiting autophagy. Thus, it has been suggested that the SUMOylation-mediated TG2 activation pathway can be an effective target for cystic fibrosis as well as other chronic inflammatory diseases, neurodegenerative disorders, and cancer. Further investigation is required to assess the role of TG2 SUMOylation in oxidative stress-driven pathology of other diseases.

14.5 ROS-Responsive Cellular Factors Associated With Regulation of TG2 Enzymatic Activity

14.5.1 Intracellular Calcium Ion Concentration ($[Ca^{2+}]_i$)

Calcium is a cofactor required for the transamidation activity of TG2, and thus for TG2 activation. In several reports, the EC_{50} of calcium for activation of endogenous or purified recombinant TG2 was estimated to be approximately 100–500 μM (Begg et al. 2006b; Bergamini and Signorini 1993; Kanchan et al. 2013; Lorand and Conrad 1984; Signorini et al. 1988). Under physiological conditions, cells maintain low levels of free calcium ion in the cytosol ($[Ca^{2+}]_i = {\sim}100$ nM) (Roderick and Cook 2008) and high levels of GTP, an intracellular inhibitor for transamidation activity of TG2 (Begg et al. 2006b). This indicates that intracellular TG2 is not active (Shin et al. 2004; Siegel et al. 2008; Zhang et al. 1998) and the increase of $[Ca^{2+}]_i$ is a prerequisite for intracellular TG2 activation.

ROS are known to increase $[Ca^{2+}]_i$ via activation of various signaling pathways and redox-sensitive calcium channels (Bogeski et al. 2011). Indeed, oxidative stresses elevated the level of $[Ca^{2+}]_i$ and calcium chelating agents such as

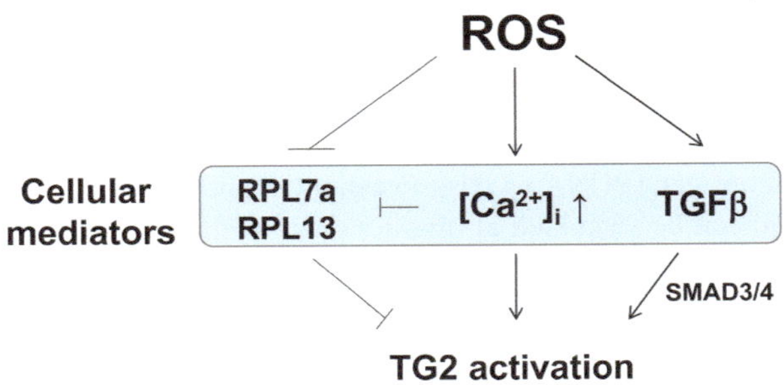

Fig. 14.5 Cellular mediators regulate ROS-dependent TG2 activation. (1) $[Ca^{2+}]_i$; TG2 is activated by ROS-induced increase of $[Ca^{2+}]_i$ (Jeong et al. 2009; Shin et al. 2008). (2) TGFβ; ROS increase the active form of TGFβ (Barcellos-Hoff and Dix 1996), which activates TG2 via SMAD3/4-dependent gene(s) expression (Shin et al. 2008). (3) RPL7a and RPL13; it has been recently suggested that RPL7a and RPL13 are endogenous TG2 inhibitor proteins (Kim 2014). They interact with TG2, leading to inactivation. ROS promote dissociation of TG2 from the ribosomal proteins, resulting in TG2 activation. *ROS* reactive oxygen species, $[Ca^{2+}]_i$ intracellular calcium concentration

BAPTA-AM and EGTA blocked ROS-induced TG2 activation (Fig. 14.5) (Jang et al. 2010; Jeong et al. 2009; Shin et al. 2008). This demonstrates that ROS activate TG2 through an increase of $[Ca^{2+}]_i$. However, it is still unknown which redox signaling pathway(s) and calcium channel(s) are responsible for the increase of $[Ca^{2+}]_i$ to activate TG2 under oxidative stress conditions. Since evidence on the involvement of $[Ca^{2+}]_i$ dysregulation in the pathogenesis of oxidative stress-related diseases has been accumulating (Hetz and Mollereau 2014; LaFerla 2002; Roderick and Cook 2008), it is expected that the verification of these $[Ca^{2+}]_i$ regulatory mechanisms will increase our understanding of TG2-related diseases.

14.5.2 TGFβ-Signaling Pathway

ROS oxidize the latent form of TGFβ into active TGFβ (Barcellos-Hoff and Dix 1996), which triggers TG2 activation (Fig. 14.5). In human lens epithelial cells and ex vivo cultured rodent lens, H_2O_2 or selenite activates TG2 in a TGFβ-dependent manner, which catalyzes the aggregation of crystallin proteins and leads to cataract formation (Shin et al. 2008). Bleomycin induces lung epithelial cells to produce active TGFβ, which promotes the accumulation of extracellular matrix (ECM) in lung fibroblasts, resulting in pulmonary fibrosis (Oh et al. 2011). Under these conditions, intracellular TG2 is activated by an unknown protein; this protein is upregulated by SMAD3/4 without a change in TG2 expression level (Shin et al. 2008) (Fig 14.5). Thus, we need to identify the TGFβ-dependent TG2 activator to elucidate the pathogenesis of TG2-mediated inflammation and fibrosis.

14.5.3 Ribosomal Proteins

Recently, ribosomal proteins, L7a (RPL7a) and L13 (RPL13) have been proposed to be endogenous TG2 inhibitors (Kim 2014). These ribosomal proteins specifically inhibit the transamidation activity of TG2, but not of other TG isotypes including Factor XIIIa, TG1, TG3, and TG4. The inhibitor proteins suppressed TG2 activity through direct interaction with β-barrel2 domain in TG2 in a mixed inhibition manner. Thus, intracellular TG2 activity was largely dependent on the expression levels of the ribosomal proteins. Intriguingly, under oxidative stress conditions, not only did the interaction between TG2 and inhibitor proteins decrease via increased $[Ca^{2+}]_i$ but also expression levels of the inhibitors decreased via ROS-mediated suppression of their promoter activity; this resulted in intracellular TG2 activation. These results suggest that RPL7a and RPL13, along with endogenous small molecule inhibitors for TG2 like GTP, cysteamine, and cystamine may participate in the prevention of aberrant TG2 activation under physiological conditions.

14.6 Concluding Remarks

ROS were thought to be essentially toxic to the cells as they cause oxidative damage to cellular components. However, a growing body of evidence suggests that ROS are involved in the regulation of many cellular processes including proliferation, differentiation, inflammation, survival, and autophagy, by eliciting redox signaling pathways (Holmstrom and Finkel 2014). The findings presented in this chapter strongly suggest that TG2 may play a role as a mediator or downstream effector in redox signaling pathways. At physiological levels of ROS, TG2 helps in the adaptation to oxidative environments. However, when cells are exposed to high levels of ROS or persistent oxidative stress, TG2 may contribute to the development of cancer, chronic inflammatory and degenerative diseases.

Interestingly, most TG2 associated disorders are age-related degenerative diseases, in which pathogenesis is closely linked to oxidative stress (Balaban et al. 2005; Finkel and Holbrook 2000; Iismaa et al. 2009) (Fig. 14.2 and Table 14.1). Moreover, TG2 is upregulated in the aging process (Lavie and Weinreb 1996; Park et al. 1999), implying that TG2 may be a key factor in age-related degenerative diseases even though the exact mechanism of TG2 activation by ROS needs to be clarified.

In conclusion, elucidation of the links between oxidative stress and TG2 will aid understanding of the pathophysiology of TG2-related diseases and development of effective therapeutic strategies.

References

AbdAlla S, Lother H, el Missiry A, Langer A, Sergeev P, el Faramawy Y, Quitterer U (2009) Angiotensin II AT2 receptor oligomers mediate G-protein dysfunction in an animal model of Alzheimer disease. J Biol Chem 284:6554–6565

Ai L, Skehan RR, Saydi J, Lin T, Brown KD (2012) Ataxia-Telangiectasia, Mutated (ATM)/Nuclear factor kappa light chain enhancer of activated B cells (NFkappaB) signaling controls basal and DNA damage-induced transglutaminase 2 expression. J Biol Chem 287:18330–18341

Balaban RS, Nemoto S, Finkel T (2005) Mitochondria, oxidants, and aging. Cell 120:483–495

Banerjee R (2012) Redox outside the box: linking extracellular redox remodeling with intracellular redox metabolism. J Biol Chem 287:4397–4402

Barcellos-Hoff MH, Dix TA (1996) Redox-mediated activation of latent transforming growth factor-beta 1. Mol Endocrinol 10:1077–1083

Basso M, Ratan RR (2013) Transglutaminase is a therapeutic target for oxidative stress, excitotoxicity and stroke: a new epigenetic kid on the CNS block. J Cereb Blood Flow Metab 33:809–818

Basso M, Berlin J, Xia L, Sleiman SF, Ko B, Haskew-Layton R, Kim E, Antonyak MA, Cerione RA, Iismaa SE et al (2012) Transglutaminase inhibition protects against oxidative stress-induced neuronal death downstream of pathological ERK activation. J Neurosci Off J Soc Neurosci 32:6561–6569

Battaglia G, Farrace MG, Mastroberardino PG, Viti I, Fimia GM, Van Beeumen J, Devreese B, Melino G, Molinaro G, Busceti CL et al (2007) Transglutaminase 2 ablation leads to defective function of mitochondrial respiratory complex I affecting neuronal vulnerability in experimental models of extrapyramidal disorders. J Neurochem 100:36–49

Begg GE, Carrington L, Stokes PH, Matthews JM, Wouters MA, Husain A, Lorand L, Iismaa SE, Graham RM (2006a) Mechanism of allosteric regulation of transglutaminase 2 by GTP. Proc Natl Acad Sci U S A 103:19683–19688

Begg GE, Holman SR, Stokes PH, Matthews JM, Graham RM, Iismaa SE (2006b) Mutation of a critical arginine in the GTP-binding site of transglutaminase 2 disinhibits intracellular cross-linking activity. J Biol Chem 281:12603–12609

Bergamini CM, Signorini M (1993) Studies on tissue transglutaminases: interaction of erythrocyte type-2 transglutaminase with GTP. Biochem J 291(Pt 1):37–39

Bhatt MP, Lim YC, Hwang J, Na S, Kim YM, Ha KS (2013) C-peptide prevents hyperglycemia-induced endothelial apoptosis through inhibition of reactive oxygen species-mediated transglutaminase 2 activation. Diabetes 62:243–253

Bogeski I, Kappl R, Kummerow C, Gulaboski R, Hoth M, Niemeyer BA (2011) Redox regulation of calcium ion channels: chemical and physiological aspects. Cell Calcium 50:407–423

Cao L, Petrusca DN, Satpathy M, Nakshatri H, Petrache I, Matei D (2008) Tissue transglutaminase protects epithelial ovarian cancer cells from cisplatin-induced apoptosis by promoting cell survival signaling. Carcinogenesis 29:1893–1900

Chung SI, Folk JE (1970) Mechanism of the inactivation of guinea pig liver transglutaminase by tetrathionate. J Biol Chem 245:681–689

Connellan JM, Folk JE (1969) Mechanism of the inactivation of guinea pig liver transglutaminase by 5,5′-dithiobis-(2-nitrobenzoic acid). J Biol Chem 244:3173–3181

Cutting GR (2015) Cystic fibrosis genetics: from molecular understanding to clinical application. Nat Rev Genet 16:45–56

D'Autreaux B, Toledano MB (2007) ROS as signalling molecules: mechanisms that generate specificity in ROS homeostasis. Nat Rev Mol Cell Biol 8:813–824

De Laurenzi V, Melino G (2001) Gene disruption of tissue transglutaminase. Mol Cell Biol 21:148–155

Falasca L, Farrace MG, Rinaldi A, Tuosto L, Melino G, Piacentini M (2008) Transglutaminase type II is involved in the pathogenesis of endotoxic shock. J Immunol 180:2616–2624

Fesus L, Tarcsa E (1989) Formation of N epsilon-(gamma-glutamyl)-lysine isodipeptide in Chinese-hamster ovary cells. Biochem J 263:843–848
Finkel T, Holbrook NJ (2000) Oxidants, oxidative stress and the biology of ageing. Nature 408:239–247
Geiss-Friedlander R, Melchior F (2007) Concepts in sumoylation: a decade on. Nat Rev Mol Cell Biol 8:947–956
Green J, Paget MS (2004) Bacterial redox sensors. Nat Rev Microbiol 2:954–966
Grek CL, Zhang J, Manevich Y, Townsend DM, Tew KD (2013) Causes and consequences of cysteine S-glutathionylation. J Biol Chem 288:26497–26504
Han BG, Cho JW, Cho YD, Jeong KC, Kim SY, Lee BI (2010) Crystal structure of human transglutaminase 2 in complex with adenosine triphosphate. Int J Biol Macromol 47:190–195
Herman JF, Mangala LS, Mehta K (2006) Implications of increased tissue transglutaminase (TG2) expression in drug-resistant breast cancer (MCF-7) cells. Oncogene 25:3049–3058
Hetz C, Mollereau B (2014) Disturbance of endoplasmic reticulum proteostasis in neurodegenerative diseases. Nat Rev Neurosci 15:233–249
Holmstrom KM, Finkel T (2014) Cellular mechanisms and physiological consequences of redox-dependent signalling. Nat Rev Mol Cell Biol 15:411–421
Hybertson BM, Gao B, Bose SK, McCord JM (2011) Oxidative stress in health and disease: the therapeutic potential of Nrf2 activation. Mol Asp Med 32:234–246
Iismaa SE, Mearns BM, Lorand L, Graham RM (2009) Transglutaminases and disease: lessons from genetically engineered mouse models and inherited disorders. Physiol Rev 89:991–1023
Jandu SK, Webb AK, Pak A, Sevinc B, Nyhan D, Belkin AM, Flavahan NA, Berkowitz DE, Santhanam L (2013) Nitric oxide regulates tissue transglutaminase localization and function in the vasculature. Amino Acids 44:261–269
Jang GY, Jeon JH, Cho SY, Shin DM, Kim CW, Jeong EM, Bae HC, Kim TW, Lee SH, Choi Y et al (2010) Transglutaminase 2 suppresses apoptosis by modulating caspase 3 and NF-kappaB activity in hypoxic tumor cells. Oncogene 29:356–367
Jang TH, Lee DS, Choi K, Jeong EM, Kim IG, Kim YW, Chun JN, Jeon JH, Park HH (2014) Crystal structure of transglutaminase 2 with GTP complex and amino acid sequence evidence of evolution of GTP binding site. PLoS ONE 9:e107005
Jeong EM, Kim CW, Cho SY, Jang GY, Shin DM, Jeon JH, Kim IG (2009) Degradation of transglutaminase 2 by calcium-mediated ubiquitination responding to high oxidative stress. FEBS Lett 583:648–654
Jin X, Stamnaes J, Klock C, DiRaimondo TR, Sollid LM, Khosla C (2011) Activation of extracellular transglutaminase 2 by thioredoxin. J Biol Chem 286:37866–37873
Kanchan K, Ergulen E, Kiraly R, Simon-Vecsei Z, Fuxreiter M, Fesus L (2013) Identification of a specific one amino acid change in recombinant human transglutaminase 2 that regulates its activity and calcium sensitivity. Biochem J 455:261–272
Kim SY (2006) Transglutaminase 2 in inflammation. Front Biosci 11:3026–3035
Kim IG (2014) The mechanism of oxidative stress induced transglutaminase 2 activation. Oral presentation at Transglutaminases in human disease processes gordon research conference, Lucca, 29 June–4 July 2014
Kuncio GS, Tsyganskaya M, Zhu J, Liu SL, Nagy L, Thomazy V, Davies PJ, Zern MA (1998) TNF-alpha modulates expression of the tissue transglutaminase gene in liver cells. Am J Physiol 274:G240–G245
LaFerla FM (2002) Calcium dyshomeostasis and intracellular signalling in Alzheimer's disease. Nat Rev Neurosci 3:862–872
Lai TS, Hausladen A, Slaughter TF, Eu JP, Stamler JS, Greenberg CS (2001) Calcium regulates S-nitrosylation, denitrosylation, and activity of tissue transglutaminase. Biochemistry 40:4904–4910
Lai TS, Tucker T, Burke JR, Strittmatter WJ, Greenberg CS (2004) Effect of tissue transglutaminase on the solubility of proteins containing expanded polyglutamine repeats. J Neurochem 88:1253–1260

Lavie L, Weinreb O (1996) Age- and strain-related changes in tissue transglutaminase activity in murine macrophages: the effects of inflammation and induction by retinol. Mech Ageing Dev 90:129–143

Lee J, Kim YS, Choi DH, Bang MS, Han TR, Joh TH, Kim SY (2004) Transglutaminase 2 induces nuclear factor-kappaB activation via a novel pathway in BV-2 microglia. J Biol Chem 279:53725–53735

Lee SM, Jeong EM, Jeong J, Shin DM, Lee HJ, Kim HJ, Lim J, Lee JH, Cho SY, Kim MK et al (2012) Cysteamine prevents the development of lens opacity in a rat model of selenite-induced cataract. Invest Ophthalmol Vis Sci 53:1452–1459

Liu S, Cerione RA, Clardy J (2002) Structural basis for the guanine nucleotide-binding activity of tissue transglutaminase and its regulation of transamidation activity. Proc Natl Acad Sci U S A 99:2743–2747

Lorand L, Conrad SM (1984) Transglutaminases. Mol Cell Biochem 58:9–35

Luciani A, Villella VR, Vasaturo A, Giardino I, Raia V, Pettoello-Mantovani M, D'Apolito M, Guido S, Leal T, Quaratino S et al (2009) SUMOylation of tissue transglutaminase as link between oxidative stress and inflammation. J Immunol 183:2775–2784

Luciani A, Villella VR, Esposito S, Brunetti-Pierri N, Medina D, Settembre C, Gavina M, Pulze L, Giardino I, Pettoello-Mantovani M et al (2010a) Defective CFTR induces aggresome formation and lung inflammation in cystic fibrosis through ROS-mediated autophagy inhibition. Nat Cell Biol 12:863–875

Luciani A, Villella VR, Vasaturo A, Giardino I, Pettoello-Mantovani M, Guido S, Cexus ON, Peake N, Londei M, Quaratino S et al (2010b) Lysosomal accumulation of gliadin p31-43 peptide induces oxidative stress and tissue transglutaminase-mediated PPARgamma downregulation in intestinal epithelial cells and coeliac mucosa. Gut 59:311–319

Maiuri L, Luciani A, Giardino I, Raia V, Villella VR, D'Apolito M, Pettoello-Mantovani M, Guido S, Ciacci C, Cimmino M et al (2008) Tissue transglutaminase activation modulates inflammation in cystic fibrosis via PPARgamma down-regulation. J Immunol 180:7697–7705

Marino G, Niso-Santano M, Baehrecke EH, Kroemer G (2014) Self-consumption: the interplay of autophagy and apoptosis. Nat Rev Mol Cell Biol 15:81–94

Mayne ST (2003) Antioxidant nutrients and chronic disease: use of biomarkers of exposure and oxidative stress status in epidemiologic research. J Nutr 133(Suppl 3):933S–940S

Mehta K (1994) High levels of transglutaminase expression in doxorubicin-resistant human breast carcinoma cells. Int J Cancer J Int du Cancer 58:400–406

Monteith GR, McAndrew D, Faddy HM, Roberts-Thomson SJ (2007) Calcium and cancer: targeting Ca2+ transport. Nat Rev Cancer 7:519–530

Nathan C, Cunningham-Bussel A (2013) Beyond oxidative stress: an immunologist's guide to reactive oxygen species. Nat Rev Immunol 13:349–361

Oh K, Park HB, Byoun OJ, Shin DM, Jeong EM, Kim YW, Kim YS, Melino G, Kim IG, Lee DS (2011) Epithelial transglutaminase 2 is needed for T cell interleukin-17 production and subsequent pulmonary inflammation and fibrosis in bleomycin-treated mice. J Exp Med 208:1707–1719

Park SC, Yeo EJ, Han JA, Hwang YC, Choi JY, Park JS, Park YH, Kim KO, Kim IG, Seong SC et al (1999) Aging process is accompanied by increase of transglutaminase C. J Gerontol A: Biol Med Sci 54:B78–B83

Penumatsa KC, Toksoz D, Warburton RR, Hilmer AJ, Liu T, Khosla C, Comhair SA, Fanburg BL (2014) Role of hypoxia-induced transglutaminase 2 in pulmonary artery smooth muscle cell proliferation. Am J Physiol Lung Cell Mol Physiol 307:L576–L585

Perkins ND (2007) Integrating cell-signalling pathways with NF-kappaB and IKK function. Nat Rev Mol Cell Biol 8:49–62

Perkins ND, Gilmore TD (2006) Good cop, bad cop: the different faces of NF-kappaB. Cell Death Differ 13:759–772

Pinkas DM, Strop P, Brunger AT, Khosla C (2007) Transglutaminase 2 undergoes a large conformational change upon activation. PLoS Biol 5:e327

Roderick HL, Cook SJ (2008) Ca2+ signalling checkpoints in cancer: remodelling Ca2+ for cancer cell proliferation and survival. Nat Rev Cancer 8:361–375

Sabharwal SS, Schumacker PT (2014) Mitochondrial ROS in cancer: initiators, amplifiers or an Achilles' heel? Nat Rev Cancer 14:709–721

Santhanam L, Tuday EC, Webb AK, Dowzicky P, Kim JH, Oh YJ, Sikka G, Kuo M, Halushka MK, Macgregor AM et al (2010) Decreased S-nitrosylation of tissue transglutaminase contributes to age-related increases in vascular stiffness. Circ Res 107:117–125

Santhanam L, Berkowitz DE, Belkin AM (2011) Nitric oxide regulates non-classical secretion of tissue transglutaminase. Commun Integr Biol 4:584–586

Sarkar DD, Edwards SK, Mauser JA, Suarez AM, Serowoky MA, Hudok NL, Hudok PL, Nunez M, Weber CS, Lynch RM et al (2013) Increased redox-sensitive green fluorescent protein reduction potential in the endoplasmic reticulum following glutathione-mediated dimerization. Biochemistry 52:3332–3345

Schofield CJ, Ratcliffe PJ (2004) Oxygen sensing by HIF hydroxylases. Nat Rev Mol Cell Biol 5:343–354

Segers-Nolten IM, Wilhelmus MM, Veldhuis G, van Rooijen BD, Drukarch B, Subramaniam V (2008) Tissue transglutaminase modulates alpha-synuclein oligomerization. Protein Sci Publ Protein Soc 17:1395–1402

Shin DM, Jeon JH, Kim CW, Cho SY, Kwon JC, Lee HJ, Choi KH, Park SC, Kim IG (2004) Cell type-specific activation of intracellular transglutaminase 2 by oxidative stress or ultraviolet irradiation: implications of transglutaminase 2 in age-related cataractogenesis. J Biol Chem 279:15032–15039

Shin DM, Jeon JH, Kim CW, Cho SY, Lee HJ, Jang GY, Jeong EM, Lee DS, Kang JH, Melino G et al (2008) TGFbeta mediates activation of transglutaminase 2 in response to oxidative stress that leads to protein aggregation. Faseb J Off Publ Fed Am Soc Exp Biol 22:2498–2507

Siegel M, Strnad P, Watts RE, Choi K, Jabri B, Omary MB, Khosla C (2008) Extracellular transglutaminase 2 is catalytically inactive, but is transiently activated upon tissue injury. PLoS ONE 3:e1861

Signorini M, Bortolotti F, Poltronieri L, Bergamini CM (1988) Human erythrocyte transglutaminase: purification and preliminary characterisation. Biol Chem Hoppe Seyler 369:275–281

Stamnaes J, Pinkas DM, Fleckenstein B, Khosla C, Sollid LM (2010) Redox regulation of transglutaminase 2 activity. J Biol Chem 285:25402–25409

Steinhubl SR (2008) Why have antioxidants failed in clinical trials? Am J Cardiol 101:14D–19D

Suto N, Ikura K, Sasaki R (1993) Expression induced by interleukin-6 of tissue-type transglutaminase in human hepatoblastoma HepG2 cells. J Biol Chem 268:7469–7473

Telci D, Collighan RJ, Basaga H, Griffin M (2009) Increased TG2 expression can result in induction of transforming growth factor beta1, causing increased synthesis and deposition of matrix proteins, which can be regulated by nitric oxide. J Biol Chem 284:29547–29558

Tucholski J, Roth KA, Johnson GV (2006) Tissue transglutaminase overexpression in the brain potentiates calcium-induced hippocampal damage. J Neurochem 97:582–594

Verma A, Mehta K (2007) Tissue transglutaminase-mediated chemoresistance in cancer cells. Drug Resist Updat 10:144–151

Wang Z, Perez M, Caja S, Melino G, Johnson TS, Lindfors K, Griffin M (2013) A novel extracellular role for tissue transglutaminase in matrix-bound VEGF-mediated angiogenesis. Cell Death Dis 4:e808

Watson WH, Pohl J, Montfort WR, Stuchlik O, Reed MS, Powis G, Jones DP (2003) Redox potential of human thioredoxin 1 and identification of a second dithiol/disulfide motif. J Biol Chem 278:33408–33415

Wu ZH, Shi Y, Tibbetts RS, Miyamoto S (2006) Molecular linkage between the kinase ATM and NF-kappaB signaling in response to genotoxic stimuli. Science 311:1141–1146

Zhang J, Lesort M, Guttmann RP, Johnson GV (1998) Modulation of the in situ activity of tissue transglutaminase by calcium and GTP. J Biol Chem 273:2288–2295

Chapter 15
Blood Coagulation Factor XIII: A Multifunctional Transglutaminase

Moyuru Hayashi and Kohji Kasahara

Abstract Factor XIII is a pro-enzyme of plasma transglutaminase consisting of two enzymatic A subunits and two non-catalytic B subunits, and platelet transglutaminase consisting of two enzymatic A subunits. FXIII plays a critical role in the generation of a stable hemostatic plug, wound healing, maintaining pregnancy, angiogenesis, apoptosis and bacterial immobilization. FXIII catalyzes intermolecular cross-linking reactions between fibrin monomers and α_2-antiplasmin. These reactions increase the mechanical strength of the fibrin clot and its resistance to proteolytic degradation. Congenital FXIII deficiency is a rare autosomal recessive disorder, most cases of which are caused by defects in the FXIII-A gene, leading to a bleeding tendency. An autoimmune hemophilia-like disease is caused by anti-FXIII antibodies. Platelet surface FXIII-A_2 is involved in fibrin translocation to lipid rafts and outside-in signaling, leading to clot retraction. FXIII-A_2-mediated protein cross-linking is associated with assembly of the extracellular matrix on a variety of cell surfaces in physiological events such as differentiation.

Keywords Factor XIII • Platelets • Surface • Clot retraction • Lipid rafts • Transamidation

15.1 Introduction

The blood coagulation cascade has evolved as a defense mechanism for maintaining hemostasis during blood vessel injury. This process is controlled by a signaling cascade consisting of 13 coagulation factors. There are two separate pathways, the intrinsic and extrinsic. The intrinsic pathway is activated by trauma inside the vascular system, and is activated by platelets, exposed endothelium, or collagen. This pathway involves factors XII, XI, IX, and VIII. The extrinsic pathway is activated by external trauma that causes blood to escape from the vascular system. This pathway involves factor VII. These eventually join together to form the common pathway. The common pathway involves factors I, II, V, and

M. Hayashi • K. Kasahara (✉)
Laboratory of Biomembrane, Tokyo Metropolitan Institute of Medical Science, Tokyo, Japan
e-mail: kasahara-kj@igakuken.or.jp

K. Hitomi et al. (eds.), *Transglutaminases*, DOI 10.1007/978-4-431-55825-5_15

X. Activation of the cascade produces thrombin (factor IIa) that activates a variety of components in the cascade. Thrombin cleaves fibrinogen to fibrin resulting in clot formation. At the end of this process, covalent bonds are introduced into the fibrin clot by the activated fibrin stabilizing factor (factor XIIIa) (Lorand 2001).

15.2 FXIII Biochemistry

Factor XIII (FXIII) is a Ca^{2+}-dependent pro-transglutaminase which cross-links proteins by catalyzing the formation of isopeptide bonds between glutamine and lysine residues (Lorand 2001). FXIII has two forms: a plasma form that is a tetramer of two carrier B subunits and two catalytic A subunits, and an intracellular form that consists of two catalytic A subunits.

The primary structures of the A subunit (FXIII-A; ~83 kDa, encoded by the F13A1 gene on human chromosome 6p24-25) and B subunit (FXIII-B; ~80 kDa, encoded by the F13B gene on chromosome 1q31-32.1) have been determined by cDNA cloning and amino acid sequence analysis (Ichinose et al. 1986a, b; Grundmann et al. 1986). FXIII-A consists of four main structural domains, the β-sandwich (amino acids 38–184), the catalytic core (amino acids 185–515), β-barrel 1 (amino acids 516–628), and β-barrel 2 (amino acids 629–731) domains, plus an NH_2-terminal activation peptide (amino acids 1–37). There are nine cysteine residues, including the active site cysteine (Cys 314), none of which forms disulfide bonds. FXIII-B consists of ten short tandem repeats, called sushi domains, held together by a pair of internal disulfide bonds.

FXIII plays a critical role in the generation of a viable hemostatic plug. Following exposure to thrombin and calcium, a zymogen is activated to FXIIIa, which cross-links proteins by catalyzing the formation of isopeptide bonds between glutamine and lysine residues of fibrin within a clot network. FXIII also catalyzes the posttranslational modification of proteins by transamidation of available glutamine residues. This reaction results in the incorporation of low-molecular-weight amines (e.g., serotonin, dopamine, and polyamine) into suitable protein substrates. Serotonin has been demonstrated to modify small GTPase by transglutaminase-mediated transamidation, during platelet activation. As a result, the small GTPase is constitutively activated, leading to irreversible aggregation and platelet α-granule release (Walther et al. 2003).

In the initial step of FXIII activation, thrombin cleaves off an activation peptide from the NH_2 terminus of FXIII-A by hydrolyzing the Arg37–Gly38 peptide bond. Then, in the presence of Ca^{2+}, the inhibitory B subunits dissociate, which is a prerequisite for the truncated FXIII-A dimer (FXIII-A_2') to assume an enzymatically active conformation (FXIII-A_2*). The conformational change of FXIII-A_2' resulting in an active transglutaminase also requires Ca^{2+}.

FXIII is responsible for the cross-linking of fibrin γ-chains in the early stages of clot formation, whereas α-chain cross-linking occurs at a slow rate. Cross-linking of the γ-chains plays a role in fibrin fiber appearance time and fiber density. Cross-linking of the α-chains plays a role in the thickening of fibrin fibers (Duval et al. 2014).

The main role of FXIII-B is to prolong the lifespan of FXIII-A in plasma. In FXIII-B-deficient patients, the FXIII-A_2 concentration in plasma is significantly decreased. Administration of recombinant FXIII-B into FXIII-B(−/−) mice accelerates fibrin cross-linking in plasma and assists the maintenance of plasma FXIII-A levels (Souri et al. 2008a). The subunits form an A_2B_2 tetramer, which circulates in plasma with a half-life of >1 week.

The A subunit is synthesized by cells of bone marrow origin and the B subunit is synthesized by hepatocytes in the liver. Formation of the FXIII-A_2B_2 complex must occur in plasma. In a previous study, enzyme-linked immunosorbent assay was carried out to investigate the affinity between the A and B subunits during their interaction, which demonstrated that their affinity constant was on the order of magnitude of $>10^{-8}$. However, this affinity constant implies that most FXIII-A_2 should circulate in the free form; however, in reality, most plasma FXIII-A_2 is present in complex with FXIII-B_2. Thus, this value must substantially underestimate the affinity between the A and B subunits.

Katona et al. (2014) demonstrated that the equilibrium dissociation constant (Kd) for the interaction between the A and B subunits was established in the range of 10^{-10} M using a surface plasmon resonance technique. On the basis of the measured Kd, it was calculated that 99 % of plasma FXIII-A_2 is present in complex with FXIII-B_2 and only ~1 % of circulating FXIII-A_2 is present as a free homodimer.

To locate the epitope on FXIII-B responsible for the interaction with FXIII-A, Souri et al. (2008b) produced various truncated recombinant FXIII-Bs. They demonstrated that those truncated FXIII-B subunits that lacked sushi domain 1 could not form a complex with FXIII-A. Katona et al. (2014) demonstrated that a monoclonal antibody recognizing sushi domains 1 and 2 of the FXIII-B subunit prevents this subunit from forming a complex with the FXIII-A subunit. However, from their subsequent experiments, they were able to localize the epitope to amino acid residues 90 to 103 on sushi domain 2. They noted that these findings are not contradictory and may reflect complex interactions between sushi domains 1 and 2 of the FXIII-B subunit.

Intracellular FXIII is present in the cytoplasm of platelets, monocytes, monocyte-derived macrophages, dendritic cells, chondrocytes, osteoblasts, and osteocytes. FXIII is activated via the nonproteolytic pathway in human platelets during activation induced by thrombin. The FXIII-A in platelets can also be activated by calpain, an endogenous intracellular protease (Ando et al. 1987).

Agonist-induced platelet aggregation in patients with FXIII deficiency and FXIII-A-knockout mice is normal. Platelet aggregation is a relatively fast process (it is completed in 5 min), whereas the cross-linking of platelet proteins is much slower. This suggests that cross-linked polymers are required only for the later phases of platelet activation, e.g., for spreading of platelets following adhesion or for clot retraction. The altered phenotype of FXIII-A-deficient platelets is characterized by a delay in their spreading (Jayo et al. 2009). Furthermore, a reduced contractile force was found in platelet-rich plasma clots from an FXIII-deficient patient (Carr et al. 2003). Clot retraction is impaired in FXIII-A knockout mice, and the addition of plasma FXIII or recombinant FXIII-A_2 to the platelet-rich plasma from FXIII-A knockout mice only partially restored clot retraction, suggesting that cytosolic FXIII also contributes to this process (Kasahara et al. 2010). To verify this idea, a plasma FXIII-free clot retraction assay is performed using washed platelets, purified fibrinogen, thrombin, and Ca^{2+} (Kasahara et al. 2013). Cross-linking of vinculin and filamin by cytosolic FXIII may be involved in platelet cytoskeleton remodeling during clot retraction (Serrano and Devine 2002).

FXIII targets a wide range of additional substrates that have important roles in health and development (Dickneite et al. 2015; Richardson et al. 2013). These include antifibrinolytic proteins, which function in cross-linking of an α_2-antiplasmin to fibrin. Clots formed in the absence of FXIII are unstable and easy to lyse by the fibrinolytic system. Proteomic approach in combination with transglutaminase-specific labeling by 5-(biotinamido) pentylamine (5BAPA) identified a total of 147 plasma FXIIIa substrates, and 48 of these were incorporated into the insoluble fibrin clot during the coagulation of plasma (Table 15.1) (Nikolajsen et al. 2014). FXIIIa cross-links a number of different proteins to clots, including factor V, thrombin-activable fibrinolysis inhibitor, von Willebrand factor, complement C3, inter-α-trypsin inhibitor, and plasminogen activator inhibitor type 2. These proteins regulate clot characteristics other than stability. The cross-linking of fibronectin to fibrin by FXIIIa has been shown to affect cell adhesion and migration in a fibrin matrix.

Cross-linking occurs between Gln398 or 399 on the γ-chain of the fibrin molecule and Lys406 on the γ-chain of another fibrin, resulting in the formation of two antiparallel isopeptide bonds that connect the D-regions of two molecules. Glutamine residues involved in the cross-linking of the α-chain are Gln221, 237, 328, and 366. Many lysine residues of the α-chain are involved as acceptor sites. Cross-linking occurs between Gln2 in the amino terminus of an α_2-antiplasmin and Lys303 on the fibrin α-chain (Richardson et al. 2013).

The amino acid sequences around the substrate glutamine site show no consensus sequence could be derived from the primary structure. However, screening for the preferred substrate sequence of transglutaminase using a phage-displayed peptide library demonstrated that QxxϕxWP (where x and ϕ represent a nonconserved amino acid and a hydrophobic amino acid, respectively) is preferred as a glutamine donor substrate (Sugimura et al. 2006).

15.3 FXIII Deficiency

Congenital FXIII deficiency is caused by defects in the F13A or F13B genes, leading to a bleeding tendency and abnormal wound healing in affected patients and spontaneous miscarriage in female patients. Several hundred autosomal,

Table 15.1 FXIII substrates cross-linked to the plasma clot

Accession no.	Protein	Reactive Gln	Cross-linking sites
P02671	Fibrinogen α-chain	8	Gln^{221}, Gln^{237}, Gln^{328}, Gln^{366}
P01023	α_2-Macroglobulin	15	Gln^{669}, Gln^{670}
P00488	Coagulation factor XIII A chain	8	
P00747	Plasminogen	20	Gln^{322}, Lys^{298}
P00734	Prothrombin	11	
P19823	Inter-α-trypsin inhibitor heavy chain H2	16	
P06727	Apolipoprotein A-IV	18	
P01024	Complement C3	27	
P02787	Serotransferrin	10	
P19827	Inter-α-trypsin inhibitor heavy chain H1	11	
P10909	Clusterin	5	
P02679	Fibrinogen γ-chain	5	Gln^{398}, Gln^{399}, Lys^{406}
P08697	α_2-Antiplasmin	11	Gln^{2}, Gln^{21}, Gln^{419}, Gln^{447}
P02751	Fibronectin	22	Gln^{3}, Gln^{4}, Gln^{16}
P02649	Apolipoprotein E	13	
P06396	Gelsolin	7	
P04004	Vitronectin	6	
Q16610	Extracellular matrix protein 1	8	
P01042	Kininogen-1	11	
P04003	C4b-binding protein α-chain	9	
P01857	Igγ-1 chain C region	3	
P23142	Fibulin-1	8	
P02760	Protein AMBP	3	
P02647	Apolipoprotein A-I	4	
Q14624	Inter-α-trypsin inhibitor heavy chain H4	5	
P04275	von Willebrand factor	12	Gln^{313}, Gln^{560}
P62736	Actin, aortic smooth muscle	3	Gln^{41}
P02765	α_2-HS-glycoprotein	2	
P01876	Ig α-1 chain c region	5	
P01871	Ig mu chain C region	5	
P69905	Hemoglobin subunit α	1	
P35858	Insulin-like growth factor-binding protein complex acid labile subunit	3	
P01834	Ig κ-chain C region	2	
P02675	Fibrinogen β-chain	6	
Q9Y490	Talin-1	4	
P0C0L4	Complement C4-A	21	
P01031	Complement C5	8	
P02768	Serum albumin	8	

(continued)

Table 15.1 (continued)

Accession no.	Protein	Reactive Gln	Cross-linking sites
Q12805	EGF-containing fibulin-like extracellular matrix protein 1	3	
P00739	Haptoglobin-related protein	1	
O43866	CD5 antigen-like	1	
P00450	Ceruloplasmin	4	
P09871	Complement C1s subcomponent	4	
P04406	Glyceraldehyde-3-phosphate dehydrogenase	1	
Q9UGM5	Fetuin-B	3	
P68871	Hemoglobin subunit β	1	
P0CG04	Ig λ-1 chain C regions	1	
P01766	Ig heavy chain V-III region BRO/TIL	1	

Nikolajsen et al. (2014)

recessively inherited cases of human FXIII deficiency are known worldwide. Most are due to loss-of-function missense mutations in FXIII-A (Iismaa et al. 2009).

Autoimmune hemophilia-like disease caused by anti-FXIII antibodies (termed AH13) or "autoimmune FXIII deficiency" is a life-threatening bleeding disorder. AH13 is rare worldwide (Souri et al. 2015). The AH13 cases are immunologically classified into three types: Aa, Ab, and B. Type Aa autoantibodies are directed against native FXIII-A and block FXIII activation. The autoantibodies not only prevent the assembly of new FXIII-A_2B_2 heterotetramers, but also remove FXIII-A from native FXIII-A_2B_2 heterotetramers by forming an FXIII-A-IgG complex. Type Ab autoantibodies preferentially bind to activated FXIII-A and inhibit its activity. Type B antibodies are non-neutralizing anti-FXIII-B subunit autoantibodies that possibly accelerate FXIII clearance. An excellent review article by Muszbek et al. (2011) gives a detailed description of FXIII biochemistry and FXIII deficiency.

15.4 FXIII Function

The pivotal role of plasma FXIII in hemostasis is well established (Richardson et al. 2013; Schroeder and Kohler 2013). In human blood coagulation, the activated form of plasma FXIII stabilizes fibrin matrix assembly during clot formation by catalyzing the covalent linkage of fibrin monomers. In addition, plasma FXIIIa enhances the incorporation of an α_2-antiplasmin into the clot network, being the principal mechanism whereby plasmin-mediated clot degradation is minimized.

FXIII plays a key role in a range of physiological functions. In addition to its role in maintaining hemostasis, these functions include (1) wound healing, (2) maintenance of pregnancy, (3) angiogenesis, (4) apoptosis, and (5) bacterial immobilization.

15.4.1 Wound Healing

Impaired wound healing was noted in a patient with congenital FXIII deficiency (Duckert et al. 1960). A key role of FXIII in wound healing has been demonstrated in FXIII-deficient transgenic mice, where healing of an excisional wound was markedly delayed compared with healing in normal mice (Inbal et al. 2005). Plasma FXIII plays a role in wound healing through a number of mechanisms: (A) FXIII enhances aggregation of platelets to the endothelium at the site of injury; (B) FXIIIa promotes the cross-linking of fibrin, thus increasing the integrity and tensile strength of a clot; (C) FXIIIa-induced cross-linking of the provisional matrix enables the infiltration of leukocytes; (D) Cross-linked macromolecules facilitate the invasion of fibroblasts and endothelial cells into the wound, enabling collagen deposition and angiogenesis (Richardson et al. 2013).

15.4.2 Maintenance of Pregnancy

Clinical findings prove that plasma FXIII is essential for carrying out normal pregnancy. Although homozygous female FXIII-A knockout mice are capable of becoming pregnant, most of them die owing to excessive vaginal bleeding during gestation (Koseki-Kuno et al. 2003). A series of histologic examinations of the pregnant animals suggest that massive placental hemorrhage and the subsequent necrosis occurred in the uteri of the FXIII knockout mice on day 10 of gestation. These findings indicate that maternal FXIII plays a critical role in uterine hemostasis and maintenance of the placenta during gestation. The cross-linking of fibrinoid components in the placenta (fibrin and fibronectin) is likely important for the sealing/anchoring effect and barrier function of Nitabuch's layer (Muszbek et al. 2011).

15.4.3 Angiogenesis

In FXIII knockout mice, the formation of new vessels into a subcutaneously injected Matrigel™ plug was significantly inhibited compared with control mice (Dardik et al. 2006a). Dardik et al. (2003) demonstrated that FXIIIa dose-dependently enhanced endothelial tube formation and endothelial cell migration.

FXIII binds to endothelial cell αvβ3 integrin, resulting in interaction between αvβ3 and vascular endothelial growth factor receptor-2 (VEGFR-2), and autophosphorylation and activation of VEGFR-2. Activation of VEGFR-2 results in phosphorylation and activation of ERK, which promotes endothelial cell survival. Activated VEGFR-2 also induces endothelial cell proliferation, through phosphorylation of ERK and upregulation of Egr-1. Finally, activation of

VEGFR-2 leads to upregulation of c-Jun, and downregulation of thrombospondin-1 (TSP-1) via the transcription factor Wilm's tumor-1, resulting in enhanced cell proliferation, migration, and survival, ultimately promoting angiogenesis (Dardik et al. 2006b). TSP-1 is one of the best characterized antiangiogenic factors. The importance of TSP-1 in angiogenesis is demonstrated by animal studies. In transgenic mice, targeted overexpression of TSP-1 in the skin was associated with potent inhibition of cutaneous tissue repair, granulation tissue formation, and wound angiogenesis (Streit et al. 2000). Therefore, the antiangiogenic properties of TSP-1 suggest the potential significance of this observation in FXIIIa-mediated promotion of angiogenesis (Richardson et al. 2013).

15.4.4 Apoptosis

Plasma FXIII targets an apoptotic molecule to downregulate its signal when cell death is initiated by an agonist. The Fas antigen is a member of the tumor necrosis factor receptor superfamily and can mediate apoptotic cell death in various cell types. FXIII inhibits apoptosis induced by a cytotoxic anti-Fas monoclonal antibody in Jurkat cells. Furthermore, an antibody against FXIII strongly accelerates the Fas-mediated apoptosis, indicating that FXIII is involved in cross-linking of Fas and downregulates Fas-mediated apoptotic cell death (Kikuchi et al. 2014).

15.4.5 Bacterial Immobilization

FXIII plays an important role in host defense against invasive bacteria. In normal plasma, the induction of coagulation by *Streptococcus pyogenes* results in immobilization of bacteria cells within a clot (Loof et al. 2011). This bacterial entrapment within a plasma clot was shown to be FXIII-dependent; cross-linking was not observed when FXIII-deficient plasma was used. Immunostaining with an antibody against N-ε-γ-glutamyl-lysine elucidated the nature of the interaction, revealing a covalent interaction between one terminal of a streptococcal M1 surface protein and a globular domain of fibrinogen. FXIII-dependent bacterial killing and cross-linking of a streptococcal M1 protein to fibrin networks were detected in tissue biopsy material from patients with streptococcal necrotizing fasciitis (Richardson et al. 2013).

15.5 Cell Surface FXIII

FXIII is present in platelets in large quantities, making the local platelet FXIII-A concentration approximately 150-fold greater than that in plasma. Platelets stabilize FXIII-depleted thrombi in a transglutaminase-dependent manner. Fluorescence confocal microscopy and flow cytometry revealed exposure of FXIII-A on an activated platelet surface (Mitchell et al. 2014). Therefore, the detection of FXIII-A that binds to activated, but not resting, platelets may be useful for the detection of pathological in vivo platelet activation (Devine et al. 1993). The Fab fragment of a polyclonal antibody against platelet FXIII inhibited the collagen-induced platelet aggregation, suggesting that FXIII-A functions on the platelet surface (Kasahara et al. 1988). However, it is unclear how platelet FXIII-A functions on the cell surface because it is not released by classical secretion mechanisms.

Lipid rafts are dynamic assemblies of sphingolipids, cholesterol, and signaling molecules such as src-family kinases that can be stabilized into platforms involved in the regulation of a number of vital cellular processes (Simons and Gerl 2010). The important functions of lipid rafts are signal transduction and membrane trafficking. Coalescence of lipid rafts on the cell surface leads to activation of src-family kinases (Kasahara and Sanai 2000). Lipid rafts are compositionally and functionally heterogeneous in the cell membrane. Platelet lipid rafts are critical membrane domains involved in physiological responses such as adhesion, aggregation and clot retraction (Bodin et al. 2003).

Cell surface FXIII-A transglutaminase is required for efficient clot retraction (Kasahara et al. 2013). Clot retraction is mediated by the interaction between the extracellular fibrin fiber and intracellular actomyosin via integrin αIIbβ3, together with the activation of the platelet contractile apparatus. Clot retraction is regulated through multiple signaling pathways. The src-dependent actomyosin contraction mediates clot retraction (Suzuki-Inoue et al. 2007). The src-dependent activation of phospholipase Cγ2 induces calcium mobilization, activation of myosin light chain kinase, and phosphorylation of myosin light chain. Flow cytometric analysis demonstrated that 5BAPA is incorporated into the surface of washed platelets by transamidation with thrombin stimulation. 5BAPA incorporation is completely impaired in FXIII-A-deficient mouse platelets. The 5BAPA-incorporated protein was identified as fibrin, which colocalizes with activated myosin in sphingomyelin-rich lipid rafts. Thrombin causes the rapid translocation of fibrin and myosin to the lipid raft fraction of washed platelets on a sucrose density gradient. The fibrin translocation to the lipid raft fraction is impaired in FXIII-A-deficient mouse platelets. These results suggest that fibrinogen is released from α-granules of platelets, converted to fibrin by the cleavage of fibrinopeptides A and B by thrombin, and translocated to sphingomyelin-rich rafts of thrombin-stimulated platelets in an intracellular FXIII-dependent manner, and activates myosin beneath sphingomyelin-rich rafts. Furthermore, raft disruption with methyl-β-cyclodextrin (which removes membrane cholesterol) inhibits outside-in signaling and clot retraction. Sphingomyelin-rich-raft-depleted platelets from sphingomyelin synthase knockout mice exhibit delayed clot retraction (Kasahara et al. 2013). These findings suggest that platelet sphingomyelin-rich rafts act as platforms where fibrin and actomyosin efficiently join via integrin αIIbβ3 to promote outside-in signal for clot retraction (Fig. 15.1). The mechanism of FXIII-dependent fibrin translocation to lipid rafts remains to be explored.

Platelets activated simultaneously by collagen and thrombin constitute a subpopulation of activated platelets. The activated platelets likely contribute significantly to thrombotic processes because they become "coated" with a number of procoagulant proteins, including factor V, fibrinogen, fibronectin, thrombospondin, and von Willebrand factor on the platelet surface. Many of these procoagulant proteins are posttranslationally modified by transamidase-mediated conjugation with serotonin, a process that increases their procoagulant activity. This mechanism may be important in the initial stabilization of the platelet plug at the site of injury (Dale 2005).

Cell surface FXIII-A is found in various cell types. FXIII-mediated protein cross-linking has been associated with extracellular matrix formation on the cell surface in physiological events such as differentiation. Myneni et al. (2014) demonstrated that cell surface FXIII-A transglutaminase acts as a switch between preadipocyte proliferation and differentiation. FXIII-A has recently been identified as a potential causative obesity gene in human white adipose tissue. It was demonstrated that preadipocyte FXIII-A forms an active transglutaminase that

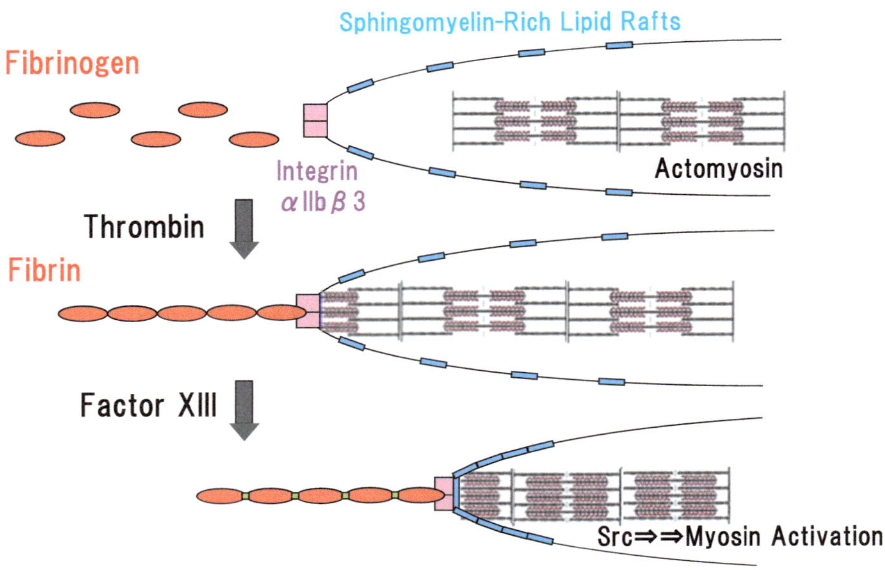

Fig. 15.1 Model of FXIII-dependent fibrin-αIIbβ3-myosin axis in platelet sphingomyelin-rich lipid rafts. Fibrin is translocated to sphingomyelin-rich lipid rafts of thrombin-stimulated platelets in an FXIII-dependent manner. Sphingomyelin-rich lipid rafts act as platforms for fibrin-mediated outside-in signaling, leading to clot retraction

translocates to the cell surface and acts as a negative regulator of adipogenesis by promoting the assembly of fibronectin from plasma into a preadipocyte extracellular matrix.

Plasma membrane FXIII-A transglutaminase also regulates osteoblast differentiation (Al-Jallad et al. 2011). The fibrillary type I collagen matrix plays a major role in regulating osteoblast activity and is required for expression of osteoblast markers. FXIII-A-deficient mice showed decreased type I collagen production during remodeling after induced myocardial infarction. In these mice, type I collagen levels were not corrected by exogenously administered plasma FXIII therapy, indicating that type I collagen matrix synthesis is regulated by cellular FXIII-A. The cross-linking activity of FXIII-A stabilizes the interaction of microtubules with the plasma membrane. Microtubule association with the plasma membrane is required for the promotion of secretory vesicle (exosome) trafficking and protein delivery to the cell surface and secretion to the matrix. Tubulin undergoes palmitoylation for membrane insertion in some cell types and it has been demonstrated to be present in lipid rafts and interacts with GM1 and GM3 gangliosides (Palestini et al. 2000; Janich and Corbeil 2007). Although the precise interaction mechanism between FXIII-A and microtubules at the plasma membrane is not clear, tubulin is a substrate of cellular FXIII-A in differentiating osteoblasts (Wang et al. 2014). Serotonin can be incorporated covalently into proteins via a transglutaminase-mediated serotonylation reaction, which in turn can alter protein function. Serotonin inhibits FXIII-A-mediated plasma fibronectin matrix assembly and cross-linking in osteoblast cultures via direct competition with transamidation (Cui and Kaartinen 2015).

One possible mechanism of FXIII-A transport to the plasma membrane could be via sphingomyelin-rich lipid rafts. Sphingomyelin-rich rafts may act as membrane transport vesicles from the trans-Golgi network to the cell surface. FXIII-A is detected in the lipid raft fraction in stimulated but not resting platelets (Kasahara et al. 2013). Alternatively, FXIII-A secretion could be mediated by Golgi matrix protein-130 (GP130), which colocalizes with FXIII-A in macrophages. GP130 functions in nonclassical protein secretion to the plasma membrane prior to excretion through membrane pores. FXIII-A is not detected in classical secretory vesicles containing trans-Golgi network protein-46 (Cordell et al. 2010). Further work is necessary to define the exact mechanism involved in FXIII-A translocation across the plasma membrane and the signaling events.

Acknowledgments We thank Dr A. Ichinose (Yamagata University School of Medicine) for helpful discussion.

References

Al-Jallad HF, Myneni VD, Piercy-Kotb SA, Chabot N, Mulani A, Keillor JW, Kaartinen MT (2011) Plasma membrane factor XIIIA transglutaminase activity regulates osteoblast matrix secretion and deposition by affecting microtubule dynamics. PLoS One 6:e15893

Ando Y, Imamura S, Yamagata Y, Kitahara A, Saji H, Murachi T, Kannagi R (1987) Platelet factor XIII is activated by calpain. Biochem Biophys Res Commun 144:484–490

Bodin S, Tronchère H, Payrastre B (2003) Lipid rafts are critical membrane domains in blood platelet activation processes. Biochim Biophys Acta 1610:247–257

Carr ME, Carr SL, Tildon T, Fisher LM, Martin EJ (2003) Batroxobin-induced clots exhibit delayed and reduced platelet contractile force in some patients with clotting factor deficiencies. J Thromb Haemost 1:243–249

Cordell PA, Kile BT, Standeven KF, Josefsson EC, Pease RJ, Grant PJ (2010) Association of coagulation factor XIII-A with Golgi proteins within monocyte-macrophages: implications for subcellular trafficking and secretion. Blood 115:2674–2681

Cui C, Kaartinen MT (2015) Serotonin (5-HT) inhibits Factor XIII-A-mediated plasma fibronectin matrix assembly and crosslinking in osteoblast cultures via direct competition with transamidation. Bone 72:43–52

Dale GL (2005) Coated-platelets: an emerging component of the procoagulant response. J Thromb Haemost 3:2185–2192

Dardik R, Solomon A, Loscalzo J, Eskaraev R, Bialik A, Goldberg I, Schiby G, Inbal A (2003) Novel proangiogenic effect of factor XIII associated with suppression of thrombospondin 1 expression. Arterioscler Thromb Vasc Biol 23:1472–1477

Dardik R, Leor J, Skutelsky E, Castel D, Holbova R, Schiby G, Shaish A, Dickneite G, Loscalzo J, Inbal A (2006a) Evaluation of the pro-angiogenic effect of factor XIII in heterotopic mouse heart allografts and FXIII-deficient mice. Thromb Haemost 95:546–550

Dardik R, Loscalzo J, Inbal A (2006b) Factor XIII (FXIII) and angiogenesis. J Thromb Haemost 4:19–25

Devine DV, Andestad G, Nugent D, Carter CJ (1993) Platelet-associated factor XIII as a marker of platelet activation in patients with peripheral vascular disease. Arterioscler Thromb 13:857–862

Dickneite G, Herwald H, Korte W, Allanore Y, Denton CP, Matucci Cerinic M (2015) Coagulation factor XIII: a multifunctional transglutaminase with clinical potential in a range of conditions. Thromb Haemost 113:686–697

DUCKERT F, JUNG E, SHMERLING DH (1960) A hitherto undescribed congenital haemorrhagic diathesis probably due to fibrin stabilizing factor deficiency. Thromb Diath Haemorrh 5:179–186

Duval C, Allan P, Connell SD, Ridger VC, Philippou H, Ariëns RA (2014) Roles of fibrin α- and γ-chain specific cross-linking by FXIIIa in fibrin structure and function. Thromb Haemost 111:842–850

Grundmann U, Amann E, Zettlmeissl G, Küpper HA (1986) Characterization of cDNA coding for human factor XIIIa. Proc Natl Acad Sci U S A 83:8024–8028

Ichinose A, Hendrickson LE, Fujikawa K, Davie EW (1986a) Amino acid sequence of the a subunit of human factor XIII. Biochemistry 25:6900–6906

Ichinose A, McMullen BA, Fujikawa K, Davie EW (1986b) Amino acid sequence of the b subunit of human factor XIII, a protein composed of ten repetitive segments. Biochemistry 25:4633–4638

Iismaa SE, Mearns BM, Lorand L, Graham RM (2009) Transglutaminases and disease: lessons from genetically engineered mouse models and inherited disorders. Physiol Rev 89:991–1023

Inbal A, Lubetsky A, Krapp T, Castel D, Shaish A, Dickneitte G, Modis L, Muszbek L (2005) Impaired wound healing in factor XIII deficient mice. Thromb Haemost 94:432–437

Janich P, Corbeil D (2007) GM1 and GM3 gangliosides highlight distinct lipid microdomains within the apical domain of epithelial cells. FEBS Lett 581:1783–1787

Jayo A, Conde I, Lastres P, Jiménez-Yuste V, González-Manchón C (2009) New insights into the expression and role of platelet factor XIII-A. J Thromb Haemost 7:1184–1191
Kasahara K, Sanai Y (2000) Functional roles of glycosphingolipids in signal transduction via lipid rafts. Glycoconj J 17:153–162
Kasahara K, Takagi J, Sekiya F, Inada Y, Saito Y (1988) "A" subunit of factor XIII is present on bovine platelet membrane and mediates collagen-induced platelet activation. Thromb Res 50:253–263
Kasahara K, Souri M, Kaneda M, Miki T, Yamamoto N, Ichinose A (2010) Impaired clot retraction in factor XIII A subunit-deficient mice. Blood 115:1277–1279
Kasahara K, Kaneda M, Miki T, Iida K, Sekino-Suzuki N, Kawashima I, Suzuki H, Shimonaka M, Arai M, Ohno-Iwashita Y, Kojima S, Abe M, Kobayashi T, Okazaki T, Souri M, Ichinose A, Yamamoto N (2013) Clot retraction is mediated by factor XIII-dependent fibrin-αIIbβ3-myosin axis in platelet sphingomyelin-rich membrane rafts. Blood 122:3340–3348
Katona E, Pénzes K, Csapó A, Fazakas F, Udvardy ML, Bagoly Z, Orosz ZZ, Muszbek L (2014) Interaction of factor XIII subunits. Blood 123:1757–1763
Kikuchi H, Kuribayashi F, Imajoh-Ohmi S (2014) Down-regulation of Fas-mediated apoptosis by plasma transglutaminase factor XIII that catalyzes fetal-specific cross-link of the Fas molecule. Biochem Biophys Res Commun 443:13–17
Koseki-Kuno S, Yamakawa M, Dickneite G, Ichinose A (2003) Factor XIII A subunit-deficient mice developed severe uterine bleeding events and subsequent spontaneous miscarriages. Blood 102:4410–4412
Loof TG, Mörgelin M, Johansson L, Oehmcke S, Olin AI, Dickneite G, Norrby-Teglund A, Theopold U, Herwald H (2011) Coagulation, an ancestral serine protease cascade, exerts a novel function in early immune defense. Blood 118:2589–2598
Lorand L (2001) Factor XIII: structure, activation, and interactions with fibrinogen and fibrin. Ann N Y Acad Sci 936:291–311
Mitchell JL, Lionikiene AS, Fraser SR, Whyte CS, Booth NA, Mutch NJ (2014) Functional factor XIII-A is exposed on the stimulated platelet surface. Blood 124:3982–3990
Muszbek L, Bereczky Z, Bagoly Z, Komáromi I, Katona É (2011) Factor XIII: a coagulation factor with multiple plasmatic and cellular functions. Physiol Rev 91:931–972
Myneni VD, Hitomi K, Kaartinen MT (2014) Factor XIII-A transglutaminase acts as a switch between preadipocyte proliferation and differentiation. Blood 124:1344–1353
Nikolajsen CL, Dyrlund TF, Poulsen ET, Enghild JJ, Scavenius C (2014) Coagulation factor XIIIa substrates in human plasma: identification and incorporation into the clot. J Biol Chem 289:6526–6534
Palestini P, Pitto M, Tedeschi G, Ferraretto A, Parenti M, Brunner J, Masserini M (2000) Tubulin anchoring to glycolipid-enriched, detergent-resistant domains of the neuronal plasma membrane. J Biol Chem 275:9978–9985
Richardson VR, Cordell P, Standeven KF, Carter AM (2013) Substrates of Factor XIII-A: roles in thrombosis and wound healing. Clin Sci (Lond) 124:123–137
Schroeder V, Kohler HP (2013) New developments in the area of factor XIII. J Thromb Haemost 11:234–244
Serrano K, Devine DV (2002) Intracellular factor XIII crosslinks platelet cytoskeletal elements upon platelet activation. Thromb Haemost 88:315–320
Simons K, Gerl MJ (2010) Revitalizing membrane rafts: new tools and insights. Nat Rev Mol Cell Biol 11:688–699
Souri M, Kaetsu H, Ichinose A (2008a) Sushi domains in the B subunit of factor XIII responsible for oligomer assembly. Biochemistry 47:8656–8664
Souri M, Koseki-Kuno S, Takeda N, Degen JL, Ichinose A (2008b) Administration of factor XIII B subunit increased plasma factor XIII A subunit levels in factor XIII B subunit knock-out mice. Int J Hematol 87:60–68

Souri M, Osaki T, Ichinose A (2015) Anti-factor XIII A subunit (FXIII-A) autoantibodies block FXIII-A2 B2 assembly and steal FXIII-A from native FXIII-A2 B2. J Thromb Haemost 13:802–814

Streit M, Velasco P, Riccardi L, Spencer L, Brown LF, Janes L, Lange-Asschenfeldt B, Yano K, Hawighorst T, Iruela-Arispe L, Detmar M (2000) Thrombospondin-1 suppresses wound healing and granulation tissue formation in the skin of transgenic mice. EMBO J 19:3272–3282

Sugimura Y, Hosono M, Wada F, Yoshimura T, Maki M, Hitomi K (2006) Screening for the preferred substrate sequence of transglutaminase using a phage-displayed peptide library: identification of peptide substrates for TGASE 2 and Factor XIIIA. J Biol Chem 281:17699–17706

Suzuki-Inoue K, Hughes CE, Inoue O, Kaneko M, Cuyun-Lira O, Takafuta T, Watson SP, Ozaki Y (2007) Involvement of Src kinases and PLCgamma2 in clot retraction. Thromb Res 120:251–258

Walther DJ, Peter JU, Winter S, Höltje M, Paulmann N, Grohmann M, Vowinckel J, Alamo-Bethencourt V, Wilhelm CS, Ahnert-Hilger G, Bader M (2003) Serotonylation of small GTPases is a signal transduction pathway that triggers platelet alpha-granule release. Cell 115:851–862

Wang S, Cui C, Hitomi K, Kaartinen MT (2014) Detyrosinated Glu-tubulin is a substrate for cellular Factor XIIIA transglutaminase in differentiating osteoblasts. Amino Acids 46:1513–1526

Chapter 16
Inhibition of Transglutaminase

Jeffrey W. Keillor

Abstract Transglutaminases (TGases) have been implicated in a number of physiological disorders, inspiring the development of TGase inhibitors as mechanistic probes and for their therapeutic potential. This chapter presents an overview of the theory and practice of enzyme inhibition, followed by a summary of recent TGase inhibitors.

Keywords Enzyme inhibition • Enzyme kinetics • Reversible inhibition • Irreversible inhibition • Transglutaminase inhibitors

16.1 Enzyme Inhibition

Enzymes are proteins that mediate the chemical reactions essential for life. These powerful biocatalysts favor improbable reactions and accelerate slow reactions, allowing chemical transformations to occur that are essential to every living organism. Normally, they show exquisite chemo-, regio- and stereoselectivity, and their activity is carefully controlled in their precisely defined biological context. However, their unregulated, aberrant activity can often lead to serious physiological disorders. As such, they represent important therapeutic targets; the vast majority of all known drugs are enzyme inhibitors. In the present era of modern medicinal chemistry, the optimization of the activity of these drugs requires a detailed understanding of the mechanism by which they inhibit their enzyme targets. Although more detailed texts may be found elsewhere, (Segel 1993; Fersht 1985; Silverman 2002) this chapter provides a very brief overview of enzyme inhibition, from a highly simplified mechanistic and practical point of view.

J.W. Keillor (✉)
Department of Chemistry, University of Ottawa, 10 Marie-Curie, Ottawa, ON K1N 6N5, Canada
e-mail: jkeillor@uottawa.ca

K. Hitomi et al. (eds.), *Transglutaminases*, DOI 10.1007/978-4-431-55825-5_16

16.1.1 Reversible Inhibition

By definition, reversible inhibitors do not permanently reduce the ability of an enzyme to catalyze its native reaction. That is, they do not irreversibly modify an enzyme; if a reversible inhibitor is removed from the environment, the enzyme of interest will recover its full activity. Typically, reversible inhibitors exert their effect through equilibrium, non-covalent binding of some enzyme species on the reaction pathway, forming a complex that does not lead to product formation. There are three general modes of reversible inhibition: competitive, uncompetitive and non-competitive. As we will see below, these modes are based on the observed effect of the added inhibitor on the kinetic parameters of the enzymatic reaction, and do not necessarily reflect a specific mechanism of inhibition.

16.1.1.1 Competitive Inhibition

Competitive inhibition is perhaps the most intuitively obvious mode. As shown in the kinetic scheme of Fig. 16.1, free enzyme can bind either substrate or inhibitor, but not both at the same time. In this way, binding of substrate and inhibitor are mutually exclusive, so the inhibitor 'competes' with the substrate for binding the same form of the enzyme.

There are several binding mechanisms that account for this kinetic behavior. In the most common mechanism (and perhaps the most intuitive), the inhibitor is bound in the substrate binding site, in the place of the substrate. More explicitly, in this mechanism both substrate and inhibitor compete for the same physical binding pocket on the enzyme. However, it should be noted that other binding mechanisms also lead to observed competitive inhibition. For example, an inhibitor may be bound in a separate site that is near the substrate binding site, such that steric hindrance from the inhibitor prevents the substrate from being bound. Alternatively, binding of the inhibitor at a separate site distant from the substrate may induce a conformational change in the enzyme that prevents the substrate from being bound simultaneously. The key feature of all of these mechanisms is that the same form of the enzyme binds both substrate and inhibitor, but binding is mutually exclusive.

Competitive inhibition manifests itself as an increase in the apparent K_M value of the related substrate. From the appropriately modified Michaelis-Menten equation (Fig. 16.2a) it can be noted that V_{max} is unaffected; the implication is that in the

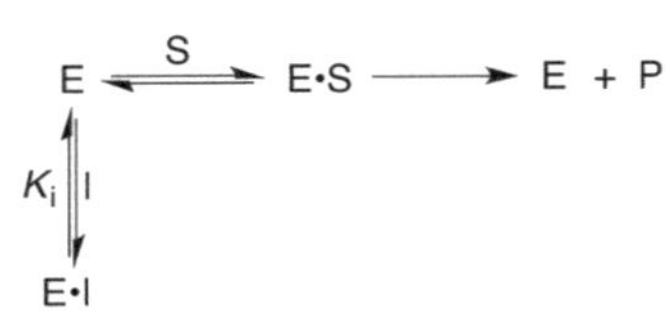

Fig. 16.1 Kinetic scheme for competitive inhibition

presence of a high enough concentration of substrate, inhibition will not be detected, as the substrate can out-compete the inhibitor for enzyme binding.

The most thorough way to determine the potency of a competitive inhibitor is by measuring its inhibition constant, K_i. This is most easily accomplished by measuring the effect of the inhibitor on the values of V_{max} and apparent K_M for a series of substrate concentrations. Non-linear regression to the hyperbolic Michaelis-Menten equation (Fig. 16.2a) is the most accurate way to do this, although some researchers still use the rather archaic and inaccurate (Tommasini et al. 1985) double reciprocal equation of Lineweaver and Burk (Fig. 16.2b). The latter plot gives lines that cross characteristically on the y-axis. The apparent K_M values can then be plotted against the concentration of inhibitor at which they were determined, in order to determine the value of K_i (Fig. 16.3a).

Another common way of determining competitive inhibitor potency is by measuring an IC_{50} value. By definition, the IC_{50} value is the concentration of inhibitor required to reduce the rate of an enzymatic reaction to half the rate that would be observed *under the same conditions* in the absence of the inhibitor. It is important to note that the IC_{50} value is therefore a *relative* measure that is highly dependent on

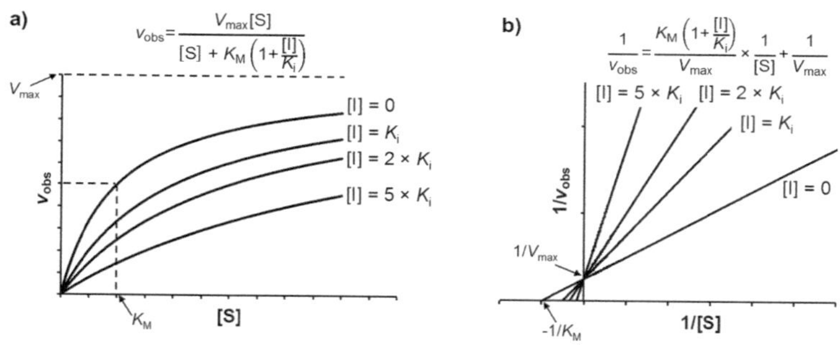

Fig. 16.2 Graphical representation of competitive inhibition: (**a**) hyperbolic Michaelis-Menten equation and plot; (**b**) double reciprocal Lineweaver-Burk equation and plot

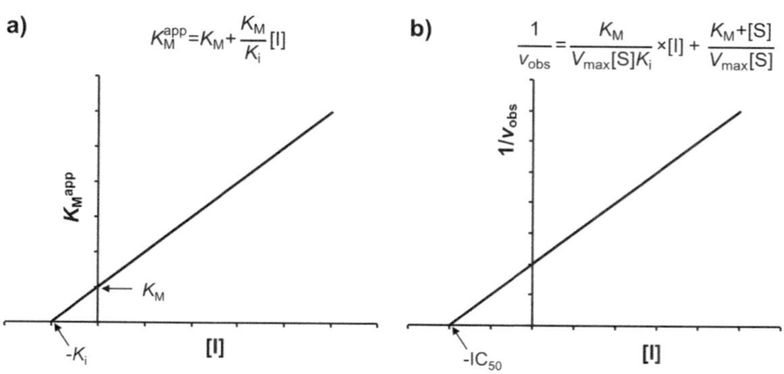

Fig. 16.3 Graphical methods for measuring competitive inhibition: (**a**) determination of the constant K_i value; (**b**) determination of the relative IC_{50} value, at a given concentration of substrate

experimental conditions, including the concentration of substrate used. For this reason, although IC_{50} values can be a useful indication of the relative affinity of two related inhibitors from the same experimental study, they cannot be compared between two different studies wherein different concentrations of substrates (or different assays) are employed. Sadly, this error is frequently committed and even published in modern medicinal chemistry literature, so it is up to the reader to interpret IC_{50} values with extreme caution. IC_{50} values can be determined graphically by plotting the reciprocal rate values measured at one concentration of substrate, as a function of varying inhibitor concentration (Fig. 16.3b). The relation between the relative IC_{50} value and the absolute K_i value can be determined by taking half the rate of the uninhibited enzymatic reaction (given by the Michaelis-Menten equation) and setting it equal to the equation for the inhibited enzyme reaction, where $[I] = IC_{50}$. For competitive inhibition, this can be reduced as shown in Eq. 16.1, also known as the Cheng-Prusoff equation (Yung-Chi and Prusoff 1973).

$$\frac{1}{2}\frac{V_{max}S}{[S]+K_M} = \frac{V_{max}S}{[S]+K_M\left(1+\frac{IC_{50}}{K_i}\right)};\ IC_{50} = \frac{K_i}{K_M}[S]+K_i \tag{16.1}$$

16.1.1.2 Uncompetitive Inhibition

A second general mode of reversible inhibition is 'uncompetitive inhibition'. This name is not as meaningful as 'competitive' inhibition, but it is intended to indicate that the substrate and inhibitor do not compete for the *same* form of the enzyme. Indeed, as can be seen by the kinetic scheme of Fig. 16.4, an uncompetitive inhibitor does not bind to the free enzyme, but rather to the enzyme-substrate (Michaelis) complex. The resulting enzyme-substrate-inhibitor complex does not lead to product formation. In one possible mechanism for uncompetitive inhibition, the binding of inhibitor distorts the Michaelis complex in such a way that the catalytic machinery cannot operate effectively.

As shown in Fig. 16.5, in the case of uncompetitive inhibition, both V_{max} and K_M are decreased by the same factor (leading to characteristic parallel lines on the Lineweaver-Burk plot). It may not be intuitively obvious why adding an inhibitor to an enzymatic reaction would *decrease* the apparent K_M of the enzyme substrate, which is usually indicative of *increased* affinity. However, looking back at the kinetic scheme, it is understandable that adding more inhibitor will favor the

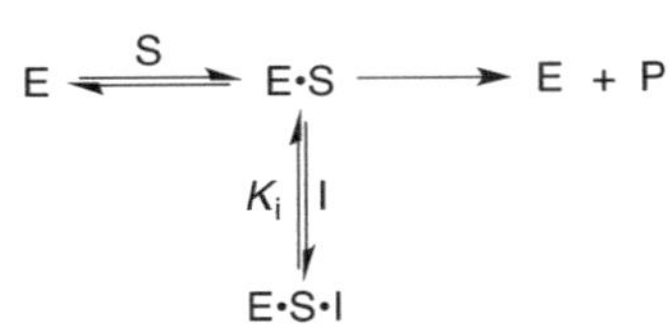

Fig. 16.4 Kinetic scheme for uncompetitive inhibition

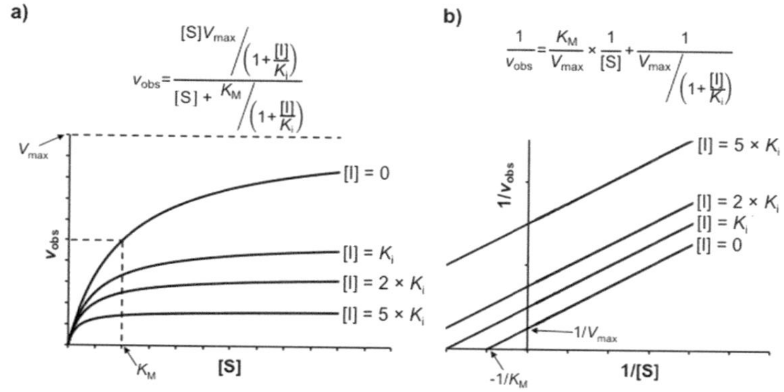

Fig. 16.5 Graphical representation of uncompetitive inhibition: (**a**) hyperbolic Michaelis-Menten equation and plot; (**b**) double reciprocal Lineweaver-Burk equation and plot

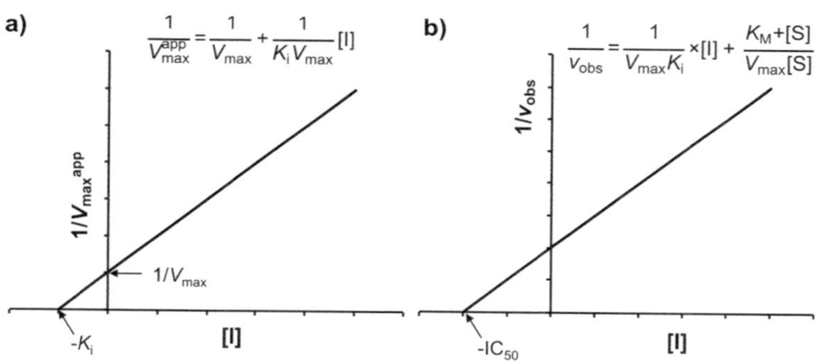

Fig. 16.6 Graphical methods for measuring uncompetitive inhibition: (**a**) determination of the constant K_i value; (**b**) determination of the relative IC_{50} value, at a given concentration of substrate

formation of the E•S•I complex, depleting the concentration of the E•S complex. The substrate binding equilibrium will then shift to the right, according to Le Châtelier's principle (Fox and Whitesell 2004), giving the impression that S is bound more tightly.

Graphically, K_i values for an uncompetitive inhibitor can be determined by plotting either $1/V_{max}^{app}$ values (y-axis intercepts from a Lineweaver-Burk plot) or $1/K_M^{app}$ values against inhibitor concentration (Fig. 16.6a). IC_{50} values can be measured by plotting the reciprocal rate values measured at one concentration of substrate, as a function of varying inhibitor concentration (Fig. 16.6b).

The relation between IC_{50} and K_i for uncompetitive inhibition is shown in Eq. 16.2. It is important to note that substrate concentration still influences the relative value of IC_{50} determined, but for uncompetitive inhibition, IC_{50} values are lower at higher S.

$$\frac{1}{2}\frac{V_{max}S}{[S]+K_M}=\frac{V_{max}S}{[S]\left(1+\frac{IC_{50}}{K_i}\right)+K_M};\ IC_{50}=\frac{K_iK_M}{[S]}+K_i \tag{16.2}$$

16.1.1.3 Non-competitive Inhibition

The third mode of reversible inhibition is generally referred to as non-competitive inhibition. This name is probably the least informative of the three, as it is intended to indicate that the observed inhibition is neither purely competitive, nor purely uncompetitive. As shown in the kinetic scheme of Fig. 16.7, non-competitive inhibition is observed when an inhibitor can bind to *both* the free enzyme and the enzyme-substrate complex. Mechanistically, this may arise from inhibitor binding at a remote site that does not affect substrate binding, but alters the conformation of the Michaelis complex, abolishing its catalytic competence.

Not surprisingly, the equation for this mode of inhibition (Fig. 16.8) appears to be the composite of the equations for competitive an uncompetitive inhibition. That is, K_M is increased by the factor of $(1 + [I]/K_i)$ but decreased by the factor of $(1 + [I]/K_i')$, while V_{max} is decreased by the factor of $(1 + [I]/K_i')$. The presence of two inhibition constants (K_i and K_i') explains why different kinds of non-competitive

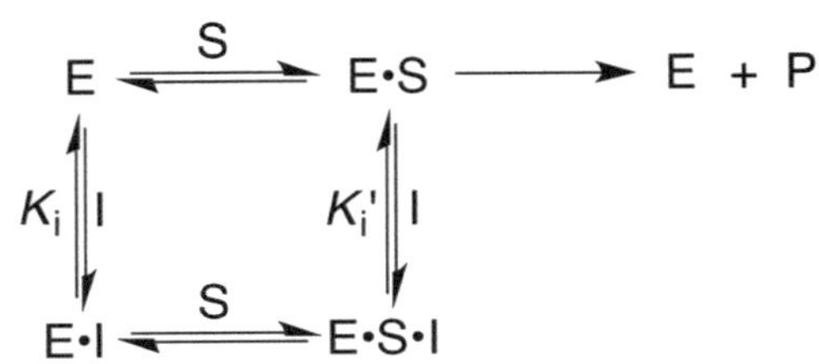

Fig. 16.7 Kinetic scheme for non-competitive inhibition

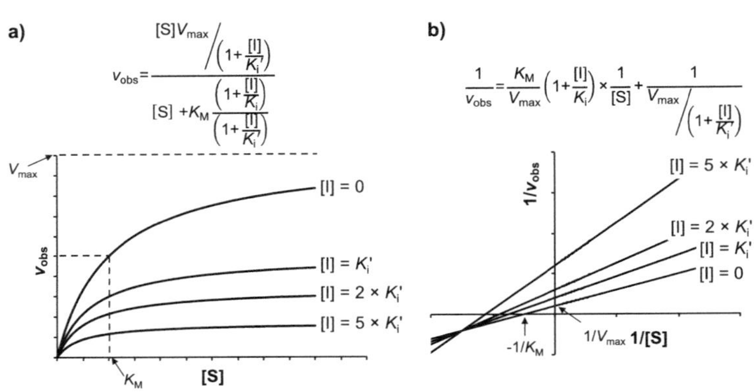

Fig. 16.8 Graphical representation of mixed non-competitive inhibition ($K_i < K_i'$): (**a**) hyperbolic Michaelis-Menten equation and plot; (**b**) double reciprocal Lineweaver-Burk equation and plot

inhibition can be observed. When K_i and K_i' are not identical (which is commonly the case), the mode of inhibition is referred to as 'mixed' non-competitive. If $K_i > K_i'$, the inhibitor is a stronger uncompetitive inhibitor than a competitive inhibitor, and K_M^{app} decreases in the presence of inhibitor. If $K_i < K_i'$, the inhibitor is a stronger competitive inhibitor than an uncompetitive inhibitor, and K_M^{app} increases in the presence of inhibitor.

To graphically determine the values of the two inhibition constants, two separate plots are necessary. The value of K_i' can be determined from the plot of $1/V_{max}^{app}$ *vs* [I] (Fig. 16.9a). The value of K_i can be determined by plotting K_M^{app}/V_{max}^{app} (the slope of Lineweaver-Burk plots, and a ratio in which the effect of $(1 + [I]/K_i')$ will cancel out) *vs* [I], as shown in Fig. 16.10.

If the values of K_i and K_i' are identical (which is very rare), the mode of inhibition is referred to as 'simple' or 'pure' non-competitive and only V_{max} is affected (Fig. 16.11).

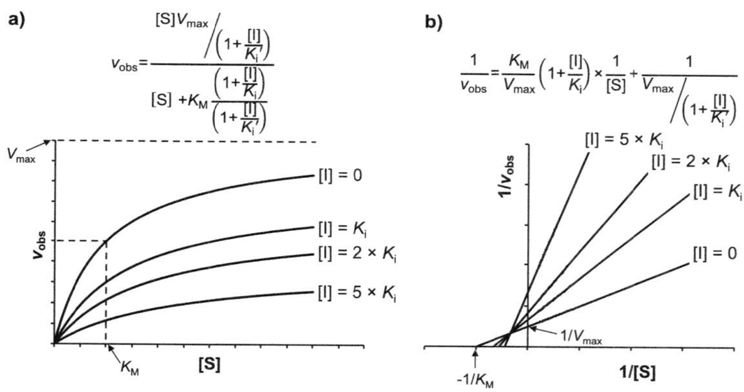

Fig. 16.9 Graphical representation of mixed non-competitive inhibition ($K_i > K_i'$): (**a**) hyperbolic Michaelis-Menten equation and plot; (**b**) double reciprocal Lineweaver-Burk equation and plot

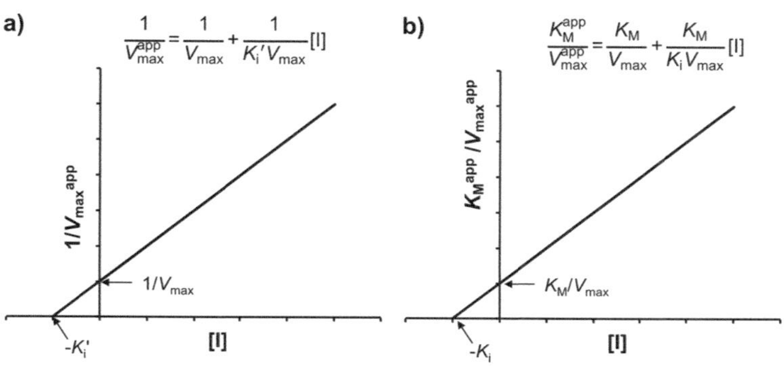

Fig. 16.10 Graphical methods for measuring the two inhibition constants of mixed non-competitive inhibition: (**a**) determination of K_i'; (**b**) determination of K_i

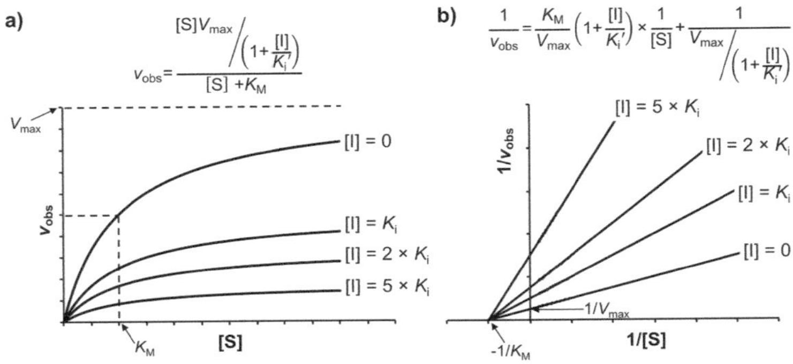

Fig. 16.11 Graphical representation of simple non-competitive inhibition ($K_i = K_i'$): (**a**) hyperbolic Michaelis-Menten equation and plot; (**b**) double reciprocal Lineweaver-Burk equation and plot

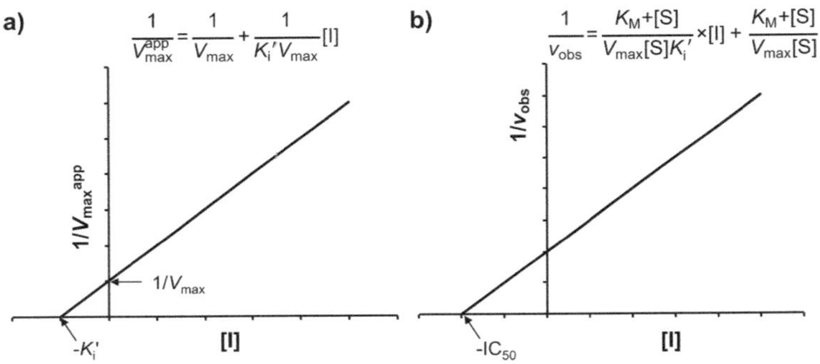

Fig. 16.12 Graphical methods for measuring simple non-competitive inhibition: (**a**) determination of the constant K_i value; (**b**) determination of the IC_{50} value, at a given concentration of substrate

To determine the value of K_i' for simple non-competitive inhibition (which is equal to K_i), one need only plot $1/V_{max}{}^{app}$ values against inhibitor concentration (Fig. 16.12a). Figure 16.12b shows the equation used to determine the value of IC_{50} for simple non-competitive inhibition, from rates measured at one substrate concentration and varying inhibitor concentration.

It is interesting to note that in the rare case of simple non-competitive inhibition, where $K_i = K_i'$, inhibition is independent of substrate binding, and $IC_{50} = K_i'$ (Eq. 16.3).

$$\frac{1}{2}\,\frac{V_{max}S}{[S]+K_M} = \frac{V_{max}S \Big/ \left(1+\frac{IC_{50}}{K_i}\right)}{[S]+K_M};\ IC_{50} = K_i \qquad (16.3)$$

16.1.1.4 Special Cases: Slow and Tight Binding

For most reversible inhibitors, the binding equilibrium is rapid and of moderate affinity. These are 'classical' reversible inhibitors. But for others, the binding equilibrium can be slow and/or particularly strong, giving rise to three other cases to consider.

16.1.1.4.1 Slow Binding

For slow binding inhibitors, it can take minutes to establish the steady-state binding equilibrium (as opposed to milliseconds for classical inhibitors). Typically this is due to very slow inhibitor binding (small k_{+i}) and inhibitor release (small k_{-i}) making up the inhibition constant ($K_i = k_{-i}/k_{+i}$). This is manifested by a time dependence of the inhibition (Fig. 16.13a). Two different experiments can be performed to demonstrate slow inhibition, relative to enzymatic reaction in the absence of inhibitor (shown as a dashed line). If the enzyme reaction is initiated by the addition of enzyme to a solution of substrate and inhibitor, the enzymatic reaction will begin at an uninhibited rate, and then slow down to the inhibited rate, as the equilibrium with the enzyme-inhibitor complex is slowly established. Alternatively, if the enzyme is first pre-incubated with inhibitor, and the reaction is then initiated by the addition of substrate, the reaction will accelerate to the steady-state inhibited rate, as the pre-formed enzyme-inhibitor complex slowly dissociates to release active enzyme.

16.1.1.4.2 Tight Binding

Some reversible inhibitors bind their target enzyme so tightly that they exhibit different kinetic behavior than classical inhibitors. For example, while classical inhibitors are usually kinetically characterized at concentrations much greater than that of the enzyme in solution, tight binding inhibitors exert their effect at concentrations approaching that of the enzyme. The effect of this near stoichiometric

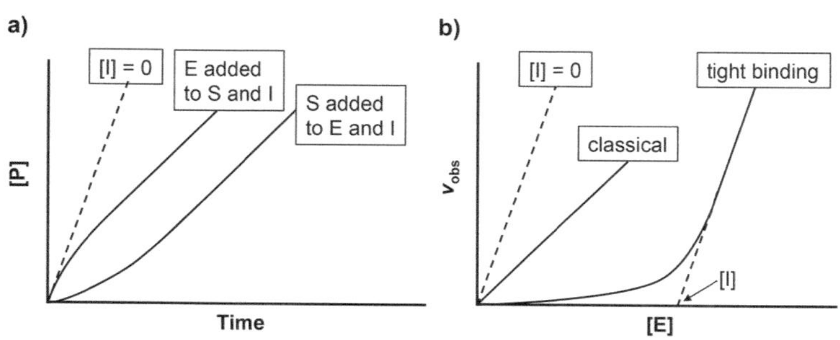

Fig. 16.13 Characteristic graphs for (**a**) slow binding inhibition and (**b**) tight binding inhibition

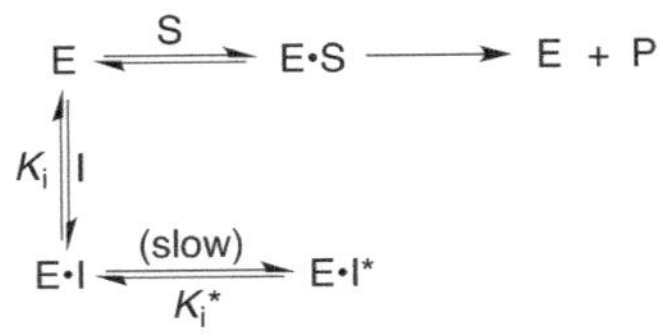

Fig. 16.14 Kinetic scheme for slow, tight binding (competitive) inhibition

inhibition can be observed in the plot of rate *vs* enzyme concentration, known as an Ackermann-Potter plot (Fig. 16.13b). Normally this plot is linear, either for the uninhibited reaction (dashed line) or for classical inhibition. But a tight binding inhibitor will suppress almost all of the activity of one equivalent of enzyme; increasing the enzyme concentration above that of the inhibitor increases the concentration of free (uninhibited) enzyme.

16.1.1.4.3 Slow, Tight Binding

A tight binding inhibitor may also have a slow binding equilibrium. For example, this is frequently observed with transition state analog inhibitors. These inhibitors bind to the enzyme in the substrate binding site, and then induce the enzyme to undergo a slow conformational change, resulting in a very stable alternative enzyme-inhibitor complex (e.g. K_i* of Fig. 16.14 can be in the pM or fM range).

16.1.2 Irreversible Inhibition

Contrary to reversible inhibition, irreversible inhibition results in the permanent modification of the target enzyme. Even if the inhibitor is subsequently removed from solution, irreversibly inhibited enzyme will not recover its activity; this is one of the most convincing demonstrations that inhibition is irreversible. First the enzyme is incubated with the inhibitor. Then the inhibitor is removed, either by high (>100-fold) dilution, or by separation (e.g. dialysis, filtration over molecular weight cut-off membrane, or chromatography). Finally, the activity of the recovered enzyme is tested and compared to a positive control of uninhibited enzyme that has been subjected to the same treatment.

Irreversible inhibition is often due to the formation of a covalent bond between the inhibitor and the enzyme, generating a stable enzyme-inhibitor *adduct* that may be detectable by mass spectrometry. However, it is worth noting that some covalent bonds are labile under physiological conditions, so the formation of a covalent bond does not always lead to irreversible inhibition.

Irreversible inhibition is usually time-dependent, also, involving the rate constant of inactivation, k_{inact}, as shown in the scheme of Fig. 16.15. However, time dependence alone cannot be used as proof of irreversible inhibition, as slow binding reversible inhibition also shows time dependence (Fig. 16.13a).

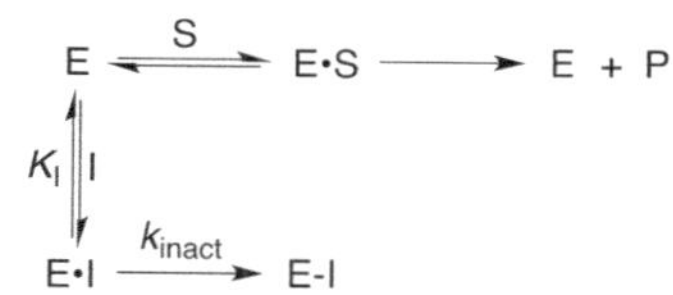

Fig. 16.15 Kinetic scheme for irreversible inhibition leading to formation of inactive adduct E-I

A detailed inspection of the time dependence is required to distinguish between these two forms of inhibition. With time-dependent reversible inhibition, some activity usually remains after the induction period, if the concentration of inhibitor is less than saturating (Silverman 2002). However, with irreversible inhibition, the final activity is usually zero, since at this point all enzyme has been inactivated.

16.1.2.1 Active-Site-Directed Inhibitors

Irreversible inhibitors may also be subdivided according to their mechanism of inhibition. Some are referred to as 'active-site-directed' inhibitors, or 'affinity labeling agents'. These inhibitors typically possess a strong electrophilic reactive group that reacts directly with a nucleophilic residue in the active site of the target enzyme, following equilibrium binding. For these inhibitors, the simple kinetic scheme shown in Fig. 16.15 is sufficient to describe their mode of action. To fully portray the overall efficiency of an irreversible inhibitor, both its affinity for the enzyme (reflected in K_I) and its chemical reactivity (reflected in k_{inact}) are required. For this reason, the ratio of k_{inact}/K_I is often used as a kind of second order rate constant for the overall reaction of an irreversible inhibitor with an enzyme.

To measure the values of these parameters, it is first necessary to measure the rate constant for disappearance of active enzyme. This can be done in a discontinuous manner, by removing aliquots from a solution of enzyme and inhibitor, removing excess inhibitor by separation or dilution, and determining the remaining activity of the enzyme present. A plot of residual activity versus time presents a mono-exponential decrease from which the rate constant k_{obs} can be determined (Fig. 16.16a). This rate constant shows hyperbolic dependence on the concentration of inhibitor (Eq. 16.4), from which k_{inact} and K_I can be determined. (Note the logical similarity between this equation, for the reaction of enzyme with equilibrium bound inhibitor, and the Michaelis-Menten equation for the reaction of enzyme with equilibrium bound substrate.) Alternatively, enzyme activity can be measured *while* irreversible inactivation is taking place, by simply adding inhibitor to a standard enzyme activity assay. In this approach, sometimes referred to as 'Kitz and Wilson conditions' (Kitz and Wilson 1962), the curve of assay product versus time can be fit to a first order equation to provide the rate constant for enzyme disappearance, k_{obs} (Fig. 16.16b). However, since the activity assay substrate most likely competes with the irreversible inhibitor for binding in the enzyme active site, the measured K_I value must be corrected by dividing by the factor $(1 + [S]/K_M)$. The apparent K_I value is increased by this factor, due to the temporary protection

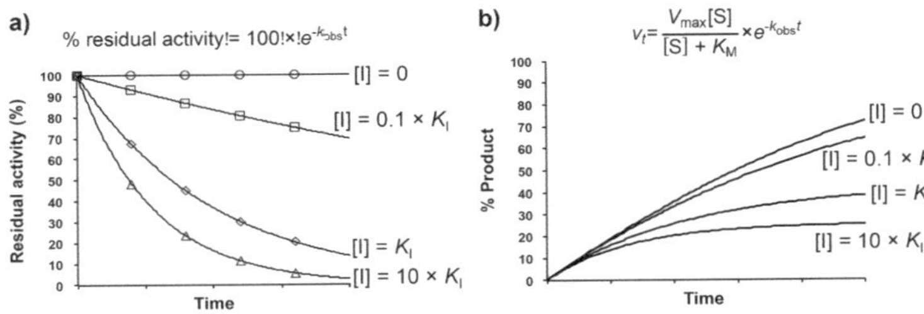

Fig. 16.16 Irreversible inhibition revealed (**a**) through discontinuous activity assay and (**b**) under continuous 'Kitz and Wilson' conditions

afforded by binding substrate in the place of inhibitor, in the same way that the apparent value of K_M is increased by a competitive reversible inhibitor (Eq. 16.5). Furthermore, it is essential to perform control experiments to show that the inhibitor being tested does not interfere with the activity assay apart from inactivating the enzyme.

$$k_{obs} = \frac{k_{inact}[I]}{[I] + K_I} \tag{16.4}$$

$$k_{obs} = \frac{k_{inact}[I]}{[I] + K_I\left(1 + \frac{[S]}{K_M}\right)} \tag{16.5}$$

As shown above, for reversible inhibition, IC_{50} values are relative values that are highly condition dependent. But for irreversible inhibition, IC_{50} values are even more relative, and even more condition dependent. If we consider that an IC_{50} value is by definition the concentration of inhibitor required to reduce enzyme activity to half that observed in the absence of the inhibitor, then it is obvious that for irreversible inhibition, this value is simply half the concentration of the enzyme! Given enough time, an irreversible inhibitor will inhibit one molar equivalent of enzyme. Therefore it is always preferable to measure constant k_{inact} and K_I values to characterize an irreversible inhibitor. However, if not enough material or time are available for measuring these constants, it may be necessary to measure a relative IC_{50} value. In this case, the precise experimental conditions must be defined explicitly, including the assay used to measure activity, and importantly, the length of *time* during which the enzyme was incubated with the inhibitor.

16.1.2.2 Suicide Inhibitors

Another subdivision of irreversible inhibitors may be referred to as 'mechanism-based inhibitors' or 'suicide inhibitors'. The kinetic scheme for these inhibitors is shown in Fig. 16.17. The important feature that distinguishes this mechanism of irreversible inhibition from that described in Fig. 16.15 is the requirement that the

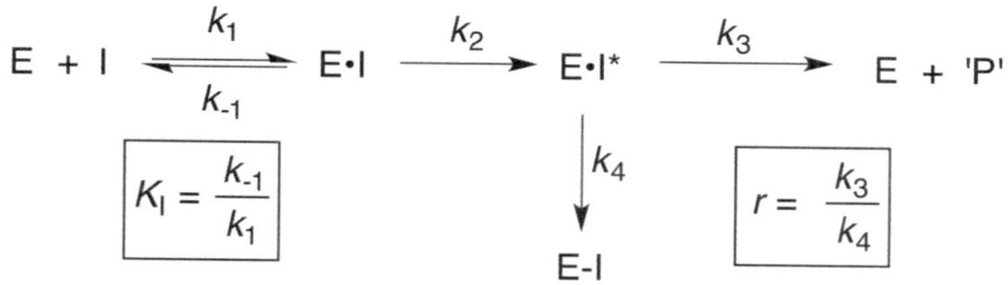

Fig. 16.17 'Suicide' inhibition, whereby the enzyme must activate the inhibitor to its competent form, I* (the ratio *r* reflects the partition ratio of regeneration to inactivation of enzyme)

inhibitor first be transformed, by the catalytic machinery of the enzyme, into its inhibition-competent form, I*. Typically, this is effected by the enzyme-mediated transformation of a weak electrophile into a strong electrophile that is able to react irreversibly with some nucleophile in the active site. Because the activation step depends on the catalytic groups of the enzyme itself, it will often show similar mechanistic characteristics (e.g. pH dependence, isotope effects, substituent effects) to the reaction of the enzyme with substrate. Another distinctive feature concerns the fate of intermediate E•I* in Fig. 16.17. This activated intermediate can often react in such a way that irreversibly inactivates the enzyme (k_4) *or* in another way by which the free enzyme is regenerated (k_3). Since neither of these steps involves any additional inhibitor binding, the ratio $r = k_3/k_4$ does not vary with inhibitor concentration. Rather, it denotes the partition ratio of the common intermediate and reports the number of molecules of 'product' that are formed upon regeneration of free enzyme, for every molecule of enzyme that is irreversibly inhibited. To fully characterize this kind of inhibition, it is necessary to measure k_{inact}, K_I and r.

16.2 Transglutaminase Inhibitors

Irreversible inhibitors of transglutaminase (TGase) were developed shortly after the enzyme was discovered, and used as probes (Folk and Cole 1966) of the enzyme mechanism. (Keillor et al. 2014) However, given the number of physiological disorders that have been proposed to be related to unregulated TGase activity, recent inhibitors have been used to interrogate the biological roles of TGase and developed with respect to their therapeutic potential. In this section we present a brief overview of recent TGase inhibitors. A more exhaustive list of inhibitors may be gathered from a collection of review articles, (Wodzinska 2005; Siegel and Khosla 2007; Keillor et al. 2011; Badarau et al. 2011; Keillor et al. 2015) but this chapter will focus on recent reversible and irreversible inhibitors of notable structures and potential.

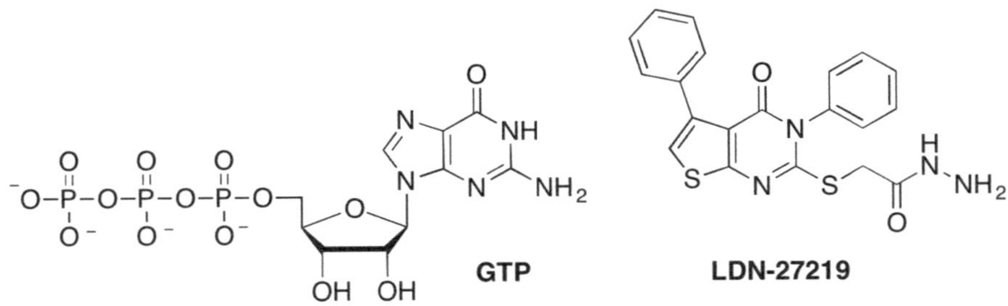

Fig. 16.18 GTP and TG2 inhibitor LDN-27219

16.2.1 Reversible TGase Inhibitors

16.2.1.1 GTP Analogs

Some of the first reversible TGase inhibitors to be reported were non-hydrolyzable GTP analogs (Lai et al. 1998) and mimics (Duval et al. 2005). One of these, LDN-27219 (Fig. 16.18), was studied in more detail and reported (Case and Stein 2007) to be a slow binding inhibitor of tissue TGase (TG2). Using an activity assay based on 8 μM *N*, *N*-dimethylated casein (NMC) as acyl-donor substrate and 16 μM α-*N*-Boc-Lys-NHCH$_2$CH$_2$NH-dansyl (KXD) as acyl-acceptor substrate, an IC$_{50}$ value of 0.6 ± 0.1 μM was measured. The authors went on to show that the inhibitor was non-competitive with respect to *both* substrates, but mutually exclusive with the negative regulator GTP.

16.2.1.2 Acyl-Donor Inhibitors

Cinnamoyl derivatives have also been reported as a class of reversible TG2 inhibitors (Pardin et al. 2008a; b). Low micromolar IC$_{50}$ values were measured, using an assay based on 54.4 μM of the chromogenic substrate *N*-Cbz-Glu(γ-*p*-nitrophenylester)Gly (Leblanc et al. 2001). One of the most potent of these inhibitors, CP4d (Fig. 16.19), was shown (Pardin et al. 2008b) to be competitive with respect to acyl-donor substrate, having a K_i of 174 ± 48 nM.

Lai and co-workers discovered other potent inhibitors by screening large compound libraries (Lai et al. 2008). The most potent of these, inhibitor ZM 449829, is also shown in Fig. 16.19. Its IC$_{50}$ value of 5 ± 5 nM was measured using the same assay used to characterize LDN-27219 (8 μM NMC and 16 μM KXD) but not the same assay used with CP4d. So although its IC$_{50}$ value cannot be compared directly with that of CP4d, it is interesting to note their structural similarities.

The same structural comparison can be done with the acylideneoxoindole inhibitors developed by Klock and co-workers (Klock et al. 2011). One of their most potent inhibitors is shown in Fig. 16.19 (compound 41). Using a sensitive fluorescence-based activity assay (Gillet et al. 2005), the K_i value of compound

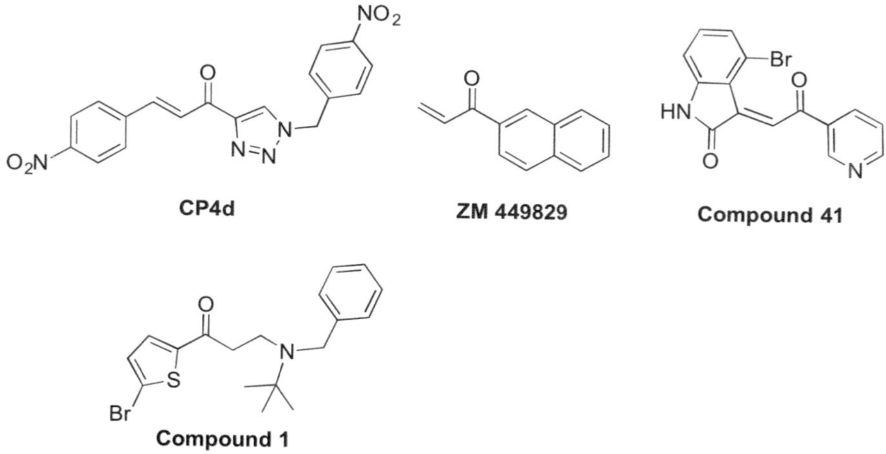

Fig. 16.19 Reversible inhibitors of TG2 that show some structural similarities

41 was determined to be 0.25 μM. This inhibition constant can be compared to the similar value measured for CP4d. Furthermore, mixed non-competitive inhibition was observed with respect to the acyl-donor substrate. The ratio of K_i'/K_i was determined to be 2.6, indicating stronger competitive inhibition than uncompetitive inhibition.

Later, Ozaki and co-workers reported (Ozaki et al. 2010) a series of aryl *N*, *N*-disubstituted β-amino ketones, whose inhibition was measured using the reaction of 20 μM monodansylcadaverine (MDC) with 0.05 mg/mL NMC. With this assay, the IC_{50} value of compound 1 in Fig. 16.19 was determined to be 81 nM. However, as discussed in the first section of this chapter, this IC_{50} value cannot be compared directly to those of other reversible inhibitors, since different assay conditions were used in the measurement.

The inhibitors shown in Fig. 16.19 were arranged in such a way to facilitate structural comparison. It is interesting to note that all of them include a carbonyl attached to an aromatic ring, and most comprise an additional α,β-double bond (see the top row of Fig. 16.19). On the other hand, the carbonyl and benzyl groups of compound 1 and CP4d can be overlaid (left column of Fig. 16.19). Although not all of these compounds can be compared for potency, since only relative IC_{50} values were reported, it is important to note that both CP4d and compound 41 were both shown to be at least partially competitive with the acyl-donor substrates, which is suggestive of a common binding site.

16.2.1.3 Alternative Inhibitors

Most recently, Kim and co-workers screened a small molecule library (Kim et al. 2014), and discovered new 3-alkynyl quinoxaline and pyrido[2,3-b]pyrazine

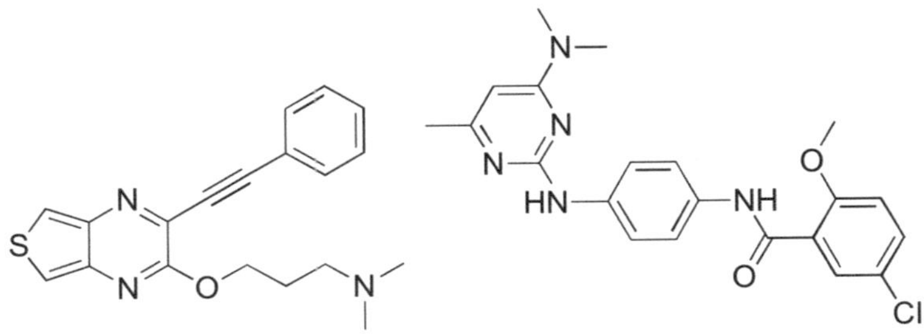

GK-428 TG53

Fig. 16.20 Additional reversible inhibitors of TG2

derivatives that showed strong inhibition of TG2. Specifically, GK-428 (Fig. 16.20) was found to have an IC_{50} value of 3.8 μM, according to a cell proliferation assay. While this assay gives more direct information regarding the biological efficacy of these inhibitors, it does not allow their mode of inhibition to be determined. However, saturation transfer difference NMR was used to verify that the inhibitor binds directly to TG2 in vitro.

An alternative approach to the discovery of TG2 inhibitors was recently demonstrated by Yakubov and co-workers (Yakubov et al. 2014). Using an AlphaLISA assay, they screened a library of 10,000 small molecules for their ability to interfere with the protein-protein interaction between TG2 and fibronectin. The inhibitor shown in Fig. 16.20, TG53, was reported to inhibit TG2-fibronectin binding with a K_i of 4.15 μM.

16.2.2 *Irreversible TGase Inhibitors*

Historically, irreversible TGase inhibitors have been developed much more than reversible inhibitors. These inhibitors are designed as analogs of the acyl-donor substrate that are competitive with that substrate (see Sect. 1.1.1) and typically active-site-directed inhibitors (see Sect. 1.2.1) (Keillor et al. 2011). In the case of irreversible inhibition, this means that binding the acyl-donor substrate can protect TGase temporarily, but when the inhibitor is bound in the acyl-donor substrate binding site, it reacts with TGase and inactivates that equivalent of enzyme. The presence of competitive substrate can therefore slow irreversible inhibition, but ultimately cannot prevent it.

Over the years, the functional groups of inhibitors that react with TGase, typically with the nucleophilic thiol group of the active-site cysteine residue, have been varied broadly. These 'warheads' have been refined for their reactivity with their target, but also for their stability in solution, allowing them to bind to their target prior to their reaction. This section presents an overview of some of the

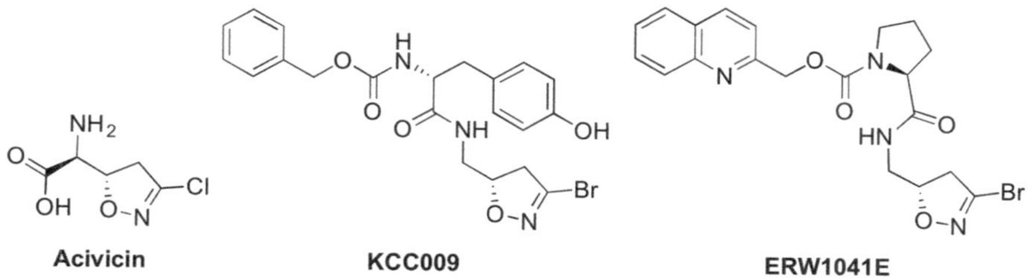

Fig. 16.21 Acivicin and analogous 3-bromo-4,5-dihydroisoxazole inhibitors

most developed irreversible inhibitors, grouped by the nature of their reactive pharmacophore.

16.2.2.1 3-Halo-4,5-Dihydroisoxazole Inhibitors

Acivicin (Fig. 16.21) is known as a natural glutamine mimic that inhibits many different cysteine-dependent enzymes (Tso et al. 1980). The Syntex group was the first to incorporate acivicin into peptidic inhibitors of epidermal (Castelhano et al. 1988) and tissue (Auger et al. 1993) TGase. Later, Khosla and co-workers studied 3-bromo derivatives that proved to be potent irreversible inhibitors with therapeutic potential (Choi et al. 2005; Watts et al. 2006; Yuan et al. 2005; 2007). For example, the enantiomerically pure form of KCC009 (Fig. 16.21) was reported (Watts et al. 2006) to have an inhibition constant (K_I) of 68 μM, an inactivation rate constant (k_{inact}) of 0.155 min^{-1} and an overall efficiency ratio (k_{inact}/K_I) of 2300 $M^{-1}min^{-1}$. KCC009 was also shown to be orally bioavailable, relatively non-toxic, and effective at rendering glioblastoma tumors sensitive to chemotherapy with *N*, *N*′-bis(2-chloroethyl)-*N*-nitrosourea (BCNU), reducing tumor size and prolonging survival in mice (Choi et al. 2005; Yuan et al. 2005; 2007). The (*S*)-proline analog ERW1041E (Fig. 16.21) was also reported to have similar inhibitory efficiency ($k_{inact} = 0.110$ min^{-1}, $K_I = 11$ μM, $k_{inact}/K_I = 10000$ $M^{-1}min^{-1}$) (Watts et al. 2006). When tested in a mouse model, ERW1041E was also found to inhibit TG2 in the small intestine, demonstrating its potential for the treatment of celiac disease (Dafik et al. 2012).

16.2.2.2 Michael Acceptors

Michael acceptors are defined as α,β-unsaturated carbonyl compounds that undergo addition reactions with nucleophiles (Michael 1886). As such, they react efficiently with active-site thiol nucleophiles. Acrylamides are one class of Michael acceptors that have proven to be stable in aqueous solution, but highly reactive with TGases (Keillor et al. 2011).

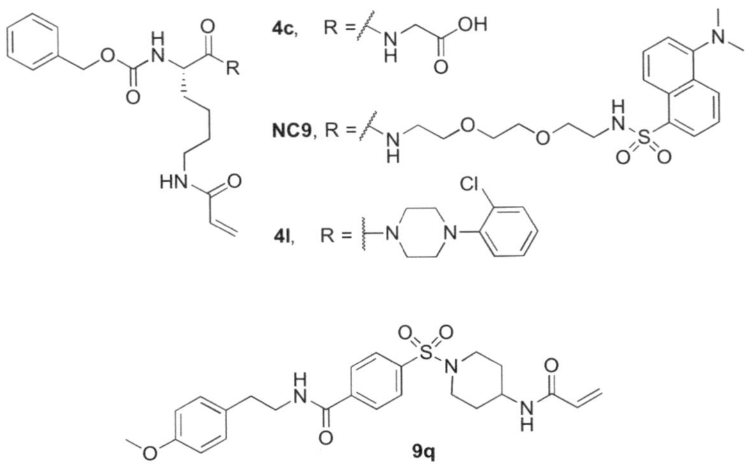

Fig. 16.22 Acrylamide-based Michael-acceptor inhibitors of TG2

In 2001, the Keillor group reported a series of derivatives, bearing an acrylamide group on the side chain of a based on the dipeptide scaffold of the commonly used substrate Cbz-Gln-Gly (de Macédo et al. 2002; Marrano et al. 2001). These were shown to be irreversible inhibitors of guinea pig liver TG2, with efficiency ratios (k_{inact}/K_I) around 10^6 M^{-1} min^{-1}. The most efficient inhibitor (compound 4c in Fig. 16.22) was shown to have a k_{inact} of 0.46 min^{-1} and a K_I of 150 nM (Marrano et al. 2001). Replacement of the glycine residue with a PEG spacer bearing a dansyl fluorophore gave irreversible inhibitor NC9 (Fig. 16.22), whose k_{inact} is 0.41 min^{-1} and whose K_I is 15 μM (Keillor et al. 2008). NC9 has since proven useful in cellular studies of TG2 inhibition (Colak et al. 2011; Gundemir et al. 2013), TG2 and Factor XIIIa localization (Al-Jallad et al. 2011), and TG2 conformation (Clouthier et al. 2012; Caron et al. 2012) (see below). More recently, a red fluorescent acrylamide inhibitor based on a more extended peptidic scaffold (RhodB-PGG-K (Acr)-LPF) was shown to be selective for TG2 over Factor XIII, and was used for staining rat aorta in a study of arterial rigidification. {Chabot, 2010, r00035}.

The CHDI Foundation also confirmed the activity of acrylamide derivatives while screening libraries for TG2 inhibitors (Schaertl et al. 2010). Subsequent derivatives were prepared and their IC_{50} values were measured by incubation for 30 min, followed by measurement of residual activity using 8 μM NMC and 16 μM KXD (Prime et al. 2012a; Wityak et al. 2012). As discussed above in Sect. 1.2.1, IC_{50} values do not allow inhibitor affinity (K_I) to be distinguished from reactivity (k_{inact}), nor do they permit comparison with inhibitors measured under different conditions (assay conditions and time). However, within the experimental conditions of one study, they can give a measure of relative efficiency (Wityak et al. 2012). To that end, derivatives 4l and 9q were reported to have low IC_{50} values (62 nM and 110 nM), good plasma stability and resistance to conjugation

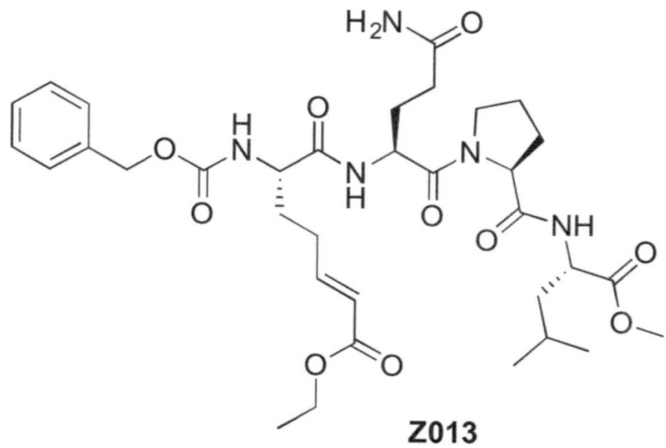

Z013

Fig. 16.23 α,β-Unsaturated ester-based Michael-acceptor inhibitor of TG2

with 25 mM glutathione over a period of 120 h in vitro (Prime et al. 2012a; Wityak et al. 2012).

The group at Zedira has also developed TGase inhibitors incorporating an α,β-unsaturated ester group as a Michael acceptor, such as compound Z013 (Fig. 16.23). This inhibitor was evaluated by Schaertl and co-workers by pre-incubating the inhibitor for 15 min, prior to measuring residual transamidation activity with 8 μM NMC and 16 μM KXD, providing an IC_{50} value of 70 nM for TG2 (Schaertl et al. 2010). Using the same assay conditions with different TGases, the authors demonstrated that Z013 shows reasonable selectivity for TG2.

16.2.2.3 6-Diazo-5-Oxo-L-Norleucine (DON) Derivatives

Another warhead that has proven effective is that of 6-diazo-5-oxo-L-norleucine (DON), arguably one of the best glutamine mimics known. The inhibitor Z006 (or 'Z-DON'), developed by the group at Zedira (McConoughey et al. 2010) is shown in Fig. 16.24. When tested under the same conditions as for Z013 (Schaertl et al. 2010) it was found to show excellent selectivity for TG2, with an IC_{50} value of 20 nM. Comparison of the structure of Z006 to that of Z013 (Fig. 16.23) reveals slight differences that may account for the difference in efficiency; however, it cannot be determined from the IC_{50} values whether this is due to a difference in affinity (K_I), reactivity (k_{inact}), or both.

Pinkas and co-workers incorporated DON in a high affinity, gluten-based peptide sequence to design the potent TG2 inhibitor Ac-P-DON-LPF-NH_2, whose K_I is 60 nM and whose k_{inact} is 0.5 min^{-1} (Pinkas et al. 2007). This inhibitor was also used to inhibit TG2 prior to crystallization, providing the first structure of a TGase in its 'open' conformation, believed to be the transamidation-active form of the enzyme.

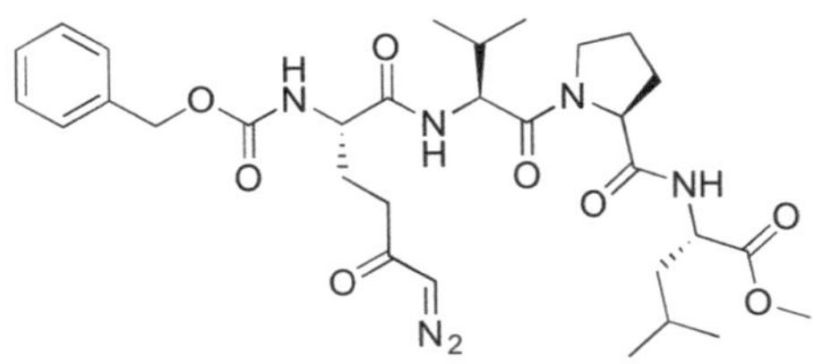

Z006

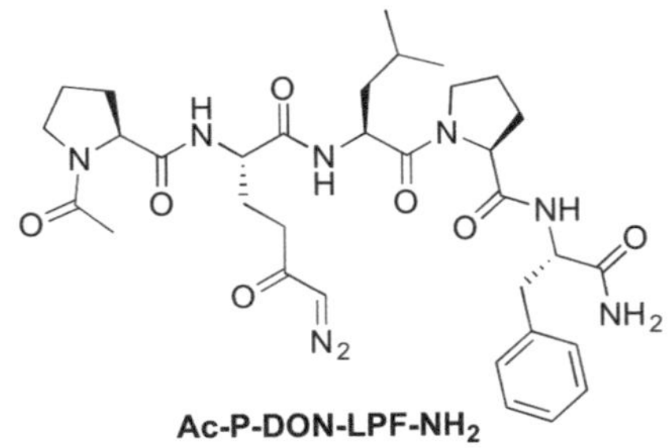

Ac-P-DON-LPF-NH₂

Fig. 16.24 DON-based irreversible inhibitors of TG2

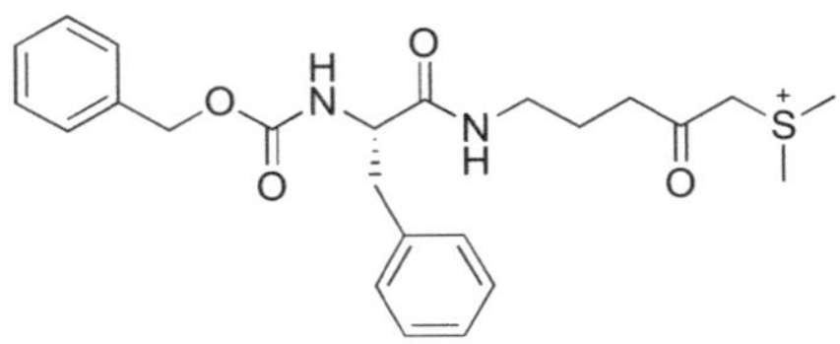

3

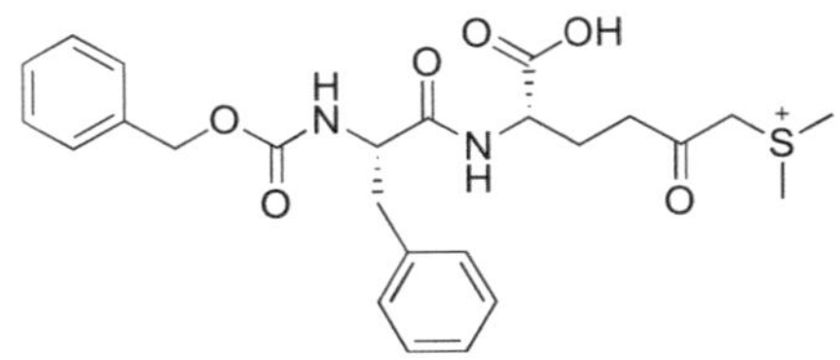

4d

Fig. 16.25 Sulfonium-based irreversible inhibitors of TG2

16.2.2.4 Sulfonium Derivatives

In 1992, Pliura and co-workers from the Syntex group reported a series of inhibitors bearing a dimethyl sulfonium warhead that were found to inhibit epidermal transglutaminase and TG2. Compound 3, shown in Fig. 16.25, inhibited TG2 with a second order rate constant of inactivation of $2.95 \times 10^4\ M^{-1}\ min^{-1}$ (Pliura et al. 1992).

More recently, the Griffin group reported a dimethylsulfonium derivative that bears an extra carboxyl group (compound 4d, Fig. 16.25), which increases its aqueous solubility and favors inhibition of extracellular TG2 (Griffin et al. 2008). The activity of this inhibitor with TG2 was determined by its ability to block the covalent incorporation of (132 μM) biotinyl-5-pentylamine (BP, acyl-acceptor substrate) into immobilized NMC (acyl-donor substrate), during 1 h incubation at 37 °C. The IC_{50} value thus obtained for compound 4d was 10 μM against TG2 and 105 μM against FXIIIa (Badarau et al. 2013). Interestingly, the methyl ester of compound 4d had less than half the aqueous solubility, with an IC_{50} value of 5 μM, suggesting that the negatively charged carboxylate does not have a strongly adverse effect on the affinity of the inhibitors. Compound 4d (also referred to as 'NTU281') has also been shown to be an effective treatment for experimental diabetic nephropathy in a mouse model (Huang et al. 2009).

16.2.3 *Perspectives*

16.2.3.1 Structure-Based Inhibitor Design

While many irreversible inhibitors are used as effective drugs, including β-lactam antibiotics and aspirin, their utility is directly related to the selectivity of their inactivation activity (Powers et al. 2002). To that end, the structure-based design of inhibitors can help enhance their affinity for their intended target (e.g. TG2) or even guide their selectivity against other targets (e.g. Factor XIII). The starting point for this approach is high resolution structural data, preferably of the target enzyme in complex with a relevant inhibitor. However, this information has been scarce for the TGases.

In 2002, Liu and co-workers reported the structure of human TG2 in complex with GDP (PDB code 1KV3) (Liu et al. 2002). Later, human TG2 was crystallized in the presence of ATP (PDB code 3LY6) (Han et al. 2010). Both of these structures show TG2 in its compact conformation and provide information of the nucleotide binding site that could be targeted to inactive the enzyme.

In 2007, the Khosla group reported a structure (PDB code 2Q3Z) of an extended conformation of TG2 that had been inactivated with their irreversible inhibitor Ac-P-DON-LPF-NH_2 (Fig. 16.24) (Pinkas et al. 2007). This provided the first structure of the 'open' form of the enzyme, and showed the acyl-donor substrate binding site as occupied by their peptidic inhibitor (Fig. 16.26). More recently,

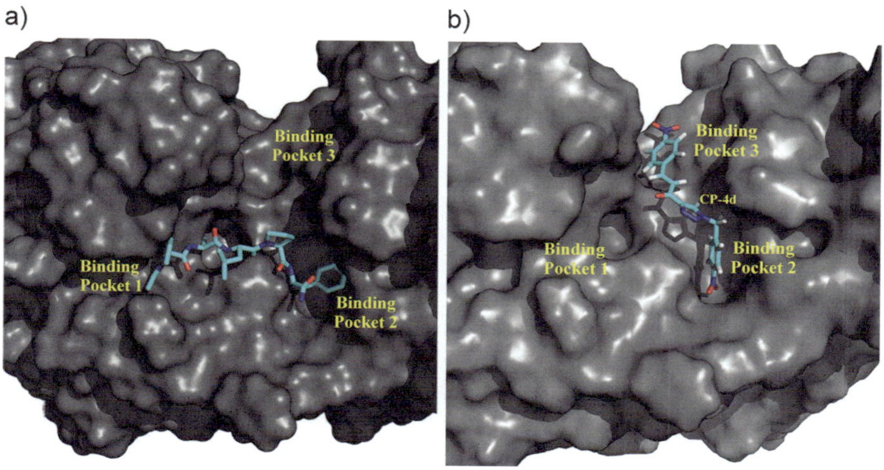

Fig. 16.26 (**a**) Structure of irreversible inhibitor Ac-P-DON-LPF-NH_2 (see Fig. 16.24) bound to human TG2 (Structure published as PDB code 2Q3Z in reference (Pinkas et al. 2007); figure adapted from reference (Keillor et al. 2015)). (**b**) Reversible inhibitor CP4d (see Fig. 16.19) docked into the acyl-donor substrate binding site from the protein structure 2Q3Z

Lindemann and co-workers deposited the structure (PDB code 3S3P) of human TG2 after reaction with irreversible inhibitor Z013 (Fig. 16.23), showing a similar 'open' conformation and the same peptide substrate binding site.

Finally, Stieler and co-workers provided a structure (PDB code 4KTY) of Factor XIII after its reaction with peptidic Michael-acceptor inhibitor 'ZED1301' (now A108 from Zedira) (Stieler et al. 2013). Importantly, this structure shows an alternative 'open' form of TGase, while providing important structural information on how Factor XIII binds its acyl-donor substrate.

To date, *no* crystallographic structure is available of TGase in complex with a *reversible* inhibitor. Therefore, in order to gain information regarding how these inhibitors may be bound, it is necessary to use computer-based modeling techniques. Homology models may be generated of any TGase for which a specific structure is not known, and minimized small molecule inhibitors can subsequently be docked into a given binding site, preferably indicated by the mode of inhibition (Pardin et al. 2009; Prime et al. 2012b). This approach has suggested the importance of the binding pockets indicated in Fig. 16.26. However, the *co-crystallization* of TGase with a tightly bound reversible inhibitor would provide genuine structural data rather than simulated information and would greatly assist structure-based inhibitor design.

16.2.3.2 Conformational Effects

TGases have long been known to exist in dramatically different conformations (Casadio et al. 1999; Di Venere et al. 2000; Mariani et al. 2000), and the conformational effects of ligand binding have been studied extensively (Keillor et al. 2014; Clouthier et al. 2012; Pinkas et al. 2007; Pardin et al. 2009; Begg et al. 2006). Recently the conformational equilibrium of TG2 has also been related to its potential function as a cytosolic scaffold protein in addition to its role as a cross-linking catalyst (Colak et al. 2011). Given the conformation dependence of all of the activities of TG2, it is important to consider the effect of inhibitor binding on the conformational equilibrium. This has been done using light scattering (Pinkas et al. 2007), capillary electrophoresis (Clouthier et al. 2012) and a FRET-based detection system (Caron et al. 2012). Interestingly, it appears that irreversible inhibitors lock TG2 in a more extended conformation (Colak et al. 2011; Clouthier et al. 2012; Caron et al. 2012), probably by disrupting important tertiary structural interactions that would otherwise stabilize the closed conformation (Keillor et al. 2014; 2015). On the other hand, it appears that some reversible inhibitors, such as CP4d (Fig. 16.19), may stabilize the closed conformation of TG2 (Colak et al. 2011; Caron et al. 2012). This underlines the potential utility of inhibitors as chemical tools for the manipulation of TG2 conformation in biological contexts, either for mechanistic studies or for therapeutic purposes (Keillor et al. 2014; 2015).

Acknowledgments JWK thanks Dr. Christopher Clouthier for the images of Fig. 16.26.

References

Al-Jallad HF, Myneni VD, Piercy-Kotb SA, Chabot N, Mulani A, Keillor JW, Kaartinen MT (2011) Plasma membrane factor XIIIA transglutaminase activity regulates osteoblast matrix secretion and deposition by affecting microtubule dynamics. PLoS One 6:e15893

Auger M, McDermott AE, Robinson V, Castelhano AL, Billedeau RJ, Pliura DH, Krantz A, Griffin RG (1993) Solid-state carbon-13 NMR study of a transglutaminase-inhibitor adduct. Biochemistry 32:3930–3934

Badarau E, Collighan RJ, Griffin M (2011) Recent advances in the development of tissue transglutaminase (TG2) inhibitors. Amino Acids 44:119–127

Badarau E, Mongeot A, Collighan R, Rathbone D, Griffin M (2013) Imidazolium-based warheads strongly influence activity of water-soluble peptidic transglutaminase inhibitors. Eur J Med Chem 66:526–530

Begg GE, Carrington L, Stokes PH, Matthews JM, Wouters MA, Husain A, Lorand L, Iismaa SE, Graham RM (2006) Mechanism of allosteric regulation of transglutaminase 2 by GTP. Proc Natl Acad Sci U S A 103:19683–19688

Caron NS, Munsie LN, Keillor JW, Truant R (2012) Using FLIM-FRET to measure conformational changes of transglutaminase type 2 in live cells. PLoS One 7:e44159

Casadio R, Polverini E, Mariani P, Spinozzi F, Carsughi F, Fontana A, Polverino de Laureto P, Matteucci G, Bergamini CM (1999) The structural basis for the regulation of tissue transglutaminase by calcium ions. Eur J Biochem 262:672–679

Case A, Stein RL (2007) Kinetic analysis of the interaction of tissue transglutaminase with a nonpeptidic slow-binding inhibitor. Biochemistry 46:1106–1115

Castelhano AL, Billedeau R, Pliura DH, Bonaventura BJ, Krantz A (1988) Synthesis, chemistry, and absolute configuration of novel transglutaminase inhibitors containing a 3-halo-4,5-dihydroisoxazole. Bioorg Chem 16:335–340

Choi K, Siegel M, Piper JL, Yuan L, Cho E, Strnad P, Omary B, Rich KM, Khosla C (2005) Chemistry and biology of dihydroisoxazole derivatives: selective inhibitors of human transglutaminase 2. Chem Biol 12:469–475

Clouthier CM, Mironov GG, Okhonin V, Berezovski MV, Keillor JW (2012) Real-time monitoring of protein conformational dynamics in solution using kinetic capillary electrophoresis. Angew Chem Int Ed Engl 51:12464–12468

Colak G, Keillor JW, Johnson GVW (2011) Cytosolic guanine nucleotide binding deficient form of transglutaminase 2 (R580A) potentiates cell death in oxygen glucose deprivation. PLoS One 6:e16665

Dafik L, Albertelli M, Stamnaes J, Sollid LM, Khosla C (2012) Activation and inhibition of transglutaminase 2 in mice. PLoS One 7:e30642

de Macédo P, Marrano C, Keillor JW (2002) Synthesis of dipeptide-bound epoxides and alpha, beta-unsaturated amides as potential irreversible transglutaminase inhibitors. Bioorg Med Chem 10:355–360

Di Venere A, Rossi A, De Matteis F, Rosato N, Agrò AF, Mei G (2000) Opposite effects of Ca2+ and GTP binding on tissue transglutaminase tertiary structure. J Biol Chem 275:3915–3921

Duval E, Case A, Stein RL, Cuny GD (2005) Structure–activity relationship study of novel tissue transglutaminase inhibitors. Bioorg Med Chem Lett 15:1885–1889

Fersht A (1985) Enzyme structure and function, 2nd edn. W H Freeman, New York

Folk JE, Cole PW (1966) Identification of a functional cysteine essential for the activity of guinea pig liver transglutaminase. J Biol Chem 241:3238–3240

Fox MA, Whitesell JK (2004) Organic chemistry. Jones & Bartlett Learning, Boston

Gillet SMFG, Pelletier JN, Keillor JW (2005) A direct fluorometric assay for tissue transglutaminase. Anal Biochem 347:221–226

Griffin M, Mongeot A, Collighan R, Saint RE, Jones RA, Coutts IGC, Rathbone DL (2008) Synthesis of potent water-soluble tissue transglutaminase inhibitors. Bioorg Med Chem Lett 18:5559–5562

Gundemir S, Colak G, Feola J, Blouin R, Johnson GVW (2013) Transglutaminase 2 facilitates or ameliorates HIF signaling and ischemic cell death depending on its conformation and localization. Biochim Biophys Acta 1833:1–10

Han B-G, Cho J-W, Cho YD, Jeong K-C, Kim S-Y, Lee BI (2010) Crystal structure of human transglutaminase 2 in complex with adenosine triphosphate. Int J Biol Macromol 47:190–195

Huang L, Haylor JL, Hau Z, Jones RA, Vickers ME, Wagner B, Griffin M, Saint RE, Coutts IGC, El Nahas AM, Johnson TS (2009) Transglutaminase inhibition ameliorates experimental diabetic nephropathy. Kidney Int 76:383–394

Keillor JW, Chica RA, Chabot N, Vinci V, Pardin C, Fortin E, Gillet SMFG, Nakano Y, Kaartinen MT, Pelletier JN, Lubell WD (2008) The bioorganic chemistry of transglutaminase – from mechanism to inhibition and engineering. Can J Chem 86:271–276

Keillor JW, Chabot N, Roy I, Mulani A, Leogane O, Pardin C (2011) Irreversible inhibitors of tissue transglutaminase. Adv Enzymol Relat Areas Mol Biol 78:415–447

Keillor JW, Clouthier CM, Apperley KYP, Akbar A, Mulani A (2014) Acyl transfer mechanisms of tissue transglutaminase. Bioorg Chem 57:186–197

Keillor JW, Apperley KYP, Akbar A (2015) Inhibitors of tissue transglutaminase. Trends Pharmacol Sci 36:32–40

Kim N, Kwak SH, Lee S-H, Juvekar V, Lee BI, Ahn H-C, Kim S-Y, Gong Y-D (2014) Novel 3-arylethynyl-substituted thieno[3,4-b]pyrazine derivatives as human transglutaminase 2 inhibitors. Org Biomol Chem 12:4932–4940

Kitz R, Wilson IB (1962) Esters of methanesulfonic acid as irreversible inhibitors of acetylcholinesterase. J Biol Chem 237:3245–3249

Klock C, Jin X, Choi KH, Khosla C, Madrid PB, Spencer A, Raimundo BC, Boardman P, Lanza G, Griffin JH (2011) Acylideneoxoindoles: a new class of reversible inhibitors of human transglutaminase 2. Bioorg Med Chem Lett 21:2692–2696

Lai TS, Slaughter TF, Peoples KA, Hettasch JM, Greenberg CS (1998) Regulation of human tissue transglutaminase function by magnesium-nucleotide complexes. Identification of distinct binding sites for Mg-GTP and Mg-ATP. J Biol Chem 273:1776–1781

Lai T-S, Liu Y, Tucker T, Daniel KR, Sane DC, Toone E, Burke JR, Strittmatter WJ, Greenberg CS (2008) Identification of chemical inhibitors to human tissue transglutaminase by screening existing drug libraries. Chem Biol 15:969–978

Leblanc A, Gravel C, Labelle J, Keillor JW (2001) Kinetic studies of guinea pig liver transglutaminase reveal a general-base-catalyzed deacylation mechanism. Biochemistry 40:8335–8342

Liu S, Cerione RA, Clardy J (2002) Structural basis for the guanine nucleotide-binding activity of tissue transglutaminase and its regulation of transamidation activity. Proc Natl Acad Sci U S A 99:2743–2747

Mariani P, Carsughi F, Spinozzi F, Romanzetti S, Meier G, Casadio R, Bergamini CM (2000) Ligand-induced conformational changes in tissue transglutaminase: Monte Carlo analysis of small-angle scattering data. Biophys J 78:3240–3251

Marrano C, de Macédo P, Keillor JW (2001) Evaluation of novel dipeptide-bound alpha, beta-unsaturated amides and epoxides as irreversible inhibitors of guinea pig liver transglutaminase. Bioorg Med Chem 9:1923–1928

McConoughey SJ, Basso M, Niatsetskaya ZV, Sleiman SF, Smirnova NA, Langley BC, Mahishi L, Cooper AJL, Antonyak MA, Cerione RA, Li B, Starkov A, Chaturvedi RK, Beal MF, Coppola G, Geschwind DH, Ryu H, Xia L, Iismaa SE, Pallos J, Pasternack R, Hils M, Fan J, Raymond LA, Marsh JL, Thompson LM, Ratan RR (2010) Inhibition of transglutaminase 2 mitigates transcriptional dysregulation in models of Huntington disease. EMBO Mol Med 2:349–370

Michael A (1886) Ueber die Addition von Natriumacetessig- und Natriummalonsäureäthern zu den Aethern ungesättigter Säuren. J Prakt Chem 35:349–356

Ozaki S, Ebisui E, Hamada K, Goto J-I, Suzuki AZ, Terauchi A, Mikoshiba K (2010) Potent transglutaminase inhibitors, aryl β-aminoethyl ketones. Bioorg Med Chem Lett 20:1141–1144

Pardin C, Pelletier JN, Lubell WD, Keillor JW (2008a) Cinnamoyl inhibitors of tissue transglutaminase. J Org Chem 73:5766–5775

Pardin C, Roy I, Lubell WD, Keillor JW (2008b) Reversible and competitive cinnamoyl triazole inhibitors of tissue transglutaminase. Chem Biol Drug Des 72:189–196

Pardin C, Roy I, Chica RA, Bonneil E, Thibault P, Lubell WD, Pelletier JN, Keillor JW (2009) Photolabeling of tissue transglutaminase reveals the binding mode of potent cinnamoyl inhibitors. Biochemistry 48:3346–3353

Pinkas DM, Strop P, Brunger AT, Khosla C (2007) Transglutaminase 2 undergoes a large conformational change upon activation. PLoS Biol 5:e327

Pliura DH, Bonaventura BJ, Pauls HW, Killackey JF, Krantz A (1992) Irreversible inhibition of transglutaminases by sulfonium methylketones – optimization of specificity and potency with omega-aminoacyl spacers. J Enzym Inhib 6:181–194

Powers JC, Asgian JL, Ekici ÖD, James KE (2002) Irreversible inhibitors of serine, cysteine, and threonine proteases. Chem Rev 102:4639–4750

Prime ME, Brookfield FA, Courtney SM, Gaines S, Marston RW, Ichihara O, Li M, Vaidya D, Williams H, Pedret-Dunn A, Reed L, Schaertl S, Toledo-Sherman L, Beconi M, Macdonald D, Muñoz-Sanjuan I, Dominguez C, Wityak J (2012a) Irreversible 4-Aminopiperidine transglutaminase 2 inhibitors for Huntington's disease. ACS Med Chem Lett 3:731–735

Prime ME, Andersen OA, Barker JJ, Brooks MA, Cheng RKY, Toogood-Johnson I, Courtney SM, Brookfield FA, Yarnold CJ, Marston RW, Johnson PD, Johnsen SF, Palfrey JJ, Vaidya D, Erfan S, Ichihara O, Felicetti B, Palan S, Pedret-Dunn A, Schaertl S, Sternberger I, Ebneth A, Scheel A, Winkler D, Toledo-Sherman L, Beconi M, Macdonald D, Muñoz-Sanjuan I, Dominguez C, Wityak J (2012b) Discovery and structure–activity relationship of potent and selective covalent inhibitors of transglutaminase 2 for Huntington's disease. J Med Chem 55:1021–1046

Schaertl S, Prime M, Wityak J, Dominguez C, Muñoz-Sanjuan I, Pacifici RE, Courtney S, Scheel A, Macdonald D (2010) A profiling platform for the characterization of transglutaminase 2 (TG2) inhibitors. J Biomol Screen 15:478–487

Segel IH (1993) Enzyme kinetics: behavior and analysis of rapid equilibrium and steady-state enzyme systems. Wiley, New York

Siegel M, Khosla C (2007) Transglutaminase 2 inhibitors and their therapeutic role in disease states. Pharmacol Ther 115:232–245

Silverman RB (2002) The organic chemistry of enzyme-catalyzed reactions, 1st edn. Academic, San Diego

Stieler M, Weber J, Hils M, Kolb P, Heine A, Büchold C, Pasternack R, Klebe G (2013) Structure of active coagulation Factor XIII triggered by calcium binding: basis for the design of next-generation anticoagulants. Angew Chem Int Ed Engl 52:11930–11934

Tommasini R, Endrenyi L, Taylor PA, Mahuran DJ, Lowden JA (1985) A statistical comparison of parameter estimation for the Michaelis-Menten kinetics of human placental hexosaminidase. Biochem Cell Biol 63:225–230

Tso JY, Bower SG, Zalkin H (1980) Mechanism of inactivation of glutamine amidotransferases by the antitumor drug L-(alpha S, 5S)-alpha-amino-3-chloro-4,5-dihydro-5-isoxazoleacetic acid (AT-125). J Biol Chem 255:6734–6738

Watts RE, Siegel M, Khosla C (2006) Structure–activity relationship analysis of the selective inhibition of transglutaminase 2 by dihydroisoxazoles. J Med Chem 49:7493–7501

Wityak J, Prime ME, Brookfield FA, Courtney SM, Erfan S, Johnsen S, Johnson PD, Li M, Marston RW, Reed L, Vaidya D, Schaertl S, Pedret-Dunn A, Beconi M, Macdonald D, Muñoz-Sanjuan I, Dominguez C (2012) SAR development of lysine-based irreversible inhibitors of transglutaminase 2 for Huntington's disease. ACS Med Chem Lett 3:1024–1028

Wodzinska JM (2005) Transglutaminases as targets for pharmacological inhibition. Mini Rev Med Chem 5:279–292

Yakubov B, Chen L, Belkin AM, Zhang S, Chelladurai B, Zhang Z-Y, Matei D (2014) Small molecule inhibitors target the tissue transglutaminase and fibronectin interaction. PLoS One 9: e89285

Yuan L, Choi K, Khosla C, Zheng X, Higashikubo R, Chicoine MR, Rich KM (2005) Tissue transglutaminase 2 inhibition promotes cell death and chemosensitivity in glioblastomas. Mol Cancer Ther 4:1293–1302

Yuan L, Siegel M, Choi K, Khosla C, Miller CR, Jackson EN, Piwnica-Worms D, Rich KM (2007) Transglutaminase 2 inhibitor, KCC009, disrupts fibronectin assembly in the extracellular matrix and sensitizes orthotopic glioblastomas to chemotherapy. Oncogene 26:2563–2573

Yung-Chi C, Prusoff WH (1973) Relationship between the inhibition constant (KI) and the concentration of inhibitor which causes 50 per cent inhibition (I50) of an enzymatic reaction. Biochem Pharmacol 22:3099–3108

Chapter 17
Substrate Engineering of Microbial Transglutaminase for Site-Specific Protein Modification and Bioconjugation

Noriho Kamiya and Yutaro Mori

Abstract Microbial transglutaminase (MTG), a robust enzyme developed initially for the manipulation of edible proteins in the food industry, has now been widely recognized as a practical protein-modifying reagent in the range of biotechnological applications. In this chapter, we introduce the potential use of MTG through our basic studies on the design of novel glutamine (Gln) donor substrates for lysine (Lys)-specific protein modification. Based on the core structure of a conventional transglutaminase substrate, benzyloxycarbonyl-L-glutaminylglycine (Z-QG), new Gln-donor substrates have been developed for the conjugation of recombinant proteins with different functionalities. The first target site for the substrate engineering was the C-terminal carboxylic group of Z-QG, which is feasibly labeled with functional moieties. For the preparation of protein-nucleic acid conjugates with novel molecular architecture, a new nucleotidyl substrate, Z-QG-(d)UTP, was created. We have also explored substitution of the N-terminal protecting group (Z) with fluorophores and biotin, and found that MTG accepts diverse functional groups at the N-terminus by inserting a short linker, leading to an increase in the utility of MTG in site-specific modification of functional proteins. Our results demonstrated how the design of (small) Gln-donor substrates of MTG can expand the scope of enzymatic manipulation in biomolecular engineering.

N. Kamiya (✉)
Department of Applied Chemistry, Graduate School of Engineering, Kyushu University, 744 Motooka, Nishi-ku, Fukuoka 819-0395, Japan

Center for Future Chemistry, Kyushu University, 744 Motooka, Nishi-ku, Fukuoka 819-0395, Japan
e-mail: nori_kamiya@mail.cstm.kyushu-u.ac.jp

Y. Mori
Department of Applied Chemistry, Graduate School of Engineering, Kyushu University, 744 Motooka, Nishi-ku, Fukuoka 819-0395, Japan

Present Affiliation: Biomass Engineering Program, RIKEN, 1-7-22 Suehiro-cho, Tsurumi-ku, Yokohama, Kanagawa 230-0045, Japan

K. Hitomi et al. (eds.), *Transglutaminases*, DOI 10.1007/978-4-431-55825-5_17

Keywords Bioconjugation • Biomolecular engineering • Fluorescent labeling • Hybrid proteins • Microbial transglutaminase • Protein labeling • Protein-DNA conjugate

17.1 Introduction

The progress of recombinant DNA technology and the knowledge increased in chemical modification strategies enable us to select a number of ways to modify proteins of interest with desired functionalities. Although many options are currently available for the expression of recombinant proteins in different types of living hosts or cell free systems, there is still room for the improvement of native functions of proteins toward specific purposes. The functional tuning of naturally occurring proteins is thus often needed in practical applications, and this goal could be achieved by suitably modifying proteins of interest (POI) by chemical, enzymatic and chemo-enzymatic approaches. From the viewpoint of site specificity in the modification of proteins, it is relevant to consider the different levels of molecular aspects, i.e., primary (amino acid residue), secondary (domain) and tertiary structures. By any means, the intrinsic function of the proteins should be retained. Chemical modification has been a major choice for covalently introducing an exogenous function into proteins because of its simplicity and variety of protocols (Hermanson 1996). For instance, covalent modifications of the α-amino group of the N-terminus and ε-amino group of Lys residues with chemically activated carboxylates have been routinely conducted, and careful adjustments of reaction conditions can ensure successful modifications. However, one of the critical drawbacks in chemical modification is the lack of site specificity due to the presence of multiple chemical functional groups on the surface of proteins, which could compete with the desired reaction. Therefore it is reasonable to seek the applicability of enzymatic post-translational modification of proteins, focused on site-specific modification of recombinant proteins.

Transglutaminase (TGase) is a potent enzyme as a protein-modifying catalyst, especially because of its site specificity and handling feasibility. In general, TGase functions in the covalent cross-linking of specific glutamine (Gln)- and lysine (Lys)-containing peptides and proteins via an ε-(γ-glutamyl)lysine bridge (Griffin et al. 2002; Lorand and Graham 2003). Accordingly, through the introduction of a specific peptide tag that works as a Gln-donor or Lys-acceptor substrate to the POI, by genetic manipulation, it would be possible to conduct an enzymatic conjugation, with a Lys-acceptor or Gln-donor substrate, establishing new chemical products (Fig. 17.1). TGase-mediated bioconjugation is thus conceptually applicable to any recombinant protein given that specific Gln- and Lys residues are available in the target protein, and this concept has actually been exemplified by a number of previous works (Rachel and Pelletier 2013; Strop 2014). It is also worth noting that the side chain amide group of Gln residue is chemically inert under physiological conditions, implying that the enzymatic activation could be an ideal solution

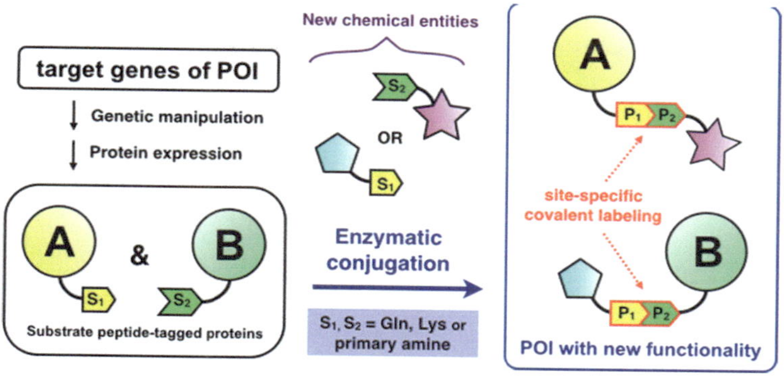

Fig. 17.1 General scheme for site-specific protein labeling catalyzed by MTG with a functional Gln-donor or Lys-acceptor substrate

for the residue-specific modification of Gln. This concept was demonstrated by the use of guinea pig TGase due to its strict substrate specificity (Coussons et al. 1991; Sato et al. 1996).

In this chapter, we focus on microbial transglutaminase (MTG) from *Streptomyces mobaraensis*, first developed for food protein engineering (Motoki and Seguro 1998; Yokoyama et al. 2004). Although it has been reported that MTG exhibits rather broader substrate specificity than mammalian TGases, its stability, its handling feasibility such as Ca^{2+}-independence, and its availability, since it originates from microbial mass production, could benefit further promotion of this enzyme in the relevant field. Interestingly, the promiscuous substrate recognition of MTG could afford the design of new Gln-donor substrates not only to label a specific Lys residue in target proteins but also to assemble labeled biomolecules with unique molecular functions and architectures.

17.2 Utility of Z-QG as a Conventional Substrate for MTG

Z-QG is a popular substrate for the catalytic activity assay of TGase (Folk and Chung 1985). The substrate specificity of MTG for primary amines was characterized by using Z-QG as a model Gln-donor substrate, and the requirement for at least three carbon linear chains between the terminal amine and substituents was shown with Lys analogues (Ohtsuka et al. 2000b). In fact, many types of functional primary amine-containing substrates (such as cadaverine derivatives) are available for MTG, allowing the labeling of a specific Gln residue in target proteins as exemplified in the functionalization of antibodies (Josten et al. 2000; Jeger et al. 2010)

Small probe molecules are useful for protein engineering in molecular biology; however, few examples have been demonstrated for a small functional Gln-donor substrate for Lys-specific labeling of target proteins (Fig. 17.2).

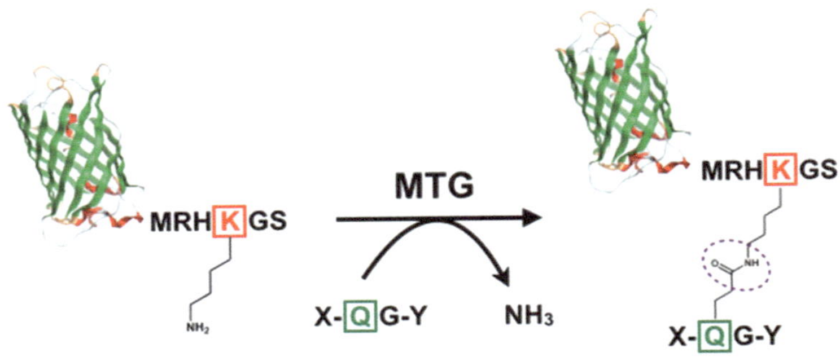

Fig. 17.2 Lys-specific protein labeling catalyzed by MTG with a small functional Gln-donor substrate

17.3 C-Terminal Modification of Z-QG

17.3.1 C-Terminal Modification of Z-QG for Protein Labeling

In general, the C-terminal carboxylic acid of Z-QG is a feasible target for introducing new chemical entities with desired function (Fig. 17.3). Given the fact that the recognition of Gln-donor substrates by MTG relies on the hydrophobic character of the N-terminus (Ando et al. 1989; Ohtsuka et al. 2000a), the C-terminal modification is reasonable to avoid steric hindrance upon the binding between MTG and a substrate. The introduction of monodansylcadaverine at the C-terminal carboxylic group of Z-QG was first studied to obtain a fluorolabeled Z-QG (Pasternack et al. 1997). In this context, C-terminal modification of Z-QG and Z-QQPL with fluorescent derivatives was explored for MTG- and human tissue transglutaminase-mediated protein labeling (Mindt et al. 2008).

17.3.2 C-Terminal Modification of Z-QG for the Preparation of Protein-Nucleic Acid Conjugates

We started from designing the C-terminal labeling of 5’-end amine-modified oligodeoxynucleotides (NH_2-ODN) with Z-QG for the preparation of protein-ODN conjugate (Fig. 17.4). This was feasibly conducted by chemical activation of carboxylate of Z-QG with N-hydroxysuccinimide (NHS). The NHS-activated Z-QG (Z-QG-NHS) was mixed with an aminated ODN, and the resultant Z-QG-ODN was conjugated with recombinant enhanced green fluorescent protein (EGFP) and bacterial alkaline phosphatase (AP). A short substrate peptide comprising six amino acids (MKHKGS, K-tag) was fused to the N-terminus of these two model proteins, allowing site-specific labeling of the peptidyl tag with Z-QG-ODN

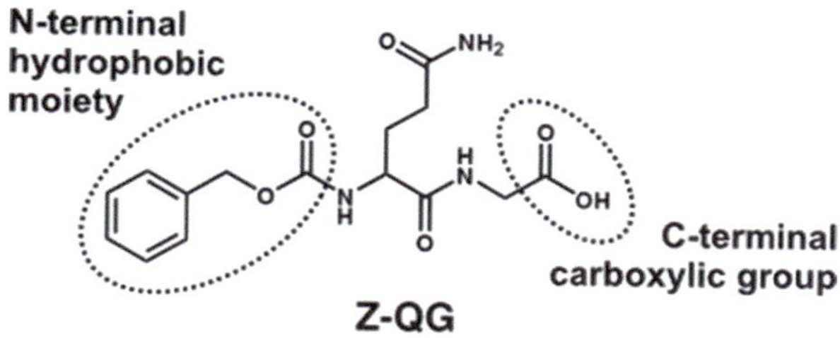

Fig. 17.3 Design concept for a functional Gln-donor substrate of MTG for Lys-specific protein labeling

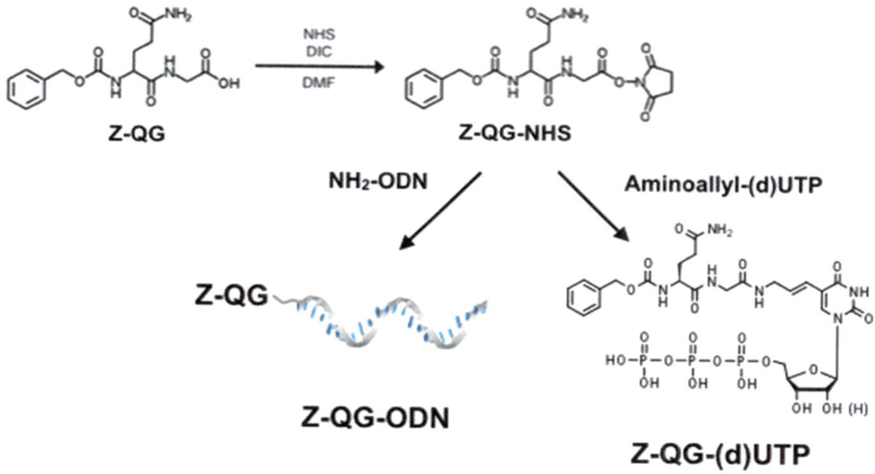

Fig. 17.4 Design of Gln-donor substrates for MTG-mediated preparation of protein-nucleic acid conjugates

(Tominaga et al. 2007). Protein-ODN conjugate is depicted as a tadpole molecule exhibiting dual functions derived from distinct functionalities of conjugated biomolecules. A potential application of this type of bioconjugates is site-specific, oriented protein immobilization through the hybridization of ODN with the complementary ODN presented on solid surface. The basic concept can be seen on a traditional in situ hybridization technique for the detection of either target DNA immobilized on membranes or mRNA on tissue sections. However, a short ODN is not suitable for those applications, which led us to develop a new enzymatic strategy for the preparation of DNA-protein conjugates (Kitaoka et al. 2011).

The design of a new nucleotidyl substrate of MTG, Z-QG-dUTP, was also based on the simple NHS chemistry (Fig. 17.4). However, in this case, we should consider the availability of the new substrate for DNA polymerase as well as MTG. Fortunately, Z-QG-dUTP was well recognized by different families of DNA polymerases (such as KOD and Taq polymerases), and a new DNA scaffold, $(Z\text{-}QG)_n$-DNA, was successfully obtained by the partial substitution of dTTP with Z-QG-dUTP in

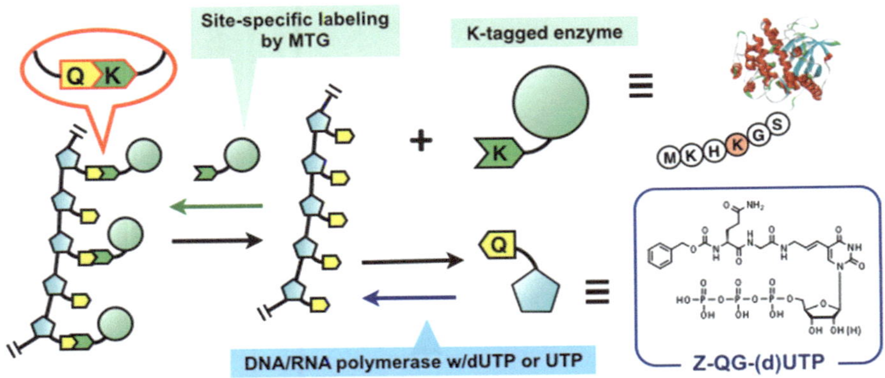

Fig. 17.5 Design of nucleic acid scaffolds with multiple Gln-donor sites for MTG-mediated labeling with a K-tagged enzyme

polymerase chain reaction. Subsequent labeling of (Z-QG)$_n$-DNA with a K-tagged enzyme resulted in a DNA-(enzyme)$_n$ conjugate with 1:n stoichiometry (Fig. 17.5). Upon the labeling of a thermostable alkaline phosphatase from *Pyrococcus furiosus*, the obtained bioconjugate with new molecular architecture enables the detection of a tiny amount of target nucleic acids having the complementary sequence in a dot blot hybridization assay under harsh hybridization conditions (Kitaoka et al. 2011). Besides the preparation of DNA-enzyme conjugates, the synthesis of Z-QG-UTP yielded RNA-(enzyme)$_n$ conjugates, which are applicable to an in situ hybridization technique for the detection of mRNA on tissue sections (Kitaoka et al. 2012). The number of enzymes labeled on the RNA scaffold impacted on the hybridization efficiency. Importantly, the potential steric hindrance, caused by the increase in the number of labeled enzymes on RNA, was controlled by simply changing the molar ratio of Z-QG-UTP to UTP on in vitro transcription catalyzed by RNA polymerases.

The synthetic strategy was applied to assemble cellulases on fully double-stranded DNA (dsDNA) to create a potent artificial cellulosome for enhancing hydrolysis of cellulosic substrates (Mori et al. 2013a). In the case of using dsDNA as a scaffold, insertion of a short ethylene oxide linker was critical for increasing the conjugation yields of DNA-(cellulase)$_n$ conjugates. The concept has been extended to label single or multiple proteins to a DNA aptamer by combining MTG and terminal deoxynucleotidyl transferase (TdT) (Takahara et al. 2013). The single protein labeling of ODN by the two-step enzymatic reaction with Z-QG-ddUTP may offer a feasible way to access tadpole molecules.

17.4 N-Terminal Substitution of Z-QG

17.4.1 Design of Fluorescent Labeling Reagents

One might imagine that the N-terminal substituent, benzyloxycarbonyl group (Z), could be a target for introducing new chemical entities (Fig. 17.3). However, the substrate preference of MTG may limit the chemical property available for the replacement of the Z moiety: it has been known that the N-terminal side of reactive Gln residue is critical for the substrate recognition of MTG. For instance, Z-QG is one of the most reactive Gln-donor substrates for MTG although GQG is not recognized as a substrate (Ando et al. 1989). As mentioned in the preceding section, most researchers were interested in the C-terminal labeling of the core structure, Z-QG. It is interesting for us that in spite of the preference of MTG for hydrophobic amino acids at the N-terminus of reactive Gln residue, it still shows rather broad substrate specificity. Therefore, we explored the idea that the substitution of the Z moiety with a fluorescent group may open new avenues for MTG-mediated protein modification with small Gln-donor substrates. In this design concept, the fluorescent group needs to play a dual role. One is for the intrinsic probe, and the other is for the substrate recognition. During the course of basic trials, we found that MTG accepts a wide range of fluorophores in the molecular size (i.e., fluorescein, dansyl and rhodamine derivatives) (Kamiya et al. 2009). Importantly, the reaction was only realized upon the insertion of a short linker (e.g., β-alanine) between the N-terminus of reactive Gln residue and the functional groups (Fig. 17.6), indicating the requirement of a linear flexible linker for the substrate recognition of MTG as was reported for primary amine substrates (Ohtsuka et al. 2000a, b; Lorand et al. 1979). The results suggest the potential utility of MTG that can accept a wider range of substrates than mammalian TGases (Sato et al. 2001).

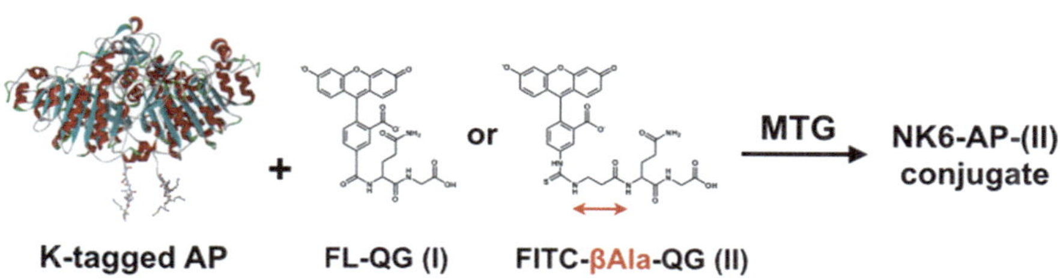

Fig. 17.6 The effect of a linker on MTG-mediated protein labeling with a fluorescent Gln-donor substrate

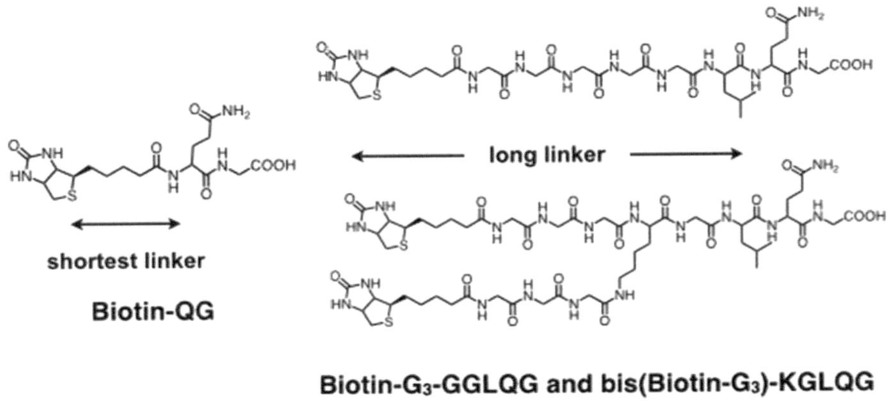

Fig. 17.7 Design of Gln-donor substrates for MTG-mediated biotinylation of proteins of interest

17.4.2 New Biotinylation Reagents and Their Applications

As another functional N-terminal substituent of the Z moiety, a small molecular ligand, biotin, was selected, which can bind to a tetrameric protein receptor, avidin. The ligand-protein interaction between biotin and avidin has been widely employed in biotechnology and bioengineering, due to its high specificity and strong affinity (i.e., $K_d = 10^{-15}$). In the specific case of Z-QG modification, we first designed a small biotinylation substrate with the shortest linker, biotin-QG (Fig. 17.7). In the structure of biotin-QG, the N-terminal biotin moiety gives hydrophobicity, and the C_4 alkyl linker, between the binding moiety and the N-terminus of reactive Gln residue, provides flexibility, which in turn facilitates substrate recognition by MTG (Mori et al. 2013b). Similarly, a larger biotinylation substrate, biotin-GGG-GGLQG, was synthesized and its C-terminal LQG tripeptides showed positive reactivity in the presence of MTG (Pasternack et al. 1997).

Like other avidin variants, avidin from *Streptomyces avidinii*, streptavidin (SA), forms a tetramer. More specifically, two SA dimers assemble, facing opposite directions. Bis-biotin can bind to the SA dimer from one face so that it behaves differently from mono-biotin. Therefore, we designed a bis-biotinylated substrate, bis-(biotin-GGG)-KGLQG, to control the vector of biotin-avidin interaction. A proof-of-concept study was conducted by the directional assembly of AP as a model protein, because of its symmetric dimeric structure (Fig. 17.8). Within a biotin-avidin system, the biotin-labeled position to a target protein should be an important factor for the formation of a supramolecular protein complex. To assess the utility of MTG combined with a new biotinylated Gln-donor substrate, we explored the internal labeling of AP to which a MTG reactive Lys-containing peptide loop (IRINRGPGKAFVT, K-loop) was inserted (Mori et al. 2011). The K-loop-inserted AP mutant maintained the original catalytic activity, and the Lys

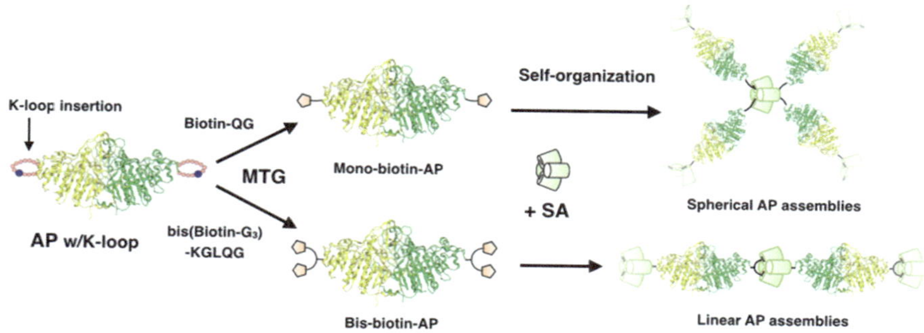

Fig. 17.8 Molecular design for assembling proteins by MTG-mediated internal biotinylation of dimeric protein

residue in the loop was recognized efficiently by MTG without impairing enzymatic activity due to the insertion of the K-loop at a suitable site in AP.

After the loop-specific biotinylation with designed new biotinylation reagents, biotinylated APs were mixed with SA to form supramolecular protein complexes (Fig. 17.8). As a result, several factors to assemble a protein complex with the biotin-avidin interaction were noticed: (1) use of a shorter biotinylation substrate and (2) arranging the position and direction of two biotins, as for example the labeled AP dimers, could prevent the intramolecular self-cyclization but promote the intermolecular interaction and the consequent enlargement of the AP-SA complexes; (3) with different types of biotinylation substrate, the shapes of protein molecular assemblies can be controlled (i.e., a spherical protein complex was formed with mono-biotin, and the bis-biotinylation of AP resulted in a one-dimensional stringy structure) (Mori et al. 2013a, b). Altogether, through the optimal design of the biotinylation substrate and biotinylation sites in a protein, the formation (i.e., growth and shape) of supramolecular protein assemblies can be designed according to each specific application. For the design of such self-assembled protein structures, protein-lipid conjugates are of interest as well, and the development of a Gln-donor substrate for MTG-mediated lipidation of a recombinant protein was demonstrated (Abe et al. 2011).

17.5 Conclusion

In this chapter, we exemplified that a small modification of the base structure of a substrate, X-QG-Y, greatly expands the utility of MTG-mediated protein manipulation in a range of biotechnological applications. Our results demonstrated that fine-tuning of artificial substrates may find new biomolecular architectures with unique functionality, and are promising for further development of MTG-mediated creation of artificial biological products. One of the next challenges would be the design of MTG mutants, exhibiting the orthogonal substrate specificities, making it possible to create one-step multifunctional modifications of different proteins and/or target sites in situ. On-going efforts in the development of new biotechnology should achieve the goal in the near future.

Acknowledgments We express our sincere gratitude to all colleagues who worked together with MTG. We also appreciate Ajinomoto Co., Inc. for providing us with the MTG sample and information on its basic characteristics. This work was supported by a Grant-in-Aid for Scientific Research (Grant Number 25289297).

References

Abe H, Goto M, Kamiya N (2011) Protein lipidation catalyzed by microbial transglutaminase. Chem Eur J 17:14004–14008

Ando H, Adachi M, Umeda K, Matsuura A, Nonaka M, Uchio R, Tanaka H, Motoki M (1989) Purification and characterisation of a novel transglutaminase derived from micro-organisms. Agric Biol Chem 53:2613–2617

Coussons PJ, Kelly SM, Price NC, Johnson CM, Smith B, Sawyer L (1991) Selective modification by transglutaminase of a glutamine side chain in the hinge region of the histidine-388 glutamine mutant of yeast phosphoglycerate kinase. Biochem J 273:73–78

Folk JE, Chung SI (1985) Transglutaminases. Methods Enzymol 113:358–375

Griffin M, Casadio R, Bergamini CM (2002) Transglutaminases: nature's biological glues. Biochem J 368:377–396

Hermanson GT (1996) Functional targets. In: Bioconjugate techniques. Academic, San Diego, pp 3–23

Jeger S, Zimmermann K, Blanc A, Grunberg J, Honer M, Hunziker P, Struthers H, Schibli R (2010) Site-specific and stoichiometric modification of antibodies by bacterial transglutaminase. Angew Chem Int Ed Engl 49:9995–9997

Josten A, Haalck L, Spener F, Meusel M (2000) Use of microbial transglutaminase for the enzymatic biotinylation of antibodies. J Immunol Methods 240:47–54

Kamiya N, Abe H, Goto M, Tsuji Y, Jikuya H (2009) Fluorescent substrates for covalent protein labeling catalyzed by microbial transglutaminase. Org Biomol Chem 7:3407–3412

Kitaoka M, Tsuruda Y, Tanaka Y, Goto M, Mitsumori M, Hayashi K, Hiraishi Y, Miyawaki K, Noji S, Kamiya N (2011) Transglutaminase-mediated synthesis of a DNA-(enzyme)n probe for highly sensitive DNA detection. Chem Eur J 17:5387–5392

Kitaoka M, Mitsumori M, Hayashi K, Hiraishi Y, Yoshinaga H, Nakano K, Miyawaki K, Noji S, Goto M, Kamiya N (2012) Transglutaminase-mediated in situ hybridization (TransISH) system: a new methodology for simplified mRNA detection. Anal Chem 84:5885–5891

Lorand L, Graham RM (2003) Transglutaminases: crosslinking enzymes with pleiotropic functions. Nat Rev Mol Cell Biol 4:140–156

Lorand L, Parameswaran KN, Stenberg P, Tong YS, Velasco PT, Jönsson NA, Mikiver L, Moses P (1979) Specificity of guinea pig liver transglutaminase for amine substrates. Biochemistry 18:1756–1765

Mindt TL, Jungi V, Wyss S, Friedli A, Pla G, Novak-Hofer I, Grünberg J, Schibli R (2008) Modification of different IgG1 antibodies via glutamine and lysine using bacterial and human tissue transglutaminase. Bioconjug Chem 19:271–278

Mori Y, Goto M, Kamiya N (2011) Transglutaminase-mediated internal protein labeling with a designed peptide loop. Biochem Biophys Res Commun 410:829–833

Mori Y, Ozasa S, Kitaoka M, Noda S, Tanaka T, Ichinose H, Kamiya N (2013a) Aligning an endoglucanase Cel5A from *Thermobifida fusca* on a DNA scaffold: potent design of an artificial cellulosome. Chem Commun 49:6971–6973

Mori Y, Wakabayashi R, Goto M, Kamiya N (2013b) Protein supramolecular complex formation by site-specific avidin-biotin interactions. Org Biomol Chem 11:914–922

Motoki M, Seguro K (1998) Transglutaminase and its use for food processing. Trends Food Sci Technol 9:204–210

Ohtsuka T, Ota M, Nio N, Motoki M (2000a) Comparison of substrate specificities of transglutaminases using synthetic peptides as acyl donors. Biosci Biotechnol Biochem 64:2608–2613

Ohtsuka T, Sawa A, Kawabata R, Nio N, Motoki M (2000b) Substrate specificities of microbial transglutaminase for primary amines. J Agric Food Chem 48:6230–6233

Pasternack R, Laurent HP, Rüth T, Kaiser A, Schön N, Fuchsbauer HL (1997) A fluorescent substrate of transglutaminase for detection and characterization of glutamine acceptor compounds. Anal Biochem 249:54–60

Rachel NM, Pelletier JN (2013) Biotechnological applications of transglutaminases. Biomolecules 3:870–888

Sato H, Ikeda M, Suzuki K, Hirayama K (1996) Site-specific modification of interleukin-2 by the combined use of genetic engineering technique and transglutaminase. Biochemistry 35:13072–13080

Sato H, Hayashi E, Yamada N, Yatagai M, Takahara Y (2001) Further studies on the site-specific protein modification by microbial transglutaminase. Bioconjug Chem 12:701–710

Strop P (2014) Versatility of microbial transglutaminase. Bioconjug Chem 25:855–862

Takahara M, Hayashi K, Goto M, Kamiya N (2013) Tailing DNA aptamers with a functional protein by two-step enzymatic reaction. J Biosci Bioeng 116:660–665

Tominaga J, Kemori Y, Tanaka Y, Maruyama T, Kamiya N, Goto M (2007) An enzymatic method for site-specific labeling of recombinant proteins with oligonucleotides. Chem Commun 2007:401–403

Yokoyama K, Nio N, Kikuchi Y (2004) Properties and applications of microbial ransglutaminase. Appl Microbiol Biotechnol 64:447–454

Index

K. Hitomi et al. (eds.), *Transglutaminases*, DOI 10.1007/978-4-431-55825-5

MIX
Papier aus verantwortungsvollen Quellen
Paper from responsible sources
FSC® C105338

If you have any concerns about our products,
you can contact us on
ProductSafety@springernature.com

In case Publisher is established outside the EU,
the EU authorized representative is:
Springer Nature Customer Service Center GmbH
Europaplatz 3, 69115 Heidelberg, Germany

Printed by Libri Plureos GmbH
in Hamburg, Germany